KU-767-695

To all the scientists, engineers, and technologists, who labor to provide people with the marvels of electronics, and who labor against the world's tendency to set these technologies up as either gods to be worshiped or demons to be exorcised.

Overview

Appendices

70084412

621.381
HoL

Surface-Mount Technology
for PC Boards

by James K. Hollomon, Jr.

0 5 OCT 2017

RECEIVED

PROMPT®
PUBLICATIONS

CARDIFF AND VALE COLLEGE

©1995 by James K. Hollomon, Jr.

REVISED FIRST EDITION—1995

PROMPT® Publications is an imprint of Howard W. Sams & Company, 2647 Waterfront Parkway, East Drive, Suite 300, Indianapolis, IN 46214-2041

This book was originally developed and published as *Surface-Mount Technology for PC Board Design* by Sams, a division of Macmillan, Inc., Indianapolis, IN.

All rights reserved. No part of this book shall be reproduced, stored in a retrieval system, or transmitted by any means, electronic, mechanical, photocopying, recording, or otherwise, without written permission from the publisher. No patent liability is assumed with respect to the use of the information contained herein. While every precaution has been taken in the preparation of this book, the author, the publisher or seller assumes no responsibility for errors or omissions. Neither is any liability assumed for damages resulting from the use of information contained herein.

International Standard Book Number: 0-7906-1060-4

Cover Design by: Suzanne Lincoln

Trademark Acknowledgments: All terms mentioned in this book that are known or suspected to be trademarks or services have been appropriately capitalized. PROMPT® Publications and Howard W. Sams & Company cannot attest to the accuracy of this information. Use of a term in this book should not be regarded as affecting the validity of any trademark or service mark.

Printed in the United States of America

9 8 7 6 5 4 3 2

Contents

4 System Design Considerations for SMT *119*

5   Printed-Wiring Layout Using SMCs   *141*

7 Quality Assurance in SMT   *211*

8 Reviewing a New SMT Design— Is It Manufacturable? 267

9 Hybrid Circuits and Multi-Chip Modules (MCMs) 299

10   Using SMCs in Military and Space-Borne Applications   *359*

C SMT Bibliography   *451*

D SMT Test Board Patterns   *465*

E Glossary of SMT Terms   *469*

Preface

Surface-Mount Technology, or *SMT*, is defined most simply as the second-level interconnection of components using planar mounting. Thus, SMT includes several circuit technologies which do not use lead-through-the-hole assembly techniques. The most widely used SMT component-packaging formats are *chip carrier*, *small-outline IC (SOIC)*, *small-outline transistor (SOT)*, *chip capacitor*, *chip resistor*, etc. *Tape automated bonding (TAB)* and some *flip chip* methods are also planar mounting techniques, and thus are classified as SMT. However, this work will concentrate on the more widely used SMT approaches, where standardization is currently well underway.

SMT holds great promise for our industry, and the race to adopt it has been described as the latest revolution in electronics. Indeed, SMT is a dramatic development, now well underway all around the globe. We can easily see why when we consider the driving forces behind it. SMT can offer 60% or greater reductions in PC board real estate, faster switching speeds, improved reliability, and simplified assembly automation. In addition, SMT often allows cost savings in materials and factory investment. These benefits, combined, are driving industry to adopt SMT as quickly as human and capital resources permit.

Surface-Mount Technology (SMT) is an *advanced manufacturing technology*. As such, the "how-to" knowledge, the human resource side, of SMT must come from the factory floor. However, the data bus between electronics designers and manufacturing personnel is either nonexistent, or, at best, difficult to access. How do designers learn what the factory really needs in order to succeed with SMT? How do manufacturing and test engineers know when to yield to the designer's demands, even if those demands create an extra burden in manufacturing or testing?

This book was written to provide a two-way bridge between the manufacturing technology and the PWB layout considerations for SMT. In it, we present standards for design and discuss why these standards exist. We look beyond legalistic adherence to design rules, and study their basis and application-specific issues that might force departure from the standards.

This book is intended for the working engineer or manager, the student or the interested layman, who would like to learn to deal effectively with the many trade-offs required to produce high manufacturing yields, low test costs, and manufacturable designs using SMT. In it, we present the information an engineer needs to both serve on an SMT project team and to appreciate and deal with the concerns of other disciplines on such a team.

Textbooks have been slow to appear since the first uses of SMT in Japan in the mid-1970s. The availability of an SMT-knowledgeable technologist (the human resource side of the SMT revolution) is generally the most significant impediment to surface-mounting progress. It is our hope that this book makes the benefits of SMT more obtainable for you.

James K. Hollomon, Jr.

Acknowledgments

Space simply doesn't permit me to credit everyone who contributed material that appears herein. All those colleagues I've worked with in the trenches have given their creativity and collected knowledge to the data gathered in this work.

I would like to give special credit to Micro Component Technology, Inc. My seven years with them blessed me richly in opportunities to learn about this new and exciting field.

There are also several who have given above and beyond the call of duty, and I feel they deserve special mention. My family, who've graciously tolerated months of being largely ignored during this writing effort. John Biancini of Supernova Design Center in Minneapolis, Minnesota, who contributed hours of boring labor, tomes of useful data, plus prayerful spiritual support and guidance for which I am eternally grateful. Martin Barton of Rockwell International, and Stacey Bilsky of Raytheon, were generous with their time as reviewers. The SMTA and my fellow committee members on the Standards committee have helped in ways beyond measure. Harold Peacock and his colleagues at the Navy EMPF in China Lake, California, were instrumental in completing the military side of this work.

And finally, my deepest gratitude to all those who have granted us permission to use their material. Since they are credited in the text, I will not list them here but will say a sincere "Thank you."

Trademarks

All terms mentioned in this book that are known to be trademarks or service marks are listed below. In addition, terms suspected of being trademarks or service marks have been appropriately capitalized. Howard W. Sams & Company cannot attest to the accuracy of this information. Use of a term in this book should not be regarded as affecting the validity of any trademark or service mark.

850 F/I and Access 2 are trademarks of Zehntel, Inc.
Band Aid is a trademark of Johnson & Johnson Corp.
Compaq is a registered trademark of Compaq Computer Corp.
Dolby is a trademark of Dolby Laboratories, Inc.

EFG is a trademark of Saphicon.

Escort and Passport are trademarks of Cincinnati Microwave, Inc.

ETA^{10} is a trademark of ETA Systems, Inc.

FEDEX and SuperTracker are registered trademarks of Federal Express, Inc.

Gumby is a trademark of Lakeside Games.

Integri-Test is a trademark of Electronic Equipment Div., Kolmorgen.

Invar is a trademark of Carpenter Technology Corp.

Kevlar is a trademark of New England Ropes, Inc.

Micro-Channel, PC-AT, and PS/2 are trademarks of IBM Corp.

Micropack is a trademark of Siemens, AG.

Plexiglas is a trademark of Rohm and Haas.

RO2800 is a trademark of Rogers Corp.

SIMM is a trademark of Wang Laboratories, Inc.

TapePak is a trademark of National Semiconductor Corp.

Teflon is a trademark of E.I. DuPont de Nemours Co., Inc.

VAX is a registered trademark of Digital Equipment Corp.

Figure 1 A new day dawning in electronics technology—the printed-circuit board. (*Courtesy Science Applications International Corp.*)

1 The Benefits and Limitations of Surface-Mount Technology

SMT, ASICs, CIM, advanced computer technologies; all are impacting the way we work and live our lives. The circuit board shown in Figure 1 is just one of the many examples we'll see of the new age in electronics manufacturing. Targeted at artificial neural network applications, this very dense card takes up just a single slot of an IBM PC AT®. Yet the diminutive card transports the simple desktop computer to unheard of levels of performance. Rated at 22 MFlops (peak), the card carries its own 12-MB RAM mobile, 32/64-bit integer, and IEEE floating-point ALU, C compiler, and macro assembler.

Surface-mount technology (SMT) usage is growing at a gallop. In 1985, SMT accounted for a small fraction of the U.S. components market. In 1987, SMT had a 10% share. By 1995, SMT usage exceeded IMC.[1] In many cases, SMT components are now less costly than their through-the-hole packaged counterparts. No small number of parts are now available only in SMT packaging. Such explosive growth is an indication that SMT has a great deal to offer the circuit designer.

1.1 What SMT Can Do for Circuitry

What are these benefits that are driving SMT's rapid growth? Understanding the rewards that come from mastery of a subject is vital to the learning experience. So, we'll begin our discussion of SMT design with a study of what it has to offer.

Figure 1-1 SMT and IMC visual comparison. This illustration is also useful for identifying SMT package styles.

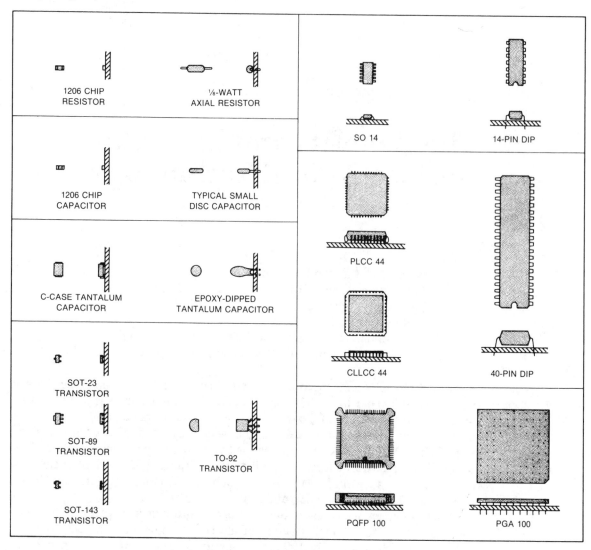

1.1.1 Size and Weight Improvements

If you were to ask any group of electronics engineers for the major advantages of SMT, it's a safe bet that "size and weight reductions" would head the list. So, let's start our discussion of advantages with a look at some specifics of circuit shrinking through SMT.

Real-Estate Reductions with SMT

Surface-mount technology is a real-estate miser for several reasons. Most obvious is the size comparison between surface-mounted components (SMCs) and typical insertion-mount components (IMCs). Figure 1-1 shows a comparison of equivalent SMT and IMC packages. For readers who are not familiar with SMT components, Figure 1-1 also provides a study tool for package identification.

While the smaller size of SMCs is a major factor in space savings, the ability to populate both sides of a circuit board often yields far greater reductions than an A/B comparison of through-hole and SMT component sizes would suggest. Additional space savings may result from the elimination of through holes and their annular rings at each component lead connection. Annular rings occupy a good deal of real estate on dense digital boards.

Figure 1-2 SMC and IMC weights and volumes compared.

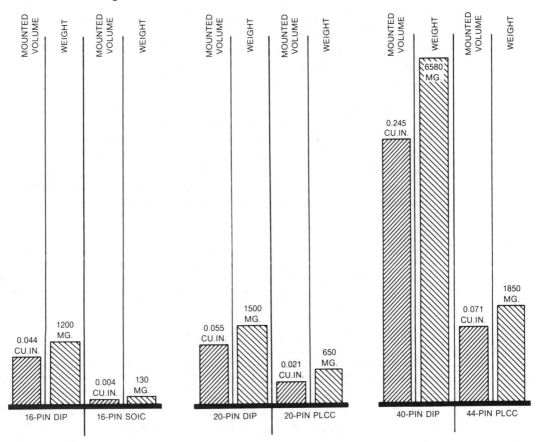

The net result is that, at best, SMT may deliver a 70% or greater improvement in density. Savings of 35% to 50% are routine. This alone would make SMT an attractive technology. Most electronics designers are struggling with space efficiency in some form. Either we need to get a given circuit into a smaller box, or we need to get more circuit into a given box. In applications where size is not important, weight or cost may be.

Weight Savings of SMT
The lower weight of SMCs is of particular interest in aerospace and portable equipment applications. Lower mass may also give SMT a distinct advantage in high vibration and shock-load environments. Figure 1-2 compares the volume and weight of equivalent SMC and IMC devices. (We will deal with design for severe environments in Chapter 4, Section 9.)

1.1.2   *Better Performance*

Looking again at Figure 1-1, it is apparent that SMCs have far smaller leads than IMCs. In fact, chip passives have no leads at all. SOICs and chip carriers have miniature or nonexistent external leads, and their internal lead paths are far shorter than a DIP of equal pin count. Shorter lead paths boost SMT performance over IMC designs in several areas.

Circuit Propagation Delays

Each additional millimeter of lead adds inductance, capacitance, and resistance (collectively called parasitics) to the signal path. Research by IBM indicates that, in high-performance computers, as much as 60% of the signal delay is attributed to interconnect parasitics.[2] Circuit performance improvements of up to 20% have been documented through changeovers to SMT.[3]

Noise Immunity

Component leads act as antennas which receive and contaminate the circuit with environmental radio-frequency noise. Compared to through-hole technology, SMT affords substantial miniaturization, particularly when fine-pitch technology is used. Miniaturization means shorter interconnectors, and thus, less coupling of RFI and EMI. The substantial reductions in lead length afforded by surface mounting give SMT a major edge in external noise immunity.

Crosstalk

Circuit noise is particularly troublesome in ECL circuits where a number of lines are switched simultaneously with very short rise times. Research by Digital Equipment Corp. and Unisys has shown that induced noise spikes on unswitched lines are primarily produced by package inductance. Therefore, SMT has a distinct advantage over IMC assembly in the reduction of coupled noise.[4]

1.1.3   *Manufacturability Advantages of SMT*

Most engineers are aware of the size and performance advantages of SMT. It may come as a surprise, however, to hear that surface mount boards are often more manufacturable than through-hole assemblies. Yet, in many applications, we find that this is the case. Let's look at some of the factors that might give an SMT design the edge over a traditional IMC implementation.

Manufacturing Floor Space

For automated assembly, IMC factories generally require one of several separate automated machines for each unique package style. A typical automated IMC factory might have a DIP inserter, a VCD axial component inserter with its associated sequencers (one sequencer for each 40 to 60 axial part numbers to be inserted in one PWB assembly set-up), a radial component inserter, and a pin inserter. Still, it's not unusual for as many as 50% of the assembly's components to require hand insertion. Typically, through-hole factories dedicate substantial floor area to hand-assembly stations.

In contrast, it takes only one flexible automated placement machine to handle the assembly of ICs, transistors, diodes, capacitors, resistors, trimmers, coils, small transformers, crystals, and other components in appropriate surface mount packages. This placement machine is often smaller than any of the previously cited through-hold machines. Thus, in applications not requiring automated assembly of mixtures of IMCs and AMCs, surface mount manufacturing generally saves floor space.

Additional floor space savings accrue from the warehousing of SMCs and Work in Process (WIP). Table 1-1 compares the dramatic differences in storage space requirements for SMCs and IMCs.

Table 1-1. SMC vs. IMC Warehousing Space†

COMPONENT	THROUGH-HOLE TYPE	SMT TYPE
Resistors	1/4-watt axials on tape and reel; 5 cubic feet.	1/8-watt 1206 on EIA RS 481A tape; 0.35 cubic feet.
Capacitors	Radials taped per EIA RS-468; 22.4 cubic feet.	1206 Chip Capacitors on EIA RS 481A tape; 0.35 cubic feet.
ICs	16-pin DIP in plastic sticks; 18.20 cubic feet.	16-pin SOICs in plastic sticks; 2.92 cubic feet.

†Space required to store 100,000 each of various components in standard shipping containers. *(Courtesy Corlund Electronics, Signetics, and R. G. Allen.)*

Where new facilities and equipment would be required for a project, many electronics manufacturers have switched from through-hole to SMT operation, because the savings in building and automation costs compelled them to do so. Others have elected to change over to increase production in a given space, thus saving the cost of having to erect new facilities.

Low-Volume High-Mix Assembly
It often comes as a shock to seasoned through-hole assemblers, but hand assembly is easier with standard pitch SMT than with through hole, and assembly rates are typically higher. Thus, SMT is not a volume manufacturing technology limited only to products built in automated factories. In fact, we often design circuits using SMT for no other reason than to facilitate low-volume hand assembly.

First-Pass Yield
Process quality varies greatly from factory to factory, and even for different products in the same factory. (Process quality is that percentage of a product that conforms to specifications as it leaves the manufacturing line and arrives at final test.) A key determinant in process quality is the degree of process control on the manufacturing line. Few SMT versus IMC comparisons involve identical manufacturing environments. Thus, it is difficult to prove that SMT offers inherently better yields than IMC. However, a great deal of statistical data is at hand to show that, with equally stringent process controls, SMT's first-pass yields can at least equal those of IMC.

As an example, let us review process quality from one SMT factory where a comparison is meaningful; that is, (1) we can compare products that are identical except that one is surface mount and the other is through-hole, (2) the products are built in the same factory with similar process controls, and (3) both versions have been produced in sufficient volumes that a strong bank of statistical data on yields has been evaluated. In this case, Cincinnati Microwave, Inc., has found SMT yields of 93% to 97% good assemblies entering final test, while through-hole yields for the identical circuit schematic run between 86% and 87%.[5] (We'll look at Cincinnati Microwave's application in detail in Section 1.5.3.) Granted that one instance is not a large enough sampling to draw any yield conclusions. It does, however, debunk the myth that SMT, as a new technology, is inherently low in process yields.

CIM Adaptability

The electronic manufacturing community is under considerable competitive pressure to maximize manufacturing cost-efficiency and quality. Both of these goals are driving us toward higher levels of automation. Computer Integrated Manufacturing (CIM) is the target of many top electronics manufacturers today, but few have hit the bull's-eye as of this writing.

Although we think of our industry as the essence of high tech, the fact is that the metals industry often leads electronics in adopting innovative manufacturing methods. This is partially because the automation for IMC assembly is primarily designed for manual set-up and as stand-alone islands of automation. Also, many of the tasks necessary to complete an IMC assembly are very difficult to automate, and off-the-shelf automation does not exist to address them.

In contrast to through-hole manufacturing, a great deal of factory-automation-compatible SMT equipment is available today. There are numerous suppliers of high flexibility machines capable of self-reconfiguration per the instructions of a host computer. The dramatic storage space reductions offered by SMT components, as shown in Table 1-1, open the door to warehousing a full complement of components for a family of board part numbers on one assembly machine. Components can then be selected under computer control, jukebox equipment targeted for the SMT factory is suitable for pass through, or smart cart handling of WIP. It is, therefore, not surprising that the majority of electronics factories with full CIM implementation are SMT factories.

Since it is a relatively new technology, there are hurdles to clear before CIM may be easily applied to an SMT factory. Software is needed for design rule checking, test engineering, test fixture development, and shop feedback to CAE/CAD. However, the solutions are now available.

1.1.4 *Reliability Increases with SMT*

Claims that SMT is more reliable than IMC may be a bit difficult to believe. After all, SMT is the new kid on the block. Nonetheless, that claim is made, and it has been substantiated by numerous companies both in their labs and on their productions lines. Discussed next are the three major factors which, in some combination, may contribute to a reliability edge for SMT over analogous IMC designs.

Reduction in Number of Solder Joints

Figure 1-3 shows how surface mounting of components, such as chip capacitors and chip resistors, reduces the number of interconnects. Since every solder connection is a potential failure point, the 50% reduction in interconnects associated with the surface mounting of these components can have a marked impact on capacitor and resistor reliability.

We should mention that the degree to which this benefit applies is dependent on several things. First, the impact of solder-joint count reduction is directly proportional to the ratio of chip resistors and capacitors to other components in the system (i.e., those components which, in through-hole versions, do not have leads attached by soldering). Second, this benefit relies on quality solder joints, proper design, and on the management of thermally induced stresses. We will discuss these topics in more detail later.

Noise Immunity

As stated earlier, each component lead and PWB trace on a circuit serves as a tiny radio antenna. Radio-frequency interference (RFI), electromagnetic interference

(EMI), and electromagnetic pulse (EMP) can sneak into circuitry through these antennae, wreaking havoc with the specified performance of the circuit. By drastically reducing the antenna area of a circuit, SMT makes designs much more noise tolerant. In Chapter 6 (Section 6.4.6), we consider the effects of radiation and electromagnetic pulse on circuitry. For a more thorough discussion of the effects of lead and line length on circuit performance, see Reference 6.

Figure 1-3 Ceramic capacitors in monolithic chip and in leaded form.

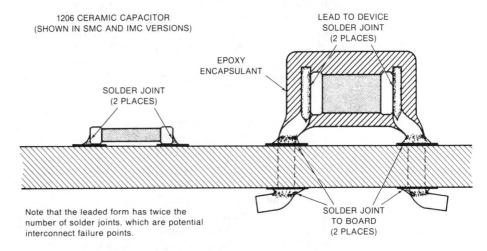

1206 CERAMIC CAPACITOR
(SHOWN IN SMC AND IMC VERSIONS)

LEAD TO DEVICE
SOLDER JOINT
(2 PLACES)

EPOXY
ENCAPSULANT

SOLDER JOINT
(2 PLACES)

Note that the leaded form has twice the number of solder joints, which are potential interconnect failure points.

SOLDER JOINT
TO BOARD
(2 PLACES)

Component Mass
Assemblies exposed to vibration and shock-induced stress benefit from the order-of-magnitude lower mass of SMCs. This is a primary factor fueling the avionics community's intense SMT interest.

1.1.5   *Cost Reductions Through Surface Mounting*

Cost is always a factor in choosing directions for electronics designs. In some industry segments, it runs third place behind performance and reliability, but more often, it is of critical importance. Therefore, the high cost of early SMT components (produced at low volumes and low economies of scale) was a delaying factor in the implementation of SMT in the United States. However, in a few instances (as early as 1983), we saw the emergence of cost as the dominant factor in choosing SMT over IMC. As SMT component volumes have increased, more commercial designs are selecting SMT because it will reduce system cost. IBM's PS/2® line of PCs is a perfect example of a cost-driven selection of SMT. Also, note this statement from a Bourns advertisement for trimmer resistors: "To survive in the marketplace, more and more products need the cost savings, space efficiency, and high performance of surface mounting."[7] Clearly, it is time to lay aside knee-jerk responses that SMT costs too much.

We will cover the details of the influence of design on cost in Chapter 4, Section 2. But now, we'll look at the key factors which may make SMT a penny pincher.

Board-Level Savings
A printed-wiring board for an SMT circuit is usually considerably less expensive than a board for the same IMC circuit. Board cost savings result from three factors.

First, SMT boards are typically 35 to 70% smaller than analogous circuits using IMCs. Thus, there's considerably less board to buy. Second, SMT boards do not require holes at every component lead for insertion of leaded components. Via holes are required only where electrical communication between layers or sides of the board must be established. So, in some designs, SMT allows a reduction in hole count, which has a direct relation to board cost.

Third, multilayer boards can often be routed with fewer layers using SMT. This is true because the reduction in through holes, mentioned above, leaves more room open for inner layer traces. Also, where via holes are required, they can be less than one half the diameter of vias on through-hole boards, where clearance for the component lead must be provided. IBM reported on a particularly high-density application. Using pin grid arrays (PGAs) and IMCs, 28 layers were required. However, the same circuit was routed in SMT in 6 signal layers.[8] Since multilayer board costs usually escalate by 7 to 15% per layer, savings in multilayer board costs can be substantial when switching from IMC to SMC constructions.

System-Level Savings

At a system level, several options are available to the designer who wishes to economize using SMT. With a median 50% reduction in real estate, it becomes possible to put the functions of two IMC boards on one SMT board of the same size. In a multiboard system, halving the number of boards required reduces the cost of backplanes and board-level interconnects, the cost of boards, and the cost of board-mounting hardware and enclosures.

In single-board systems, system cost may be reduced by downsizing the board and the housing for the board. Another approach is to keep the housing the same size and add more functions through surface mounting. This may permit giving customers greater value added, and can justify a higher selling price, which equates to a more profitable position for the manufacturer. Adding functions to aging products may revitalize a tired line, saving substantial new product development costs and allowing the continued amortization of previous investments.

Cost of IMC lead prep, sequencing and insertion often is higher than the cost of placing the SMT version of the same component. This is true both in automated and in-hand assembly. Whenever comparing costs, it is vital to look at cost-of-ownership on the board, not just the raw cost of the parts involved.

Plant Investment Savings

As we discussed in Section 1.1.3, for similar product volumes, automated SMT production requires less equipment and floor space investment than automated IMC assembly manufacturing. Where brick and mortar investment is required for a new product, the savings in front-end capitalization may be a major factor in system costing.

1.2 Applications Where SMT May Not Make Sense

To date, no single circuit-assembly technology satisfies all applications. Some needs are best served by SMT. Other applications beg for custom large-scale silicon integration, hybrids, and yes, even IMC methods. Before we begin a detailed analysis of SMT design, let's review application issues that might speak for some other technological approach.

1.2.1 Component-Driven Disadvantages for SMT

Price of Components
Some components cost more in surface-mount packaging than in through-hole style. Other SMCs currently enjoy a price advantage over their IMC analogs. Note that, even in applications where SMT components are somewhat more expensive than through-hole components, the other surface-mount cost savings, discussed earlier, may still make SMT the most economical choice. Whether SMT component costs are a burden or an advantage is entirely a function of the mix of component styles, and the weight of the noncomponent savings for a given application. To arbitrarily refuse to look at SMT "until component price parity is reached" may be false economy.

Availability of Components
Certain component types are widely available in SMT styles. Other families of components are either difficult to find, or are not offered at all in surface-mountable packages. Components, such as large coils, transformers, power resistors, high-capacitance or high-voltage capacitors, high-power transistors, large rectifiers, and other large components, are of limited availability as SMCs. In applications where most components must be through hole, the slight real-estate gain offered by changing a small percentage of the population to SMT may not be justifiable.

1.2.2 "Support Concerns" as a Delaying Factor

Even in situations where SMT has a clear cost, size, and performance advantage, there are additional considerations to weigh when determining if a change to surface mount is a sound business decision. Next, we'll study several issues that have led some managements to delay entry into SMT, even though the advantages at the product level were clear.

Field Repair of Assemblies
SMT assemblies are relatively simple to repair and rework. Components are generally removed and replaced in special workstations, using heated air or heated nitrogen to desolder and resolder. Repair technology is straightforward, and the necessary workstations are not expensive. Several, like the one shown in Figure 1-4, are available for around $2500. However, workstations aren't free. Also, transporting them for field service would be difficult.

For companies with field-repair strategies for embedded electronics in heavy equipment, repairing SMT might mean a considerable investment in training and equipment. Getting SMT repair gear to work sites would become a continuing burden. In such cases, it might be wise to step back from SMT and develop new strategies for repair before changing manufacturing technologies. Instead of field repair, management might opt for a game plan involving the isolation of faults to the board level and the field exchange of boards.

Through-Hole Installed Base
Having a large installed base in the field makes some companies think twice about committing to SMT. To support such an installed base, some through-hole capability must be maintained, or new SMT designs of the same form, fit, and function as the existing IMC boards may be developed.

Figure 1-4 Typical SMT repair station. *(Courtesy Nu-Concept Systems, Inc., Colmar, PA.)*

1.2.3 *Amortization of Existing Through-Hole Equipment*

Announcing to the directors that the new million-dollar-plus automated insertion line is obsolete before it's fully operational can have a profound and undesirable effect on a career. The fear of having to make such an announcement has certainly delayed development of SMT in some companies.

In reality, changing to SMT is very seldom an overnight, 100% proposition. Many firms have started into SMT by adding a few pieces of equipment to an existing through-hole factory. Therefore, the fear that tightens the vocal cords of the manager who must request new equipment may be misguided. But the fact that those vocal cords do tighten has certainly stopped more than one management from talking about SMT.

Eventually, such managers will probably get stuck between the pain of asking for new capital and the pain of explaining a loss of competitive edge and market share, if they don't ask for new capital. Such an avoidance/avoidance conflict may needlessly delay an important technological move. A proactive measured development of new technology, without obsoleting existing strength, should be examined closely if you feel that your insertion investment is too new to retire.

1.3 Common False Perceptions Regarding Adopting Surface-Mount Technology

There are several widely held misconceptions about SMT. These false impressions have delayed SMT acceptance in some companies that desperately needed the competitive edge it would give them. We'll cover these SMT fallacies in enough detail that you can separate any misapprehensions from valid concerns in selecting good SMT applications.

1.3.1 SMT Components Must Reach Parity with Through Hole

In every single SMT Overview talk I've attended or given, someone has asked, "When do you think SMCs will reach parity with IMCs?" In examining the person's motives for this question, it generally emerges that they plan to ignore SMT until the answer comes back, "SMCs are now cheaper than IMCs in all instances." As discussed above, this attitude completely discounts system cost, which is what we're really trying to reduce. It ignores the marketing value of offering a lighter, smaller product with more functions. We know of instances where this attitude was responsible for allowing competitors to obtain an irreversible technological lead.

Some products are not particularly cost sensitive. Some are. For those that are not cost sensitive, SMT's size and performance benefits should be the primary factors evaluated in deciding what technology should be chosen for a given design. However, for price-sensitive products, it is imperative that system cost and potential manufacturing savings be carefully weighed. The shortcut of comparing the price of a few components is a dangerous false economy.

1.3.2 Standards Have to Be Developed First

When SMT first began to catch the interest of leading-edge manufacturers, there was a distinct problem with lack of standards. Works, such as IPC-SM-782, and books such as this were not available. The only way to SMT proficiency was to develop it in an R&D lab. The situation has changed. There are JEDEC, EIAJ, and IEC standards on components. The IPC has issued standards on design, workmanship, soldering, and various facets of testing. Phil Marcoux of International Quality Technology, Santa Clara, California, has correctly quipped, "There's no lack of standards. There are plenty of standards. Just pick one." It's our hope that this book may serve as a guide in this standards selection process.

1.3.3 You Have to Be in High-Volume Manufacturing to Consider SMT

Certainly, the high-visibility SMT applications, the ones we read about in trade journals, are generally high-volume operations. This doesn't mean there aren't hundreds and thousands of small- to medium-volume SMT shops. In fact, a quick review of our customer files shows that the 80/20 law is in operation in SMT volume. (The 80/20 law states that 80% of the output of a given endeavor results from 20% of the effort. As an example, in most companies, 80% of the sales usually come from 20% of the customer base. Also, 80% of the sales usually come from 20% of the sales force. This law will be found to apply to many diverse situations.) About 80% (77.8% exactly) of the companies on file are producing low to moderate volumes of surface-mount assemblies.

1.3.4 *The Equipment Costs Are Prohibitive*

Again, the high-visibility operations that are written up in the trade press tend to be highly automated factories. These are neat to read about, so that's no surprise. These aren't the only SMT factories in the world, however. I lived within a few minutes drive of an electronics business that's recently entered SMT. Essex Technologies, of Carpinteria, California, makes automobile alarm systems. They invested around $50,000 in equipment to set up what they call their prototype SMT line. All this equipment was new, not used. In the first full month of operation, their prototype line shipped 3000 double-side-populated boards with a total of 60 SMCs per board.[9]

1.4 SMT Manufacturing Styles and Their Effect on Benefits

The broad spectrum of SMT can be divided into three manufacturing styles for study purposes. These styles are called simply, Type 1, Type 2, and Type 3 SMT. Jim Hall of Dynapert HTC developed this classification in 1983 to help understand the soldering technology of SMT.[10] Since that definition was established, a legion has followed, trying to establish their mark by redefining the definition. None have gained wide acceptance, so I suspect the definition should stand.

Each style is distinct in the soldering approach(s) it requires. Since the manufacturing style chosen for a design has an impact on the mix of SMT benefits achieved by the design, we'll discuss manufacturing style from a benefits/ limitations point of view in this chapter. Later, we'll discuss design for manufacturing, relative to each of the assembly styles.

1.4.1 *Type 1 SMT*

Type 1 SMT is pure surface mount. Type 1 uses SMCs only. No through-hole mounting of components is used in Type 1. (*Note*: It is possible to lead prep through-hole components and surface mount them on a Type 1 board.) Boards may be populated on one or both sides.

In regard to soldering approaches, Type 1 assembly is characterized by the use of reflow soldering. While it is conceivable that an exclusively SMT board could be produced by wave soldering, only a very small percentage of Type 1 production departs from the reflow soldering rule. Type 1 SMT is the only style that typically uses reflow soldering as the sole soldering method. Because it uses reflow soldering, Type 1 manufacturing allows the use of all styles of surface-mounted components. Some SMCs are not adapted to wave or immersion soldering.

Advantages
Type 1 SMT affords the maximum real-estate gains possible with surface mounting (in most applications). The only instance where this might not be true is in those

circuits which have a large contingent of hairpin-mounted resistors, or single in-line package (SIP) components.

Also, because Type 1 SMT generally involves reflow soldering, all types of SMCs may be used, including those intended exclusively for reflow soldering. In some other SMT approaches, the requirement for passing component bodies through a soldering wave places limits on the types of components allowed. Generally, leaded and leadless chip carriers, SOT-89s, aluminum electrolytic capacitors, connectors, and switches should not be passed through molten solder. Given proper processing, however, all such components are available in SMT styles appropriate for reflow soldering.[11]

Type 1 SMT is relatively straightforward from a manufacturing process viewpoint. Single-side-populated Type 1 boards require only one reflow soldering pass. (Double-side-populated Type 1 boards are generally soldered one side at a pass.) In contrast, Type 2 boards require both a reflow and a through-hole soldering process. Type 1 boards do not require adhesives. Type 3, and some Type 2, boards use adhesives to hold SMCs on the solder side of the board.

Finally, for product lines requiring a new automated-equipment investment, Type 1 manufacturing requires no investment in insertion equipment. Since several different insertion machines, each dedicated to one single package style, are required to automate the typical insertion line, Type 1 SMT may bring substantial savings in both capital equipment and floor space.

Limitations

Since Type 1 SMT allows no use of IMCs (except by special lead prepping), the most glaring disadvantage of this approach is in component availability. Even where all components for a design are available in SMC types, the designer may prefer to use an IMC in some instances for cost considerations. Type 1 SMT complicates such a decision. Types 2 and 3 do not place this restriction on the designer.

Type 1 SMT also requires learning about solder pastes, silk screening (or some solder-paste dispensing method), and reflow soldering—technologies foreign to many through-hole assembly shops. These technologies are well understood within the hybrid microcircuit industry, however, and a wealth of data is available to support the learning of them. But, the simple fact of the necessity of learning a new process must be viewed as one reason to avoid any unnecessary change. Type 2 SMT requires the same new processes, but Type 3 avoids both reflow soldering and the use of solder pastes.

Best Applications

Type 1 SMT is well suited to high-density memory applications. Here, the required components are readily available as SMCs and are produced in sufficient volumes to be quite cost-competitive with IMC versions. Board-size reductions, and the reductions in numbers of layers resulting from SMT conversion, generally make surface-mount methods more cost-effective than through hole for such applications.

Avionics has made good use of the very high densities, low weight, and reliability of Type 1 SMT. Table 1-2 lists data collected by Texas Instruments Incorporated on system weight and volume savings made by several users of TI copper-clad Invar substrates. Photographs of these applications are shown in our discussion of military and aerospace applications of SMT—presented at the end of this chapter.

Table 1-2. System Weight and Volume of Military SMT vs. IMC Design[12]

Military Electronics Manufacturer	Program	Reduction DIP/SMT	
		Volume	*Weight*
Martin Marietta	Avionics Memory Board	8:1	5:1
Eaton AIL Division	B1B Countermeasures	4:1	2:1
Rockwell Collins	JTIDS	8:1	3:1

Any application which requires that several boards be sandwiched closely together in a minimum housing volume should profit from a Type 1 approach. By sticking exclusively to SMCs with their low profiles, boards may be mounted much more closely than IMC heights would allow. Figure 1-5 shows the close-order stacking of both surface-mount and through-hole assemblies, with a typical mix of components. The example shown compares boards with SMT and IMC. The DIP ICs illustrated are 14- and 40-lead types vs. SO-14 and PLCC-44, chip vs. teardrop tantalum, chip vs. axial resistors, and SOT-23 vs. TO-92 transistors.

Figure 1-5 Close-order stacking of SMT boards vs. IMC assemblies.

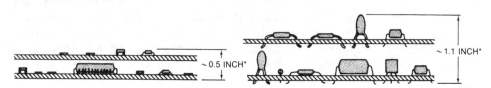

*Dependent on component heights, board thickness, and clearance required.

1.4.2 *Type 2 SMT*

Type 2 SMT is defined as a surface-mount assembly method where both SMCs and IMCs are mixed together on at least one side of the board. As this definition implies, Type 2 boards may be populated on one or both sides.

Type 2 SMT is distinct from Type 1 and Type 3 in that two separate soldering approaches are used. Where SMCs and IMCs are integrated on the same side of the board, the SMCs are generally reflowed and the IMCs are wave soldered (except where hand soldering is used in lieu of one or both automated soldering processes). Using hand soldering for the second-side IMCs, assemblies have been built with surface-mount and through-hole components mixed on both sides of the board. However, before suggesting such a construction, have a very solid justification in mind to placate the manufacturing and test people.

Type 2 SMT covers a wide variation of assembly styles. Boards may be populated on one or both sides. As long as at least one side combines surface attachment with insertion mounting, the assembly is Type 2.

Advantages

Where component availability or cost premiums are a barrier to converting an IMC assembly to SMT, Type 2 methods may provide a convenient path around the obstacle. Type 3 SMT also provides a way to integrate IMCs and SMCs in one assembly. However, Type 3 usage constrains the designer to using only immersion-solderable SMCs. Type 2 excels in applications where the designer wishes to take advantage of SMCs, which are not immersion solderable, and where he also wishes to use some IMCs for cost or availability reasons.

It's also worthwhile to note that certain IMCs, such as SIP or ZIP packages for resistor networks and multichip memory modules, are very real-estate efficient in single-board systems where board stacking height is not a concern. So, maximum use of available board square area is another factor that may suggest Type 2 usage. Just such an application is depicted in Figure 1-6, a very-high-density, 18-MHz, VME Bus, single-board, Unix Platform from Mizar, Inc.

Figure 1-6 An 18-MHz, VME Bus, single-board, Unix Platform. *(Courtesy Mizar, Inc.)*

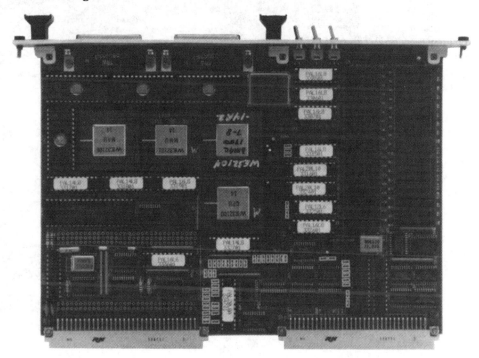

Limitations
The advantages of Type 2 SMT are high density, component availability, and component-cost shopping flexibility. In short, Type 2 SMT seems almost too good to be true. One might suspect that there's a catch. Manufacturability is that catch. The dual soldering-process requirement can complicate manufacturing, add process steps, and push manufacturing costs up while depressing first-pass yield. Test accessing may prove a problem. There are design rules to minimize the impact of these problems. We'll discuss these in detail when we cover manufacturability and design for test. For now, we will just note that the benefits of Type 2 SMT come at a price.

Best Applications
Type 2 SMT is a component-driven technology. It should be chosen when the benefits of SMT are very attractive, but component availability or component cost would shut the door on SMT and Type 3 won't open that door.

Figure 1-7 shows the Micro-Wand II™, a miniature hand-held computer/barcode scanner, built with the use of Type 2 SMT. Federal Express uses a Hand Held Products, Inc. SUPERTRACKER™ to keep track of the 840,000 packages they move in the United States in an average workday. The SUPERTRACKER provides FEDEX™ the means to tell you, in virtual real time by phone, where your package is and exactly when it will arrive at its destination. This product is small, hand-holdable, shirt-pocket sized, and is packed with features to do its intended job. And, to survive in its environment, it is rugged. It is designed to withstand a 7-foot drop onto concrete. This is a product which couldn't exist without Type 2 SMT.

Figure 1-7 Micro-Wand II™, a computer/barcode scanner. *(Courtesy Hand Held Products, Inc.)*

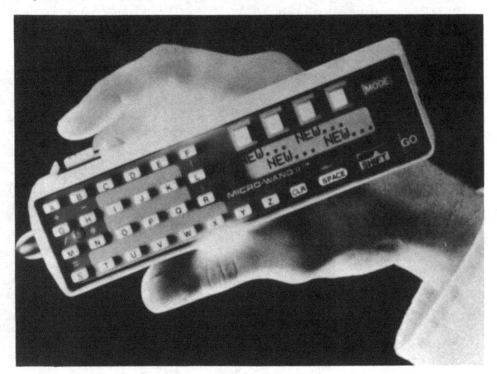

1.4.3   *Type 3 SMT*

Surface-mount technology, as a major commercial force, originated in Japan in the early 1970s. In the mid-1970s, Panasonic was granted patents on SMT pad designs (not enforced to deter SMT usage). Early Japanese surface mounting, and the bulk of surface mounting in Japan as late as 1990, was Type 3. Type 3 SMT was an innovative use of existing building blocks in the miniaturization of typical consumer products.

Type 3 SMT is the second alternative to mixing SMCs and IMCs. Because of its similarity to Type 2, some writers have suggested calling Types 2 and 3 just one type. This would not, however, simplify the understanding of the soldering technologies, because, unlike Type 2, Type 3 utilizes a single-pass soldering approach and no reflow soldering. In Type 3 assembly, all SMCs are glued to the solder side of a board. IMCs are inserted on the classical component side. The board is then immersion soldered, attaching components on both sides of the assembly in a single automated soldering operation.

Advantages

For a shop familiar with through-hole manufacturing, Type 3 presents a more manageable learning assignment than Types 1 or 2. There is no need to develop any solder-paste expertise, reflow soldering technology, or new solder-materials knowledge. However, in no way am I to be understood to say that learning to produce quality, high-yield, Type 3 assemblies is a cake walk. The Type 3 process book has less chapters than the other types, but each chapter is plump, and it must be fully digested if good results are to be obtained.

Type 3 SMT may well satisfy the need to adopt surface-mount technology while protecting an investment in insertion automation. Type 3 SMT integrates into an existing IMC automated line with less impact than either Type 2 or Type 1. You need only find space in your line for a placement machine, with its associated adhesive dispense and curing ancillaries. No additional soldering steps are required. (Note that standard single-wave soldering systems probably will not give quality soldering results on SMCs without major modifications. Generally, SMT adoption dictates that through-hole wave-soldering machines be replaced with SMC specialized equipment.)

Component engineers find Type 3 assemblies a boon. The types of SMCs most commonly used in this approach—chip capacitors, resistors, and SOTs—are the most widely available SMCs. They are also reasonably priced. SOICs (wave-solderable glue chip packages) are typically near parity with or less expensive than DIPs. Chip capacitors are usually less expensive, and chip resistors are usually very close to parity with their leaded cousins.

Limitations

Type 3 SMT constrains the designer in several areas where Types 1 and 2 give great flexibility.

The first significant constraint is in the area of components allowed. Any SMCs selected for Type 3 assembly must be immersion solderable. While a few shops are experimenting with wave soldering of J-lead plastic-leaded chip carriers (PLCCs), our investigation has failed to locate anyone using this technique in production. If any high pin-count components (>40 pins) are required in a design, a style of SMT that allows use of chip carriers might be in order. DIP packaging for >40 pins requires excessive board real estate. Also, the long internal lead lengths of large DIPs becomes a deterrent to their use. While PGAs may present an answer, these packages impose a severe cost premium over PLCCs and ceramic chip carriers (CCCs). SOLICs are usually limited to 28-pin devices or less. Higher pin counts in SO packaging present serious problems to silicon designers. It becomes difficult to locate all the bonding pads on the IC die in such a manner as to avoid bonding wire crossovers when the die is interconnected to the lead frame.

PWB designers are also more rigidly constrained in the component location of SMCs for immersion soldering than they are for reflow soldering. Components may not be located so close that solder bridges from one termination to another. Also, the component layout must take into account possible eddy currents, or shadowing in the solder wave, which might cause solder misses on terminations. These constraints are well understood. We will deal with them later in our discussion of Type 3 design. However, the constraints place limits on the benefits the designer can reap from Type 3 SMT.

Best Applications

Good applications for Type 3 SMT are determined by the mix of components required for the circuit. The top candidates have no very high pin-count devices (>40 pins), and have a large number of discretes. If an IMC circuit has a large contingent of capacitors and resistors, such that ≥60% of its components can be converted to immersion-solderable SMCs, the chances are good the circuit will be a cost-effective Type 3 changeover candidate. The circuit shown in Figure 1-8, an automobile radio, is an excellent example of Type 3 assembly.

Figure 1-8 Automobile radio in Type 3 SMT. Bottom side showing SMC passives. *(Courtesy of Delco Electronics Corp.)*

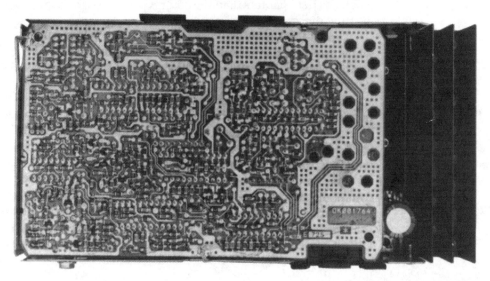

Since Type 3 assembly requires that most or all of the ICs on the circuit be inserted DIPs, with leads clenched, the ability to handle DIP insertion in the projected volume of a product line is a determining factor in the choice of Type 3 technology. A new product line, requiring new facilities, might require less capital equipment and building space if implemented in Types 1 or 2 SMT. Type 3 is a reasonable direction when altering an existing assembly line. And, where Type 3 SMT fits the application, costs and yields of the process can match world-class manufacturing standards.

1.5 Actual Examples of SMT Benefits and Limitations

1.5.1 *Automotive Applications*

The automotive industry of 1970 was only peripherally interested in electronics. The simple AM car radio was the most sophisticated electronic instrument aboard the average car. Now, electronics plays a vital role in producing cars and trucks that meet consumer and regulatory demands. Engine operation is computer-controlled to optimize mileage and reduce exhaust emissions over the full range of operating conditions that the car may see. On-board electronics may control the braking system, thus preventing wheel lock-up which causes skidding. Other systems may report fuel mileage, estimated time of arrival, and engine status to the driver. Some cars even have computer-synthesized voices which remind passengers to buckle up for safety, and, occasionally, announce that "A door is ajar."

The future holds even more in electronic benefits to the driver. Satellite positioning systems, in conjunction with electronic map databases, will give drivers a real-time display of their location and route to destination. Electronic automatic transmissions and anti-collision systems will also soon be available, according to Jeff Brenner of Chrysler.

In view of the exhaust emission restrictions and fuel economy laws, it is safe to predict that a car will soon pack more electronic computing horsepower than engine horsepower.

Automotive electronics application are typically driven by cost, reliability, and size. Weight reduction is also an important element in fuel economy. Reliability is particularly challenging in automotive applications because of the environment in which circuits must function. Interior auto electronics must be designed to operate over a temperature range from -50° F (-45.6° C) to +125° F (+51.7° C). Shock loads from road hazards and a constant vibration further complicate the designer's job. In the engine compartment, ambient temperature extremes may swing from -50° F to +250° F (+121.1° C). Engine-mounted electronics must also withstand an atmosphere of severe vibration, dust, and oil.

Automobile radios, like the one shown in Figure 1-8, are a perfect example of the drive to get more function into a given space. The opening in the dashboard is fixed. But yesterday's simple AM radio is no longer appealing to many customers. Features like FM stereo, auto tuning, automatic frequency control (AFC), stereo cassette player, CD player, Dolby™ noise suppression, etc., which were once found exclusively in packages two orders of magnitude larger than a car radio, must somehow now fit into the car's dash. And this compression of function must be carried out while simultaneously moving from the sedate environment of the home living room to the bouncing, rattling, temperature-cycling chamber of a car's interior.

Figure 1-9 Automotive fuel-injector control board. *(Courtesy of Delco Electronics Corp.)*

All three major U.S. auto makers have turned to SMT as one of the technologies needed to help them meet these challenges. A side benefit is that SMT greatly simplifies their current shift to pass-through automation of PWB assembly. Automotive after-market manufacturers and foreign auto makers also rely heavily on surface mount for miniaturization and reliability in a challenging environment.

Figure 1-9 illustrates one use of Type 2 SMT, which shrinks and makes durable a fuel-injection controller, allowing on-engine installation. Direct engine mounting allows EPA certification without having to transport for testing one of each auto body style with each available engine option. The available space and harsh environment in the crowded engine compartment led GM to choose SMT for this application.

Designed and manufactured by Delco Electronics Corp., the electronic fuel-injection (EFI) system shown in Figure 1-9 features the latest in surface-mount technology. Mounted accurately to within 0.001 of an inch, the board employs multilayered technology. Its four layers of circuitry contain more than 350 surface-mount devices, including plastic-leaded chip carriers (PLCCs), small-outline integrated circuits (SOICs), crystals, resistors, capacitors, and diodes.

Off-road equipment presents a truly brutal environment for electronic circuits. In Figure 1-10, we see an engine control circuit for a diesel road grader. Here, SMCs are used for their superior vibration resistance.

Figure 1-10
Printed-circuit assembly (vehicle controller) using surface-mount technology.
(Courtesy Case/IH and Magnavox.)

In laboratory tests of SMT, Caterpillar engineers have been able to vibrate boards until the leads bent and ultimately fractured on J-lead PLCCs, yet no damage was observed in the solder joints.

1.5.2 *Computers and Peripherals*

Mainframe computers have used SMT and other high-density technologies for years. But the advent of desktop and briefcase personal computers has greatly intensified the computer industry's interest in high-density circuitry approaches. Each year we see more and more functions crammed into a few cubic feet of desktop computer. Since 1975, desktop personal computers have progressed from rudimentary 4- and 8-bit machines to 32-bit systems capable of addressing more than 4 gigabytes of RAM and virtually unlimited off-line storage. As of this writing, desktop computers have invaded the performance territory of larger computers, such as the VAX™. In fact, with transputer boards installed, they're outperforming them. And a whole new class of computers, the truly portable laptop computer, has come to life and now rivals the desktop unit in power.

Figure 1-11
Briefcase-size PC
AT®-compatible
computer.
*(Courtesy of Toshiba
America, Inc., Irvine,
CA.)*

Mass-storage devices have followed this same progression curve. The 8-inch (20.3 cm) floppy disk of a few years ago was replaced by the 5¼-inch (13.3 cm) disk. The 3½-inch (8.9 cm) disk is becoming popular today. Amazingly, with each decrease in size came an increase in storage capacity. The hard disk of 1977 was a appliance the size of a dishwasher, and probably cost more than its operator's house. In 1987, a 760-Megabyte full-height 5¼-inch (13.3 cm) hard-disk drive cost less than a decent used car and it fit into one little corner of a desktop computer.[13] Technology advances on many fronts have contributed to this dramatic price and performance improvement. Next, let's look at some actual computer uses of SMT and see the role that surface mount has played.

Extensive use of surface-mount technology helped create the powerful laptop computer shown in Figure 1-11. The Toshiba T3100/20 personal computer packs the punch of an IBM PC AT® into a package that is less than ⅓ cu. ft. (8.8 liter) in size, and weighs 15 pounds (less than 7 kg). The 80286 processor runs selectably at either 4 or 8 MHz. Standard memory is 640 KB, and can be extended to 2.6 MB. The package shown includes a built-in 20-MB hard disk, and a 720-KB 3¹/₂-inch floppy disk drive.

Figure 1-12 IBM PS/2 Personal Computer motherboard. *(Courtesy IBM.)*

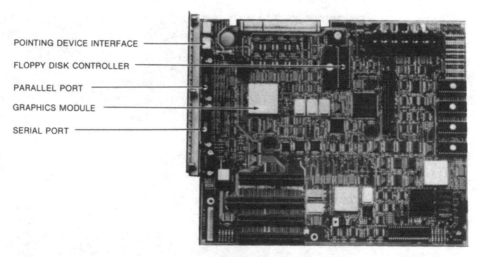

POINTING DEVICE INTERFACE

FLOPPY DISK CONTROLLER

PARALLEL PORT

GRAPHICS MODULE

SERIAL PORT

Figure 1-6 illustrated the very-high-density use of Type 2 SMT to impliment a single-board, single or multiuser, VME bus, Unix platform. The 32-bit WE32100 18-MHz Central Processor Unit (CPU), WE32101 Memory Management Unit (MMU), WE32106 1.3M Whetstone math coprocessor, and WE32104 32-bit DMA controller are in 100- to 125-pin ceramic PGAs. One megabyte of on-board zero-wait-state, 70-nanosecond, random-access memory (RAM) is provided using ZIP memory modules. These space-efficient ZIPs use PLCC memories but package them on tiny daughter boards, with pins coming off one side for through-hole mounting on the main board. The SOLICs at the rear of the assembly take care of the bus interfacing. An additional 63 SMCs are located on the bottom of the board.

Figure 1-13 A hard disk and controller card. *(Courtesy Plus Development.)*

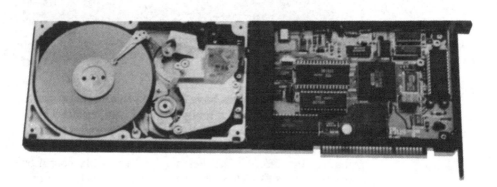

IBM has combined SMT and application-specific ICs (ASICs) to put a number of system functions on the motherboard of its PS/2™ line of 32-bit high-performance desktop PCs. High-resolution color graphics, the Micro-Channel™ input/output system, a video clock, and a disk controller are all built into custom circuits. Over 85% of the active and 80% of the passive devices in the PS/2 are

SMT.[14] With all its additional power and built-in features, the PS/2 has seventy-three fewer chips than the PC AT. Two ASICs have replaced seventy-five of the AT's chips.[15]

The card shown in Figure 1-13 uses a single expansion slot on a PC, leaving the front panel opening available for additional drives. On the card is a 40-Megabyte hard disk and its controller. Type 2 technology was selected for this application because some components were more available and less costly in IMC packages. Note the efficient use of the available space in this design.

1.5.3 *Consumer Electronics*

Since consumer electronics was the first application area for the widespread use of surface mounting, we can expect to find abundant examples of the use of SMT in today's consumer electronic products. As the following application reviews show, our expectations are on target. Note how the size- and cost-driven consumer designs differ from the performance- and size-driven computer applications we've just studied.

As mentioned earlier, one-for-one SMT/IMC circuit comparisons are not easy to find. When changing from IMCs, someone usually thinks of some additional things to put in the space that is saved. We'll start our look at actual consumer SMT applications with one of the few true part-for-part changeovers available for comparison.

Figure 1-14A shows the before and after views of a dedicated-frequency radio receiver. In Figure 1-14B, the receiver circuitry of both boards is the same. In one case, it is implemented in through-hole technology. In the other case, it's built of SMCs. Both products were built in the same factory using essentially the same process controls.

The products receive radio transmissions on two narrow frequency bands, 10.525 and 24.150 Gigahertz. These frequencies just happen to coincide with those used by X- and K-band radar speed detectors. There is a certain degree of popular interest in owning such a radio receiver. This popular appeal has blessed the manufacturer, Cincinnati Microwave, Inc., with sufficient sales to allow them to collect a valuable bank of heuristic data on both designs.

Several interesting contrasts emerge from a study of the statistical data on these two product lines—data collected by Cincinnati Microwave, Inc.

The surface-mount line delivers a significantly higher percentage of good product to final test. The first-pass yield of the IMC line varies between 86 and 87%, while the process quality of the SMT line runs from 93 to 97%. The wider process window seen on the SMT line is due to printed-wiring-board (PWB) quality. One PWB vendor's shop supplies the boards that produce the 97% yield. But this small shop cannot supply sufficient quantities of boards to support CMI's production requirements. The available second sources have not been able to match the small shop's board quality. Certainly, PWB quality varies in similar fashion for both lines, but the impact to the SMT line is far more significant.

Reliability of the surface-mount Passport detector is also significantly better than the IMC Escort detector. CMI is the sole repair agent for the product, so they have been able to track field failures closely. For an equal number of units in the field for an equal time, the through-hole product is three times more prone to failure. This is probably due mainly to the high-vibration automotive environment putting more stress on the IMCs, which typically have masses an order of magnitude greater than their SMC counterparts.

Figure 1-14
Passport™ and
Escort™ radar
detectors.
*(Courtesy of
Cincinnati
Microwave, Inc.)*

(A) Completed detectors.

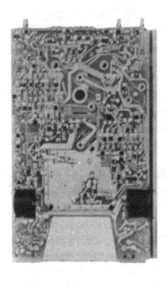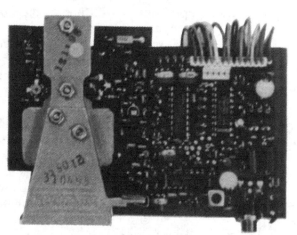

(B) Detector circuit boards.

Finally, of great interest to CMI and their shareholders, in their first full year of production of the SMT Passport detector, net corporate profits were up 54%. The IMC Escort receiver was a mature product, with 6 years in the market and 1.5 million units sold. Yet, the excitement generated by the pocket-size Passport and the manufacturing savings of SMT were able to generate this stellar bottom-line performance.

In Figure 1-15, we see a keyless automobile lock system. As shown, the electronics packaging must be dense, to fit all the required functions in the available space. And, SMT stands up in the rugged environment. This product is manufactured by Essex Technologies in the Carpinteria facility we discussed earlier. They serve as a subcontractor for ARA Manufacturing Co., the company that developed and markets the device.

Figure 1-15 A keyless auto lock system. *(Courtesy of ARA Manufacturing Co., Grand Prairie, TX.)*

(A) The main electronics module, showing Type 2 SMT technology.

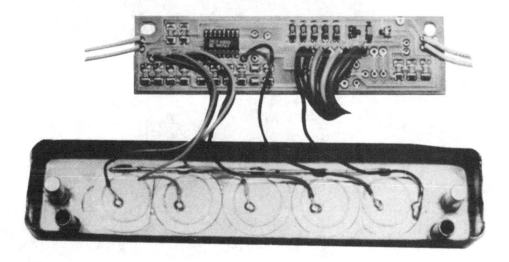

(B) The keypad module, showing Type 1 SMT (except for the hand-soldered wires and one DIP capacitor).

1.5.4 *Contract Assembly*

Figure 1-16 shows the SMT assembly area at Corlund Electronics, a contract electronic assembly company in Newbury Park, California. The compact placement system shown handles low-volume jobs and rapid changeovers. Another placement system, equally compact, automates high-volume passive and active-element assembly. According to Corlund Vice President Bruce MacDonald, flexibility is essential in a contract SMT assembly line. "We have to be ready to handle what the customer specifies, and handle it without massive tooling costs," says MacDonald. "This system allows us the flexibility we need to meet our customer's requirements." A typical contract assembly job, a disk controller assembly for Tandon, is shown in Figure 1-17.

Figure 1-16 An SMT assembly area at Corlund Electronics, showing the flexible SMT assembly equipment. *(Courtesy Corlund Electronics.)*

Figure 1-17 A typical product of a contract SMT assembly service. *(Courtesy of Tandon and Corlund Electronics.)*

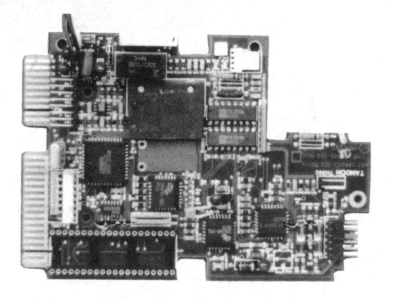

According to Mr. MacDonald, the SMT section of the plant occupies only 30% of the floor space, but it typically produces 60% of the contract assembly revenue at Corlund. Notice the contrast with the manual stuffing area at the same plant, shown in Figure 1-18.

Figure 1-18 A medium-volume IMC assembly area at Corlund Electronics. *(Courtesy Corlund Electronics.)*

1.5.5 *Military and Aerospace Applications*

Figures 1-19, 1-20, and 1-21 show some typical military avionics applications of SMT. Here surface mount is critical in the packaging of the needed high-level functions into the smallest space available, thus keeping assembly weight low for flight systems, as required in modern warplanes. These are the weight-savings examples shown in Table 1-2.

Figure 1-19 An avionics board from LANTIRN. *(Courtesy Martin-Marietta.)*

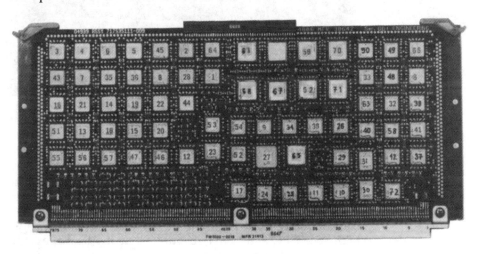

Figure 1-20 A
B1B countermea-
sures assembly.
*(Courtesy AIL
Division, Eaton.)*

Figure 1-21 A
JTIDS assembly.
*(Courtesy Rockwell
International, Inc.)*

1.5.6 *Industrial Applications*

The drive toward higher frequencies in electronic systems is echoed by a drive for
higher performance in the test equipment that must test these systems. Figure 1-22
shows the use of SMT for a performance-driven advanced board tester. This timing
control board, from Teradyne's L290 VLSI module functional test system, uses
high-speed ECL and SMT to implement the board-level test of complex VLSI
modules (with hundreds of pins) at pattern rates up to 40 MHz.

Figure 1-22 A timing generator board from Teradyne's VLSI functional tester. *(Courtesy Teradyne, Boston, MA.)*

Figure 1-23 A hand-held digital multimeter. *(Courtesy John Fluke.)*

(A) The Fluke 77 digital multimeter.

(B) Top side of circuit board, showing IMC technology and one hybrid module.

(C) Backside of circuit board, with a Sharp SM5 microprocessor at top and a Fluke A/D convertor at the bottom, both in 60-pin flat packs.

Figure 1-24
Hand-held
controller card
for a warehous-
ing system.
*(Courtesy MSI
Data.)*

Figure 1-25 A
heart pacemaker
which automati-
cally responds to
body movement.
*(Courtesy
Medtronic, Inc.)*

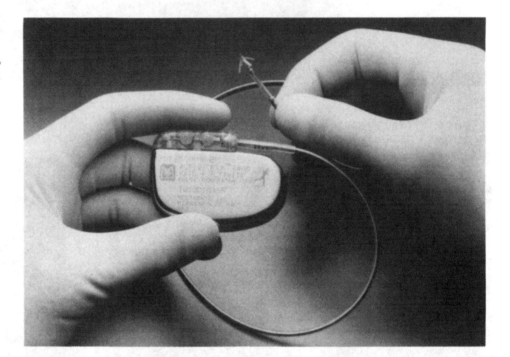

(A) The new Activitrax™ Pacemaker.

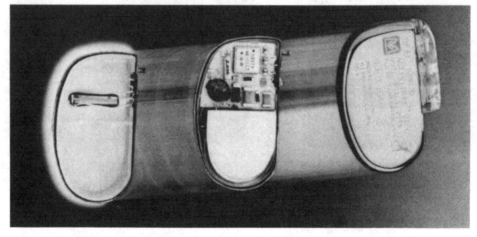

(B) Activitrax™ Pacemaker, showing the pulse generator, circuitry, and sensor.

Large test electronics is not the only instrumentation market segment turning to the advantages of SMT. Small instruments, like the ones shown in Figures 1-23 and 1-24, can be hand-held. But, customers are seeking ever increasing functions within the tiny box. SMT is often the answer in designing hand-held instruments.

1.5.7 *Medical Uses of SMT*

A piezoelectric sensor is shielded deep inside the solid titanium casing of the new Activitrax™ pacemaker made by Medtronic, Inc. When this sensor is slightly stressed or bent by the patient's body activity, the pacemaker automatically increases or decreases the heart rates. The pacemaker will react to as little as ten one-millionths of an inch of movement. It will increase heart rates as high as 150 beats per minute, and will decrease them as low as 60 bpm for sleeping or resting. Pacemakers are usually implanted in a simple 20- to 30-minute surgical procedure.

Both the size and reliability benefits of SMT are highly valued in the medical application shown in Figure 1-25. People's lives are being saved each day by the use of surface-mount technology in such devices.

Not only is SMT used to give people a new lease on life, it is applied in medicine to make life more bearable for the living. The circuit shown in Figure 1-26 is used to electronically stimulate nerve centers, thus blocking pain. Trans-subcutaneous Electronic Nerve Stimulation (TENS) technology has proved effective for some chronic pain sufferers where all other resources, short of heavy pain-killing drugs, had already failed.

Figure 1-26
Trans-subcutaneous Electronic Nerve Stimulation (TENS) unit. *(Courtesy StaoDynamics, Longmont, CO.)*

The device in Figure 1-27 improves the quality of life of patients who need regular injections for chemotherapy, pain control, infertility treatment, or neonatal care. The micro pump shown in Figure 1-27A is small enough to be carried in normal daily activity, thus freeing people to lead a useful life, rather than constantly being required to visit a clinic or be hospitalized for injections.

Figure 1-27 A miniature infusion pump and the pair of circuit boards which control it. *(Courtesy Pharmacia Deltec, Inc., St. Paul, MN.)*

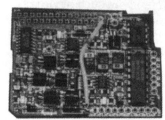

(A) The CADD-LD micro infusion pump and two circuit boards.

(B) Close-up of the microprocessor board. Note unique billboard mounting of capacitor at upper right.

Figure 1-28 SMT
telecommunica-
tions hybrid
circuit board.
*(Courtesy Mitel
Semiconductor.)*

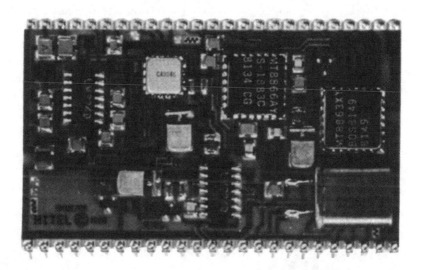

1.5.8 *Telecommunications Applications of SMT*

The telecommunications industry is committed to SMT for its reliability, small size, and low cost. The circuit shown in Figure 1-28 is a hybrid which uses SMT mounting of ICs to avoid the yield and cost problems associated with chip and wire techniques.

As telecommunications moves from a stationary to a mobile-based technology, SMT becomes even more necessary for weight and packaging-density savings. Figure 1-29 shows a portable 800-MHz PC programmable radio which has both trunk and talk-around capability. Type 3 SMT density and weight savings play a vital role in miniaturizing this radio. And, as telecommunications switching becomes more sophisticated, more circuitry must be packed into control hardware at the switching office. Figure 1-30 shows how SMT is applied to meet this challenge.

1.6 References

1. Compare to Gnostic Concepts, *The Impact of Surface Mount Technology, A Multi-Client Study*, San Mateo, CA, 1984.
2. Kolias, John T., "Packaging Impact on System Performance," *IEPS Surface Mount Technology Compendium of Technical Articles Presented at the 1st, 2nd, and 3rd Annual Conference*, International Electronics Packaging Society, Inc., Glen Ellyn, IL, 1984, pg. 363.
3. Hutchins, Dr. Charles, "VLSI Prompts Designers to Ponder Packaging," *Electronic Design*, Hasbrouck Heights, NJ, December 22, 1983.
4. Moore, Robert and Weaver, Bruce, "Electrical and Mechanical Considerations in Design of LCC for Higher Performance Applications," *IEPS SMT Compendium of Technical Articles from 1st, 2nd, and 3rd Annual Conference*, International Electronics Packaging Society, Inc., Glen Ellyn, IL, 1984, pg. 529.

5. Kuk, Donald, Advanced Technology Development Manager, Cincinnati Microwave, Cincinnati, OH; Telephone interview, 19 November 1986.

6. Lewis, E. T., "Interconnection Designs Considerations for VLSI Multichip Packaging," *1986 Proceedings—International Symposium on Microelectronics*, International Society for Hybrid Microelectronics, Reston, VA, October 1986, pg. 722.

7. Bourns, Inc., Advertisement, *Electronic Component News*, July 1987, pg. 13.

8. Caswell, Greg, et al, *Surface Mount Technology*, International Society for Hybrid Microelectronics, Reston, VA, 1984, pg. 6.

9. Moon, Roger, Manufacturing Engineering Manager, Essex Technologies, Carpinteria, CA; Telephone interview, 25 September 1986.

Figure 1-29 A mobile 800-MHz trunk radio. *(Courtesy E. F. Johnson.)*

(A) The basic radio. Unit can be programmed using a personal computer.

(B) Radio with cover removed and top circuit board set aside showing the IMC technology.

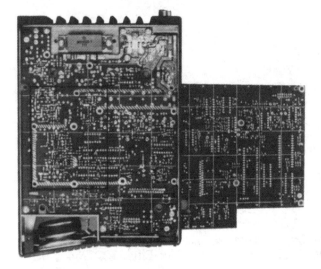

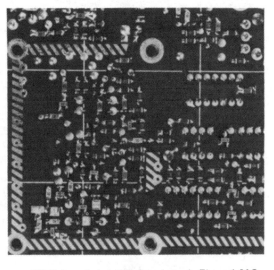

(C) Bottom view of circuit boards shown in Figure 1-29B. Note Type 3 SMT technology.

(D) Enlarged view of the boxed area in Figure 1-29C, giving a better view of the SMCs.

10. Hall, James W., "Soldering Techniques for Manufacturing Surface Mounted Circuits," *International Journal for Hybrid Microelectronics*, Volume 6, Number 1 (currently available from ISHM) Reston, VA, October 1983.

11. Bruno, Terry, "Update—Surface Mount Aluminum Electrolytics," *SMTA Newsletter*, SMTA, Edina, MN, November 1986.

12. Texas Instruments Incorporated, El Paso, TX; and SMTA, Edina, MN. From material presented at EXPO SMT, 1 October 1986, Las Vegas, NV.

13. Jason, Warren, Telecommunications and Computer Applications Consultant, Integrated Data Concepts, Los Angeles, CA, Telephone interview, 28 November 1986.

14. Francis, David, President, Micro Process Technology (Newsletter), San Jose, CA, Telephone interview, September 1987.

15. Alster, Norm, "PS/2: Surface Mount Makes It an Inside Job," *Electronic Business*, 1 September 1987, pg 44.

Figure 1-30
Modern telecommunications controls circuit board. (Courtesy Rockwell International, Inc.)

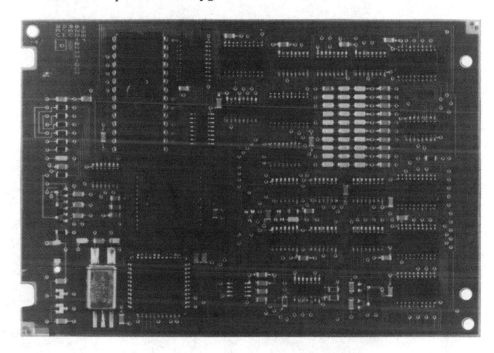

Figure 2 Components—a critical variable in the SMT equation.

(A) Assorted SMCs, the building blocks of SMT circuits.

(B) PLCC socket on a dedicated PC board. (Courtesy Loranger International Corp.)

2 Surface-Mount Components (SMCs)

2.1 SMT Passives
2.2 SMT Actives
2.3 Connectors and Sockets
2.4 Switches
2.5 Other Components

The starting point of SMT is the components. Figure 2 shows advances that have been made in component miniaturization technology since the advent of surface mounting.

This book was written for the working engineer who must continually refer to component specifications and land-dimension drawings. In an attempt to serve the engineer's needs, in the first edition, published in 1989, we included detailed component data and land recommendations in Appendix A. In the ensuing years, as we have watched the proliferation of SMT device packages, we have recognized the limitations of this strategy. In our 1995 update, we are adding considerable component information, but please be cautioned to check EIA and JEDEC standards and consult component vendors for package outline information in order to ensure that you are working with "all the news that's fit to print."

In the chapter, we'll discuss component selection and specifications for an SMT project. In component selection, one important factor to consider, in addition to electrical characteristics, is the manufacturing equipment capabilities. The placement system used to assemble boards will restrict both the types of components which can be automatically handled, and the type of component carriers in which the parts must be purchased. Teamwork from all departments should be used in component selection so as to avoid costly surprises on the production and test floor.

Note: While every effort has been made to include the latest data on parts standards, new parts and the documents controlling them are rapidly changing. For generalized, up-to-date information on component packages, always refer to the Electronic Industries Association standard, EIA-DPD-100, JEDEC Registered, and Standard Mechanical Outlines for Electronic Parts.

2.1   SMT Passives

Let's begin our study of surface-mounted components with *passives*. Passives are circuit elements (components) which do not change their basic character when an electrical signal is applied. Passives include such elements as capacitors, resistors, inductors, etc.

2.1.1   *Capacitors*

Capacitors are relatively mundane components in through-hole assemblies. Certainly, the design engineer may have to do some homework occasionally to find a part with suitable voltage ratings, capacitance, dissipation factor, etc. But, for most through-the-hole uses, capacitor selection is a nearly rote activity.

This is not the case with SMT. Surface-mount package constraints place practical limits on the capacitance and voltage ratings available. SMT processing may further constrain component selection, since the component may be required to survive a trip through a reflow soldering furnace or even direct immersion in a molten solder bath. Fortunately, there is a rapidly growing collection of parts that answer some or all of these needs. In the following, we'll analyze each of the major SMT capacitor families available today.

We have not attempted to catalog every available SMT component. Even if we had, new offerings appear daily. Only common parts are covered herein. We mention this in order to caution you that you should not be deterred from a search for a specific SMT component which meets a particular requirement merely because we have not mentioned it in this work.

Ceramic Chip

The *ceramic chip* is basically the same animal you're used to in the leaded world of ceramic capacitors. It is the building block of the ceramic leaded capacitor, only no leads have been soldered to its end terminations, and it hasn't been encapsulated. Beyond those fundamental differences, SMT ceramic capacitors have end terminations designed to yield quality results in the required soldering process (at least, they had better have such).

Figure 2-1 A cutaway view of an MLC chip capacitor.

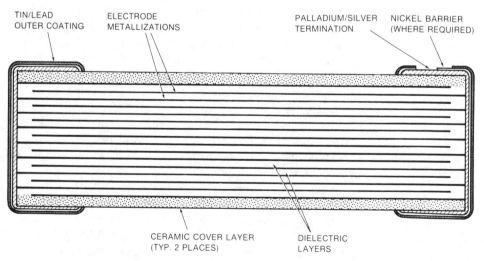

TIN/LEAD OUTER COATING   ELECTRODE METALLIZATIONS   PALLADIUM/SILVER TERMINATION   NICKEL BARRIER (WHERE REQUIRED)

CERAMIC COVER LAYER (TYP. 2 PLACES)   DIELECTRIC LAYERS

Ceramic chip capacitors are the workhorse capacitors of SMT. They go by various names, such as Chip Cap, Multilayer Ceramic Chip (MLC), and Ceramic Cap. Figure 2-1 shows the construction details of an MLC cap. As explained in the next few paragraphs, that designation is descriptive of the component.

MLC chips are built up of multiple-laminated layers of a partially metallized green ceramic material (Figure 2-1). Each layer's metallization extends all the way to one end of the device, but stops just short of the other end. The layers are arranged such that the terminated end alternates. In green form, the layers are then laminated by heat and pressure. The green monolith is fired for up to 16 hours at around 1200°C. End terminations are then added to the fired monolith and it is again fired at around 800°C for about 15 minutes. A nickel barrier layer may then be plated or sputtered over the base termination---for protection in the soldering process. Some parts are made with a palladium-based termination. Unless process engineering develops a foolproof method for soldering of such parts, they are best avoided. They may be slightly cheaper than capacitors using a precious-metal system, but the savings in component cost will almost certainly be offset by higher processing expenses brought on by the poor solderability of the palladium metallization.

Of particular concern in specifying chip capacitors, end-termination materials and methods have a dramatic impact on soldering yields. Most chip caps are terminated with thick-film mixtures of silver, palladium, and glass grit. Silver readily goes into solutions with tin, the major ingredient in most electronics solders. This silver/tin solubility poses the danger that the silver in the termination will enter into solution with the tin in the solder, leaving the end termination damaged, and destroying its cohesive bond to the capacitor (the metallization going into solution with the solder is a phenomenon called *leaching*).[1] Nickel barriers, with an over-coating of tin/lead for solderability, are the most common approach to solving this problem from the component side of the equation.

Leaching is a time/temperature based problem, as shown in Table 2-1. As long as time-at-temperature is kept within safe limits, as indicated by the

Table 2-1. Leach Resistance of Typical MLC Capacitors[2]

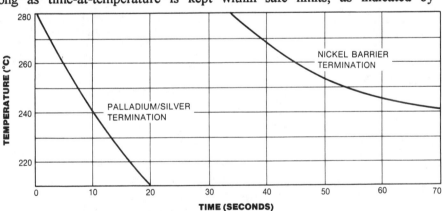

component vendor, leaching should not be a problem. However, remember that this time/temperature consideration must include any rework or repair that the component may undergo in its lifetime.

Since leaching was identified as an SMT problem several years ago, the tendency has been toward specifying the use of nickel barriers on all parts to avoid any possible problems. According to capacitor manufacturer AVX, several sophisticated users are now bucking this trend and have returned to palladium silver. They prefer to eliminate leaching through process control because the palladium-silver termination is more stress resistant and poses no tin/nickel intermetallic dangers. We'll look at the pluses and minuses of nickel barriers in more detail in Chapter 9, Section 8.2.2, and in Appendix A, Section A.1.1.

Figure 2-2 Standard Tape & Reel packaging dimensions (see Table 2-2).

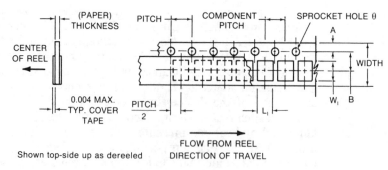

(A) Punched carrier tape.

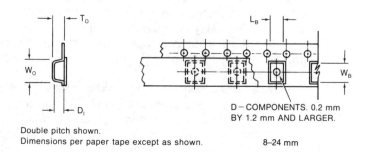

(B) Embossed tape (double pitch shown).

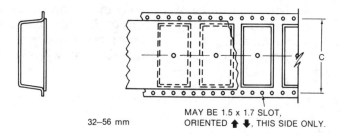

(C) Large-pocket embossed tape.

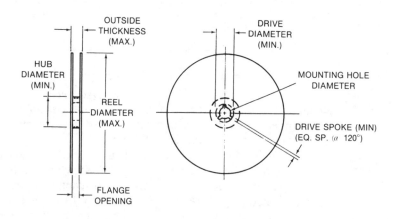

(D) Reel dimensions (see Table 2-3).

Carrier Packaging—Chip capacitors have been used in the hybrid industry for many years. Typically, they were supplied in bulk containers and were either placed on assemblies by hand, or sorted and fed to automated placement equipment by a bulk vibratory feeder equipped with tooling to suit the component's size and aspect ratio. High-volume application of chip caps to PWBs brought pressure for higher speed and reliability in handling methods. Today, the most widely accepted solution is Tape & Reel packaging. Figure 2-2 shows the basic dimensions of SMT tape and reels (Tables 2-2 and 2-3). For more details, please refer to the referenced specifications.

Table 2-2. Standard SMT Tape & Reel Packaging Dimensions (per EIA-RS-481A Specifications)

Width	Pitch	Component Pitch	Thickness (see Note 2)	Sprocket Hole φ	A	B	C	D	D_I Embossed Only	L_B Embossed Only	L_I Punched Only	T_O Embossed Only	W_B Embossed Only	W_I Punched Only	W_O Embossed Only	Z (see Note 3)
8	4	4	1.1 (max)	1.5	1.75	3.5	—	1.0 (min)	Note 1	Note 1	Note 2	2.4 (max)	Note 1	Note 2	4.2 (max)	25
12	4	4, 8	—	1.5	1.75	5.5	—	1.5 (min)	Note 1	Note 1	—	4.5 (max)	Note 1	—	8.2 (max)	30
16	4	4, 8, 12	—	1.5	1.75	7.5	—	1.5 (min)	Note 1	Note 1	—	6.5 (max)	Note 1	—	12.1 (max)	40
24	4	12, 16, 20, 24	—	1.5	1.75	11.5	—	1.5 (min)	Note 1	Note 1	—	6.5 (max)	Note 1	—	20.1 (max)	50
32	4	16,24, 32	—	1.5	1.75	14.25	28.5	2.0 (min)	Note 1	Note 1	—	10.0 (max)	Note 1	—	25.6 (max)	50
44	4	24,32, 40	—	1.5	1.75	20.25	40.5	2.0 (min)	Note 1	Note 1	—	10.0 (max)	Note 1	—	37.6 (max)	50
56	4	48	—	1.5	1.75	26.25	52.5	2.0 (min)	Note 1	Note 1	—	10.0 (max)	Note 1	—	49.6 (max)	50

Notes:

1. D_I, L_B, W_B, and W_O are determined by the component size. Dimensions are selected to yield a cavity with a 0.05 (min) to 0.50 (max) clearance on each axis, and to prevent a component rotation of more than 10° within the cavity.

2. Material thicknesses, L_I and W_I, are determined by the component size. Dimensions are selected per Note 1.

3. Tape with components must withstand bending around a mandrel of radius Z without damage to tape, cover seal, or components.

4. All dimensions are in millimeters.

Table 2-3. Tape reel dimensions.

Tape	Reel Diameter (max)	Flange Opening	Outside Thickness	Hub Diameter	Mounting Hole	Optional Drive Diameter	Drive Spoke
8	330	8.4	14.4	50	13.0	20.2	1.5
12	330	12.4	18.4	50	13.0	20.2	1.5
16	330	16.4	22.4	50	13.0	20.2	1.5
24	330	24.4	30.4	50	13.0	20.2	1.5
32	330	32.4	38.4	50	13.0	20.2	1.5
44	330	44.4	50.4	50	13.0	20.2	1.5
56	330	56.4	62.4	50	13.0	20.2	1.5

Standards—For chip capacitors, the standard sizes shown in Figure 2-3 and Table 2-4 are recommended by the EIA.

Table 2-4. EIA Chip Capacitor Standard Sizes. **Note:** Miniaturized parts in 0603, 0504, and 0402 sizes are now also available and are being widely used in high density applications.

EIA Size Designation	Dimensions (mm/in) Length	Width	Thickness (Max)	Terminal
0805	1.8–2.2 (0.070–0.087)	1.0–1.4 (0.040–0.055)	1.3 (0.050)	0.3–0.6 (0.012—0.024)
1206	3.0–3.4 (0.118–0.134)	1.4–1.8 (0.055–0.070)	1.5 (0.060)	0.4–0.7 (0.016—0.028)
1210	3.0–3.4 (0.118–0.134)	2.3–2.7 (0.090–0.106)	1.7 (0.067)	0.4–0.7 (0.016—0.028)
1812	4.2–4.8 (0.166–0.190)	3.0–3.4 (0.118–0.134)	1.7 (0.067)	0.4–0.7 (0.016—0.028)
1825*	4.2–4.8 (0.166–0.190)	6.0–6.8 (0.236–0.268)	1.7 (0.067)	0.4–0.7 (0.016—0.028)

* Not included in SMTA Component Standards Group recommended sizes.

Figure 2-3. Chip capacitor standard sizes.[3]

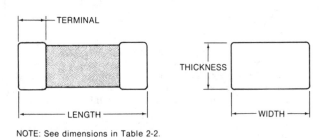

NOTE: See dimensions in Table 2-2.

Design Considerations—End-termination specifications should be written based on the required assembly processing requirements. For reflow soldering, components with end terminations as shown in Figure 2-4 are recommended for maximum soldering yield. However, components with a greater tin content may provide a longer shelf life. The outer tin-lead coating should never have a reflow temperature more than 5°C (9°F) higher than the melting point of the solder paste to be used in reflow attachment. For wave soldering, end terminations with a tin content matching that of the solder in the wave-machine pot are recommended.

Figure 2-4 Chip capacitor end terminations.

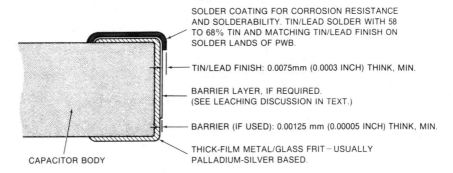

SOLDER COATING FOR CORROSION RESISTANCE AND SOLDERABILITY. TIN/LEAD SOLDER WITH 58 TO 68% TIN AND MATCHING TIN/LEAD FINISH ON SOLDER LANDS OF PWB.

TIN/LEAD FINISH: 0.0075mm (0.0003 INCH) THINK, MIN.

BARRIER LAYER, IF REQUIRED. (SEE LEACHING DISCUSSION IN TEXT.)

BARRIER (IF USED): 0.00125 mm (0.00005 INCH) THINK, MIN.

THICK-FILM METAL/GLASS FRIT – USUALLY PALLADIUM-SILVER BASED.

CAPACITOR BODY

Because of their leadless design, chip capacitors have a lower equivalent series resistance (ESR) and equivalent series inductance (ESL) than their IMC cousins. Therefore, they are a boon in high-speed circuitry design.

Tantalum and Aluminum Electrolytic Capacitors

Tantalum capacitors are typically specified when an extremely stable capacitor is needed. They are the most stable of the electrolytic capacitors. SMT electrolytics, with other electrolyte materials, should be specified in the same manner as that discussed here for tantalums.

Figure 2-5 Typical tantalum "brick" capacitors. *(Dimensions are given in Table 2-5.)*

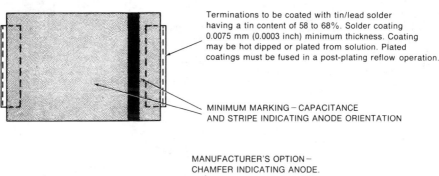

Terminations to be coated with tin/lead solder having a tin content of 58 to 68%. Solder coating 0.0075 mm (0.0003 inch) minimum thickness. Coating may be hot dipped or plated from solution. Plated coatings must be fused in a post-plating reflow operation.

MINIMUM MARKING – CAPACITANCE AND STRIPE INDICATING ANODE ORIENTATION

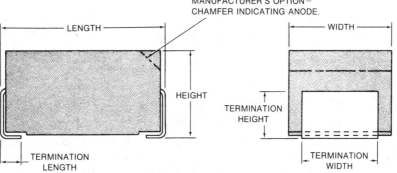

MANUFACTURER'S OPTION – CHAMFER INDICATING ANODE.

LENGTH

WIDTH

HEIGHT

TERMINATION HEIGHT

TERMINATION LENGTH

TERMINATION WIDTH

Table 2-5. Typical Tantalum "Brick" Capacitors (See Figure 2-5)

EIA RS-228 Size	Size Code/Standard Capacitive Range (mm/in)			
	A Case	*B Case*	*C Case*	*D Case*
Length	3.0–3.4 (0.118–0.134)	3.3–3.7 (0.130–0.146)	5.7–6.3 (0.224–0.248)	6.8–7.6 (0.268–0.299)
Width	1.4–1.8 (0.050–0.070)	2.6–3.0 (0.102–0.118)	2.9–3.5 (0.114–0.138)	4.0–4.6 (0.157–0.181)
Height	1.4–1.8 (0.055–0.070)	1.7–2.1 (0.067–0.083)	2.2–2.8 (0.087–0.110)	2.5–3.1 (0.098–0.122)
Termination Length	0.5–1.1 (0.020–0.043)	0.5–1.1 (0.020–0.043)	0.5–1.1 (0.020–0.043)	0.5–1.1 (0.020–0.043)
Termination Width	1.1–1.3 (0.043–0.051)	2.1–2.3 (0.083–0.090)	2.1–2.3 (0.083–0.090)	2.3–2.5 (0.090–0.098)
Termination Height	0.7 Min (0.028)	0.7 Min (0.028)	1.0 Min (0.040)	1.0 Min (0.040)

The dimensions for four standard sizes of this package, recognized by the EIA, are given in Table 2-5. The package outline is illustrated in Figure 2-5. Four additional EIA sizes form an "extended range" for high capacitors or voltage requirements. Their outline is shown in Figure 2-6 and dimensions listed in Table 2-6. See the transistor specifications (discussed later in the chapter) for comments on end-termination treatments for use on component procurement drawings. Other package formats are available, but often present difficulties in PWB assembly.

Figure 2-6 Extended-range tantalum "brick" capacitors. (Dimensions given in Table 2-6.)

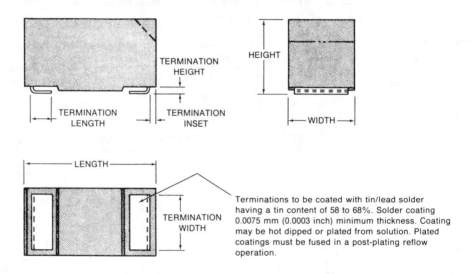

Terminations to be coated with tin/lead solder having a tin content of 58 to 68%. Solder coating 0.0075 mm (0.0003 inch) minimum thickness. Coating may be hot dipped or plated from solution. Plated coatings must be fused in a post-plating reflow operation.

Construction---While construction details for tantalum capacitors vary from manufacturer to manufacturer, these traits are common. A solid tantalum dielectric material is employed, from which these capacitors draw their name. End-termination leads are welded to the plates of the capacitor, and a protective encapsulation of epoxy resin plastic may be molded around the capacitor body such that the end terminations are left exposed for attachment to a PWB.

Carrier Packaging---Like chip capacitors, tantalum capacitors are available in Tape & Reel packaging. This is the preferred handling method. In small quantities, tantalum bricks may also be purchased in plastic sticks or in bulk. Bulk packaging is not recommended, however, except for very small quantities where hand placement will be the assembly method. This is because tantalum capacitors are polarized components, and typical polarizing markings vary from being difficult to being impossible to automatically orient. Also, we should note that many automatic placement systems do not have standard accessories that permit them to deal with tantalums in plastic sticks. Therefore, Tape & Reel packaging is the preferred handling method for these parts. Table 2-7 shows tape sizes for standard tantalum bricks.

Table 2-6. Extended-Range Tantalum "Brick" Capacitors (See Figure 2-6)

| EIA RS-228 Size | Size Code/Extended Capacitive Range (mm/in) | | | |
	3518 Case	*3527 Case*	*7227 Case*	*7257 Case*
Length	3.3–3.7 (0.130–0.146)	3.3–3.7 (0.130–0.146)	6.9–7.5 (0.272–0.295)	6.9–7.5 (0.272–0.295)
Width	1.6–2.0 (0.063–0.079)	2.4–3.0 (0.095–0.118)	2.4–3.0 (0.095–0.118)	5.2–6.2 (0.205–0.244)
Height	1.7–2.1 (0.067–0.083)	1.7–2.1 (0.067–0.083)	2.5–3.1 (0.098–0.122)	3.0–3.7 (0.118–0.146)
Termination Length	0.6–1.0 (0.024–0.040)	0.6–1.0 (0.024–0.040)	0.8–1.2 (0.031–0.047)	0.8–1.2 (0.031–0.047)
Termination Width	1.6–1.8 (0.063–0.071)	2.4–2.6 (0.095–0.102)	2.4–2.6 (0.095–0.102)	5.4–5.8 (0.213–0.228)
Termination Height	0.7 (0.028)	0.7 (0.028)	1.0 (0.040)	1.2 (0.047)
Termination Inset	0.4–0.6 (0.016–0.024)	0.4–0.6 (0.016–0.024)	0.6–0.8 (0.024–0.032)	0.6–0.8 (0.024–0.032)

Table 2-7. Tape & Reel Sizes for Tantalum Bricks

| Standard Brick Size | Tape Specifications | | Extended Brick Size | Tape Specifications | |
	Width	*Pitch*		*Width*	*Pitch*
A Case	8 mm	4 mm	3518	8 mm	4 mm
B Case	8 mm	4 mm	3527	8 mm	4 mm
C Case	12 mm	8 mm	7227	12 mm	8 mm
D Case	12 mm	8 mm	7257	12 mm	8 mm

2.1.2 *Resistors*

Leadless resistors are, as of this writing, the most widely used component in the SMT stable. They are to SMT what the $1/4$-watt carbon or metal-film axial resistor is to insertion-mount processing. Unfortunately, while SMT resistors may offer real-estate and reliability benefits over their IMC relatives, penny for penny, they don't often rival them for performance. Typical $1/4$-watt metal-film resistors are ±1% tolerance, have a thermal coefficient of resistance (TCR) of 100 parts per million (PPM), and a 1% stability over 10,000 hours. A similar $1/8$-watt chip resistor, priced in the same ball park, will have a 5% tolerance, 200 PPM, and 2% stability over

10,000 hours.[5] Tighter specifications in SMT parts are available, but they may carry a price premium. Just remember, it's lower system cost we're after, and we "may have to pay a cent here to get two there."

On the positive side of the scales, chip resistors are like chip capacitors in that the absence of leads reduces parasitic inductance and allows a faster, more noise-tolerant operation.

Chip Resistors

The 1/8-watt chip resistor is the most commonly specified SMT resistor in U.S. design. Commercial chip resistors are typically offered in ±5% and ±1% tolerances. Standard resistance values, as defined by the EIA, are listed in Table 2-8.[6]

Table 2-8. Standard Resistor Values

SERIES E24 (±5%)		SERIES E96 (±1%)							
10X	33X	10.0X	13.3X	17.8X	23.7X	31.6X	42.2X	56.2X	75.0X
11X	36X	10.2X	13.7X	18.2X	24.3X	32.4X	43.2X	57.6X	76.8X
12X	39X	10.5X	14.0X	18.7X	24.9X	33.2X	44.2X	59.0X	78.7X
13X	43X	10.7X	14.3X	19.1X	25.5X	34.0X	45.3X	60.4X	80.6X
15X	47X	11.0X	14.7X	19.6X	26.1X	34.8X	46.4X	61.9X	82.5X
16X	51X	11.3X	15.0X	20.0X	26.7X	35.7X	47.5X	63.4X	84.5X
18X	56X	11.5X	15.4X	20.5X	27.4X	36.5X	48.7X	64.9X	86.6X
20X	62X	11.8X	15.8X	21.0X	28.0X	37.4X	49.9X	66.5X	88.7X
22X	68X	12.1X	16.2X	21.5X	28.7X	38.3X	51.1X	68.1X	90.9X
24X	75X	12.4X	16.5X	22.1X	29.4X	39.2X	52.3X	79.8X	93.1X
27X	82X	12.7X	16.9X	22.6X	30.1X	40.2X	53.6X	71.5X	95.3X
30X	91X	13.0X	17.4X	23.2X	30.9X	41.2X	54.9X	73.2X	97.6X

Note: X in the table is a power of 10 multiplier yielding the following ranges: E24 ±5%, 10 ohms to 2.2 megohms; E96 ±1%, 49.9 ohms to 2.2 megohms.

Construction---Chip resistors are generally built in a batch process. Many individual resistors are fabricated on a single ceramic substrate. The process derives from thick-film hybrid manufacturing methods. First, conductive metal terminations are screened on the top surface of the substrate where individual resistor ends will be. These terminations are then fired. Next, conductive ink, generally based on ruthenium oxide, is screened between the terminations and fired. The resistors are then laser trimmed to tolerance. The substrate is cut into individual resistors, which are terminated with a palladium-silver metallization. A nickel barrier is usually applied over the base termination to prevent silver leaching, and a final solderable coating, generally tin/lead, is co-plated or dip-applied over the barrier.

Figure 2-7 shows the standard construction of a chip resistor and details the critical dimensions for standard sizes. Table 2-9 lists the dimension values.

Within the U.S. market, the 1206 resistor has been the dominant size, but there is a trend toward greater use of the 0805 for high-density designs. The 0603 (nominally 0.060" by 0.030" wide) has gained considerable market share in the Far East. Parts as small as 0402" (nominally 0.040" long by 0.020" wide) are now used in growing numbers.

Table 2-9. Chip Resistor Dimensions

Size Code	Dimensions (mm/inch)			
	Body Length	*Width*	*Thickness*	*Termination Length*
RC0805	1.8–2.2 (0.070–0.087)	1.0–1.4 (0.040–0.055)	0.3–0.7 (0.012–0.028)	0.3–0.6 (0.012–0.024)
RC1206	3.0–3.4 (0.118–0.134)	1.4–1.8 (0.055–0.070)	0.4–0.7 (0.016–0.028)	0.4–0.7 (0.016–0.028)
RC1210	3.0–3.4 (0.118–0.134)	2.3–2.7 (0.090–0.106)	0.4–0.7 (0.016–0.028)	0.4–0.7 (0.016–0.028)

Figure 2-7 Diagram of a typical chip resistor.

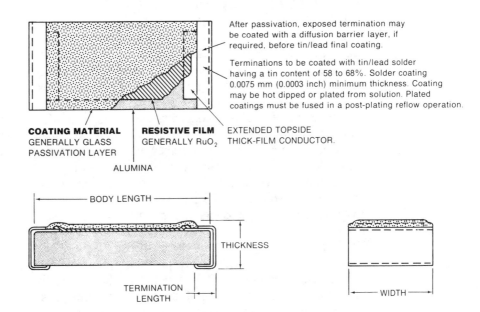

After passivation, exposed termination may be coated with a diffusion barrier layer, if required, before tin/lead final coating.

Terminations to be coated with tin/lead solder having a tin content of 58 to 68%. Solder coating 0.0075 mm (0.0003 inch) minimum thickness. Coating may be hot dipped or plated from solution. Plated coatings must be fused in a post-plating reflow operation.

COATING MATERIAL GENERALLY GLASS PASSIVATION LAYER

RESISTIVE FILM GENERALLY RuO$_2$

EXTENDED TOPSIDE THICK-FILM CONDUCTOR.

ALUMINA

BODY LENGTH

THICKNESS

TERMINATION LENGTH

WIDTH

Standards---In 1989, the EIA completed amendments to EIA/IS 30. The new specification, EIA/IS 30A, adds considerable data on solderability and leach resistance and it tightens physical tolerances.[7] Refer also to EIA 575 for standard SMT resistors and EIA 576 for precision parts.

Carrier Packaging---Tape & Reel is the preferred handling medium for chip resistors. Both the 0805 and 1206 sizes fit in 8mm tape on 4mm pitch.

Specifications---Resistor parameters that must be specified are resistance, resistive-tolerance, power rating, mechanical size, TCR in PPM, 10,0000-hour stability, and end-termination material and format. To prevent tombstoning and other solder defects, end terminations should be leach resistant and of equal length on both ends of the part. It is also important that the termination project a minimum of 0.25 mm (0.010 inch) in from the end of the part. Components with little or no top and bottom terminations (terminated on the ends only) are much more prone to tombstoning. End terminations should be uniform in thickness. Any lumps or globs on the terminations will likely cause significant process problems during the assembly and soldering operations.

For wave-soldered consumer products which will not be reworked (low enough in cost and/or high enough in yield to allow scrapping of those assemblies that fail final test), eliminating the nickel barriers will cut cost. However, where this cost savings is used, care must be taken with soldering time/temperature profiling to avoid leaching and/or silver contamination in the wave-machine solder pot.

Cylindrical Resistors

Cylindrical resistors are commonly called MELFs (Metal-Electrode Face-bonded). MELFs are typically supplied in two sizes. The MLL-34 is an 1/8-watt resistor, and the MLL-41 handles 1/4-watt at 70°C. There is a larger package that is both called and rated as "1/4 watt." As with chips, MELF wattage capacities must be derated per the manufacturer's suggested curve if they are to be operated in high-temperature ambient conditions.[8]

The MELF is not widely used in the U.S. market, but is very common in Asian SMT. An understanding of regional preferences is useful in deciding when to specify chips and when MELFs are a better choice. The Asian electronics market is biased toward consumer-electronics manufacturing. While there are numerous ultrahigh-tech applications, such as space instrumentation and super-computers in Japan, consumer electronics is a major force in determining the market share of the various component styles there.

Consumer-electronics PC boards generally are relatively small, low in cost, have few high-pin-count ICs, and have a large number of passive components. Because consumer products are so price driven, even small-cost differentials between components are likely to influence designers in their component selection. Whereas U.S. designers often reject MELFs from a fear of the war stories they hear about MELFs rolling off pads in reflow soldering, Japanese consumer process engineers may feel compelled to develop methods and equipment that allow them to take advantage of the lower cost of MELFs.

A quick view of Japanese consumer-electronics manufacturing shows that many of the PC boards are not reflow soldered. A substantial percentage are wave-soldered in a Type 3 process. Since all the SMCs are glued firmly in place, there is no concern regarding part movement. However, the fact that several major U.S. manufacturers use MELFs very successfully, with reflow soldering and no adhesives, states that the soldering method is not the only important consideration. Another factor in the Japanese success with MELFs is in the use of placement stations, or heads, optimized for the handling of cylindrical parts of a certain diameter, which are presented in a given way. Low-volume high-product-mix manufacturing environments would likely find the MELF a penny-wise and pound-foolish choice, but high-volume-optimized lines may well take advantage of its low cost and adaptability to bulk feeding.

Construction---In general, MELF resistors are constructed in similar fashion to metal or carbon-film axials, except that no axial leads are attached. Some manufacturers attach a metal cap to the component body and others apply metallizations similar to those on chip capacitors or chip resistors.

Standards---No specific EIA Standard exists for MELF construction. Most available parts conform to JEDEC DO-35 form, but without leads, as in Figure 2-

8. This basic form factor is used for cylindrical components of all types. Capacitors, resistors, inductors, and diodes are available in this form factor. Table 2-10 lists the dimensions for each case style.

Table 2-10. MELF component dimensions. Note: Additional case sizes 0805, 1206, 1406, and 2309 have been added since the first edition of this book.

Case Style	Dimensions (mm/inch)		
	Length	*Diameter*	*Termination*
MLL-34 (SOD-80)	3.3-3.7 (0.130-0.146)	1.5-1.7 (0.059-0.067)	0.29-0.55 (0.011-0.022)
MLL-41 (SOD-87)	4.8-5.2 (0.189-0.205)	2.44-2.54 (0.096-0.100)	0.35-0.51 (0.014-0.020)

Figure 2-8. MELF component construction.

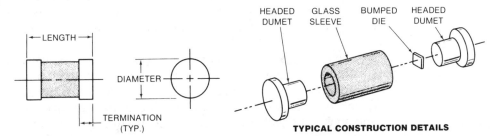

TYPICAL CONSTRUCTION DETAILS

Carrier Packaging—For low- to moderate-volume production, Tape & Reel is the preferred handling method. Certainly, it is favored for MELF diodes and other polarized components. However, some high-volume placement equipment is available with bulk hopper-style feeders for nonpolar MELFs of a given body size. This handling approach minimizes component cost, and has proven very reliable.

Specifications—MELF resistors are specified in the same manner as chip resistors. Particular attention should be paid to both the end terminations, per the specifications for capacitors, and the termination bond integrity. Also, MELFs, more than rectangular components, require process-engineering support. The process engineer should be involved in the decision to use MELFs. The engineer should understand the importance of the advantages of MELFs since their use may require the investment of considerable process-engineering resources. And the process engineer should be involved from the beginning of the selection process so that last minute process development does not become a stumbling block to production start-up.

Trimmer Potentiometers

The good news here is that there is a wealth of different trimming potentiometers available in SMT packages. The bad news is that the proliferation of available styles is due to a lack of standards, and some parts might have difficulty meeting any specification that a rational group would develop. The watchword is caveat emptor. Be sure you understand how the potentiometer is constructed, and how your process and field use will affect it, before you decide to abandon leaded trimmers.

Construction—SMT trimmer construction details obtained from different manufacturers vary greatly. Many are built on 96% alumina substrates with attachment metallizations similar to those of chip resistors. Others are constructed on epoxy-glass substrates. Organic substrate types employ a wide range of metallizations and resistive inks. Trimmers may be plastic encapsulated, cermet body, or open frame.

We can make one safe generalization with regard to construction. Whatever the materials and construction details, worthwhile trimmers must be built to survive the soldering, cleaning, and rework operations of the manufacturing environment. For sealed units, this implies protection from contamination that might enter the part during soldering or cleaning, and suggests that the seals around the adjustment-shaft entry are an area of critical concern. Where applications permit, an open construction may provide a low-cost part without such a convenient hiding place for contaminants and corrosives. The decision between use of sealed and open constructions involves considering what contaminants might remain in an open part after processing, and then assessing the protection available for the open part from environmental contaminants, while in service or storage.

Standards—Work is under way in the EIA, but not complete as of this writing, on standards covering trimmer lead spacing and size, maximum height, and time/temperature limits for processing.[9] Until such work is complete, the designer must devote time to becoming trimmer literate, or use the mixed-technology approach and stick with leaded trimmers which are through-hole mounted. For in-house certification, consider humidity testing per MIL-STD-202, Method 103 for 100 hours with a minimum insulation resistance of 10 megohms. Leak testing in 85 °C perfluorinated hydrocarbon fluid may be used to test the integrity of sealed units.

Carrier Packaging—Tape & Reel packaging, where available and compatible with placement equipment feeders, is the preferred carrier package. However, many trimmers do not fit the tape-thickness limits of EIA-RS-481A tape standards. Therefore, the tape may not fit dereeler feeders designed to handle this standard. Check your carrier format options and involve the manufacturing-engineering department in the decision.

Specifications—Beyond the benchmarks and cautions stated earlier, specify the resistance range, resistance tolerance, number of turns, physical description, TCR, absolute minimum resistance (where the desired part should approach zero resistance), insulation resistance, dielectric strength, resistance stability (over 10,000 hours, or after high-temperature exposure, whichever is more appropriate), temperature range, load life, processing-environment survival characteristics, environmental requirements, and mechanical life.[10]

Resistor Networks
As space savers, SMT resistor networks are easy winners when compared to through-hole or SMT discrete resistors. But SMT resistor networks usually come in a poor second in board area consumed, when compared to IMC SIP or ZIP packages. Let's not rule against SMT networks too hastily, though. Where boards must be sandwiched together closely, or the component side must be kept low profile for packaging considerations, SMT networks should be considered. Also, the small board-area premium required by the SMT part may be more than offset by processing considerations where a design would be pure surface mount were it not for a few SIP resistor networks.

Construction—Construction varies among manufacturers. In general, networks are fabricated using thick-film resistive techniques on 96% alumina substrates. Networks are available in packages similar to leadless and leaded chip carriers, and in packages resembling gull-wing and J-leaded SOICs. Of course, similarity to IC packaging goes no further than the outside form factors—if it goes that far. There is no die inside the package. Ceramic substrates are often protected by encapsulation instead of lid attachment. Also, some network packages are not as tall as the IC packages they mimic. And, a number of manufacturers supply pseudo SOIC parts. These wider than standard, and narrower than wide-body SOs, are called Small-Outline Medium (SOM) by some vendors. Be forewarned, vendors may make scant mention of the fact that their SOM doesn't fit IPC SOIC land-pattern dimensions.

Standards—Resistor network standards are progressing as EIA project number PN 1905. Currently, pseudo standardization is achieved by following, or sort of following, either the LCC, PLCC, SOIC, or SOJ outline. In dual-in-line formats, the SOJ is less popular than its gull-wing counterpart. However, it is gaining some momentum because it maximizes resistor network substrate area for a given slice of PWB real estate. Also, J-lead solder joints are easier to visually inspect than gull wings despite the common perception to the opposite. See more on lead-form inspectability in Chapter 7.

Carrier Packaging—Tape & Reel and plastic tubes, or "sticks," are popular carrier packages. Sticks are appropriate for small quantities and repeated material changes on the production line. Tape & Reel, if available, is better for long production runs. Of course, the choice is fundamentally dependent on the handling capabilities of the placement machine.

Specifications—Surface-mount resistor networks are specified in a fashion similar to through-hole networks. The package style, number of pins, electrical configuration (isolated, bussed, or dual terminator), and resistance specifications are typically called out. As with all SMT devices, the solderability of the lead finish is of critical importance, and bears far more attention than you might give for a through-hole device. For leaded SMT networks, see comments on lead finish given in Section 2.2.2 under "specifications" of transistors.

2.1.3 *Inductors and Coils*

Tuning Coils, Transformers

Construction—Surface-mountable inductors and coils are available in open designs or shielded designs (to reduce magnetic coupling between adjacent components). The winding cores are usually either ceramic or ferrite. The internal joints are generally soldered with high-temperature alloys that are impervious to process environments. Protected types are postmolded with an epoxy resin or other appropriate molding compound. Bare beryllium-copper end terminations are available on some parts, but solder-coated terminations with 60 to 63% tin (near eutectic tin/lead solder) should be specified to ensure acceptable process yields.

Standards---Standards activity is under way in the EIA, under its P-3.8 committee for inductors and coils. As of this writing, there is no issued standard for molded parts. As a work-around, most users choose parts conforming to the package outlines for molded brick-style tantalum capacitors (Figures 2-5 and 2-6). However, dimensions for inductor bricks may follow their own directions.

Open-style inductors, tuning coils, and transformers vary widely in package outline. A careful analysis of the available offerings, and a well-defined package on the specification control drawing are the order of the day with these components.

Carrier Packaging---Parts are generally available in bulk or in Tape & Reel shipping packages. Coils and transformers are often problem parts for vibratory feeders, so automated handling usually requires that taped components be specified.

Specifications---Specifications for SMT coils should include all the electrical considerations typical of through-hole components, plus a close control of the physical package and attention to solderability and soldering process resistance. For inductors with actual metal-tab terminations rather than fired-on metallizations, see the comments on lead finish given in Section 2.2.2 (transistor specifications). For fired-on terminations, refer to the comments given under multilayer ceramic chip capacitors (Section 2.1.1).

2.2   SMT Actives

SMT seminars and papers have brought forward a great deal of data on the physical dimensions and processing methods for various SMCs. Data, such as package thermal characteristics and package-related propagation delays, are much less widely available. As of this writing, most component specification sheets do not distinguish between leaded and SMC packaging except as to physical dimensions. The component manufacturer may be able to help, where data are needed for critical applications. Regarding propagation delays, a rule of thumb for less critical designs is that SMC packaging will allow safe operation at the high limit of the specifications, for the same part in an IMC package.

2.2.1   *Diodes (2-Pin Packages)*

Construction

Diodes are commonly packaged in MLL-34 or MLL-41 MELF bodies, or in plastic-encapsulated parts based on the SOT-23 outline. Some dual diodes are also available in SOT-143 packages. Construction details for the MELF are similar to a DO-35 diode, except that the leads are replaced with face-bonded metallizations for surface mounting. SOT-packaged construction is the same as for SOT transistors. There is a trend toward dual diodes in SOT-23 packages and toward diode arrays in SOIC packages. SOT-23s are available with two diodes, with a common-anode, common-cathode, and series connection; or as a single diode, with one lead of the SOT unused. There is an alternative package with a rectangular body and two pins roughly following the format of the SOT-23, except having its two pins at the opposing ends of the part. This package is called the SOD-123. It is available with pins in gull-wing format, a la SOT-23, or with terminations formed under the body in the fashion of the "brick" style tantalum.

Standards

The mini MELF MLL-34 (SOD-80) and MELF MLL-41 follow roughly after JEDEC DO-35, and specifically after the EIAJ standards for MLL designations. The SOT-23 outline is controlled by EIA TO-236. The SOT-143 package closely follows the body outline of the TO-236, but it has four pins and slightly different dimensions from the SOT-23.

Carrier Packaging

For most applications, the preferred carrier is 8 mm tape on 4 mm pitch.

Specifications

The major difference in specifying diodes in SMT packages vs. leaded diodes is in the resistance to soldering. Where nonleaded termination systems have a precious-metal base layer, a barrier layer may be specified to prevent diffusion of the precious metal during soldering, as stated earlier in the discussion on capacitors. For MELFs with bonded end caps or leaded device specification, follow the advice given next under specification of transistor terminations.

2.2.2 *Transistors (3-Pin Packages)*

Construction

SMT small signal transistors are typically packaged in the SOT-23 format. Higher-current devices are packaged in SOT-89, DPAK, TO-252, or modified insertion-mounted cases. Plastic-case transistors in SOT packages are constructed like TO-92 leaded transistors. The die is eutectic or conductive epoxy attached to the die flag on the lead frame. The die bonding pads are interconnected to the leads of the lead frame by wire bonding, and then the package is postmolded with epoxy novalac or a similar encapsulant. Finally, the individual devices are trimmed away from the lead frame runner, lead-formed, tested, sorted, and packaged for shipment.

Figure 2-9 Component variations allowed by EIA TO-236 specifications.

Standards

The SOT-23 format is controlled by EIA TO-236 within the U.S. and Europe. SOT-89 specifications are per TO-243. Note that the SOT-23 outline is a rough envelope.

All the components shown in Figure 2-9 fit the specifications. These parts are also covered by EIAJ specifications, which vary from the EIA outlines. So caution is the word when substituting vendors, and purchase decisions best not be made on price alone!

Carrier Packaging

The SOT-23 format is packaged in 8 mm tape with 4 mm pitch, while the SOT-89 is packaged in 12 mm tape with and 8 mm pitch. For packaging of other lead-formed transistors, consult the vendor.

Specifications

As shown in Figure 2-10, there are three board standoff heights available within the TO-236 outline. The industry trend is toward a medium-profile part with a 0.08 mm to 0.13 mm (0.003 inch to 0.005 inch) standoff, since this part works in both reflow and wave soldering and provides sufficient clearance for cleaning.

Figure 2-10   Standoff specifications for the SOT-23.

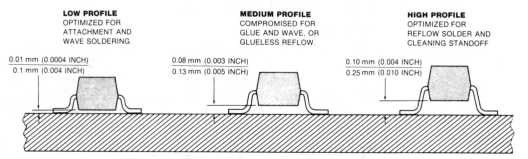

Terminations should be solder coated with solder that is close to the 60/40 tin/lead eutectic. Fifty-eight percent to 68% tin is allowed by specifications. The higher tin contents may exhibit improved solderability after a long storage period. However, tin/lead contents matching those of the solder paste and substrate metallization tin/lead ratios promote high reflow soldering yields. Solder coatings may be either hot-dip applied or electroplated. Hot-dipped terminations must be free from lumps, icicles, or other solder surface irregularities, and must not show any dewetted or pinholed areas. Plated terminations must be reflowed after plating to fuse the solder to the base metal. The tin/lead finish should be at least 7.5 microns (0.0003 inch) thick. *Note*: This termination treatment applies to all leaded SMCs and therefore, this discussion will not be repeated under each category.

2.2.3   *Transistors (4-Pin Packages)*

The SOT-143 format may be used to package unijunctions and gated devices requiring four leads, and can also be used for RF switching transistors, where the extra lead improves high-frequency performance. Another case style, the SOT-223 provides a more robust package with an enlarged lead on one side and three smaller pins on the opposite side. The enlarged lead makes this package useful for high-dissipation semiconductors.

Construction

Construction of the four-lead SOT-143 is similar to the SOT-23 package discussed earlier.

Standards
The SOT-143 format basically follows the outline of TO-236, except that it has a fourth extra-large lead. Virtually every dimension differs slightly from the SOT-23 package, however.

Carrier Packaging
The recommended carrier package is 8 mm tape with 4 mm pitch.

Specifications
See comments under "transistor specifications" above.

2.2.4 *ICs (8- to 28-Pin Dual-In-Line Packages)*

ICs have been surface mount packaged in DIL flatpacks for years. Hermetic flatpacks are available, and this package was popular for some time in military and hi-rel design. However, the package is expensive. It requires handling for lead trimming and forming operations, and its leads are very fragile and easily damaged. Because of these disadvantages, flatpack production has been on a downward trend since 1984. Three-tenths of 1% of all ICs sold worldwide will be packaged in flatpack in 1987, and production volume is shrinking by 15% per year.[12] Clearly, some alternative is necessary for new SMT designs.

The SO family of ICs provides one alternative. Their small size (about one-eighth of the material of a DIP package) and their substantial volume usage make them very price competitive with DIP packages. In a recent study, we found merchant-marketed ICs in SO at parity with, and up to 10% cheaper than, DIPs. The leads of the SOIC are about 5 times thicker and about 2.5 times shorter than flatpack leads. Thus, the SOIC is far more forgiving of manufacturing handling.

Construction
The SOIC and its wide-bodied cousin, the SOLIC, are manufactured like DIPs. Dies are generally attached to a bonding flag on the lead frame and interconnected electrically to the lead frame by wire bonding. Alternatively, dies are inner-lead bonded to TAB tape which is then outer-lead bonded to the lead frame. In either case, the assembly is then postmolded with a protective epoxy covering.

Standards
JEDEC has standardized a 0.150 inch (3.8 mm) nominal-width body SOIC in 8-, 14-, and 16-pin sizes, and a 0.300 inch (7.6 mm) nominal-width body in 16, 20, 24 and 28 pins.

There is also a standard for a nominal 0.300 inch body-width part having J-leads. The standard covers 14- to 28-pin devices in 2-pin increments. This package has been targeted for memories, where its J-leads allow very close-order spacing for high-density designs. Center pins may be omitted in selected part numbers to facilitate busing of a wide power trace.

Carrier Packaging
For short production runs, plastic tubes are the packaging format of choice. For longer runs, where suitable tape feeders and taped components are available, reeled components are recommended. One standard plastic tube carries 48 16-pin SOICs. A 16 mm tape with 8 mm pitch holds at least 2500 (see Table 2-11) of the same device. (Many customers hesitate to fill the reels to the capacity because they

cannot consume the full reel in a single production run.) So a placement machine with one tube at a feed station would have to be reloaded 52 times for every 1 reload required with the tape. Table 2-11 details reel sizes for various SO family parts.

Table 2-11. 13-Inch (330-mm) Tape and Reel Sizes for SO Family ICs

| Package Type | Tape Size | | Reel Width | Typical Component Count Per Reel** |
	Width	*Pitch*		
SO-8	12 mm	8 mm	18.4 mm	2500
SO-14	16 mm	8 mm	22.4 mm	2500
SO-16	16 mm	8 mm	22.4 mm	2500
SOL-16	16 mm	12 mm	22.4 mm	1000
SOL-20	24 mm	12 mm	30.4 mm	1000
SOL-24	24 mm	12 mm	30.4 mm	1000
SOL-28	24 mm	12 mm	30.4 mm	1000

*Based on partially filled reels; may be more at user request.

Specifications

There are some areas where caution is called for in specifying SOICs. First, be aware that there are several standards competing in the world. EIAJ parts, for instance, come in a variety of body widths. To make matters worse, JEDEC parts are controlled by an inch-based lead pitch of 0.050 inch (1.27 mm). Some, but not all Japanese parts are controlled by a metric pitch of 1.25 mm (0.04921 inch). In an 8-pin device, this is of no concern. However, for 28-pin devices, the incremental error because of pitch discrepancy is a substantial 0.010 inch, enough to make bridging between leads likely if a board is laid out for inch parts and metric parts are used, or vice versa.

There are also a number of components built in what the manufacturers call an SO package, which do not fit standard pads for either the JEDEC SO or SOL devices. Resistor networks and crystals are among the prime offenders. A word to the wise is for the user to check the manufacturer's outline drawing. Nonstandard parts will generally work just fine as long as you've laid out your PWB for the part you are getting.

The basic comments regarding the specifying of lead finish for transistors will apply equally to SO packages.

2.2.5   *ICs (20- to 124-Pin Packages)*

There is such a diversity of package alternatives in this category that some explanation is in order. If nothing else, this exercise may ease the pain of dealing with the package proliferation. At best, it may help you make better package selection decisions.

The SOIC solved many problems inherent in the flatpack format. However, the SO series left some other important areas unresolved. There is a trend toward increasing I/O in ICs, driven by greater levels of integration on silicon. Devices of ≥ 68 pins grew from a minuscule 0.03% of the worldwide packaging market in 1983, to 1.35% in 1990 (about 400% growth).[13] Beyond 1990, as ultra-large-scale integration and the use of very-high-pin-count fine-pitch parts began to gather steam, this growth has become exponential. Unfortunately, the dual-in-line SO package is both space inefficient and very difficult to wirebond without crossovers (potential short circuits) in packages of 28 pins and up. Even if the silicon designer can locate bonding pads to avoid crossovers, the potential for shorts produced by the mold-wash of bond wires makes the SO a poor choice for high-pin-count device packaging. Above 40 pins, DIPs are also a poor packaging alternative. Figure 2-11 compares board area and internal lead length for a 64-pin DIP and a 68-pin 50-mil chip carrier. The DIP requires 3.2 times the board area and the longest internal lead is 3 times the length of the chip carrier. Clearly, for high I/O silicon and VLSI, a four-sided device is needed.

Figure 2-11 DIL and quad IC packaging compared.

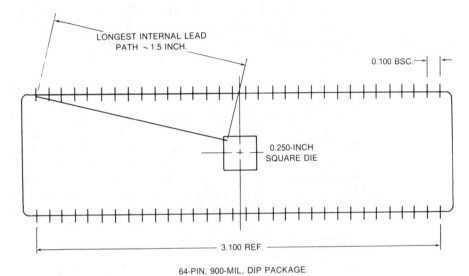

The first package to fill this need in production volumes was the quad flatpack. However, it suffered from all the negative qualities of its DIL sister. And, as if these woes were not sufficient, it posed serious design and assembly automation problems for users. Various pin counts had unique lead pitches, ranging from 0.040 inch (1.016 mm) to 0.026 inch (0.66 mm). It's no surprise that the quad flatpack is not the Model T of the packaging industry.

From initial design efforts funded through the Manufacturing Technology (MANTECH) program of the U.S. Air Force, the CLLCC was born in the late 1970s.[14] This hermetic package solved all the concerns listed for the quad flatpack, but it brought new concerns regarding cost and solder-joint integrity. Ceramic carriers are much more expensive to produce than postmolded plastic DIPs of equal pin count. And, because there are no leads to absorb stresses of thermal expansion, the CLLCC package demands special attention in the matching of the thermal coefficient of expansion (TCE) between the ceramic package body and the substrate. Common board materials, like epoxy/glass, expand at about three times the rate of the alumina ceramic of the chip-carrier body, and so, makes a poor substrate for CLLCCs. Most materials that match alumina's TCE also match or exceed its cost (about three times FR-4).

To allow chip-carrier use on organic PWBs and reduce packaging cost, a number of package formats were developed using leaded construction, plastic postmolding, and varying lead forms. The plastic leaded chip-carrier (PLCC), developed around 1980, became the dominant style in the U.S. market. PLCCs provide well-protected leads to form flexural elements, thus relieving expansion stresses that might otherwise damage solder joints. This package met the needs of a large segment of the PWB assembly market through the 80s, and continues to dominate the SMT package market in 28- to 68-pin devices where fine-pitch is not a requirement.

Many semiconductors are available in either SO or PLCC packages, so some suggestions for selection are in order. (Where package selection involves choosing between lead forms, see the lead discussion given later under CCCs and in Chapter 8.) The 20-pin SOIC requires slightly more board real estate than the 20-pin PLCC. The 14- and 16-pin SOICs take slightly less space than the PLCC. The SOIC is significantly lighter than the PLCC. Also, its DIL format makes surface-layer routing of bused designs far easier. It is a snap to route memory with DIL packages. Thus, in some applications, the SOIC plus its routes may take less space than the PLCC with its tracks. At around 24 to 28 leads, the PLCC begins to have a clear advantage, and it is rapidly gaining its packaging market share in the 44- to 68-pin bracket.

However, the PLCC doesn't meet the military market's demand for a hermetic, leak-testable package. And, even though the PLCC provides four sides for I/O, the 1.27 mm (0.050 inch) lead centers of the package limit it to ≤84-pin applications. A 100-pin 50-mil PLCC suffers from questionable solder reliability and takes up too much PWB real estate.

We've already established that the CLLCC provides a hermetic package for military and hi-rel applications. And, TCE concerns are routinely addressed by using substrates of similar TCEs. Doesn't this take care of the hermetic package segment? The answer is a firm NO. Two concerns remain unmet by the 50-mil and 40-mil CLLCC.

1. All available substrate materials matching the TCE of ceramic are expensive and place other severe limits on the design, assembly, or reliability of board assemblies. For a detailed discussion of these concerns, see Chapter 5, Section 7.

2. As I/O increases to or beyond 84 pins, CLLCC solder-joint reliability drops off dramatically, even when devices are mounted on substrates matching the TCE of the carrier. This is because heat dissipated within the device under operation causes a thermal gradient between the package and the board, leaving the carrier relatively hotter than the board. In large packages, the differential in thermal expansion produced by this thermal gradient is severe enough to work-harden and, eventually, break solder joints.

In answer to both these concerns, a family of ceramic leaded chip-carriers (CLDCCs) was developed. Thus, a wide variety of packages are in the hermetic chip-carrier family, as detailed in Figure 2-12.

Construction

JEDEC Type-A PLCCs are built using dies attached to lead frame and wire bonding, or TAB interconnect to the lead frame and postmolding, in the same manner that SOICs or plastic DIPs are made. A small percentage of PLCCs are premolded parts designed to be socket-attached to board assemblies. Premolded parts may also be reflow attached. JEDEC postmolded specifications cover only 28-, 44-, 52-, and 68-pin parts. Square premolded JEDEC PLCC specifications detail 8 sizes—from 20 to 124 pins. But most 20- and 84-pin PLCCs are actually postmolded. This is because the original JEDEC Type-A standard left to interpretation many details of part construction. There are also rectangular premolded PLCCs in 5 styles—from 18 to 32 pins.

Figure 2-12 The hermetic chip-carrier family tree.

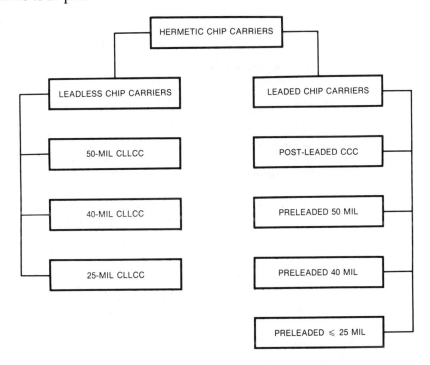

There are four main 50-mil CLLCC families registered with JEDEC for square parts. Types A, B, C, and D differ somewhat in construction details, but they share the following points in common. All are built on a ceramic substrate (base carrier) having metallizations for die attach and interconnection to the outside world. Each is covered using a lid, which is generally glass-to-metal sealed for hermeticity.

Type A is designed for lid-down mounting in a socket, placing its heat-dissipating side up for air cooling (particularly forced air) rather than conductive heat removal through the board. Its cavity-down orientation makes it well suited to heat-sink mounting on its top side. However, it does not provide for solder connection to boards. It is not widely used today.

Types B and C are designed for lid-up mounting via reflow soldering to TCE-compatible materials, or by socketing on noncompatible substrates. The lids on both packages extend above the CCC body. The Type-B lid is metal (such as gold-plated Kovar), and Type C's lid is ceramic. Type C dominates today's CCC marketplace.

Type D is similar to Type B except that its metal lid is recessed so that its top surface is flush. The package is designed for cavity-down mounting, making it, like

Type A, suitable for air-cooled applications. However, unlike Type A, this package allows the option of reflow or socket attachment to board assemblies.

Figure 2-13
Typical lead configurations for CLDCCs.[26]

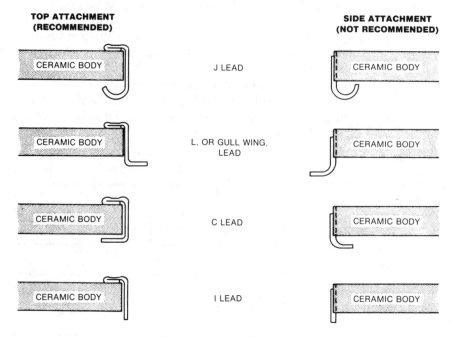

NOTE: While directed at CLLCC lead frames, these comments, except those regarding side and top attachments, apply equally to other devices with these lead forms.

(A) The lead configurations.

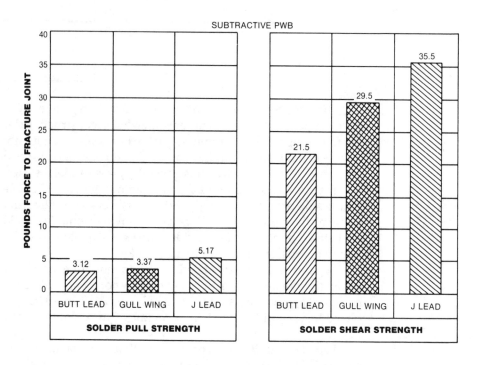

(B) Pull and shear test results.

In addition to the four "square" families, there are Type E and **Type F** "rectangular" standards. Rectangular parts have not captured a major share of the component market, however.

True preleaded CCCs are manufactured with a thin metal-foil lead frame enclosed by two ceramic body halves. The body is joined around the lead frame with hermetic glass-to-metal seals in CERDIP fashion. The specifications for such parts are covered in JEDEC MS-044. They have J-lead forms. However, the specification provides only for cavity-up 68- and 84-pin parts.

To meet the need for other lead counts and cavity-down orientations, manufacturers may post-attach lead frames to MS-003, MS-004, or MS-005 parts. This produces a chip-carrier modeled after side-brazed ceramic DIPs. The leads are brazed or thermocompression-bonded to the top surface of the carrier. In some cases, the leads are attached to side castellations. Since these carriers are supplied to the user with leads in place, it is customary to speak of them as preleaded, even though leads are attached as a secondary operation. Figure 2-13 shows the typical lead configurations used. We can see in Figure 2-13 that side-attached leads provide less of a flexural element for a given component standoff. Thus, top-attached leads are preferred, to maintain the lowest possible component profile for a needed flexural capability in the leads.

Figure 2-13 also provides a convenient point to evaluate the relative strength of solder joints produced by various lead forms. The chart in Figure 2-13B shows the results of pull and shear tests of J-lead, gull-wing, and butt-lead joints. Clearly, both the J-lead and gull-wing formats provide higher initial tensile and shear stress resistance than the butt-lead. While joint strength may have some relationship to long-term reliability, the equation for this relationship has not been demonstrated. Also, joint reliability is probably impacted by a large collection of variables outside the lead form. The chart in Figure 2-13C comes closer to the point of comparing the joint reliability of the J-leads vs. the I- or butt-leads. The chart shows the results of extended thermal cycling tests of J- and butt-leaded parts.[16]

Where preleaded CCCs are unavailable, users may add a lead. Postleaded CCCs are built per standard CLLCC types B, C, and D. After fabrication and lidding, the leads are soldered on by the component vendor or by the end user. Additionally, post-lead vendors can attach wire leads to the CLLCC castellations using a high-temperature solder, so that the lead attachment remains unaffected by subsequent soldering during PWB assembly or rework. In general, leads are of the two types shown in Figure 2-14.

Figure 2-13
(continued)

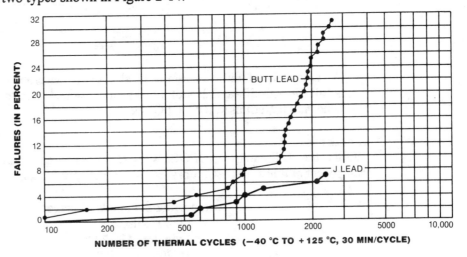

(C) Results of extended thermal cycling tests.

Quad plastic flatpacks (PQFPs) are an emerging answer to the need for an IC package with over 50 pins and up to about 250 pins. The JEDEC JC-11 committee has registered 8 packages in this family. Currently, these registrations range from 52 pins through 244 pins, on a 0.635 mm (0.025 inch) lead pitch. EIAJ had previously registered a PQFP, but loose tolerances, unprotected leads, and concerns about solder integrity led JEDEC to set a separate course.[17]

Figure 2-14 Lead types for postleaded chip carriers.

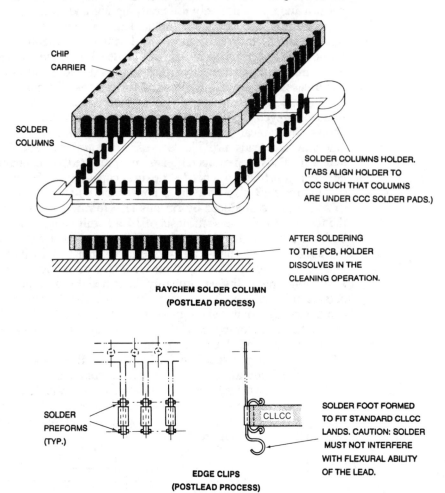

A ceramic quad flatpack family is also under development within JEDEC. Plans call for both 50- and 25-mil pitches, dimensioned such that the leads will fit the same land pattern used for the PQFPs.[18]

Standards
Within the U.S. market, JEDEC's JC-11 committee handles standards for component outlines. JEDEC, a branch of the EIA, can be contacted at:

Engineering Standards Manager
2500 Wilson Blvd.
Arlington, VA 22201
(703) 907-7500

EIA Japan activity is also of interest, and is important to assess when selecting Japanese suppliers for components.

We hope the preceding discussion will silence those critics of SMT who claim there aren't enough standards before somebody listens to them and generates an order of magnitude more. Lack of internal workmanship documents and process-control specifications within a given company may be an obstacle to SMT. But there is no dearth of industry-wide standards to draw them from. In fact, the real problem is just the opposite. How do you sort through the confusion and pick the right standards. Books such as this one are intended to help you successfully negotiate the maze of standards and process approaches.

Carrier Packaging

With leaded carriers, coplanarity of leads, as defined by the sketch of Figure 2-15, is of critical importance in maintaining high manufacturing yields and in keeping solder defects in check when using SMT. Carrier packaging for other SMT components serves primarily to maintain the parts in an oriented position for delivery to a placement machine. With leaded chip carriers, this concern is overshadowed by the need to protect the component leads from bending or damage during the handling operations. Leadless chip carriers must be protected from smashing together on edge while in tubes. Thus, carrier tapes and matrix trays are popular for these parts. Carrier tape specifications for PLCCs and CLLCCs are shown in Table 2-12.

Figure 2-15
Coplanarity variance is defined as the total deviation of leads from a single plane.

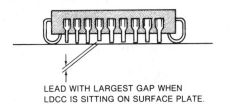

LEAD WITH LARGEST GAP WHEN
LDCC IS SITTING ON SURFACE PLATE.

Table 2-12. 13-Inch (330-mm) Tape and Reel Sizes for Chip Carriers

Package Type*	Tape Size		Reel Width	Typical Component Count Per Reel**
	Width	*Pitch*		
PLCC-20	16 mm	12 mm	22.4 mm	1000
PLCC-28	24 mm	16 mm	30.4 mm	750
PLCC-32	24 mm	16 mm	30.4 mm	750
PLCC-44	32 mm	24 mm	43.4 mm	500
PLCC-52	32 mm	24 mm	43.4 mm	500
PLCC-68	44 mm	32 mm	55.4 mm	250
PLCC-84	44 mm	32 mm	55.4 mm	250

*PLCCs shown. CLLCCs and LDCCs are the same.
**Based on partially filled reels. This number may be increased to suit user needs.

The new JEDEC PQFP has tabs protruding from its four corners. These tabs extend beyond the perimeter of the leads so that a device banging around in a tape carrier cavity is afforded some protection from lead damage. These tabs make it possible to ship PQFPs in plastic tubes, as well.

Specifications

In specifying leaded or leadless chip carriers, pay close attention to the lead finish and see that it is compatible with your soldering process. Leaded chip carriers should have lead finishes as detailed earlier for transistors.

Leadless carriers generally are constructed with a trimetal termination of gold, nickel, and a refractory metal, such as tungsten or molybdenum.[17] Gold is fond of forming extremely brittle intermetallics with the tin in solders. Such intermetallics are a sure source of failure for solder joints, which, under stress, will crack right along the intermetallic and solder junction. This problem is corrected by selecting tin-free solders, by keeping the time at solder-reflow temperatures short, or by overcoating the gold.

Another concern is the solderability of parts. Aggressive fluxes would provide an easy answer for through-hole devices. But, leadless chip carriers allow very little cleaning space between their bellies and the board. Therefore, it's not good practice to use fluxes that might harm the circuit if the cleaning process doesn't do a perfect job. To ensure (and test) solderability, it's a common practice to dip-solder coat the CLLCC terminations.

CLLCCs are available with (relatively) low-cost glass-frit sealed lids or with more expensive solder-sealed lids. Solder attachment allows a lower temperature sealing than the CERDIP-style glass-frit seal. And solder seals meet all the process requirements of MIL-M-38510. Therefore, solder-sealed parts are required for some military and hi-rel applications.[19]

So, for CLLCCs, engineers must determine how the parts will be soldered, who will solder-dip the leads, what the end use dictates in the seal selection, and whether heat will be primarily dissipated into the PWB or into the air (cavity up or cavity down, respectively).

Figure 2-16
Comparison of IC package sizes.
(Courtesy National Semiconductor.)

2.2.6   ICs (Fine-Pitch Packages)

In 1989, when the first edition of this book was printed, fine-pitch devices (≤ 0.635 mm/≤ 0.025 inch) were the new frontier of SMT. As we predicted then, the early 1990s were dominated by activity in fine-pitch surface-mount technology. The enormous miniaturization afforded by fine-pitch will certainly drive interest in it until even more compact approaches gain sufficient acceptance to sweep this technology aside. Figure 2-16 compares a 40-pin DIP, a 44-pin PLCC, and a miniature package showing the great size advantage of reduced lead spacing.

Japanese engineers, searching for packaging formats to break the 84-pin barrier and conserve board space, have developed a family of 36- to 100-pin fine-pitch devices. These packages use standard lead frames, wirebond techniques, and postmolding. Leads are on centers ranging from 0.432 mm (0.017 inch) to 0.635 mm (0.025 inch). These packages have very fragile leads, and are limited to about 100 pins. At higher pin counts, mold-wash (the tendency of the postmolding operation to move and short-circuit the bonding wires) lowers component manufacturing yields to unacceptable levels.[20]

In Europe, Siemens has developed a variation of the tape automated bonding (TAB) procedure, called Micropack™. Micropack is a TAB direct-on-board technique that uses a three-layer copper Polyimide tape with outer terminations typically on 0.508 mm (0.020 inch) centers. Three-layer tape can be used for very close center inner-lead bonding on the chip. Wire bonds on the IC die are generally set to a 0.254 mm (0.010 inch) minimum pitch to allow room for the bonding tool and to prevent bond wire shorting. A 6-mil pitch (0.152 mm) is about the limit for current bonding technology. Three-layer tape has been successfully used in production for bonds on 0.102 mm (0.004 inch) centers, and 0.051 mm (0.002 inch) is the current limit. This technology can handle production devices with over 300 leads and, in studies, has handled 600 I/Os. Special Micropack-style designs provide excellent thermal paths for very high thermal dissipation.

Disadvantages of the TAB on-board approach are the cost of three-layer tape, the requirements for specialized equipment for full parametric test and burn-in of components before placement, and a lack of protection for the die.

In the United States, National Semiconductor has developed the TapePak™ approach for SMT use of TAB. TapePak uses a 0.071 mm (0.0028 inch) single-layer copper-tape "lead frame" to form interconnections to a die. The device is postmolded with a protective package around the die. A protective plastic ring is simultaneously molded outside the attachment leads of the device, as shown in Figure 2-17. The leads for attachment to the circuit are on 20- or 10-mil (0.508-0.254 mm) centers. But, the protective ring supports an outer perimeter of fanned-out leads on 1.27 mm (0.050 inch) centers, allowing package handling with current burn-in and test technology. The disadvantages of TapePak are the limits of lead spacing and bond-pad centers on the die which are imposed by single-layer copper tape, and the requirement for specialized excising stations to remove the outer protective ring and lead form the component before placement on pick-and-place machines.[21]

Several large vertically integrated companies worldwide are also working with Flip Chip technology. Flip Chip involves a direct to the board upside-down (from which the name is drawn) mounting of dice, which have tiny hemispherical solder bumps grown on their bonding pads.

Their advantages are extremely high density and an excellent thermal path. The disadvantages are a limited supply of the Flip Chips, a lack of protection for the die, a very limited availability of the necessary high-accuracy placement equip-

Figure 2-17 The
TapePak device.
*(Courtesy National
Semiconductor
Corp.)*

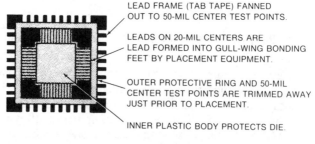

LEAD FRAME (TAB TAPE) FANNED
OUT TO 50-MIL CENTER TEST POINTS.

LEADS ON 20-MIL CENTERS ARE
LEAD FORMED INTO GULL-WING BONDING
FEET BY PLACEMENT EQUIPMENT.

OUTER PROTECTIVE RING AND 50-MIL
CENTER TEST POINTS ARE TRIMMED AWAY
JUST PRIOR TO PLACEMENT.

INNER PLASTIC BODY PROTECTS DIE.

(A) Device elements

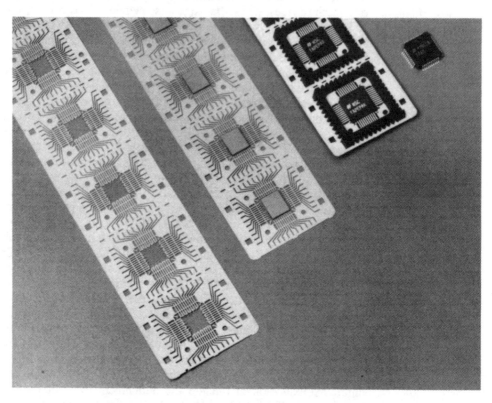

(B) Steps in device construction.

ment, limits on device testing prior to placement, and a difficulty in inspecting
solder joints.

All of these technologies have found growing uses. In space-constrained
applications such as notebook computers and hand-held electronics, TAB, Flip
Chip, and COB are methods of choice. These applications have spawned rapid
growth in equipment, sources of known-good-die, service infrastructure, and
standards for these advanced manufacturing technologies.

Figure 2-17
(continued)

(C) 124- and 40-pin devices showing size.

2.3   Connectors and Sockets

There are numerous and varied connector offerings now in the SMT arena, and the family is expanding rapidly. Many connectors (particularly true of the pure SMT types) are low, or zero, insertion/extraction force designs for use in minimizing the forces applied to surface-mount solder joints during use. If moderate-to-high insertion/extraction forces are anticipated (such as in designs involving large pin or tin-plated parts, or in designs requiring repeated insertions and removals), it is best to choose a interconnect device providing some form of mechanical fastening to the board. Typical mechanical hold-down schemes include screw connection, heat staking of bosses, and snap-in bosses.

Construction
Construction of SMT connectors and sockets differs from inserted interconnect construction in several ways. First, since the solder connections are on the component side of the board, many SMT connectors employ gull-wing lead forms to provide for visual inspection of solder joints. Gull-wing mounting requires more board real estate than typical insertion styles. Second, special plastics are often used as dielectrics for SMT connectors, because the parts must withstand the full heat of the soldering-process environment.

Standards
Activity to provide standards is under way in a subcommittee of EIA P-5.1.

Carrier Packaging
Carrier packaging varies dependent on the assembly methods and equipment to be used, and on the manufacturer's available offerings. The most common approaches are specialized carrier tapes for high-volume robotic applications, matrix trays or tubes for medium-volume automation, and bulk bagging for small runs where hand assembly will be used.

Specifications

Beyond the considerations involved in specifying through-hole PWB-mounted connectors, SMT brings a concern about the solderability of lead finish, the survivability in soldering and rework/repair processes, and the mechanical fastening needs.

2.4   Switches

Construction

Switches are not yet fully standardized for SMT. The variety of jobs that switches must address makes standardizing to a limited number of designs difficult, if not impossible. Most SMT switches are of the small slide or rotary- type. A few push-button types are also available. Outwardly, many of the SMT switches currently available look like gull-winged versions of through-hole designs. However, their materials and sealing may be quite different in order to suit the relatively hostile process environment that SMT components may encounter.

Standards

Consult EIA P-13 for the current status of work under way in P-13.1 on rotary switches, P-13.2 for slide, rocker, and toggle switches, and P-13.3 for the push-button and keyboard switches.

Carrier Packaging

Generally, switches are supplied in bulk or in plastic tube, but some are available in specialized tapes. Selection of the carrier package should be determined by the availability from the switch suppliers and the requirements of assembly automation.

Specifications

As with all leaded SMCs, surface finish of the leads should be controlled. Refer to the discussion of transistor specifications (Section 2.2.2) for details. Also, the specifications should detail the board-attachment method to use for stress relief, if any, and the carrier packaging needed.

2.5   Other Components

The following is just a sampling of the support components available in SMT packaging today. Except for very large and high-power devices, virtually all component types are now offered in SMT style.

2.5.1   Crystals

An early objection to surface-mount packaging for crystals was that their size and bulk didn't blend with 2-pin SMC packaging. As a solution to this, larger device packages have been adapted, as shown in Figure 2-18. Crystals are available in chip carrier and SO packages, and also in metal cans with the two wire leads formed in gull-wing fashion for surface mounting.

Figure 2-18 SMT crystals. *(Courtesy Q-Tech Corp.)*

2.5.2 *Relays*

The first SMT relays on the market, gull-wing lead-formed DIP relays, are still widely available today. The alternatives include plastic rectangular packages, with the leads trimmed and formed in "C" fashion around each end, and some variations on standard surface-mount IC packages.

2.5.3 *Jumpers*

Construction
Jumpers are basically shorting bars constructed like chip resistors, but with conductive inks, instead of resistive inks, fired on the top surface. Chip jumpers are used to avoid double-sided, or multilayer, signal traces on boards where only a few tracks cannot be routed without crossovers.

Standards
There are no chip jumper standards in the U.S. or Europe as of this writing. In Japan, chip jumpers follow the CR21 and CR32 resistor outlines (0805 and 1206, respectively).

Carrier Packaging
Carrier tape is the preferred shipping format, except where specialized placement equipment dictates some other choice.

Specifications
Jumpers are specified by component outline, termination material, resistance to the soldering process, and the maximum voltage/wattage rating. A CR21-size jumper will handle roughly one order of magnitude more power than the equivalent-size chip resistor, or 1 ampere at 70 °C, while the CR32 handles twice that amount.

2.5.4 *Variable Capacitors*

A variety of caseless and enclosed variable capacitors are available for surface mounting. Construction types include Teflon™ dielectric and air trimmers. Care must be taken in selection in order to consider the entire manufacturing process, and choose components that will still be trim caps after assembly. See comments in the earlier discussion regarding trimmer resistors.

2.5.5   Thermistors

Chip thermistors are available with form factors roughly following those of chip capacitors, as shown in Figure 2-19. However, there are many sizes of thermistors outside the five recommended by the EIA for MLC chips. Refer to the discussions of the chip-cap specification process (Section 2.1.1) for guidance in specifying the end terminations for surface mounting of thermistors. Figure 2-19 shows two formats. Surface-mount thermistor chip wraparound terminations for PC board mounting are shown on the right, and hybrid wire bondable terminations are on the left. For volume uses, 8- or 12-mm carrier tape is the preferred packaging format. Where tape is not available from the component supplier, but desirable for manufacturing, service bureaus exist to reel parts.

Figure 2-19 SMT thermistor chips. *(Courtesy Dale Electronics, Inc.)*

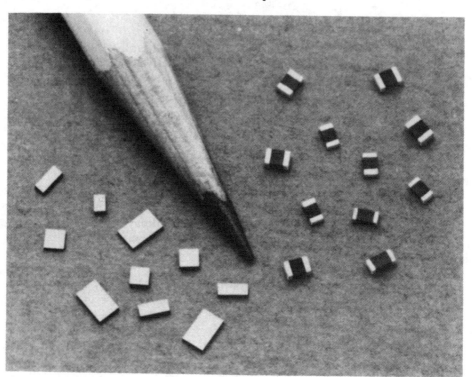

2.5.6   Inductors

Construction

SMT inductors are generally wire-wound around a ceramic or ferrite core. Some are left bare, some are conformally coated, and some are encapsulated, as shown in Figure 2-20, for protection and to improve automated handling characteristics. Bare chip inductors are difficult to handle with automated equipment. A molded part with a flat top is important for pick-and-place equipment.

Coils are available in both unshielded or shielded form. Shielded designs reduce the magnetic coupling between components. Internal soldered connections should be welded or high-temperature soldered to withstand SMT assembly-process environments.

Figure 2-20 Chip
inductors in
preferred
packaging.
*(Courtesy Dale
Electronics, Inc.)*

Standards

Two case sizes have been proposed by the SMTA components committee and are specified in ANSI/IPC SM-782.[22] However, our 1987 survey of component availability indicated less than an enthusiastic support of any standard from suppliers. Therefore, it is wise to allow time for shopping when purchasing inductors.

Carrier Packaging

The two standard types listed above come in 8-mm/4-mm pitch, and 12-mm/8-mm pitch tape. However, those companies that are battling to resist the standard are just as determined to package parts in innovative ways as they are committed to creativity in package size. Again, check early with prospective suppliers and be aware of those services which will tape components. Note, taping may add some lead time as the taping house must purchase suitable tape and tooling for some nonstandard components.

Specifications

For parts with end terminations fired onto the body, refer to the chip capacitor specifications (Section 2.1.1) for end termination treatment. For molded parts with metal lead terminals, refer to the specifications of SOT-23 for termination specifications.

2.5.7 *Optoelectronics*

Construction

Packaging optocouplers for SMT use is a challenge. Typically, the package bodies are too small to provide the space required for the needed 2500–4000-volt isolation. Three avenues are being used for SMT as of this writing. Typical DIP parts are sometimes lead formed to gull-wing configuration or are trimmed to butt-lead form for surface mounting. In a more generically SMT approach, Siemens offers a high-body version of the 50-mil SOIC, as shown in Figure 2-21. Motorola plans to second source this package. The critical footprint is the same as a JEDEC SOIC, and the larger body height needed to meet isolation specifications does not interfere with most automated assembly operations. For SMT applications requiring hermeticity, TRW offers a CLLCC variant with two pins each on two sides.

Figure 2-21
Optocouplers in
50-mil SOIC and
100-mil DIP
packages.
*(Courtesy Siemens
Components,
Cupertino, CA.)*

Standards
As of this writing, no standard exists for optocouplers. The Siemens/Motorola package shown in Figure 2-21 was developed by those suppliers, based on input from an ad-hoc committee of users and assembly equipment manufacturers.

Carrier Packaging
The SOIC- and CLLCC-style parts can be supplied in Tape & Reel for large production runs. However, the SOIC's nonstandard height violates EIA tape-thickness specifications. This may impact feeding of the tape in some de-reelers. SOICs may also be supplied in plastic sticks for short runs or small lot jobs. Matrix trays are available for small quantities of the CLLCC.

Specifications
For parts with end terminations fired onto the body, refer to the specifications for CLLCCs for end termination treatment. For molded parts with metal-lead terminals, refer to the specifications for SOT-23 (Section 2.2.2) for termination specifications.

2.6   References

1. Manko, H. H., *Solders and Soldering*, 1979, McGraw-Hill, pg. 119.

2. *Surface Mounted Components Catalog and Applications Manual*, MuRata Erie North America, Inc., 1987, pg. 30.

3. "Fixed Electrolytic Tantalum Capacitors," *EIA Specification RS-228*, EIA, Washington, DC.

4. ANSI/IPC-SM-782, *SMT Land Patterns*, IPC, Lincolnwood, IL, 1987, pg. 4.

5. Carter, G., Corporate Director Q&A and Chairman of the EIA Engineering Council, Dale Electronics, Inc., Telephone interview, July 27, 1987.

6. "Preferred Values," *EIA Specification 385*, EIA, Washington, DC.

7. Carter, op. cit., ref. 5.

8. Marcoux, Phil, Ed. *AWI's SMC Library*, SCI/AWI, Inc., Sunnyvale, CA, 1986.

9. *The Trimmer Primer IV*, Bourns Trimpot, Riverside, CA, 1986.

10. Ibid.

11. ANSI/IPC-SM-782, op. cit., ref. 4.

12. *VLSI-ERA IC Packaging—1986 Update*, Electronic Trends Publications, Cupertino, CA, pg. 10-2.

13. Ibid.

14. Tsantes, J., "Technology Update, Leadless Chip Carriers Revolutionize IC Packaging," *EDN*, May 27, 1981, pg. 49.

15. Derfiny, Dennis and Dody, Glen, "On Optimizing the PLCC Leadform," *Proceedings of NEPCON West 1987*, Cahners Exposition Group, Des Plaines, IL, February 1987, pg. 251.

16. Braden, J. S., "New Surface Mountable Packages for VLSI Devices," *MEPPE Journal*, MEPPE, Palo Alto, CA, Vol. 1, No. 2, May 1987.

17. Ibid.

18. Schoenthaler, D., Hymes, L., and Miller, W., Chairmen IPC Component Mounting Subcommittee, *Printed Board Component Mounting*, ANSI/IPC CM-770C, The Institute for Interconnecting and Packaging Electronic Circuits, Lincolnwood, IL, March 1987, pg. 37.

19. Ibid.

20. Levi, M., "Advanced Packaging for High-Lead-Count Integrated Circuits," *Electronic Engineering Times*, February 24, 1986, pg. T14.

21. Ibid.

22. ANSI/IPC-SM-782, op. cit., ref. 4.

Figure 3 A highly flexible CIM SMT factory. *(Courtesy Allen-Bradley Industrial Computer & Communication Group, Cleveland, OH.)*

(A) Automated handling marks this flexible SMT line.

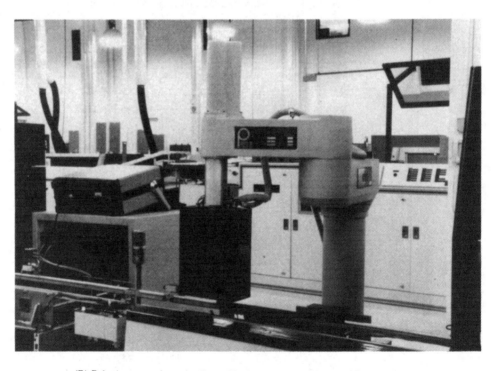

(B) Robotic arm performs work outside the scope of standard assembly equipment.

3 SMT Manufacturing Methods

High flexibility and high levels of computer control were the engineering watchwords that guided the design of the facility shown in Figure 3. Traditional high-volume production lines deliver low manufacturing cost through economy of scale. This pass-through system's economy of scope makes cost-effective on-shore manufacturing of a large product mix possible.

Engineers who have recently entered the SMT arena often disagree about fundamental points of design methods. But there is one area where we find amazing unity. Experienced engineers agree that SMT is a manufacturing process-driven technology.

Manufacturing methods have such an impact on design directions that we are devoting an entire chapter of this book to the manufacturing side of SMT. Let's start with an overview of the manufacturing process.

3.1 An Overview of the SMT Manufacturing Process

Basically, SMT manufacturing involves six process steps:

- Attachment-media dispensing
- Component placement
- Attachment-media curing
- Soldering
- Cleaning
- Test

77

After these steps, boards passing final test are shipped. Boards failing final test are disassembled to remove the noncomplying areas, and then reassembled, using essentially the same process steps. Of course, this is a gross over-simplification of the actual manufacturing process flow, but this uncomplex view is very useful in understanding SMT processing.

3.1.1 Attachment-Media Dispensing

By attachment media, we mean that material which will hold the SMC in position—after placement and prior to reflow soldering. The attachment media may or may not continue to hold the component after soldering.

Attachment media for surface-mounted components is generally either solder paste, liquid flux, or an adhesive. In any case, the material must be viscous enough to hold components in place until further processing steps permanently attach them.

Reflow Soldering

SMCs on reflow-processed assemblies are generally placed in preapplied solder paste. *Solder paste* is a mixture of tiny spherical solder particles, flux and its activators, and the carrier vehicles. The vehicles used for solder paste are generally thixatropic (such a material becomes less viscous and flows under pressure, but regains viscosity after the pressure is removed, preventing unwanted spreading of deposits).

Thixatropic solder pastes may be dispensed by screen or stencil printing, and by syringe or pump dispensing. Solder pastes are also occasionally dispensed by offset printing using transfer fingers. Figure 3-1 shows a small screener in the process of depositing solder paste.

Figure 3-1 Silk screening of solder on an SMT board. *(Courtesy DeHaart.)*

Reflow-soldered assemblies may also be produced by placing pre-tinned parts in sticky flux on pre-tinned boards, or by using a combination of viscous liquid flux and solder preforms. These methods are not widely used, but have been proven worthwhile in solving certain application challenges.

Wave Soldering

Surface-mounted components on wave-soldered boards are generally placed in adhesive paste. After being cured, the adhesive holds the SMCs in place as they are passed through a solder bath, which makes electrical connections between the components and board, and which reinforces the mechanical bond supplied by the adhesive.

Thixatropic adhesives may be applied by all the methods used with solder pastes. Nonthixatropic adhesives are syringe or pump dispensed, or are offset printed.

3.1.2 *Component Placement*

For engineering prototyping and very-low-volume assembly, or for the assembly of boards with very low population counts, hand assembly is often used. Some low-volume assembly is also handled by semiautomated placement, using a machine to present the correct components and point to the placement locations for a hand-assembly operator. However, almost all SMT production, with even moderate volumes, is done by automated assembly. Figure 3-2 shows an automatic placement head in the process of adding SMCs to a PWB assembly.

Figure 3-2 The head of an SMT placement machine placing a component on a board.

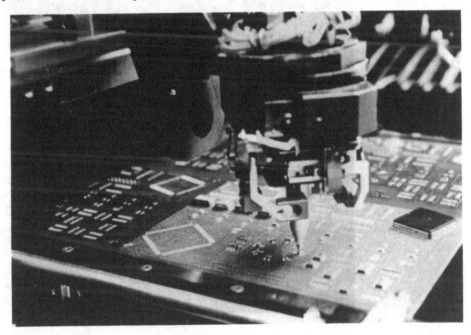

Automatic pick-and-place machines programmably select, orient, and place on the board a variety of components. Pick-and-place machines are available to fill virtually any application requirement. Machines can handle a manufacturing volume ranging from a thousand components to upwards of two million components per day. Some machines handle only a restricted range of similar components. Others handle a very wide range of component styles and sizes. Pick-and-place machines also differ greatly in their level of machine intelligence (their ability to sense, and adapt to changes in the process environment). Finally, a matter of little surprise, equipment varies widely in price. Automatic placement equipment starts in the $20,000 range, and goes upward to well over $1,000,000. In general, prices increase with the four desirable features: throughput, intelligence, flexibility, and reliability (accuracy, repeatability, and consistent operation). Unfortunately, even for a megabuck, there currently is no machine that delivers high marks for all four desirable features.

3.1.3   *Attachment-Media Curing*

In some processes, after component placement, the attachment media (solder paste or adhesive) must be cured before the soldering step.

Solder Paste

Curing of solder paste is accomplished by heating the assemblies, with the components placed in wet paste. Temperatures and soak times vary dependent on materials, but they generally run between 80 °C and 150 °C for 15 to 60 minutes. Where curing times, temperatures, or assembly materials dictate, curing may be done in a nitrogen or forming-gas atmosphere to prevent excessive oxidation of assembly materials and/or solder. Figure 3-3 shows an infrared furnace used for solder paste curing.

Figure 3-3 IR furnace used for solder paste curing.

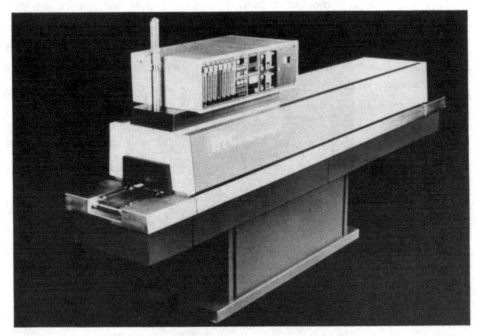

Curing evaporates the volatiles from solder paste, preventing a rapid boil-off which might create defects in certain soldering processes. Higher-temperature curing may also begin flux activation with certain pastes.

Curing has been shown to dramatically reduce solder defects in vapor-phase soldering, where steep temperature gradients can cause rapid outgassing of flux volatiles in uncured paste. Escaping gas bubbles have been shown to increase tombstoning, produce solder balls, and leave voids in solder joints.[1]

We apologize for the many qualifiers in the above discussion. To be accurate, yet brief, we cannot cover specific data on all the solder paste formulations used in surface mounting. Pastes may be made from varying combinations of tin, lead, silver, indium, gold, palladium, etc. Rosin, organic acids, halides, and synthetic materials are used as fluxes. The activity levels of the fluxes can be varied from unactivated to highly active. And, thixatropic vehicles are made of a wide range of materials, in varying viscosities and with distinct properties. To further complicate matters, there are infinite permutations and combinations of these basic materials. To cover them all would require a separate and exhaustive book. Solder paste and process equipment vendors are an excellent source of data on the materials you'll be using.

Adhesives

Since very few air activated, or anaerobic, adhesives are used in SMT component attachment, curing is a common step used in the wave-soldering assembly process to accelerate cross-linking of the epoxy adhesives. Some adhesives are cured by heat, some by exposure to ultraviolet radiation, and some by a combination of the two.

3.1.4 *Soldering*

After placement (and after curing where it is required), the next major process step to consider is soldering. Below, we'll briefly cover the common methods used in SMT soldering. We will not enter into any discussion of which is best. Your application will determine the soldering-method choice. Process selection is such an important fact of design that the full SMT team should participate in the selection.

Figure 3-4 SMT assembly in a vapor-phase soldering system.

Reflow Soldering

Reflow soldering describes the application of sufficient heat to an assembly to melt pre-applied solder. The solder is usually in the form of paste, but it may be in preforms or pre-tinned components and boards. Figure 3-4 shows an assembly immersed in the vapor blanket of a vapor-based soldering oven.

The term "*re*flow" probably stems from early hybrid methods that involved pre-tinning (flowing), and heating ("*re*flowing"). It's less clear why the term was applied to the melting of solder pastes and preforms. Indeed, metal powders for solder pastes are made by atomizing melted solder into an inert atmosphere. Likewise, solder is melted in the process of making preforms. But solder is also melted to cast the bars used to fill a wave-solder pot, and wave soldering is called flow soldering.

Whatever its genesis, the term reflow clearly delineates soldering by the heating of preapplied solid solder, as opposed to the application of molten solder to assemblies (flow or wave soldering).

Reflow soldering may be accomplished in several ways. The most common reflow approaches are conductive, convective, nonfocused IR (panel and T-3 lamp), focused IR (laser and spot-focused lamp), and vapor-phase soldering (VPS). Each of these methods is characterized by a somewhat different heat-transfer mode. For a more complete discussion of these soldering approaches, see References 1 through 5

Flow Soldering

SMT places new requirement on flow (or wave) soldering, and it generally requires specialized equipment to meet these requirement. SMT flow-soldering systems must solve two antipodic problems inherent with the soldering of SMCs on a board's bottom (solder) side. These are voiding (too little solder) due to gas entrapment and eddy currents around the SMC, and bridging (too much solder) between adjacent SMC terminations.

A variety of solutions to SMT voiding and bridging have been incorporated in flow-soldering systems for SM. Today, well-controlled surface-mount flow-soldering systems can produce very low defect rates.[6]

3.1.5   *Cleaning*

Component "standoff" distances provide clearance, allowing the cleaning solvents space in which to flow easily under components and remove harmful contaminants. Many SMCs mount very close to the board. This makes cleaning under the components a special concern in the engineering of the SMT cleaning process. Leadless components are particularly troublesome, since they have no leads that can be used to provide standoff. And, leadless chip carriers are the most difficult, because they have little standoff, cover a large board area, and have interconnections that form a picket fence around all four sides, which seriously retards cleaning-agent flow.

In the early days of SMT, cleaning was a challenge, but the approach was nearly a no-brainer. Everybody knew that you used solvent cleaners to get the soils from under SMCs with their tiny cleaning clearances. The solvent of choice was almost universally CFC based. Of course, concern over the depletion of the earth's ozone layer by CFCs has changed all this. As the Montreal Accords clicked into full effect, there was considerable confusion and debate over the proper way to clean SMT assemblies[7]; but the one nearly-universal given was that the CFC route was not it.

Cleaning Strategies

Electronics manufacturers today have numerous cleaning strategy choices available to them. They can select from:

1. Aqueous cleaning with saponifiers.
2. Aqueous cleaning without saponifiers.
3. No-clean solder/flux systems.
4. Semi-aqueous cleaning.
5. Solvent cleaning with ozone-friendly chemistries.

Which one is best? If I were to assert one or the other, I would risk disaster on two fronts. One, I would incur the ire of those people whose application dictates a different method than the one I advocated. Two, I would be at risk of lynching by a mob of angry vendors whose methods were bypassed. Not desiring either of these outcomes, I will pass on the chance to take a firm stand, and just provide a list of things to consider in your selection process.

Aqueous cleaning with saponifiers can be used with RMW and synthetic-activated (SA) fluxes. The bonus of this approach is that a wide range of solder/flux formulations may be selected, active SA fluxes can be used where poor solderability is a problem, and finished work can be of excellent, clean, shiny-solder appearance. The downside is that an increasing number of jurisdictions impose stringent environmental regulations on the contaminants allowed in water dumped into their sewers. In many cases, the water you return to the sewer must be cleaner than the water the locality supplies you from the tap. The result is that very costly waste-water-treatment plants or closed-loop water systems may be a requirement. Finally, water molecules are relatively small items, and it would seem they would fit into the tiniest of spacers. However, water has a very high surface tension, and that property of water tends to inhibit its penetrating capacity. The saponifier is added to reduce the surface tension of the water and increase its penetration and, thus, its cleaning action in the tight spaces so prevalent in SMT assemblies. Unfortunately, cleanliness requirements often force us to get the saponifiers off the assembly, and we cannot do this with saponified water. Deionized water is used, and, again, we are up against the surface-tension problem. The answer to the apparent conundrum is in design of assemblies with sufficient component standoff to permit cleaning.

Aqueous cleaning without saponifiers follows the discussion immediately above, but adds limitations on the types of flux formulations permitted. If no saponifier is used, the water solubility of the flux residue must be adequate to ensure full removal during the cleaning operation.

No-clean soldering is an attractive method. Where there is no very-high-impedance circuitry on a board, and where use environments are reasonably dry, it is often possible to simply forego cleaning. In such cases, a mildly activated rosin flux is used, and the solder formulation is carefully selected for a balance of printing, wetting, fluxing, and low residue characteristics. The flux chemistry should be such that residues are very low in ionic content, and that any ionics that remain after soldering are rendered harmless by the encapsulating effect of the rosin base of the flux. The benefit of the no-clean method is that the cost and difficulty of thorough cleaning are eliminated. Downsides of this approach are that no-clean solders tend to do poorly unless the solderability of all parts and boards is very good, and the residue that remains may interfere with test probing or contaminant probes. Also, the circuit still looks as if it needs cleaning, and this bothers some veteran inspectors. Caution: do not clean a no-clean for the sake of looks. In doing so, you may defeat the encapsulating effect of the flux residue and leave exposed ionic contaminants in a position to harm the circuit materials.

Semi-aqueous cleaning makes use of a solvent-water mixture where the solvent does the cleaning and the water serves as a carrier to flush away the dissolved wastes. Semi-aqueous cleaners generally are closed-loop in design. Contaminants are purged from the cleaning vehicle, and it is recycled through the system. Advantages are very high levels of cleanliness and the freedom to use more active fluxes for difficult-to-solder materials. The penalties are the cost of equipment and chemicals, and the fact that some of the semi-aqueous methods require the use of highly flammable chemicals.

Solvent cleaning with ozone-friendly chemistries is an obvious and much perused alternative to CFC cleaning. The challenge remains finding a chemistry that approaches the cleaning power, low cost, and handling ease of CFCs without bringing its own set of environmental concerns. Some of the early contenders are already on the list of substances identified as ozone depleters.

When Is an SMT Circuit Clean?
Cleanliness testing is often done by measuring the conductance of a 75/25 percent-by-weight solution of isopropanol and water. This method is complicated by the same standoff problems that make the cleaning process difficult. Contaminants hiding under a tightly clinging SMC are as difficult to dissolve for conductometric measurement as they are difficult to remove in a cleaning machine. Research has shown that many SMT assemblies, which are reflow soldered with RMA flux solder paste, will pass MIL-P-28809 cleanliness testing without any cleaning. The problem is that MIL-P-28809 tests for ionic contaminants and it may not detect rosin contamination, or any ionics entrapped in rosin on the assembly.[8]

A destructive sampling program (including the removal of low standoff components to facilitate contaminant extraction) may be required to ensure cleanliness on an SMT line. Turbidity testing involves a light-scattering test to determine how much organic contaminants are precipitated from an isopropanol wash solution when a dilute aqueous acid is added to the wash. Turbidimeters measure the light-scattering angle at 0°. They are accurate only when measuring a fairly turbid solution. Nephelometers measure light scattering at an angle to the incident beam. For nearly clear solutions, they are much more sensitive than turbidimeters, and thus perform better in SMT testing. Nephelometers are available through scientific instrument distributors for around $1000.

Ultraviolet spectrometric analysis shows the abietic acid level in an isopropanol wash solution by measuring the sharp peak absorption of abietic acid at 241 nanometers. Abietic acid is the principal component in rosin, and gives an accurate measurement of rosin presence down to 1 PPM in the wash solution. Ultraviolet spectrometers range between $4000 and $10,000, dependent on versatility. They, also, are available through scientific instrument distributors.[9]

3.1.6 *Testing*

Testing is often identified by "experts" as a great stumbling block for surface-mount technology. I placed experts in quotes because I've learned, in my engineering career, that the true experts are not those who can tell you how impossible something is, but those who can tell you how to do the thing. So, let's look at what is different in the SMT vs. IMC inspection and test procedures, with an eye on how to test SMT assemblies.

The Challenges of SMT Testing
Inspection and test of SMT boards must accommodate the close lead spacings (1.27 mm/0.050 inch, or less) of SMCs, the tight geometries of SMT assemblies, and the potential of assemblies populated on both sides. Each of these SMT characteristics pose a challenge for visual inspection, in-circuit testing, and repair/rework. SMT also facilitates packaging high-pin-count ICs. Devices with a large number of I/O pins may complicate testing, since many are untestable by in-circuit means.

Close Lead Spacings

Bed-of-nails testing of bare and populated IMC boards is a simple and well-understood science. The standard 2.45 mm (0.100 inch) lead centers of IMCs lend themselves to probing, and probe cards built on such generous center distances are reasonably priced and rugged.

But what happened when we introduced SMT with 1.27 mm, 1.016 mm, 0.635 mm (0.050 inch, 0.040 inch, 0.025 inch), and finer-pitch devices? Typical of technological advances, we stumbled along for a while. Then engineers developed spring-loaded probes for fine pitches. At first, these probes were expensive, but competitive pressures quickly brought them into line with coarser-pitch probes. Today, SMT pitches are readily addressed by probe cards. The cost per point for 1.27 mm (0.050 inch) probe cards is reasonable. Finer-pitch probe cards do carry a cost-and-ruggedness penalty, however. The moral is "Design no finer than is absolutely required."

Dense Assemblies

Planning is the testability watchword introduced by increasing circuit density. Pay very close attention to device clearances (particularly around large and tall components). Make all test nodes accessible. And consider the case for built-in testing. A little investment here will pay enormous benefits in manufacturing and test costs in a production environment.

Double-Sided Population

Provided proper design rules are followed, double-sided boards, with only passives and a few transistors on the bottom side (Type 3 technology), are not much of a test problem. However, SMT allows the building of complex Type 1 and 2 assemblies with dense IC populations on both sides, and these can spell trouble for test engineers. Test nodes for top-side components often are not accessible from the bottom of the board because a bottom-side component obscures them or because they connect through a buried or hidden via or one obscured by a bottom-side component. Solutions to this include built-in testing, active partitioning for functional testing, double-side probing for functional or in-circuit testing, or some combination of these approaches.

Please note that functional testing generally requires access to many or all the circuit nodes in order to reduce test time and provide the most accurate troubleshooting guidance from testing. Therefore, planning to functionally test a board is no carte blanche for covering up test access points.

High Pin Counts

One of SMT's primary benefits is that, by reducing pin spacings and allowing I/O on four sides of a device instead of the DIP's two, SMT facilitates much higher I/O counts than does through-hole packaging. While this is great news for circuit designers looking for ways to replace 50 glue logic chips with one ASIC, it places a burden on test engineers. High-pin-count custom ICs and microprocessors often can't be fully tested in-circuit. Functional and/or built-in testing may be a requisite with high-pin-count devices.

The Answer

Testable and manufacturable SMT boards occur by chance about as often as a group of monkeys types out the great books of the western world in random play with typewriters. More often, testable complex assemblies are the result of rigorous design efforts guided from the inception by a multidisciplinary team that includes process, test, manufacturing, and quality engineers. We'll discuss specific design-for-testability issues in detail in Chapters 4 through 7. For now, let's review key differences between SMT and IMC assemblies, in preparation for a study of various manufacturing approaches to building SMT circuits.

3.1.7 *Key Differences Between SMT and IMC Assembly*

The manufacturing of circuit boards using surface-mounted components differs from through-hole manufacturing in several important ways. We have touched on these differences in the preceding chapters, but we will recap them here before beginning a detailed discussion of the manufacturing processes.

Differences in Engineering
Aside from the obvious adaptations to new component packaging and mounting techniques, team engineering is a major change being pushed by SMT. In our earlier discussion of testing, we stated that a team approach is vital to the success of SMT engineering efforts. While team (or simultaneous) engineering may be employed to benefit in just about any manufacturing operation, it's unlikely you'll manage SMT's complexity without teamwork.

This was not the case in IMC manufacturing. Manufacturing and test engineers in electronics assembly operations have traditionally been largely ignored during the design phase of product development, and are then assigned the thankless task of solving whatever riddles design engineering might "throw over the wall" to them.

The new emphasis on complete teamwork is elevating the status of manufacturing, process, test, and quality engineering in the electronics industry. On the downside, early involvement in the design operations removes the hedge a manufacturing group used to enjoy. When manufacturing participates in the design, it can't refuse to build a new product on the grounds that it is unmanufacturable.

Differences in Component Attachment
The primary difference in SMT component attachment is the lack of the mechanical connection to the board that is supplied by the through-hole-mounted leads of IMCs. The SMT solder joint is not only the electrical connection for the device, it is the major or sole mechanical connection as well. Therefore, solder-joint integrity is of critical concern in producing reliable SMT assemblies.

Operating in opposition to the need for high-integrity solder connections, SMT joints are often more difficult to inspect than IMC solder connections. The solder joints are smaller, on closer centers, often obscured under the component they mount, and are not inspectable from the board's opposite side. Thus, a very-high-yield soldering process, tightly monitored, and under excellent control, is a vital part of production SMT manufacturing.

Differences in Soldering Methods
Solder difficulties can't be resolved by simply turning up the temperature in the solder pot, or by passing boards through for a second or third try in the solder wave. Component-lead and board-metallization solderability must be near perfect.

SMT soldering also differs from IMC soldering in the control of the amount of solder applied. With limited process attention, wave soldering of inserted components in properly sized through-plated holes results in the proper amount of solder being deposited. And the solder deposit is relatively easy to visually inspect on an IMC assembly. In contrast, reflow soldering requires a close attention to design of the solder land, the solder print area, the solder deposit thickness, the solder paste consistency, and the printing operation in order to produce joints with just the right amount of solder applied. And, wave soldering of SMT circuits involves resolving the trade-off between defects produced by too much solder and defects caused by too little solder. Both these defects can and do occur in the same process setup, as shown in **Figure 3-5**. Thus, SMT soldering, far more than IMC soldering, requires an attention to the amount of solder applied.

Figure 3-5. A wave-soldered SOIC showing the application of both excessive and insufficient solder.

Differences in Component Testing

As we stated earlier, SMCs have much closer lead centers than IMCs. Also, SMT boards may have components on both sides. These differences complicate the accessibility of SMT assemblies for testing. Matters are made worse by the fact that, unlike IMCs, probing on device leads is prohibited in SMT. This is because the probe pressure might press an open lead into contact with a land and give a false pass. Also, misalignment with a lead might damage fragile fine-pitch probes.

SMT boards may also have very-high-pin-count devices, many of which are not testable by in-circuit means. The functional testing of VLSI is just as dependent as in-circuit testing on test node access.

The differences in SMT and IMC tests are much more obvious than the way to resolve them. Therefore, test engineering must be part of the initial design of SMT boards. Test access requires space and often conflicts with the miniaturization efforts. Access vs. real-estate-savings trade-offs must not be resolved in a kangaroo court.

Differences in Repair Methods

In their most basic form, SMT and IMC repair is the same. Both involve the taking off of nonconforming parts or circuitry as necessary, the cleanup of the rework area (including removal of solder where components were extracted), and the repetition of the original process steps to replace those components. In some instances, the lack of through-hole clenched leads makes SMT repair simpler than IMC repair. But the higher densities of the surface-mount and double-sided component population both add complexity where SMT is concerned. And, because of the time required to reflow all of the leads of a large SMT IC package, the shielding of adjacent circuitry from conductive and convective heat is a unique challenge of surface-mount repair.[10]

As with testability, repairability for SMT assemblies must be a goal of the initial design. Repair department personnel should be a part of the design team.

3.2 Manufacturing Styles of SMT

Manufacturing style or type (Types 1, 2, and 3), more than any other single factor, determines the process flowchart used to assemble a given SMT product. Armed with our understanding of the SMT process from the previous overview discussion, we will now study how process steps are varied to suit different SMT manufacturing styles.

3.2.1 *Type 1 SMT Manufacturing*

Type 1 Defined

To review, Type 1 SMT is an exclusive surface-mount method, with components mounted on one or both sides. All styles of surface-mount components are used in Type 1 processing. Type 1 components are virtually always reflow soldered. A typical Type 1 process flowchart is shown in Figure 3-6.

Figure 3-6. A typical Type 1 SMT process flowchart.

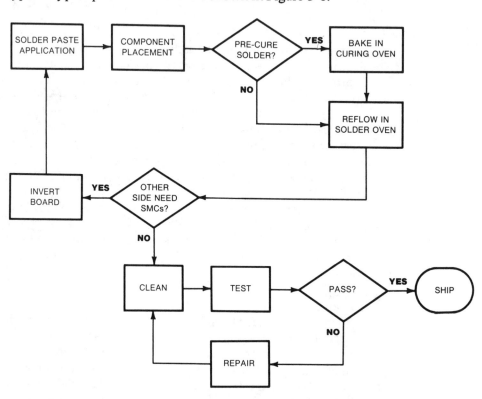

Figure 3-6 shows that double-sided populated boards are reflow soldered twice. On the second pass, the board is inverted and the top-side components are placed upside down in the reflow oven. New SMT engineers often stagger at this. They see visions of all the upside-down components winding up on the floor of the furnace. In fact, with all standard SMCs, solder surface tension will hold upside-down components in place, even though they are reflow soldered on the second pass. The ratio of surface-tension attractive force to component weight determines whether a component will remain on the bottom of a board when held only by molten solder. Surface-tension attractive force is determined by the termination surface area, and can be calculated as illustrated in **Figure 3-7.**

The equation for calculating the device-to-substrate clearance (for above board mounting) is given in Equation 3-1.

$$X = \frac{\delta A_p \sin (\theta_2 - 90)}{F_w + \delta B_1 \cos \theta_1 + \delta B_2 \sin \theta_2} \qquad \text{(Eq. 3-1)}$$

where,

X = Clearance, device to substrate,
δ = Surface tension,
A_p = Total area of chip-carrier bottom pads,
F_w = Weight of chip carrier,
B_1 = Wetted perimeter of chip-carrier castellation,
B_2 = Wetted perimeter of chip-carrier bottom pads,
θ_1, θ_2 = Wetting angle.

The equation for the solder surface-tension holding force (for below board mounting) is given in Equation 3-2.

$$F_p = F_w + \delta B_1 \cos \theta_1 + \delta B_2 \sin \theta_2 - \frac{\delta}{r_o} (A_p) \qquad \text{(Eq. 3-2)}$$

where,

F_p = Pull-off force,
r_o = Radius of curvature at the bottom pad location.

At the point of pull-off, the expression reduces to

$$F_p = F_w + \delta (B_1 + B_2) \qquad \text{(Eq. 3-3)}$$

Figure 3-7. Factors for calculating surface-tension attraction and stand off height.[11]

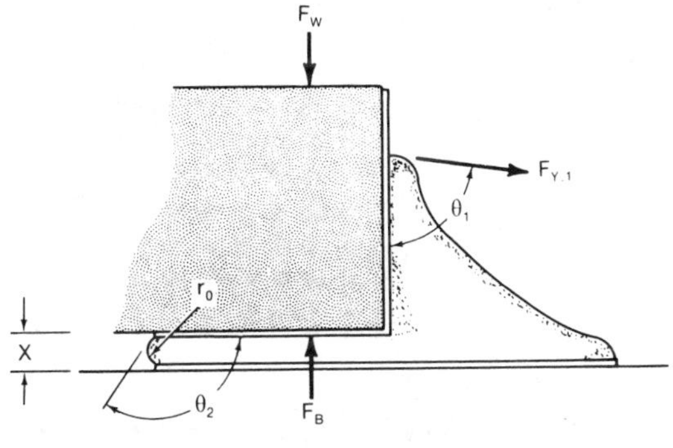

DEVICE TO SUBSTRATE CLEARANCE (ABOVE BOARD MOUNTING) $X = \dfrac{\delta A_P \sin(\theta_2 - 90)}{F_w + \delta B_1 \cos \theta_1 + \delta B_2 \sin \theta_2}$

SOLDER SURFACE-TENSION HOLDING FORCE (BELOW BOARD MOUNTING) $F_P = F_w + \delta B_1 \cos \theta_1 + \delta B_2 \sin \theta_2 - \dfrac{\delta}{r_o} (A_P)$

In practical terms, each lead of a J-lead PLCC adds about 0.08 gram of attractive force in the bottom-side reflow of previously soldered joints. Thus, a 44-pin PLCC needs 24 leads soldered to hang upside down by molten solder (an error margin of 45% if all leads are soldered). However, a 84-pin PLCC must have 71 leads attached, leaving a margin of only 15%. The obvious conclusion is that it is better to locate large ICs on the last reflow side.[12]

Please note that while molten solder may support a bottom-side component during reflow, fresh solder paste most certainly will not. The flux and thixatropic vehicles will lose viscosity in solder preheating long before the solder reflows. Unless they're glued or otherwise fixtured, upside-down components must have already been reflowed in order to remain on the board during soldering.

Type 1 Characterized

Type 1 SMT excels in circuit miniaturization, simplicity of process flow, reliability, and use of high-pin-count devices. On the negative side of the equation, Type 1 usage limits component selection to surface-mountable varieties or special SM lead-formed IMCs. And, Type 1 manufacturing requires the mastering of solder-paste application and reflow soldering techniques. The benefits and limitations of Type 1 technology are shown in more detail in Table 3-1.

Table 3-1. Benefits and Limitations of Type 1 SMT Assembly

Features	Advantages	Limitations
SMT components only.	Best miniaturization.	Component availability.
PLCCs/LCCs/µPaks.	Simple automated process.	Total new line required.
TAB and Flip Chip.	Adapts to batch of 1.	Reflow and screening learning curve.
SO and SOL ICs.	Single solder process.	TCE match concerns (where leadless hermetic component packages are used).
Chip capacitors and resistors.	Hi-reliability possible.	
Connectors and miscellaneous.	Potential high yield.	
SMT lead-formed IMC.	Potential lower cost.	
Very high density.	Improved electrical performance.	
High reliability.	Noise immunity.	
One or both sides populated.	Speed.	
	Allows uses of high-pin-count packages.	

3.2.2   Type 2 SMT Manufacturing

Type 2 Defined

Reviewing Type 2 SMT, we recall that it involves a mixture of SMCs and IMCs on at least one side of the board. Type 2 assemblies may be populated on one or both sides. There are a multitude of ways to construct Type 2 boards, and each manufacturing method has its individual characteristics as to what component styles may be used in what combination(s). However, several generalizations can be made regarding Type 2 SMT. On at least one side, all styles of surface-mount components may be used, just as in Type 1 processing. And that side (the one with all styles of SMCs) will be reflow soldered. Several typical Type 2 process flowcharts are shown in **Figures 3-8 and 3-9.** Many variations of these charts could be developed from permutations of the SMT and IMC process steps. The ones shown are offered as food for thought, not as a required blueprint.

Figure 3-8. Type 2A SMT process flowchart.

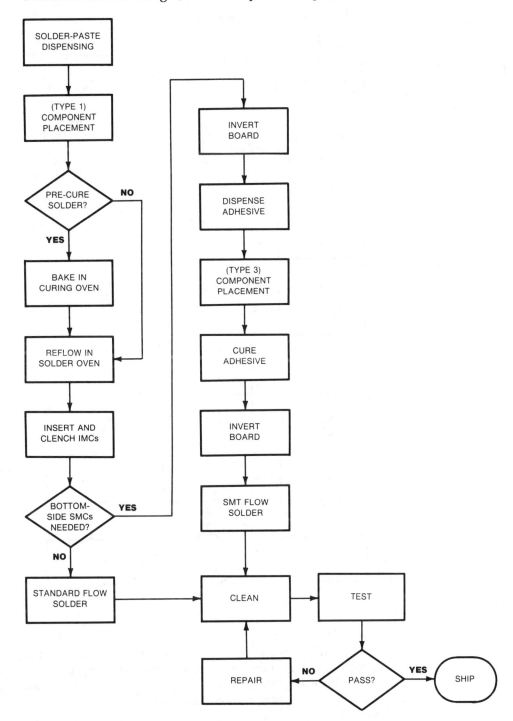

Figure 3-9. Type
2B SMT process
flowchart.

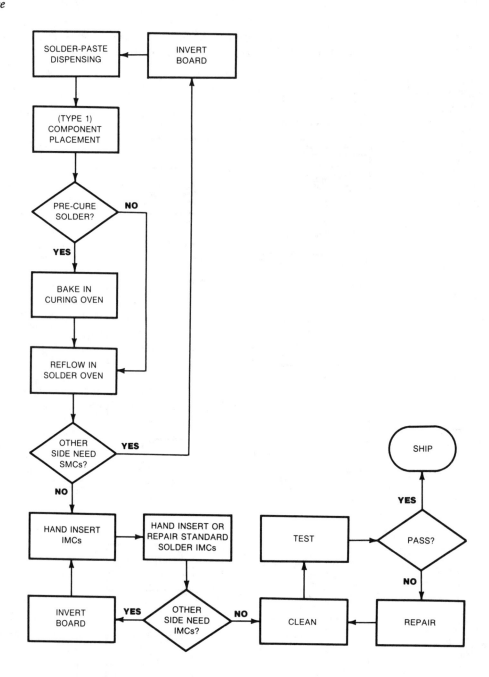

Type 2 Characterized

Type 2 SMT allows impressive circuit miniaturization and the use of high-pin-count devices, coupled with the flexibility of using insertion-mounted components. On the negative side of the Type 2 equation, the mixture of reflow and flow soldering may significantly complicate both assembly, test, and repair of circuits. And, like Type 1, Type 2 manufacturing requires the mastering of solder-paste application and reflow soldering technologies.

Type 2A combines the potential of using high-pin-count chip carriers on the top side of a board with the manufacturing efficiencies of Type 3 SMT for mounting of passives and discrete transistors on the bottom side. Thus, Type 2A would be attractive where a circuit with a high R/C population count also used a few high-pin-count (≥64-pin) ICs. The passives could be bottom-side mounted à la Type 3, and the high I/O actives reflowed topside in efficient SMC packaging.

Type 2B comes closer to Type 1 technology, but allows through-hole mounting by hand or semiautomated means. Because of the labor-intensive approach used for IMCs in Type 2B technology, it is best suited to SMT circuits where only a few through-hole parts will be needed.

The benefits and limitations of Type 2 technology are listed in more detail in Table 3-2.

Table 3-2. Benefits and Limitations of Type 2 SMT Assembly

Features	Advantages	Limitations
Type 2 Generalities	*Type 2 Generalities*	*Type 2 Generalities*
Mix of IMC and SMT components. Some reflow solder. All SMCs per Type 1.	Component availability. Component price shopping. Good miniaturization. High-pin-count packages can be used.	Reflow and screening learning curve. Multiple soldering processes. Complicated test and repair.
Type 2A Peculiarities	*Type 2A Peculiarities*	*Type 2A Peculiarities*
Type 1 SMT processing (top side only). Type 3 processing (bottom side SMCs). Automated IMC processing.	Good density in circuit with high percentage of resistors and capacitors. Relatively automated process. Plus all of the general advantages.	IMC components on top side only. Plus all of the general limitations.
Type 2B Peculiarities	*Type 2B Peculiarities*	*Type 2B Peculiarities*
Type 1 SMT process (top or both sides). Manual IMC processing (top or both sides).	High density in circuit with few IMCs. Allows IMCs and SMCs on both sides. Plus all of the general advantages.	Labor-intensive IMC process. Requires operator finesse. Plus all of the general limitations.

3.2.3 *Type 3 SMT Manufacturing*

Type 3 Defined

To recap, we've said that Type 3 SMT is exclusively IMC on the classical component side and exclusively SMC on the bottom side. Only flow-solderable styles of surface-mount components are used in Type 3 processing.

Type 1 is clearly distinct from Type 3 in processing, since it is fully reflow soldered. Types 2 and 3 may seem alike, as both involve mixed SMC and IMC populations. However, they differ in two key areas. Type 2 mixes IMCs and SMCs on the same side of the board. Type 3 separates them on different sides of the board.

And Type 2 involves two different soldering technologies—reflow and flow soldering. Type 3 is exclusively one-pass flow soldered. A typical Type 3 process flowchart is shown in **Figure 3-10**.

Figure 3-10. Typical Type 3 SMT process flowchart.

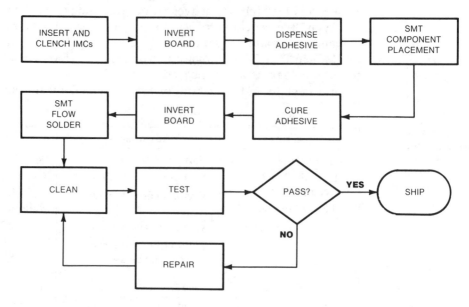

Type 3 Characterized

Type 3 SMT allows some miniaturization, particularly for circuits having a high percentage of discrete components. Type 3 uses a simple automated process flow with only one soldering step. However, Type 3 limits SMT component selection to flow-solderable varieties, which rules out the use of space-efficient chip carriers and high I/O packaging. And, Type 3 manufacturing requires the mastering of adhesive application and curing, with its inherent reliability concerns. The benefits and limitations of Type 3 technology are given in more detail in Table 3-3.

Table 3-3. Benefits and Limitations of Type 3 SMT Assembly

Features	Advantages	Limitations
IMCs on topside only.	Simple 1-solder process.	Only flow-solderable SMCs allowed.
SMCs on bottom side only.	Automated process.	
	Available components.	Not compatible with high I/O SMC styles.
Flow-solderable SMCs.	Good miniaturization of circuits with a high percentage of resistors and capacitors.	Adhesives required.
Chip capacitors and resistors.		
SOTs.		Adhesive and SMC flow-solder learning curve.
SOICs.	Relatively simple test and repair procedure.	

3.3 Process Steps and Design

In the preceding manufacturing discussion, we looked at the SMT manufacturing process in very broad terms, we covered differences between the IMC and SMC manufacturing technologies, and developed a basic understanding of the process flows for various styles of SMT assemblies. All of this will serve as a background for the meat of this chapter, which follows. How do manufacturing requirements impact design decisions? We'll start the discussion by analyzing this question, process by process.

3.3.1 Solder Application

Solder-Paste Screening

Screening and stencil printing are two widely used techniques for applying solder paste to assemblies for reflow soldering. Silk screening and stencil printing have been used for many years for solder application in the manufacturing of hybrid circuits.

Silk-screen printing is a very old art, so named for the fact that the printing screens were originally made of silk stretched tautly on a frame. Areas to be printed were left as open mesh silk, whereas the areas that were to be free of ink on the printed image were filled in on the screen so that ink would not flow through them. Printing was accomplished by forcing ink through the screen's open areas by wiping the ink across the screen with a squeegee.

Today, most industrial screens are made of a more durable stainless-steel mesh, rather than silk. And images are formed on the screen photographically, using an emulsion coating which washes away in the print areas after the photo exposure. But the basic process hasn't changed dramatically since its inception.

Screen printing of solder requires a planar circuit surface. Even a very small lead protruding from an opposite-side IMC would destroy a printing screen. Therefore, screening is generally done before through-hole insertion. Also, there is a lower limit on the size feature that can be printed with screened solder paste. For solder-paste application, a comfortable rule of thumb is to print nothing smaller than 0.635 mm x 1.27 mm (0.025 inch x 0.050 inch) with a silk screen.

With very little investment, a screen-making operation can be set up in a corner of your shop, delivering reasonable cost and quick turn-around on screens.

Solder-Paste Stenciling

Stencils act like silk screens in defining a printed image for squeegee applied paste. Like screens, solder stencils require a planar printing surface. Since stencils do not have screen wires partially obscuring the print area, they are suitable for printing fine features below the typical limits of screens. Solder paste dots well under 0.254 mm (0.010 inch) square can be stencil printed reliably.

Stencil manufacturing is beyond the scope of most small shops, and is typically done by outside services. Both the cost and lead time for stencils are considerably higher than for screens. However, stencils are much more durable, easier to clean, and require less on-line maintenance. Thus, stencils are generally preferred over screens for high-volume printing jobs.

Solder-Paste Dispensing

Solder paste may also be applied to the lands by dispensing. Dispensers generally use a timed pulse of pneumatic pressure to extrude paste from a syringe, or draw paste from syringes and dispense it using a positive displacement pump. Dispensing is inherently a sequential process, usually being done one solder site at a time. Hand dispense guns are often used for prototype and short-run assembly.

Since dispensing heads can be computer controlled on X–Y motions, dispensing allows the programmable selection of solder-application patterns. While it is inherently a slow sequential process, and screening is a fast batch process, the flexibility afforded by programmable dispensers is attracting many manufacturers to this method of solder application. The relatively slow sequential process is often no slower than sequential placement, and thus may not form a bottleneck in a production line.

The primary advantages of the time-pressure dispense method is the lower cost of dispensing equipment. The disadvantage is accuracy and repeatability. Solder paste is 85 to 95% metal powder by weight. Under the heavy pressure necessary to extrude solder paste through a small dispensing nozzle, the more fluid elements of the paste (flux, vehicles, and thixatropic agents) may tend to separate out and be dispensed in disproportionate volume. When this condition occurs, the increase in metals content in the paste soon clogs the nozzle. Positive displacement dispensing is said to overcome this problem by applying dispensing stress to only a very small volume of paste, which is dispensed in a single shot.

Figure 3-11. A tombstoning solder defect.

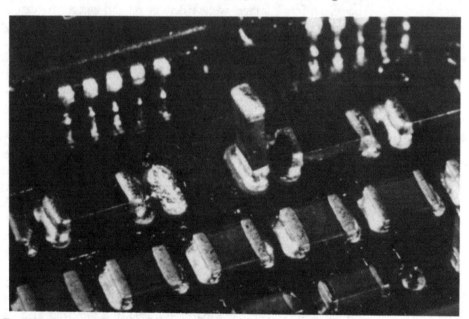

Process-Driven Design Rules for Solder-Paste Application

Only certain pastes are suitable for a given application process. Generally, the viscosity and metals content of the pastes are varied to suit the application methods. Very viscous pastes with a high metals content stencil well. Pastes of moderate viscosity and metals content are better for screening. And pastes of lower viscosity and metals content are indicated for dispensing. Solder paste and equipment vendors can furnish exact guidance in the selection of pastes appropriate for a particular process and application.

In reflow soldering, the amount of solder forming the solder joint is determined by the solder print area, the print thickness, and the paste metals content. The standard device land patterns suggested herein are tailored, in the case of 50-mil-pitch SMT, for screen or stencil printing of 7 to 10 mils of wet solder paste, and will provide the proper amount of solder to the joint under such print conditions. For dispensed solder, the dispensing volume should normally be equal to the volume in a 0.18 to 0.25 mm (7- to 10-mil) deposit exactly covering the land area. For fine-pitch components, it may be necessary to reduce the solder-print thickness to 4-to-7 mils. In order to provide sufficient solder to standard-part leads on a fine-pitch assembly, the print pattern for the standard pitch parts may be expanded by up to 25%. The resulting overprint will not cause bridging, but will withdraw into the solder bond during reflow, driven by the cohesive and wetting forces of the solder.

Solder volume is sometimes varied from the above amounts for certain design and manufacturing objectives. By reducing solder volume, we can reduce tombstoning, the solder defect shown in Figure 3-11. Within limits, lowering the solder volume also increased the joint ductility. Conversely, adding solder will, within limits, increase the joint strength. However, the goal of assembly is not to produce the strongest possible joint, but to produce a sound ductile joint capable of absorbing stresses. Studies have shown that excessive solder contributes to component cracking and termination failures as shown in Figure 3-12.

Figure 3-12. Solder joint is too strong - termination has pulled away.

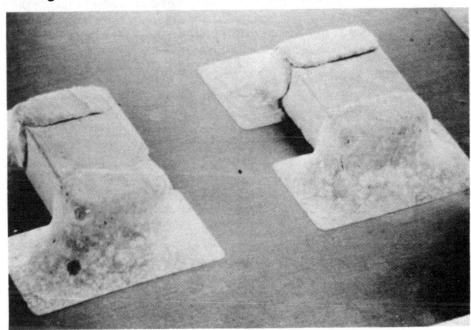

3.3.2 *Adhesive Application*

For flow soldering, SMCs must be held in place by an adhesive until they are soldered. The adhesive is applied to the component sites prior to SMC placement, and is cured immediately after placement. Manufacturing speed is dependent on the adhesive being quickly cured.

SMT flow soldering subjects the SM components to one or more passes through molten solder while hanging from the bottom of the board. Therefore, it is important that the cured adhesive have sufficient holding power for the task at hand, and retains that holding power at the elevated temperatures seen during the soldering process. For the sake of rework, it is equally important that the adhesive not form so strong a bond that components can't be readily removed.

Adhesives are generally left in place after the soldering process. Thus, long-term reliability depends upon the adhesive being fully cured (having no surface tackiness which could pick up contaminants). Adhesives must also be free from voids, which may entrap solder during the flow operation and introduce a reliability concern.

Adhesives are usually applied by screen or stencil printing, dispensing, or transfer (offset) printing.

Adhesive Screening and Stenciling
Adhesives are screened and stenciled in the same manner described earlier for solder pastes. In order to be applied by these methods, adhesives must be thixatropic in nature, with an appropriate rheology.

As with solder printing, adhesive screening or stenciling must be done on a planar surface. Therefore, it is not an acceptable application method for bottom-side attachment on circuits where through-hole components are already inserted. For auto insertion, SMCs are usually added after the inserted components to allow maximum bottom-side clearance for cut-and-clench tooling, and to protect the SMCs from the pounding of the auto-insertion process. Therefore, screen printing finds limited application in adhesive processing.

Adhesive Dispensing
Like solder-paste dispensing, time-pressure methods with syringes, and positive displacement pumping, are typical dispensing methods. Many X–Y placement systems designed for assembly of Type 3 discrete components have an upstream station, running in slave format to the placement station, where a single dot of adhesive is dispensed in parallel to each placement move. This arrangement takes advantage of the X–Y motion used in placement, giving the dispensing operation a free ride. However, unless equipped to implement nonslave motion, the slave station can only dispense in the exact pattern that the placement head follows. As we'll see later under design considerations, other dispensing patterns may be desirable or even mandatory.

Adhesive Transfer Printing
Transfer, or *offset*, printing provides a batch printing method comparable with screening in speed, but suitable for the application of adhesive on nonplanar surfaces. Offset adhesive printers use a finger, or multiple fingers, to pick up a metered layer of adhesive from an adhesive tray and transfer it to the printing site(s). A doctor blade is passed across the tray to reestablish a smooth adhesive layer of correct thickness after each printing pickup.

Offset printing may involve one finger running programmably in an X–Y motion, or multiple fingers individually set in a carrier plate at proper dimensions for each printing location. Single-finger printers operate sequentially and are relatively slow, but highly programmable. Hard-tooled multiple-finger printers can print a large number of sites simultaneously. While they are much more setup intensive than the programmable type, their high throughput makes them attractive for large production-lot offset printing.

Process-Driven Design Rules for Adhesive Application
Again, Rule 1 is to choose a material suitable for the given application process. Epoxy and equipment vendors can furnish exact guidance in selecting the adhesives appropriate for a particular process and application.

Figure 3-13.
Determining the
proper glue-
deposit volume.

MINIMUM STANDOFF. DO NOT DEPOSIT SO MUCH
ADHESIVE THAT EXCESS COULD BE SQUEEZED
ONTO LAND AREAS.

MAXIMUM STANDOFF. ENOUGH ADHESIVE MUST
BE DEPOSITED TO ENSURE CONTACT BETWEEN
COMPONENT AND GLUE DOME.

(A) Standoff variance of component.

MAXIMUM DOT SIZE. SHOULD NOT FOUL LANDS
UNDER A COMPONENT WITH MINIMUM STANDOFF.

MINIMUM DOT SIZE. MUST CONTACT AND ADEQUATELY
ATTACH A COMPONENT WITH MAXIMUM STANDOFF.

(B) Deposit volume variance of applicator.

Figure 3-14.
Solder defect
caused by adhesive
contamination.

ADHESIVE HAS FOULED
THE LAND AND CAUSED
A SOLDER OPEN.

Another important process design consideration using adhesives is how much glue to apply. Proper adhesive volume is determined by two variables. First, the height of component standoff (if any) dictates how high a droplet you'll need to ensure contact between the component body and the dome of glue deposited on the PW board. Second, the rheology of the adhesive and dispensing-volume tolerance of the application equipment determines the quantity that must be dispensed to assure formation of the required height in that dome of glue. Figure 3-13 illustrates how these factors are weighed in determining, by empirical methods, adhesive dispensing specifications for a given application.

Never deposit more adhesive than necessary to satisfy the above test, as excessive adhesive greatly increases the likelihood of the adhesive fouling the solder lands, and causing resultant solder defects. Where components have no standoff, the glue deposit should be kept as small as equipment and adhesion considerations allow. Adhesive contamination of the solder is most likely to occur under components with no standoff, as shown in Figure 3-14.

Designing for adhesive application also includes the determination of where to put the glue dot(s) for a given component. With relatively small components and thermal-set epoxies, this is straightforward. The dot goes under the center of the component. However, for UV-cured epoxies, centering the dot would block the light in the curing operation and result in disaster. Figure 3-15 gives dot-location recommendations for thermoset and UV applications. Dot-location data should form part of the CAD component library database.

Figure 3-15.
Adhesive dot-location recommendations.

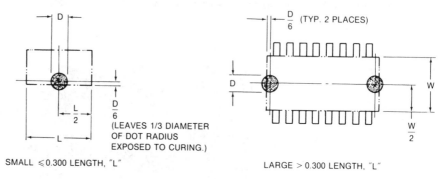

(A) UV-curing rules.

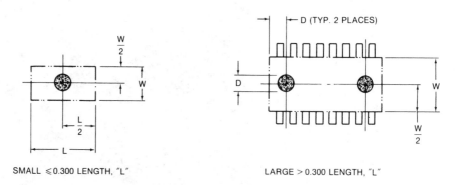

(B) Thermoset rules.

3.3.3 *Placement of Components*

There is probably enough attention focused on robotics today, so we will start our discussion of placement by noting that components can be placed by hand. In fact, at times, components should be placed by hand in an SMT shop. Even the simplest SMT automated placement equipment requires considerable setup attention before addressing a new task. The engineering prototyping of one or two circuits, and the repair operations, are often much more efficiently done by hand. But this can only happen when someone has developed the necessary finesse. Note: It is well to select as SMT assembly technicians those people who are gifted by God with the patience of Job and with hands as steady as Gibraltar. If this is done, you will soon reach a pleasant surprise. The day will come when, for purposes of easy hand assembly and prototyping, you will prefer SMC to IMC assembly.

Placement Equipment Defined

To assemble an SMT board, the proper components must be selected, oriented, and correctly located on the assembly. Placement equipment may be broadly defined as machines which assist in the assembly of SMT boards by helping select, and/or orient, and/or place components.

Placement Equipment Categories

Beyond the simple manual placements of SMCs, there is a wide range of equipment for SMT assembly. Equipment can be categorized by its mode of operation, degree of flexibility, machine intelligence, and level of automation.

Let's start by looking at the *mode of operation* categories. At the low end in cost and throughput are the semiautomated sequential systems, so named for their step-by-step assembly of one component at a time. At the low-volume end of sequential equipment are X-Y machines built around pen plotter motions. Then there are the moderate-volume sequential machines, many of which incorporate the latest in robotics technology, such as vision and tactile feedback, to improve their placement reliability. And there are the high-volume sequential machines, often having multiple heads on a turret. Some of these machines are arranged so that several heads are performing separate functions simultaneously. There are very-high-volume machines which use batch techniques, placing a large number of components in a single machine cycle. Mass-placement machines are also called simultaneous units. Finally, there are the hybrid machines, which combine the various features of both sequential and simultaneous placement.

On the low end of the *flexibility* ladder, we find equipment that does very few tasks, and concentrates on doing them very efficiently. At the top are the true robot, which can exchange end effectors, allowing them to address an extremely wide range of assembly tasks. In general, increased flexibility comes at the expense of throughput and reliability of operation for any one task. However, robotics technology is making great strides toward a marriage of these three benefits of automation.

Categorizing placement equipment by *intelligence* may sound a bit silly to anyone who's spent a few minutes watching a robot assemble SMT boards. I have seen robotic assembly machines get one step out of sequence and place hundreds of components where their next door neighbor belonged. I was a witness when a robot forgot to index in the X and Y axes, and started building a skyscraper of components at one side. Many's the time I've seen a machine drop a component in transit but go on to stuff its empty vacuum nozzle down in gooey solder paste. And I haven't spent much time as a robotics observer.

One might justifiably ask if the phrase "machine intelligence" is a contradiction in terms, just like "jumbo shrimp" or "political integrity?" The answer is "No," but only when we agree on the definition of intelligence in this sense. Intelligence here refers to a machine's ability to sense and react to its own operation and to the environment around it.

At the low end of the IQ scale, equipment has no internal or external feedback. The machine that built the component skyscraper, described earlier, was in this category. If it had been provided with internal feedback to verify whether or not the X–Y motion was occurring properly, it would have sensed the motor failure and called for help. Internal monitoring means sensing that the required actions within the system are occurring. Good robotic design practice demands that no machine action be inferred. All actions should be confirmed by feedback loops (monitored by encoders, etc.).

On the next step up the IQ scale is equipment with internal, but no external, feedback. The machine that dropped a component and still put its nose down into solder paste was not properly equipped to sense component presence on the pickup head. In fact, this particular machine did have a sensing mechanism, but the sensor didn't always work. The sensing of external assembly conditions is an area where advances in robotics are rapidly improving the performance of automated machinery. Coupled with computer hardware and software advances (expert systems and artificial intelligence), sensing technology is dramatically extending the reach of machine intelligence today. Table 3-4 details some of the external sensing technology available in placement equipment/systems as of this writing.

Table 3-4. Types of External Sensing in Placement Equipment

Sensor Category	Placement Task(s) Addressed by This Type Sensor
Delta Pressure	Component presence on vacuum pickup tip.
Tactile Feedback	Component presence/size on pickup tip.
	Placement pressure when placing a component.
Component Testing	Verification of proper component-feeder loading.
	Verification of polar component orientation.
	Limited parametric testing of components (R/L/C).
Machine Vision	PWB orientation and referencing.
	Skipping rejected sections of multiboard panels.
	Solder paste or adhesive dispensing inspection.
	Placement site locating.
	Component form inspection and orientation.
	Completed placement inspection.
Bar-Code Readers	PWB recognition.
	Component carrier recognition.
	Setup vs. program verification.

Finally, as with all manufacturing equipment, we can separate placement machinery by *level of automation*. At the bottom of this staircase are the semiautomated workstations which simply serve to guide a manual operator through placement moves. A step above these, but still in the broad category of islands of automation, are the various stand-alone systems for populating boards. We say "stand-alone" to indicate that these systems provide no means of automatically loading workpieces from upstream processing stations, nor do they have means of passing completed workpieces to downstream machines. Next come the pass-through systems with automated board handling. And, at the top of the heap in degree of automation, we have the self-configuring robotic systems. These are systems which, to a marked degree, are able to automate the setup process. They may scan the bar code on an incoming bare board and then call up the proper assembly sequence and interchangeable tooling needed to build the given part number. These machines may be pass-through types, or may be structured for robotic loading and unloading.

Process-Driven Design Rules for Placement

There are numerous areas where placement-equipment requirements should influence design. We'll discuss the more generic of these. Wisdom dictates that you quiz your placement-system vendor or assembly house for additional equipment-specific design rules.

The clearances required between adjacent components vary, dependent on placement-equipment accuracy and, where required, tweezer clearances. Using the variables illustrated in Figure 3-16, the SMT team should establish minimum clearance requirements for each component part number. This information should form part of the CAD component library database. As adjacent components are located on the board, these clearance requirements should rule how closely they are arrayed.

Figure 3-16.
Clearances for
collision-free
automatic
placement.[13]
(Reprinted from
Focus on SMT
Design, Courtesy
AMTI, Chesapeake,
VA.)

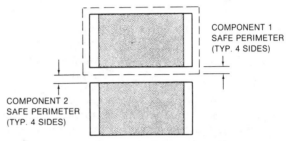

The Safe Perimeter extends an equal distance on all sides from the nominal dimensions of the component (see text).

As illustrated in Figure 3-16, the minimum clearance requirements dictate the *Safe Perimeter* for each component. The Safe Perimeter extends an equal distance on all sides from the nominal dimensions of the component.* To this end, the Safe Perimeter equals some of the factors, A + B + C + D, where

A = The tolerance over the nominal component width or length, whichever is greater.

B = The component placement accuracy. (Note: Many placement machines specify this as ±0.001 inch or ±0.002 inch. Actually their specifications seem to apply to the accuracy of their X–Y motion at the lead screw, not to the total error budget, which includes centering-jaw inaccuracies; Z-axis roll, pitch, and yaw inaccuracies; and theta inaccuracies. Few equipment suppliers can or will discuss this matter in detail. Actual observation and testing may be in order and may show that the total error budget at 3 sigma is closer to ±0.008 inch or ±0.010 inch.)

C = The centering or jaw-carrying allowance, if these will be in contact with the component as it touches the board. (Some equipment breaks the contact of their centering jaws above the board, and extends a vacuum nozzle only to place the component on the board. If this is the case, dimension C can generally be ignored.)

D = The tooling hole or optical registration error. (This item applies only to clearance of components loaded in a previous tooling setup.)

---

* Reprinted from *Focus on SMT Design*, Courtesy AMTI, Chesapeake, VA.

Placement machines with automated board handling use a multitude of PWB indexing schemes. Each system imposes its unique dimensional standards on assemblies. Areas of design and layout concern for automated board handling include, but are not necessarily limited to, those given in Table 3-5. Since the first publication of this book in 1989, one great advance has occurred on this front. The Surface-Mount Equipment Manufacturer's Association (SMEMA) has issued a standard design to allow integration of machines with automated board-handling features. Before buying machinery for a pass-through line, the reader should obtain the most recent version of this standard.

Table 3-5. Design Concerns for Automated PWB Assembly Handling

Area	Concern(s)
Edge Clearance	Size of component free area on PWB edges for conveyor handling.
Registration Feature	Tooling hole and/or fiducial mark size, shape, and locations.
Top/Bottom Clearance	Highest allowable projection from the top and bottom surfaces of board.
Flatness	Maximum allowable board warpage, including any sag produced by component weight. Avoid installing heavy parts until after placement.

Both the variety of component types and the quantity of different component part numbers are limited by the placement machine. It is advisable to keep to a minimum the time from beginning an SMT assembly to the final soldering and cleaning. Time limits vary, dependent on the solder pastes or adhesives used, but any restriction speaks against multiple passes through the placement system. Where designs require component varieties and counts beyond the equipment capabilities, consider breaking the assembly into several boards or mother/daughter cards.

Vision systems may pose a number of restrictions on design. These include clear areas around features to be oriented or inspected, and the required contrast between feature and background (which impacts materials selection).

The programming of placement equipment can be a tedious and time-consuming task, and errors can be introduced during manual programming. Therefore, engineering should provide as much help as possible to the manufacturing personnel responsible for programming. Depending on the placement system used, this might range from simple assembly details to post-processed CAD data suitable for direct downloading to the manufacturing line.

Simple assembly details should include a drawing showing the component center distance from board reference (tooling features), and the placement programming data from the CAD component library data base. The component library might include the clearance perimeter, placement pressure, pickup tooling, centering-jaw sequence and pressure, adhesive dot location and volume, and the component shipping-carrier information.

Post-processing for CAD downloading might include a conversion to placement machine language. Also, post-processing may involve the addition of details which are necessary to placement but are absent in the CAD data base, such as those listed for the component library, feeder-location assignment, optimum-assembly sequence determination, data on component rotation from feeders, and the conversion of component center distances from the abstract 0-0 point of the CAD system to the concrete 0-0 of the placement machine. Some placement-equipment software includes the rudimentary artificial intelligence necessary to simplify program development. Optimization routines might take care of some of the preceding steps, and various CAD systems differ in the level of detail which can be extracted for program generation. If CAD downloading is part of your manufacturing plan, start early and obtain the resources to develop the necessary post-processor. Your CAD system and placement machine vendors may offer some help. (Be warned, however. Some will only put forward obstacles to protect their proprietary software.) But remember, very few suppliers have off-the-shelf solutions to your CAD downloading needs.

3.3.4 *Curing of the Attachment Media*

As we established earlier, solder paste sometimes is cured, and adhesives are almost always cured, before the reflow soldering operation. Solder-paste curing is done by a prolonged bake in a low-temperature oven. Adhesive curing generally involves baking, ultraviolet-light exposure, or a combination of the two.

Solder-Paste Baking
Curing of solder paste is common prior to vapor-phase and laser soldering. Without curing, the rapid temperature rises of these processes can cause an explosive outgassing of the volatiles in uncured paste. For small production requirements, the assemblies may be baked in a batch oven. For higher volume manufacturing, belt-type IR or convection furnaces are typically used. Where bake times are extended, temperatures are particularly high, or oxidation-sensitive materials are used, a nitrogen or forming gas atmosphere may be specified. In general, however, ambient atmosphere works fine and keeps the curing costs to a minimum.

Adhesive Curing
Thermoset adhesives are cured in the same manner as that described for solder-paste curing. UV-curable adhesives are cured under ultraviolet light. Small belt and batch UV machines are available for curing printing inks, and can be readily adapted for low production requirements. For higher production, belt machines combining UV and IR are typically used, since heat greatly accelerates the polymerization of UV epoxies. Indeed, some UV adhesives demand elevated curing temperatures.

Design Considerations for Curing
In laying out boards that will require thermal exposure for curing, it is good practice to distribute the heat-sink masses evenly across the assembly. Large variations in thermal mass in an assembly produce hot and cold spots during a baking operation. Where uneven heat-sink masses can't be avoided, the assemblies may be processed by lowering the temperature and increasing the soak time. Longer soak times allow temperature equalization across the assembly.

Assemblies that are to be UV cured should be designed so that ultraviolet light can easily reach its required target. Avoid any features with a parasol effect.

3.3.5 *Soldering of SMCs*

SMT Soldering Categorized
The first-order division of SMT soldering processes is between flow and reflow. Within each of these categories are various methods for accomplishing the soldering process. SMT flow soldering may be done on double-wave equipment, agitated wave machines, hollow waves, and specialized drag-soldering units. Reflow soldering is commonly done using conductive belt, convection, infrared, laser, and vapor-phase heating sources.

SMT Soldering Reliability Concerns

Many SMT soldering processes expose components to higher temperatures and steeper thermal gradients than what wave soldering places on IMCs. Therefore, it is reasonable to be concerned about component survival and reliability after processing. Table 3-6 shows the reliability data developed by Texas Instruments Incorporated on boards processed through vapor-phase soldering (VPS). The boards were processed using a 4-minute cycle in batch VPS, with 30 seconds at 215 °C.[14]

Table 3-6. Reliability Impact of the VPS Process

Number Of Reflow Cycles	Life Test (125 °C)		85% Relative Humidity (at 85 °C)		Temperature Cycle (-65/+150 °C)		Autoclave	
	Test	*Fail*	*Test*	*Fail*	*Test*	*Fail*	*Test*	*Fail*
NONE	125	2	100	0	100	0	65	1
1	125	3	100	0	100	2	65	1
2	125	1	100	0	100	0	65	1
3	125	5	100	1	100	0	65	2
4	125	0	100	0	100	0	65	1
5	125	0	100	1	100	2	65	0

The test used to obtain the data given in Table 3-6 approximates the process stress that SMT boards might encounter in double-sided reflow and repair soldering cycles. (A control group of boards was not processed through the reflow cycle.) As the numbers show, well-controlled SMT soldering processes pose no threat to appropriate types of SMCs. Thus, as a designer, your job need not include any worry about soldering-process-induced failures. Your job is to select the components appropriate for the soldering methods your assemblies will see.

Soldering-Process Impact on Design

First and foremost in design for solderability is the management of thermal mass. For flow soldering, this includes avoiding any large thermal vias which can connect a flow-type solder joint to a large heat sink. Such connections drain heat from the joint area and can produce poor wetting, and open or marginal joints. Figure 3-17 shows how a thermal connection to a ground plane is reduced to avoid this condition.

We mentioned thermal-path control in relation to flow-soldered boards because vias within SMC lands are permitted in flow-solder design rules. However, the same comment applies to reflow soldering, where special construction rules are followed to allow vias in lands, or wherever lands have a direct thermal path to a nearby heat sink.

For reflow soldering, thermal-mass management also covers the board layout necessary to produce a relatively balanced thermal mass across the entire board surface. Having a large heat sink in one area of a board causes uneven heating in any mass reflow process. The solder lands near the heat sink do not reflow as quickly as those more distant from it. We know that tombstoning is aggravated when one end of a device reflows before the other. So, at best, large variations in thermal-mass distribution may cause components to pop wheelies. At worst, the low thermal-mass areas may exceed the safe time/temperature limits of assembly materials before the heavy thermal mass can be brought up to reflow temperatures.

Figure 3-17.
Reduce any
thermal paths to
large heat sinks.

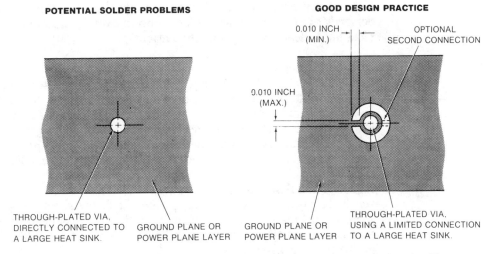

NOTE: This rule applies to vias connected directly to solder lands. Thermal paths may also be reduced by
locating the via external of the land, with a 0.010-inch-wide trace that is 0.010-inch long (min.).

If possible, leave high thermal-mass objects off the assembly until after reflow soldering. If you must put a large heat-sinking device on an assembly prior to reflow, locate it near the edge of the card in a low population area. And don't place potential tombstones near it. In all reflow processes, the card edges heat more rapidly than the center (assuming equal thermal-mass distribution) because they have less thermal coupling and more exposed area for heat absorption.

For vapor-phase soldering, avoid creating low spots or cavities that might trap condensed soldering vapors. Perfluorinated soldering fluids are extremely expensive, and poor designs that create drag-out from the soldering operation can be exceedingly costly.

3.3.6   *Cleaning*

Cleaning Equipment Categorized
Cleaners can be divided along two distinct lines. From a strictly process viewpoint, we can classify cleaners by the solvent used. Aqueous cleaners use water, usually with a saponifier (a wetting agent) as the cleaning solvent. Solvent cleaners use an organic-based solvent. Combination systems use an organic-based solvent followed by a water wash. From a materials-handling view, we can separate cleaners into batch and in-line styles. Both batch and in-line systems are available in aqueous or solvent types, so our division produces a cleaner classification matrix.

Cleaning is basically a time/process-dependent equation. In other words, an aggressive cleaning process will accomplish a given task quickly, whereas some other process may yield equally clean assemblies if we are willing to wait a few years. Remember, just a little water and blowing sand was able to clean away millions of tons of solid rock and expose what we now recognize as the Grand Canyon. However, electronics manufacturing operations usually must reckon time in minutes or seconds, not geologic periods.

Batch Cleaners
Batch cleaners are generally devices for dipping assemblies into a cleaning solvent. The solvent may be in liquid or vapor form. In addition, sprays, heating, or agita-

tion of liquid may be used to enhance cleaning. Excepting ultrasonic immersion cleaners, few batch-style machines clean rapidly and repeatedly enough to meet the demands of SMT. With the military concern over the safety of ultrasonic cleaning, most SMT cleaners are in-line style machines.

In-Line Cleaners

In-line cleaners employ belt conveyors to transport the assemblies through a series of cleaning stages. These stages may include vapor baths, hot sprays, heated fluid immersions, ultrasonically agitated immersions, or combinations of these, followed by a drying stage at the end. Specialized machines, with aggressive cleaning processes such as ultrasonic immersion or very high-pressure sprays, have been developed to meet the challenges of cleaning surface-mount assemblies.

Aqueous and Semi-Aqueous Cleaners

Batch-style water-based cleaners operate like dishwashers. They may be far more sophisticated than the home appliances in their process methods and controls, but their basic operation is the same. In-line aqueous systems follow the description of in-line cleaners, given above.

Arguments in favor of aqueous cleaning include the low cost of solvents, and environmental or safety concerns which are less than those associated with some organic solvents. The negative arguments include contaminated waste-water treatment; the concern that water, even with a saponifier, may not penetrate under the minimal standoff of many SMCs; and the fact that special flux systems must be used for water cleaning. Certain water-removable flux residues are of more concern than rosin flux, if not thoroughly removed in cleaning. Thus, homework is required in solder-paste selection.

Semi-aqueous systems bring a similar set of concerns to those listed above for the aqueous cleaner. In addition, some of the chemistries used are highly flammable, and have a low flash point. Thus, automated fire control (again, without the use of ozone depleting chemicals) may be a requirement.

Solvent Cleaners

Solvent cleaners use organic fluids, often CFCs, chlorinated, or fluorinated hydrocarbons. Such solvents are particularly suited to removing organic soils, such as rosin flux. Because some polar contaminants may be entrapped in organic soil, stabilized organic/polar azeotropes are often used as solvents to remove both types of soils.

Batch solvent cleaners are typically either vapor degreasers or ultrasonic baths. In-line machines operate as described in our earlier discussion of in-line equipment. Most SMT solvent cleaners are of the in-line style because batch types generally require unacceptably long cleaning cycles to produce desirable results.

Combination Cleaners

Combination cleaners are in-line machines that use a series of solvent stages, followed by a final aqueous stage, to address the problem of removing both organic and polar soils.

3.3.7   *Testing of Finished Assemblies*

Testing is a topic which deserves a book all its own, and therefore cannot be fully addressed in a few paragraphs herein. Because of space limitations, we will present an overview of testing methods in this section. Then, in Chapters 4 through 7, we will talk in more detail about testability design. For detailed discussions of test engineering, we suggest you see Reference 15 through 21 in the bibliography given at the end of this chapter, and also refer to the additional references in Chapters 4 through 7.

Test Methods Categorized

Broad categories of testing can be drawn around the type of defect or parameters being inspected. We test for electrical defects and function, using manufacturing-defects analyzers, in-circuit testers, functional testers, burn-in systems, and mock-ups. Assembly integrity and design-rule adherence are checked using various manual and automated vision approaches. Solder-joint quality is inspected using certain vision systems and using laser thermometric analyzers. Assemblies are checked for harmful contaminants in cleanliness testing. And all of the above, plus basic engineering, may be tested using *destructive physical analysis (DPA)*. We'll now briefly cover each of the types of testing we've identified.

Manufacturing-Defects Analyzers

Manufacturing defects analysis (MDA) is a fast test for shorts, opens, and reversed-polarity components. Boards are probed using "bed of nails" fixtures. MDA does not confirm assembly or component operation, but it gives a rapid feedback on some areas of the manufacturing process. MDA testing is usually coupled with other tests to fully qualify the assemblies.

In-Circuit Testers

Ideally, in-circuit testers will fully parametrically test each and every circuit element by isolating each element and testing it as an individual unit. To reach this ideal, every circuit element's I/O must be able to be fully probed in a "bed of nails" fixture, and every circuit element must be completely in-circuit testable. SMT works against both these requirements. Dense circuit populations (the goal of most SMT design programs), close I/O pin spacings, and bottom-side devices, can make the probing of all I/Os, or even all circuit nodes, difficult to impossible. And the increasingly complex high-pin-count devices facilitated by SMT often can't be fully tested in-circuit. Imagine testing for the entire truth table of a 32-bit micropro-cessor and all of its support chips.

The benefit of in-circuit testing, as opposed to MDA testing, is the finding of component as well as manufacturing defects. Thus, in-circuit testing may provide a guide to repair without costly troubleshooting by technicians. Of course, this benefit is limited to designs that allow in-circuit analysis of a major portion of a circuit's elements.

Functional Testers

Functional testing looks at circuit operation *in toto*, and/or in definable chunks. A chunk of a circuit may include one or more elements. Functional testers access the circuit, using assembly I/O, such as edge connectors or cabling, and by the probing of circuit nodes, on a bed of nails. Full access to circuit nodes is essential so a functional test can isolate a fault and furnish detailed troubleshooting guidance. Where probe access must be limited, the design team should devote attention to active partitioning and built-in tests, so as to reduce the costs of troubleshooting on the bench.

Functional testing may be paired with other testing so that the functional test looks only for areas of circuit operation not verifiable in other tests, while manufac-turing defects, component failures, and troubleshooting guidance are derived primarily from outside the functional test.

Assembly and System Burn-In Testers

Burn-in testing involves the monitored operation of the assembly or system at elevated temperatures for a defined period of time. Some companies call a simple on-line operation of a system, without elevated temperature, *burn-in*. More sophisticated burn-in tests can spot the time of failure and can provide significant detail regarding the failure mode, when out-of-specification operation occurs. Like other board tests, access to circuit nodes determines the level of detail that can be obtained in burn-in testing.

As with burn-in testing of individual components, assembly and system burn-in are aimed at catching those functioning components which will fail (under power) early in the circuit's life. Such failures are called *infant mortalities*. Components which do not succumb to infant mortality tend to survive to a ripe old age. Therefore, eliminating weak sisters by operating under temperature stress will greatly enhance circuit reliability. Full-board or -system burn-in will also catch heat-related failures which might occur in the field when the system is under power at elevated ambient temperatures.

Mock-up Tests

Mock-up testing may be as simple as powering up a system to see if it performs its specified functions. Or, mock-up testing may involve plugging the system into a black box to speed up testing and/or capture more information than would be obtained by just operating the system. Mock-up testing provides a low capital investment approach to determining if an assembly is a live one. However, it is best applied to testing systems with only a few operating modes and a limited number of response patterns. Complex systems, with many permutations of responses for various stimuli, would take an enormous amount of mock-up test time. Mock-up testing may also fall short in providing troubleshooting guidance.

Visual Testing

Vision is used in every facet of electronics for quality verification. Visual testing runs from simply "eyeball" inspection of a board to passing assemblies through sophisticated automated vision or X-ray systems. Simple operator inspection, with or without magnification, is used to detect missing or grossly misaligned components. Microscope inspection may be directed at solder quality and fine alignment. Automated vision systems using visible light are high production tools which, with carefully developed programs, will rapidly detect missing and misaligned components and some solder defects. X-ray systems may be used to spot hidden defects, such as voids in solder, delaminations in PWB material, cracked components, and broken internal wire bonds in IC packages.

Laser Solder-Inspection Systems

Laser thermometry uses infrared laser radiation, directed at an individual solder joint, to rapidly heat the joint. The inspection system then turns off the laser and monitors the radiated heat from the joint. Good solder joints provide a solid thermal path into the board, so the built-up temperature quickly dissipates. Defective joints are flagged because they retain the heat longer. Like X rays, laser thermometry finds both visible and hidden solder flaws.

Cleanliness Testing

Conductometric testing has been the mainstay of PWB cleanliness testing in through-hole technology. A mounting body of evidence suggests that new methods, as discussed earlier in Section 3.1.5 of this chapter, will be useful in the cleanliness testing of SMT assemblies.

Nondestructive and DPA Testing

Finally, the physical analysis laboratory provides a powerful testing tool. *Destructive physical analysis (DPA)* is particularly useful in evaluating early SMT designs and detecting the causes of failures. DPA includes microsectioning, scanning electron microscope (SEM), and chemical analysis. Non-destructive techniques include hot-spot detection, X-ray radiography, particle impact noise detection (PIND) of packaged circuits, and high-powered microscopic observation.

By definition, DPA is not a production test, but it may be used to sample a small percentage of boards from a production line. Such sampling can yield early warnings of impending process problems.

3.3.8   *Rework and Repair*

Unless assemblies are very low cost, you will probably want to salvage those that fail test. The good news is that SMT boards can be reworked. In fact, you may find a 68-pin chip carrier easier to remove than a 64-pin DIP with clenched leads. The bad news is there are new and difficult challenges to meet in SMT repair.

Let's begin our repair discussion by looking at some of the challenges SMT presents. Obviously, one is the localization of repair reflow on densely populated boards to avoid damage to areas adjacent to the repair. A subset of this is that removal of SMT ICs is done by reflowing all leads simultaneously. While this can be accomplished quickly using a hot-air repair station, the PWB will remain heated much longer than it would if you were to desolder a DIP, one lead at a time. Prolonged high-temperature exposure can cause delamination or measling in PWBs and can cause metal migration between board plating and solders. Another SMT repair problem is the visual identification of components which are small, may have no marking, and are easily confused with one another.

Now, we'll look at typical repair processes and see how design can enhance their effectiveness.

Heated Tweezers and Special Repair Irons

These simple repair tools are soldering irons with tips that are shaped to cause the simultaneous reflow of all terminations of a certain size SMC. The tweezer variety can actually close on a component to grip it during the repair process. The specialized irons have no gripper, but are shaped to heat all leads of the target device simultaneously. The cohesive bonds of the solder and wetting forces of solder on the iron cause the reflowed component to come away from the board when the iron is lifted.

Both tweezers and tong-tipped irons are usually used to remove and replace small components with leads on two sides. To remove a faulty device, the repair technician places the heated tongs of the iron on the solder joints of the component, waits for the solder to reflow, and then lifts the part with the iron. After removing a component, the joints should be cleaned of all solder, and fresh material is then used for replacement. Otherwise, solder contamination, particularly from the dissolution of traces from the board, may produce a marginal joint or worse. Replacement of components, using fresh solder and flux, is done with the same iron, for reflow soldering.

Convective Reflow Stations

Convective reflow repair stations use either heated gas or heated air, which is directed in a focused stream from above and/or from below the repair areas to melt the solder for rework. Solder joints of faulty components are heated. The operator removes the part with tweezers or a vacuum tool. Alternatively, some repair stations provide a vacuum head, which is used to remove components. After solder removal and cleanup, new solder and flux are added, and a new device is reflowed to the board using the same convective heating procedure.

Design for Specific Repair Procedures

First, the self-evident. Repair operations are always easier on single-side populated assemblies, and on boards with a little elbow room between the components. Adequate clearance between components is critical for rework when using tweezers or shaped irons. Plenty of clearance also is the key when shielding neighboring parts in a convective reflow operation. It is possible to convectively rework double-sided assemblies. Top-side heating only should be used. However, conductive heat transfer through the board will probably bring the bottom of the rework area over the reflow temperature. Solder surface tension will hold all standard SMCs on an upside-down board, so components won't likely fall off unless they are upset externally while the solder is molten. The problem is that metal migration and solder contamination will occur, both during component removal and replacement. Thus, the part directly under a convectively reworked area may look fine, but may have seriously weakened solder connections.

3.4 Degree of SMT Automation

To define degree of automation, let's imagine a degree of automation scale. At the low end of the scale is a total hand-assembly operation. On the high end is a machine that has a raw-material hopper at its input and finished, bagged, and boxed assemblies coming from its output. This topic has enormous impact on the manufacturing engineers of the team, particularly when we approach the high end of the scale. However, we will limit our discussion to an overview of design for automation in this text.

3.4.1 *Prototype Assembly by Hand*

The tiny, often unmarked components, dense layouts, and fine lead pitches of SMT can all spell headaches for assembly technicians who are learning to build surface-mount assemblies by hand. But, manual assembly frees the designer from many of the process design constraints listed earlier. Let's look at the design constraints that pertain to hand assembly.

Some designs would be severe burdens on automated lines because they would require backtracking for multiple passes through certain process steps. Such designs may be no particular problem for manual assembly. Humans are much more adept than machines at jumping back and forth between tasks.

Manual assembly requires sufficient component clearance so unsteady human hands can make placements without upsetting adjacent parts. Tweezer clearance must be considered. And, you should provide enough room so that technicians can clearly see how to orient miniature device leads on PWB lands and can visually inspect the finished work.

Manual assembly eliminates the concerns of automated registration and the handling of boards in placement equipment. Tall components, heavy components, irregularly shaped components—all these pose less concern in manual handling than in automated assembly. However, the process-dependent rules set forth in Section 3.3 for solder-paste application, adhesive application, curing of adhesives, soldering, and cleaning, apply to all degrees of automation.

3.4.2 Islands of Automation

A word of explanation is in order before beginning our discussion of this topic. Islands of automation, as used here, means stand-alone machinery for the various process steps. We refer to equipment which is not able to, or is not used to, automatically handle and index assemblies between various process steps. Factory automation often uses stand-alone islands, called *work cells*, to do specific process steps. Robotic carts or smart conveyors may index work between these "islands." However, since we will consider these arrangements separately under factory automation, we'll restrict our view here to the narrow definition given above.

When all workstations are stand-alone units, it's entirely possible that each will use a different approach to orient and handle the workpieces. Board clearances, indexing, and orienting features should be designed to accommodate all the pieces of process equipment that will be required for the assembly.

Islands of automation can accommodate multiple passes through a given process step. However, minimizing work in process (WIP), avoiding multiple setups, and the rapid completion of assemblies, will all speak against pin-ball-machine process flows. If possible, design the assemblies to flow in an orderly fashion, one time through the manufacturing line, just as if the line was connected by conveyors.

3.4.3 Line Automation

Line automation describes process machines (and steps) connected by fixed conveyors. In line automation, workpieces must visit each process step in the line, whether any work is needed at that step or not.

Line automation dictates close attention to design to facilitate board handling. Free areas are usually required on the board or panel edges. Tooling holes and/or fiducial targets may be needed for alignment. Thus, design rules should be developed around the handling specifications of the line.

The design rules should also take into account the minimums and maximums for each process step during a pass through the line. In other words, if the placement system only allows feeders for 100 unique component part numbers, limit the assemblies to that number. It generally doesn't make economic or quality sense to make multiple trips across a pass-through line. Multiple passes, requiring setup variations, would be an economic disaster.

Manufacturing efficiency of line automation requires considerable manufacturing-engineering input from the earliest stages of system design. This comment is especially true when high-volume manufacturing is expected, which is often the reason for selecting a pass-through manufacturing approach. High-volume pass-through manufacturing adds emphasis to all the design-for-manufacturability comments discussed above, and in Chapter 4 (Section 6) and Chapter 6. Line automation doubles the importance of the manufacturing group's input to the SMT team.

3.4.4 *Factory Automation*

Factory automation is used here to mean flexible automation which requires little or no operator intervention to build a wide variety of PWB assemblies. Factory automation integrates the computer and machine intelligence with machine vision and sophisticated sensing of the manufacturing environment. The ultimate goal of this integration is to build a self-configuring flexible factory capable of throughputs which rival line automation. Such a flexible factory would be said to offer "economy of scope" whereas the pass-through line boasts "economy of scale."

The "design for manufacturing" comments made earlier, regarding line automation, are only amplified by factory automation. And the complexity of software and process control, required for successful factory automation, dictates the starting of projects far in advance of delivery dates and the allocation of substantial resources for prototyping the process and developing the line. We would not recommend factory-automated SMT be undertaken by any organization not thoroughly familiar with surface-mount design and manufacturing.

3.5 Manufacturing Volume of SMT

As we mentioned earlier, manufacturing volume has a significant impact on design directions. Next, we'll look at various production levels, and see how the SMT team adapts processes to suit production capacity.

3.5.1 *Low-Volume Production*

We will arbitrarily define low volume as the placement of less than 25,000 components per average month. By this definition, a placement system or technician placing about 175 components per hour could handle the required output in a single shift. Such a rate is reasonable for manual, semiautomated, or pen-plotter style placement. Note: Since placement is usually the pacing process in an SMT line, it makes a convenient benchmark against which we can gauge the manufacturing volumes.

Low-volume manufacturing shifts some of the SMT team's attention from design for producibility to parts procurement and quality issues. At low volumes, manual techniques can cover for many manufacturability sins. But low-volume manufacturing robs an organization of clout with component suppliers and distributors. This can spell trouble for parts availability and quality. And low volumes make statistical process control somewhere between impractical to impossible. The SMT team should develop a strategy to deal with the unique challenges that low volume may pose.

Where low volumes are complicated by a high mix (a requirement to manufacture many different assembly part numbers), a simple primarily manual approach should be the target. Where possible, avoid excessive setup and programming burdens. For instance, bench troubleshooting will probably cost far less than the development of sophisticated tests to provide detailed fault analysis.

3.5.2 Medium-Volume Production

We'll say that medium volume extends from 25,000 components to around 300,000 component placements per month. And, 300,000 placements per month equates to about 2150 parts per hour, in a 20-day single-shift operation. There is a wide range of flexible placement equipment that is capable of supporting rates of 25,000 to 300,000 placements per month. In reference to placement rates, we'll offer a word to the wise. There is a large troupe of manufacturing personnel who've been disappointed by equipment that wouldn't deliver even 300K parts per month when the specification sheets promised 3 times that! What we're talking about here is day in and day out production on real world assemblies, not an idealized laboratory test. And it's our experience that with one flexible placement machine, 300,000 parts per one-shift month is a high number in the real world.

Our comments on "design for process considerations," made earlier, are aimed at just such a manufacturing volume. The only additional comment necessary is that a high manufacturing mix amplifies demands on the SMT team to develop solid design rules for the organization, so that each successive design is manufacturable on the existing line.

3.5.3 High-Volume Production

For our discussion, high-volume manufacturing will begin at over 300,000 component placements per month. We chose this volume because it corresponds to just over 2000 parts per hour in single-shift mode. And, 2000 parts per hour is a significant breaking point for flexible-placement machinery. Only a select few of the legions of flexible P&P machines available can sustain rates very far above 2000 parts per hour over a long haul, what with all the downtime and setup times considered.

Lines that are set up for significantly higher rates will often have several placement machines. In high-volume production lines, one of the machines may be a "chip shooter" (a very-high-rate machine optimized for a narrow range of tasks, such as the placement of passives from 8-mm and 12-mm tape only).

All the comments in this chapter and in the rest of this book regarding design-for-manufacturability, test, and repair will apply tenfold when high volumes are planned. Special emphasis should be placed on standardized panel sizes and handling features, so that any board may pass through the line without causing tooling changes. Standardized inventory will also pay off, particularly if the part-number count can be kept low enough for stocking on the line. Multiple-pass manufacturing should not even be considered. If necessary, avoid it by routing problem boards to special work cells. A better approach would be to simplify the assembly by using standardized modules on daughter cards.

The SMT team should be up to speed on surface mounting at the prototype and moderate-volume levels before applying that knowledge to high-volume manufacturing of SMT. Likewise, a solid knowledge of manufacturing automation is a precious asset for any group launching into high-volume automated SMT assembly.

In closing, remember that all the rules and caveats given above are for guidance. After looking them over, if you feel you must violate one or more, that's all right. Many others have successfully done so. This book was not written to stifle engineering creativity. We hope the discussion of manufacturing rules will serve instead to guide those creative efforts.

3.6 References

1. Charbonneau, Richard A., "Infrared Vs. Vapor Phase," *Circuits Manufacturing*, September 1986, pg. 27.

2. Hutchins, Charles L., Ph.D., *Surface Mount Technology—How to Get Started*, D. Brown Associates, Warrington, PA, September 1986.

3. King, Scott, "SMT Module Yield Improvement with Infrared Reflow," *Proceedings of NEPCON West 1986*, Cahners Exposition Group, Des Plaines, IL, February 1986, pg. 995.

4. Hollomon, James K. Jr., "Soldering for Present and Future Surface Mount Applications," (In three parts), *Electri·Onics*, July, August, and September 1985.

5. Lichtenburg, Larry, "Reflow Soldering Assembly," *Hybrid Circuit Assembly*, March 1984, pg. 21.

6. Sedrick, A. V., "Design Guidelines for Achieving High First-Time Solder Yields with Mixed Component Technology," *Proceedings of NEPCON West 1986*, Cahners Exposition Group, Des Plaines, IL, February 1986, pg. 797.

7. Johnson, Kathryn L., "Ultrasonic Defluxing of Military Electronic Assemblies—A Reevaluation, Part 1: The Test Plan," *Proceedings of NEPCON West 1986*, Cahners Exposition Group, Des Plaines, IL, February 1986, pg. 389.

8. MacLeod, Norman, "MIL-P-28809: Testing in the Presence of SMDs," *Proceedings of Circuit Expo West 1986*, World Wide Convention Management, Libertyville, IL, 1986, pg. 77.

9. Archer, Wesley L., Cabelka, Tim D., and Nalazek, Jeffrey J., "Quantitative Determination of Rosin Residues on Cleaned Electronic Assemblies," *11th Annual Electronics Manufacturing Seminar Proceedings, Electronics Manufacturing Programs Office, China Lake Naval Weapons Center*, China Lake, CA, pg. 19.

10. Crawford, John A., "Surface Mount Device Repair Study—Miniature/Microminiature (2M) Electronic Repair Program, Naval Sea Systems Command (SEA-060M)", *Proceedings of the 1987 SMTA Technical Symposium*, SMTA, Edina, MN, October 1987, pg. 3-15.

11. Spigarelli, Donald J., "Design and Process Considerations for Soldering Surface Mounted Components," *IEEE Eurocon 1982 Proceedings*, pg. 365.

12. Hutchins, Charles, Ph.D., *Surface Mount Technology—How to Get Started*, D. Brown Associates, Warrington, PA, 1986, pg. 2-2.

13. Hollomon, James K. Jr., *Focus on SMT Design*, Anatrek, Chesapeake, VA, 1988, pg. 3.2.4.2.

14. Hutchins, Charles, Ph.D., and Ganden, Howard, "Pitting SMT Against Standard Mounting," *Electronic Engineering Times*, October 19, 1987, pg. T8.

15. Pynn, Craig, *Strategies for Electronics Test*, McGraw-Hill, New York, NY, 1987.

16. Hroundas, George, "PCB Test Strategies for Manufacturing Yield Improvement," *Proceedings of NEPCON West 1987*, Cahners Exposition Group, Des Plaines, IL, February 1987, pg. 911.

17. Lussier, Paul V., "Developing a Strategic PCA Test Plan," *Proceedings of NEPCON West 1987*, Cahners Exposition Group, Des Plaines, IL, February 1987, pg. 920.

18. Stubendorff, James M., "Design for Testability," *Proceedings of NEPCON West 1986*," Cahners Exposition Group, Des Plaines, IL, February 1986, pg. 336.

19. "SMD Testing," *Evaluation Engineering*, December 1987, pg. 26.

20. Turino, Jon, "Test/Evaluation Outlook for '88," *Evaluation Engineering*, December 1987, pg. 69.

21. Tustin, Wayne, "EE Vibrations," *Evaluation Engineering*, December 1987, pg. 15.

Figure 4 An SMT team that has tackled the thorny problems of system design.

4 System Design Considerations for SMT

This book is aimed at a technical subject and intended for use by a technical audience. But it is important to never lose sight of the people-side of bringing the technical wonders of SMT to light. As Figure 4 suggests, we'll cover some of the human facets of successful design project management in this chapter. That's me in the lower left, talking about SMT to a crowd back in 1983.

SMT is a complex multivariable design problem. As such, good SMT designs require good systems engineering. In this chapter, we'll develop a plan for an SMT systems-engineering effort. We'll begin development of this plan by ranking the benefits that actual SMT users say they achieve from surface-mount technology. Our plan will follow an engineering approach in order to extract the best of what SMT has to offer.

Throughout 1987, my company conducted a study of the electronics business in the upper Midwest. The following material from the data we collected may prove useful in guiding system design teams who are entering SMT for the first time. In our survey, we asked actual users of SMT to rate ten potential SMT benefits on a scale of 1 to 10, showing which had the greatest importance in their application. They were instructed that 10 equals the most important and 1 is the least important. Figure 4-1 details the averages for each benefit from among the 104 sites that responded.[1] Remember, these are ratings from actual SMT users.

Figure 4-1 How users rated SMT benefits in importance.

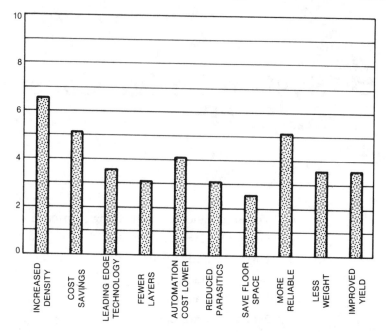

NOTES:
1. SMT benefits are rated on a scale of 1 to 10, with 10 being highest.
2. Results shown are the averages of responses received from 104 sites currently using SMT.

In Figure 4-1, we see some of the benefits others have achieved using SMT. As your team begins systems engineering on a new project, remember these ratings and fight compromises in design that would fall short of the full promise that SMT offers. With our benchmark thus set, let's develop an organized attack plan for the systems engineering of SMT products.

4.1 The Simultaneous Engineering Approach

"Simultaneous Engineering," as it is called at General Motors, is one name for the team approach to product development. Team engineering has recently come into vogue in the USA, probably as a result of cross-cultural fertilization from Japan. We've been deeply impressed by Japan's industrial success, taking its economy from post-war ruins to world-class performance in just a few decades. The U.S. auto industry, in particular, has studied the Japanese model of automotive development and manufacturing. Chrysler, Ford, and GM all have team engineering programs underway today. Let's begin our discussion of team engineering by looking at its application by the Ford Motor Company, where it is called "The Team Concept."

The Ford Taurus, shown in Figure 4-2, has been a runaway success with both critics and consumers. Introduced in 1985, it was the first U.S. car built by team engineering. Simultaneous engineering, according to Cadallic engineer Dave Mattis, works as follows.

"In a nutshell, we start by involving every department that will ultimately bear responsibility for the product. Representatives from design, engineering, manufacturing, marketing, finance, and sometimes even suppliers, all work together, right from the start."

Figure 4-2 The Taurus---first U.S. car produced by Team Concept. An interesting footnote: In 1995, 6 years after publication of our original edition, the Taurus is number 1 in its class in sales. *(Courtesy Ford Motor Company.)*

How does this contrast with business as usual? In the sequential mode, the design department would develop plans for a new model. The plans would then go to engineering, where the available components, such as transmissions and engines, would be fitted into the new enclosure dreamed up by design. Often, friction would arise when existing hardware wouldn't fit. The result would be a compromise that might leave each department somewhat frustrated. Next, the design would go to finance, where the cost of each component, engineering choice, and design direction would be called into question. Finance would usually find something objectionable, with the result of more compromises.

When design, engineering, and finance finally came to an agreement on the new car, plans would go to the manufacturing division. Here, assembly line managers might say, "We can't build it," or "We can only build it if you give us half a billion dollars for new tooling." New and painful compromises would be the likely result. Finally, and only after all these parties came to a truce and decided to build the new model, outside suppliers were called in to provide new specified parts. Because of the compromised design, suppliers might be muscled into bidding so low that either their profits or their quality, or both, were seriously eroded.

The car that finally emerged from such a cumbersome design-by-negotiation process generally was too late to market, too poor in quality, and too far from what marketing needed to suit customer demands. Engineering lead time was up to six years, far too long for marketing to accurately predict customer needs and wants in today's fast-changing world.

Recognizing that something had to be done to regain competitiveness, Ford's current chairman, Don Petersen, hired Dr. W. Edwards Deming in the early 1980s. One of Dr. Deming's most quotable slogans is, "Quality is a management decision." He means quality comes from top management commitment, not from a forgotten few inspectors at the end of the line. In other words, quality demands that all departments in a manufacturing organization work as a team with quality as a major goal. Out of this quality approach, simultaneous engineering has emerged as a powerful tool to improve engineering and manufacturing efficiency. The result? Ford's vice president for design, Jack Telnack, says they hope to bring their normal engineering lead time down from 54 to 45 months. The resulting savings will be enormous, but the standard to measure against is even more impressive. Mazda boasts a 24- to 36-month cycle using simultaneous engineering.[2]

Team engineering is here today, and will undoubtedly become the accepted way of engineering any complex product. It is already the accepted way in all of the successful SMT operations that our consulting work has uncovered. From the preface of this book forward, we have spoken of the SMT team. At this point, let's look at what this team is. How does team engineering work? What proven track history does team engineering carry in the electronics industry? Below, we'll answer those questions. Let's start by looking at some SMT team applications from industry today.

4.1.1 Applications of the Team Approach

Delco Hughes Electronics Corporation

Delco purchased their first chip placement machine for SMT in 1978, but didn't enter into SMT production in volume until 1983. A team of 12 to 15 players from diverse disciplines, such as design engineering, manufacturing, purchasing, production, and maintenance, studied SMT for 5 years before turning on volume production. Delco took this time to study surface-mount technology even though they already had the benefit of some SMT experience from years of hybrid manufacturing.

According to Daniel K. Ward, assistant superintendent of manufacturing engineering at Delco's Kokomo production facility,

> "Purchasing and parts engineering groups obtained samples and did preliminary testing on component reliability. We also researched the best way to design our boards for SMT. We put both of these together in our products."[3]

Delco's facility at Kokomo, Indiana, shown in Figure 4-3, is now reported to be the highest-volume SMT factory in the world.

Steve Hinch, Hewlett-Packard's corporate SMT program manager, says,

> "It's important to get purchasing involved as soon as you know you're going to do surface mount. We've found a wide quality difference among suppliers. It's clear that the selection of surface-mount component suppliers must be a partnership effort between the design engineers and purchasing."[4]

Western Digital Corporation

WDC found SMT and team engineering an ideal marriage. According to spokesman Jim Elliot,

> Right from the start, WDC realized that 'SMT was not just another way to melt solder.' The objective was high quality, high flexibility, and high

volume at predictably lower costs. This they achieved through SPC and discipline in all areas of SMT.

Figure 4-3
Delco's production line in Kokomo, Indiana. *(Courtesy Delco Hughes Electronics, Inc.)*

WDC has one high-volume SMT plant in Irvine, California, as seen in Figure 4-4. The line pictured, originally installed in 1984, was being extensively modernized as this was being written. Unfortunately, pictures of the new line were not available at press time. WDC SMT plants are also operating in Cork, Ireland, and in Puerto Rico. As of this writing, Western Digital is just bringing Plants #4 and #5 on line. And they're consuming millions of components a day.

Disk controllers are a major portion of Western Digital's business. Today, other disk-drive manufacturers are getting into SMT or are getting out of the disk business. But WDC is no Johnny-come-lately. They committed over $3,000,000 in capital equipment to SMT in 1984.

Unlike mainframe-driven computer companies, WDC has tied its success to the personal computer. The increasing use of VLSI and SMT is packing enormous horsepower into the little desktop boxes, and they're showing up in every corner. As WDC chairman Roger Johnson says, "At last, the inmates are running the asylum."

In the roller-coaster marketplace of PCs, WDC decided early that high-flexibility manufacturing was more likely to preserve profitability than shrewd long-range market forecasts. SMT was an important element in developing the manufacturing flexibility they needed. That flexibility philosophy has recently passed the litmus test. WDC sustained their sales growth through a 40% decline in hard-disk-controller prices and a precipitous 80% fall in business from their largest customer, IBM. They finished 1987 at around $500,000,000 in revenues.[5]

Figure 4-4 West-
ern Digital's
Irvine SMT Line.
*(Courtesy Western
Digital Corp., ©Bill
Varie, 1987.)*

4.1.2 *The Need for the Team—a Complexity-Driven Problem*

Once upon a time, a single skilled craftsman could turn out a beautiful and well-crafted product like a pair of boots. Most products were no more complex than a pair of boots. If you needed more boots, you employed more craftsmen. But, as manufacturing volumes grew and product complexity multiplied, the place of the generalist craftsman gave way to the technical expert, one who is highly skilled in a single facet of the creation of a product.

This "Age of the Specialist," coupled with an exponential increase in complexity of manufactured goods, has made teamwork an essential element in product development and production. This fact is particularly true in SMT. Design oversights often can't be taped together on the production floor. If testability, repairability, manufacturability, etc., aren't designed into an SMT board, the results may run from poor (higher manufacturing costs, lower yields, and more field failures) to abysmal (product not manufacturable within required costs, multimillion dollar write-offs, and a declaration of bankruptcy).

What happens when a company adopts surface mounting? Which of the corporate operations are likely to be affected? SMT will impact management, finance, purchasing, receiving inspection, materials handling/storage, basic circuit-schematic design, circuit-board layout, PWB manufacturing, assembly methods and tooling, test engineering and equipment, repair and rework methods, mechanical enclosures and board interconnect layouts, and warranty repair strategies, to name just a few areas. Even the dust from the janitor's broom may be an issue. And SMT will go beyond the corporation in question to impact its suppliers, as well.

Therefore, the single most important factor in successfully entering the SMT field is a smoothly operating team approach. To set up an SMT implementation team, do all of the following:

1. Establish top management involvement, understanding, and support.
2. Form a design team with all the functional areas involved, including the key vendors.
3. Give the team the time and resources to do the job.
4. Set and fully communicate realistic goals.[6]

4.1.3 *How Simultaneous Engineering Works*

There is no absolute answer to the question, "How should team engineering work?" The New York Yankees differ from the Minnesota Twins. Each team is shaped by the culture of its town, the personalities of its players, and the attitudes of its managers. But both have demonstrated they can be winning teams. In the same way, engineering teams are molded by the culture, members, and management of their companies. But, there are a few absolutes to guide team formation. And there are some starting points we can suggest for your consideration of the optional points. We'll cover the firm rules below.

Absolutes
Successful team engineering breaks down departmental battle lines. Team members no longer wear just one hat. They're not just designers, or manufacturing managers, or controllers. They can't care only about the interests of one department. They put on a second hat of "team member." The team's assigned task is the responsibility of each and every member. The team either wins, or it loses. And Most Valuable Player awards don't often go to losing teams. So the first absolute is that team engineering works by enforcing the Golden Rule,

> "Therefore all things whatsoever ye would that men should do to you, do ye even so to them."[7]

When trade-offs must be made, each player must consider the good of all the departments on the team rather than trying to protect the sacred turf.

Rule 1—Whatever the form your company chooses for simultaneous engineering, it should be designed to preserve the Golden Rule effect.

Rule 2—Don't leave any involved department off the team. If territorial squabbles interferred with design compromises before team formation, imagine the cooperation level when six departments gang up on one. And imagine further that the one department feels both picked on and left out of the important work. Leaving out important members will absolutely not save money. It will waste money.

Remember, the whole reason for simultaneous engineering was to boost efficiency.

Rule 3—Give the SMT program every chance to succeed by providing the team with all necessary resources.

Rule 4—Management support and involvement is critical in an SMT start-up. A substantial level of man hours and corporate resources will be required for success. Commitment of such resources will involve top management. A middle manager trying to start SMT as a skunk-works project will have to cut too many corners, and will certainly fall short of achieving SMT's full promise. The unbridled organizational backing needed to tackle surface-mounting technology only comes when top management is informed and excited about SMT.

Aside from providing the necessary resources, the other reason that top management support is vital is for communication. Since SMT impacts all functional areas of a company, the SMT program will run smoothly only with the support of all departments. And such support isn't likely to come out of one excited engineer talking up SMT to finance, tool engineering, test, etc.

Rule 5—Management must assess what surface mount can offer the organization, set realistic goals for the adoption of SMT, and then communicate the goals and benefits of achieving them to the full population.

Options

All right, we've got a team. Every functional area is represented on the team. We've even invited key suppliers to give their input to our first design effort. Now, how does the team proceed to develop those first few critical designs. In Sections 4.2 through 4.10, we'll outline a systematic approach to steer the initial design decisions. Mr. W. C. Clark, a consummate engineer and a cousin of General Mark Clark, once said, "All engineering involves painting yourself into a corner. You just want to be sure it's the one with the door." Systems engineering is a navigational tool for finding the corner with the door.

In systems engineering, we will create a specification and drawing package which will guide each future step in the product's manufacture. This documentation package will grow as we move through the following listed steps. Each successive stage will build upon the past. And each will be guided by the previous steps. It is entirely possible that an advanced step like cost analysis or manufacturability consideration may point to a fatal flaw in the early planning. This can send the whole project back to Square 1. A really fatal flaw could even force management to abandon the project. Of course, we hope to avoid this by thinking each early move through, à la championship chess. But even Bobby Fisher has been beaten. It's far better that system engineering should highlight a serious problem than that the problem be found on the manufacturing floor. So move ahead, endeavoring to win on the first roll, but don't feel that having to make a second pass is a failure; it's money saved.

4.2 System Performance

Step 1 in systems engineering is the development of a basic statement of system performance. This begins at the low level of simply describing just what the new product will do. We then define, in increasing layers of detail, how the system must perform to do the function(s) required.

SMT is an exciting development to marketers in electronics because it offers product-development possibilities that can solve problems and improve people's lives (to a good marketer, people and customers have the same meaning). In areas such as system performance, cost, and form/fit/function, the marketing side of the SMT design team should have a great influence on product development.

4.2.1 *Product Performance to the Customer*

We in high-tech industries have a tendency to lose sight of the customer's needs. Perhaps this comes from our instinctive engineer's love for the technology involved in building our products. We must remember that most customers don't care what circuit technology we used to build the computer, calculator, or modem that we sell them. They care what it will do, how long it will continue to do it, how easy it is to use, and how much it costs.

Engineering-intensive industries are not alone in the love of their product. Even low-tech businesses have occasionally forgotten what they're really trying to deliver to their customers. Perhaps, a look at several real-life examples of this will serve to bring this vital facet of systems engineering into better focus.

Once upon a time, the Hollywood studios thought they were in the movie business. And, they loved the movie business. After all, the movies were an important part of American culture. Why, the Pentagon even hired them to make morale-boosting movies during the great war.

When television was invented, the movie moguls figured it had absolutely nothing to do with Hollywood, because Hollywood was in the movie business. By the time that video recording was perfected for home use, and satellite TV was developed, the Hollywood studios finally realized that these technologies had something to do with them and their customers. They were alternatives which might attract customer dollars from the movie theaters. The studios actively fought to suppress these "bad" technologies. They fought other entertainment, and hung on to being in the MOVIE business until they were almost a bankrupt industry. Then, the brighter studio heads figured out that they were in the entertainment business. They began to seek ways in which their expertise could be applied to delivering what their customers wanted, regardless of the medium. The result is today's healthy studios—churning out TV movies, satellite programming, video tapes, and yes, even box-office successes at the movie theaters.

In like manner, railroads once upon a time were the way that freight was moved over the land. The railroad companies were in the railroad business. Their executives had models of steam engines, like the huge, single-expansion, articulated, Union Pacific 4-8-8-4, displayed on their desks. When the first few packages were moved by truck, the railroads had no interest in this. And the suggestion that sensible men would ship freight in those newfangled flying machines brought guffaws on the railroad's mahogany row on more than one occasion. The railroads had to fade into near-economic ruin before they realized that their customers didn't care about freight trains, they wanted freight transportation.

Nobody in America knew more than the rail industry about transporting large quantities of freight. But the railroads had to come to understand their strengths, in relation to solving customer problems, before they could find creative ways to solve their own marketing dilemma. When they realized that they were in the transportation business, they begin to acquire well-positioned complementary businesses. They developed containerized cargo, interfacing the rails with sea, truck, and air transportation. And, they revived their industry.

Interesting though they be, what does the recovery of the railroad and movie industries have to do with SMT? Well, both effected a recovery by creatively applying their available skills and expertise to deliver what their customers wanted. Successful electronics companies succeed because they do the same thing. And SMT is a powerful tool that you can use to deliver what customers want and need.

For instance, computer users, myself included, want more storage space for large database applications. We want the storage-unit size to be small for use in desktop applications. And, we want to easily interface the storage device to whatever computer we're using. Control Data Corporation's 8-inch disk drive, pictured in Figure 4-5, was the first 8-inch (20.32-cm) drive introduced into the market having greater than a 1-Gigabyte capacity. The 1.236-Gigabyte drive uses SMT to pack all its control electronics into a "half a rack" space. SMT is also used to build three interchangeable interface cards, allowing the user to select the drive with SCSI, SMD, or IPI-2 interfaces for easy connection to existing hardware.

Figure 4-5 An 8-inch high-capacity Winchester Drive. From the vantage of our second edition, looking back at this picture highlights the speed of change buffeting today's electronics engineers. *(Courtesy Control Data Corp.)*

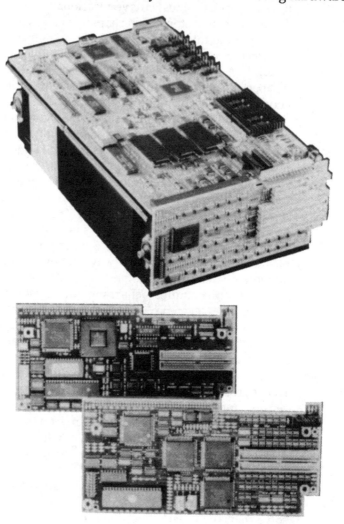

4.2.2 *High-Speed Circuitry*

Where performance means speed, the reduced lead capacitance and inductance of small SMCs are a driving force in their selection for many products. Smaller packages with shorter internal lead paths give faster rise times. And shorter traces on down-sized boards mirror these speed gains. But SMT performance benefits are not without a cost. Existing designs that ran perfectly in IMC may develop timing problems when switched to surface-mount technology.

Testing or modeling of SMT performance is a requisite for the timing of sensitive designs. Manufacturer's specification sheets typically make no distinction between the performance of a device in DIP or SMC. And, to further complicate timing-test issues, the breadboarding of SMCs, using wire-wrap and sockets, may add enough parasitics to invalidate the test.[8] SMT requires a development of design confidence because designers must breadboard on the PWB. And, the cut-and-weld method of fixing a PWB error gets tricky with 6-mil lines and 6-mil spaces.

Simulation on a CAE workstation is recommended where available. Since component specification sheets don't tell the SMT timing story, noncritical designs generally model SMCs at the fast limit, rather than the design median, of the DIP part's rise time. Where this is not accurate enough, you must "factor in time" for rise-time testing in the design process.

4.3 Form, Fit, and Function

Step 2 in systems engineering is the form, fit, and function decision. While we usually write these three items in this order, we'll actually consider function first, fit second, and form third. The topics flow much more naturally from one another when taken in this reverse order.

4.3.1 *Function*

Function is the most basic description of what we plan to develop in our system-engineering effort. What will the new product be? What will it do? How can the product be designed to suit the customer's desires? Function is the stage where we formalize our preceding performance statement into a specification integrated with the beginnings of how we'll shape the actual product hardware.

In considering function, we must decide, at a macro level, how each thing the product does will be implemented. For example, let's say the task is to create a timer for the remote control of household appliances. Will the customer want it to be an electronic or a mechanical timer? Once the electronic functions are identified, the team can move on to define how each will be implemented on a micro level. If the timer is to be fully electronic, would the customer's interests best be served by a monolithic approach, by a hybrid approach, an SMT board, etc.

The output of the function stage of the systems-engineering process is a detailed written description of all the required functions for the product, and a plan for the basic implementation of these functions.

4.3.2 *Fit*

Fit is the way in which a product interfaces with the immediate outside world. The appliance timer we envisioned above might need to be connected between a 115-volt wall outlet and things like coffee pots or floor lamps. It might be situated on a kitchen counter or under an end table in a living room. And the user would need to have some feedback as to its settings and operational status. These considerations would come under "fit."

Under fit, we would also consider how it will connect with other elements in its intended environment. Depending on task, this process may be driven internally or externally.

For a self-contained and self-sustaining product, it may be that no compelling fit constraints exist. The job is then a relatively simple inwardly driven one. We work outward from the circuit, determining what fit best suits its intended function. If the product's relationship to external elements is important, we work inward to fit our product to its expected environment. For example, suppose we are developing an SMT card to replace a through-hole board in a 19-inch rack-mounted VME-bus computer. The new board must have edge connectors and cable connectors in exactly the same locations as its predecessor, and it must mount from the same hardware.

The information we derive on circuit fit is added to the function description produced in Section 4.3.1.

Fit comes before form (size and shape) because this order of progression allows important external factors, which influence form, to be considered in advance of choosing the form. For instance, a circuit which must fit in a half-height space for a disk drive, coexisting with mechanical drive hardware, could be physically very different from a circuit of an identical function which is to fit in a 19-inch rack mount.

4.3.3   *Form*

Form refers to product size, shape, and weight. Form is, in many instances, responsive to fit. Our example of a household-appliance timer would have a very different form factor from a timer for OEM heavy-equipment manufacturers.

As with fit, form may be externally dictated by the required fit and function. Or, in the absence of compelling forces, form may be internally determined by circuit requirements, cost considerations, etc. Externally dictated miniature form factors are often the genesis of surface-mount programs.

Consider our preceding illustrations. The half-height disk controller would have a rigid fit and form factor imposed upon it. Most small-computer disk-drive manufacturers are turning to SMT and VLSI to crunch needed functions into the tiny envelope available. The household timer would have a form and fit largely determined by pleasing outside package appearance, low cost, and minimal size. Miniaturization and cost reductions would be probable drivers of SMT for such an application. And the 19-inch rack-mounted VME card would have some latitude in form, but its fit would be dictated by the rack-mount hardware and the electrical interfaces required. Surface mounting would probably not be a size factor, but would be a cost, reliability, or function-driven choice in such an application.

Documentation of the form determined for our product is added to the preliminary specification. Our form, fit, and function documentation will be the nucleus for the entire engineering project. At each successive stage, it will guide investigation. And, each additional step toward final product plans may amend the original form, fit, and function write-up.

System Packaging

The basic form, fit, and function description can be further refined now by going to a deeper level of detail on product design. Selection of the form, fit, and function will guide the system packaging design effort. How will each board be shaped, interconnected, and housed? This information forms the next concentric circle of data in our design specification.

System Interface

With system packaging understood, we move on to specify how the system will interface with the outside world. This area covers system I/O and user interfaces, such as switches, indicators, keyboards, etc. This forms the next growth ring on the project data tree.

Material Selection

Next, the materials, such as PW boards and solder, are specified. By this point, the team must have a reasonably solid idea of how the board will be built, because certain materials will not be compatible with certain processes.

Component Selection

Right along with the board materials comes component selection. The two are so interrelated that they should be done together.

As SMT technology matures and grows in usage, component sourcing difficulties shrink exponentially. However, finding needed components for conversions occupied a considerable quantity of my time and energy in 1987, and I'll be pleasantly surprised if the problem evaporates in the next 5 years. Therefore, early and diligent team attention (including purchasing) to supply is an obvious step. Also, SMC styles impact manufacturing method selection, and so deserve consideration at the very beginning of a project.

Are SMCs the Only Way?

The first important component decision for the team is, "Will this be an all SMT board?" If the answer is yes, the component selection process requires very early attention. If, however, one or two components must be through-hole mounted, there is much to be said for relaxing the "SMCs wherever possible" rule. Resistor networks, R/C networks, and multichip modules in SIP or ZIP packaging are very space efficient, and are often readily available at competitive pricing. They deserve due consideration on any board which will require wave soldering.

Components, a Multiple Choice Question

In addition to team effort, engineering can often break component availability impasses by selecting alternatives which are electrically suitable to replace specified parts. According to Steve Leibson of EDN, you should investigate pin- and function-compatible devices from second sources, equivalent devices from related-device families (i.e., replace a 74LS part with a 74S, 74HC, or 74F part), and superset compatible parts (such as using a $2K \times 8$-bit memory in lieu of a $1K \times 8$-bit unit).[9]

Multipackage Land Patterns

Another strategy that may ease component supply crunches is the incorporation of multiple-land layouts on a PW board. Figure 4-6 shows several variations of multipackage pads integrated on a board.

Figure 4-6 Multi-package lands ease SMC availability concerns.

NOTE: *These two examples are simply offered as food for thought. The same strategy has been used in through-hole technology for years. For instance, say we need a 1000 mF Radial-Lead Aluminum-Electrolytic Capacitor rated at 10 Volts. It has a lead spacing of 5 mm (0.197"). However, experience has shown that, at times, a 16 volt part, with a lead spacing of 7.5 mm (0.295") is either cheaper or more readily available. In such cases, space permitting, putting redundant patterns on the board is good design practice.*

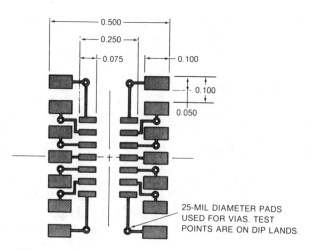

NOTE: Sizes are in inches.

25-MIL DIAMETER PADS USED FOR VIAS. TEST POINTS ARE ON DIP LANDS.

(A) A 14-pin DIP, gull-wing lead form, and 14-pin SOIC integrated circuit (note non-IPC lands).

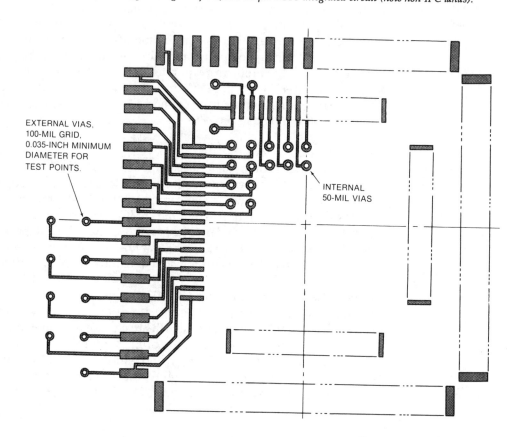

EXTERNAL VIAS, 100-MIL GRID, 0.035-INCH MINIMUM DIAMETER FOR TEST POINTS.

INTERNAL 50-MIL VIAS

(B) A 68-pin, 50-mil PLCC and 68-pin, 25-mil PQFP integrated circuit (showing two via alternatives).

4.4 SMT Manufacturing Style

In the product-specification phase discussed earlier, the team will probably have developed a good idea of the SMT style that best fits the new creation. At this point, we'll need to formalize thoughts on manufacturing style. The output of this step (Step 3) is a manufacturing flowchart showing each process step in order. As the flowchart develops, the team should review all available resources in light of the processes required. Planning must include both the flowchart and the manufacturing plan to execute the flowchart.

4.5 Manufacturability

Armed with our preliminary outline of a product and the process flow to build it, it's time to turn to manufacturability (Step 4). At this step, we'll look for ways to make our product easier and less costly to manufacture. Strategies for yield improvement, cost reduction, work in process reduction, etc., will be part of this investigation.

For discussion, we've considered manufacturability, SMT manufacturing style, and manufacturing cost separately. In practice, however, they are taken together as a product manufacturing review. The topics are so closely related that they must be considered as a unit.

In Chapters 5, 6, and 7, we'll discuss ways that design can improve SMT manufacturability. The team will use such information at this point to hone the manufacturing plan from the rough flowchart, discussed in Section 4.4, to a tuned plan for efficient production. A review of design manufacturability is covered in Chapter 8, Section 8.2. Data on manufacturability and process details should be added to the process-flow documentation.

4.6 Manufacturing Cost

In this phase of system engineering, the team should perform a careful cost analysis of the new product. This phase starts with a critical look at each piece part and value-added step. How can value be amplified? How can cost be reduced? After value engineering comes the tallying of predicted costs. The value analysis and cost estimate data are added to the product documentation package.

Both manufacturability (discussed earlier) and cost analysis at this stage, produce their share of project U-turns. The team shouldn't look at a reversal as a failure. Rather, it is a success of system engineering in preventing a costly failure in production. Quite possibly, the lessons learned in coming to the U-turn will steer a second effort to a stunning success.

4.7 Capital Investment Evaluation

Product plans which prove manufacturable and cost effective must now be reviewed as a prospective investment for the corporation. Each corporate finance department will have a standard yardstick it applies to this evaluation process.

Usually, there are a number of investment alternatives competing for corporate resources in any fiscal period. The purpose of capital investment evaluation is to determine which of all the opportunities before the corporation deserve the investment of available funds.

The output of this stage of system engineering is the cornerstone of the business plan for the new product.

4.8 System Reliability

From both our own projects and our market research in analyzing other companies' SMT successes, we're convinced that surface-mount technology can actually be a vehicle to improved quality and reliability. It is equally clear that SMT, improperly applied, can wreak havoc with circuit reliability. Chapter 7 will be devoted to quality issues. Next, let's look briefly at quality-related differences between SMT and through-hole products, and then turn our attention to the goals of the system reliability step of system engineering.

4.8.1 How SMT Reliability Differs from IMC

SMT is generally selected, among other reasons, for its density advantages. But SMT miniaturization can be a mixed blessing. While it allows the packing of more circuit function into a given space, this also means the packing of more power dissipation into that space, which can have a devastating impact on system reliability.

Through years of familiarity, many designers have developed a sixth sense of thermal management with DIP packages. Be aware that SMT throws seat-of-the-pants thermal engineering into the ringer. You should allow time in the design process for thermal modeling. Tried-and-true TTL designs might benefit from some experimentation with advanced CMOS when SMT changeover time comes around.[10] Thermometric photography of a breadboard may be in order. And, breadboarding is not as simple with SMCs as it was with through-hole proto boards.

4.8.2 The Output of System Reliability Analysis

In the system-reliability study, the team should calculate predicted reliability and B-10 life, lay plans for maximizing reliability, and develop a detailed test strategy to ensure that the reliability predictions become reality on the production floor. Test planning should not focus on final test, but on process control. The general wisdom that a poorly designed and manufactured product can't be tested into excellence applies firmly to surface-mount devices.

4.9 The Manufacturing, Storage, and Operating Environment

To build maximum reliability into electronic assemblies using SMT, or any other technology, the design team must consider the full range of environmental conditions that assemblies and component parts will see during manufacturing, warehousing, and operation.

In this section, we will provide an overview of the issues to consider in design for environmental stress. The team should review environmental issues and document the intended manufacturing, storage, and operational environment as part of the product data package.

4.9.1 *The Manufacturing Environment*

ESD Control

In recent years, ESD has been identified by destructive physical analysis (DPA) laboratories as the cause of a large percentage of all electronic equipment failures. Management often questions the worth of ESD programs when asked to invest in static protection. Table 4-1 provides statistical data gathered by Sperry Flight Systems, which can be used to calculate the likely cash benefits of ESD control.

Table 4-1. Potential Savings from ESD Control

Quality Problem	Percentage Attributed to ESD
Component Failures	5% to 25% Caused by ESD
DOA Components	50% ESD Caused
Infant Mortality	>50% Caused by ESD

Typical operating costs are between $50.00 and $100.00 in finding and replacing one failed component in the factory. Field-failure costs vary somewhere between 10 to 100 times higher than factory-failure costs. The exact multiplier is dependent on the repair strategy used. The net result of all this is that in most manufacturing situations, ESD control will produce a 25% to 50% improvement in output.[11]

Manufacturing Processes

Aside from external stresses, such as ESD, assemblies must survive the sometimes brutal environment of the manufacturing floor. In Chapter 3, we talked about the vibration and shock of component insertion, the heat of SMT soldering, and the action of aggressive cleaners. The team should review the process flow and ensure that all materials and components in the assembly are going to survive the stresses of the manufacturing environment. Any unusual care required in the manufacturing process should be clearly identified and should become part of manufacturing process instructions.

4.9.2 *The Storage Environment*

Operating environments attract attention. Some products, like spacecraft electronics, face incredibly severe and wildly varying environments. Others, like office automation, live in relatively sedate and desirable habitats. However, designers considering thermal cycling must reckon with storage conditions as well as operating conditions. In the warehouse, or even in the office on weekends, the office automation product could see temperatures well below zero or above 100 °F. The team should plan for the product to emerge breathing from the storage environment. The acceptable storage-environment parameters should be added to the documentation library for the product, and this should serve as a guide in writing storage instructions, both for distribution and end users.

4.9.3 *The Operating Environment*

Finally, the team must consider the operating environment for the new product. What impact will temperature extremes, RFI, EMI, radiation, contaminants, moisture, shock, and vibration have on circuit operation and reliability? We list this item last only because it's the first and, often, the only thing we instinctively consider when the word environment is mentioned. The operating environment is far from last in importance, however. And scant data on SMC performance under environmental stress may make this phase a time-consuming part of the total systems-engineering job. Testing will almost certainly be in order if the environment is at all harsh.

The operating environment specifications are added to the data package, and they are the key to generating safe-limit literature for users and for setting testing parameters for QC.

4.10 Design Feedback

By design feedback, we mean the formalized approach planned for analyzing and correcting the anomalies uncovered after production begins. A well-conceived design-feedback system provides a wealth of data to improve the teams' skills, the design guideline library, corporate efficiency, and corporate profits. Design feedback is the cornerstone of a "solution-oriented problem-solving approach." Solution-oriented problem solving is a powerful new tool being directed at manufacturing efficiency improvement today.

4.10.1 *Suggestions for Feedback Implementation*

We now offer a few suggestions toward setting up the design feedback system. Feel free to customize these and innovate.

1. All employees involved with the product in any form should be encouraged to contribute suggestions and they should be solicited for input about things they see going wrong. Nobody knows more about a product than the people who handle it day in and day out. When suggestions or comments are received, they should get immediate attention before the team, and a report of action should be given the originator. The absence of a response, taken as evidence that management doesn't care, will quickly crush any worker's constructive participation.

2. Develop a mechanism to unmask process-control difficulties that are responsible for recurring line, test, and field failures. Add to that mechanism the apparatus to do something about the root problem. Communicate these efforts to the workers so they know reliability and efficiency are important to management.

3. Have a scheduled way to contact a random sampling of customers for their input about product performance. Each of us has had experience with some gizmo that would need a miracle to perform its specified

function. Do you remember thinking that the guys who made that thing must never have used their own product, or they'd know better than try to sell you one? Remember how long it was before you bought anything else from the responsible (really, irresponsible) company?

The preceding are tried and true methods of improving manufacturing efficiency. These and other methods have been described in depth in works by such quality gurus as Deming, Juran, Crosby, et al. They will work in any manufacturing operation, but they are essential for complex problems like surface-mount technology. The important point here is to review all design-significant feedback that is set before the team, and see that the SMT design database is a constantly growing and improving library.

4.10.2 *Likely Areas for Concern*

In our study of the upper Midwest, respondents rated the following areas, shown in Table 4-2, as particularly troublesome in their SMT programs. Table 4-2 represents answers from over 100 companies currently involved in SMT. Forewarned is forearmed. Your system approach should detect and prevent these "gotchas."

Table 4-2. Users Rate Their Unmet Needs in SMT[12]

SMT Issue	Percentage Identifying Unmet Needs	
	Unmet in 1987	*Expected 1990*
Component availability	23%	21%
Standards/Design guidelines	20%	7%
Automation in-house	13%	9%
Competitive manufacturing cost	11%	9%
Material inspection/control	5%	9%
Prototyping ease	5%	2%
All SMT needs unmet, Leaded ceramic CC available	4%	0%
Design resources for SMT	4%	5%
Ability to rework/repair	2%	7%
Better PWB materials, Local assembly houses available methods for fine-pitch parts	2%	5%
Faster deliveries, Lack of expertise, P&P equipment better	2%	2%
Better market anticipation	2%	0%
Better design service, Better cooling of fast ICs, Ceramic board fabrication, Flexible equipment	0%	2%

In closing, I will warn that, while SMT design is not impossibly complicated, it is definitely not business-as-usual for those familiar with through-hole. Achieving SMT design excellence can be compared to walking a tightrope over the Grand Canyon. So long as you make no false steps, you reach great heights and enjoy an utterly breathtaking view. However, mistakes can be very costly indeed. The safety net is the team, and the dedicated attention to detail in engineering. Often, in the rush to move a product swiftly from inception to market, engineering is short-changed. This is always unwise, and doubly so when we are dealing with a complex integration of electronics and manufacturing technology such as SMT. Engineering typically represents well under 10% of the total investment required to deliver a product throughout its lifetime. Yet well over 90% of the decisions that determine the product's cost are cast in concrete during the engineering phase. Over 90% of the lifetime cost of a product is generally in cost of materials, manufacturing, distribution, and service. Yet these areas are usually able to contribute only minor reductions in product cost. The same equation holds true when we look at product quality and reliability. Engineering is in a highly leveraged position to contribute to product excellence. In order to improve product quality and lower costs, management attention must be focused on excellence in R&D.

That is not to say that sales, marketing, manufacturing, all the other departments have an insignificant role. Certainly, unless they do their jobs excellently, cost and quality may suffer indelibly. However, their function is to actualize the potential cost savings and quality inherent in a design. Only engineering and R&D are in a key position to generate that potential.

4.11 References

1. Hollomon, James K., *Advanced Manufacturing Technology in the Upper Midwest—A Research Report*, Anatrek, Norfolk, VA, 1988.

2. Jeanes, William, "The Idea That Saved Detroit," *Northwest Magazine*, Minneapolis, MN, September 1987, pg. 15.

3. Duensing, S., "An Inside Look at Delco's Surface Mount Process, Part 1," *Surface Mount Technology*, October 1987, pg. 16.

4. Russel, John F., "Surface Mount Technology: Standards Finally Fall into Place," *Electronics Purchasing*, September 1987, pg. 52.

5. Hollomon, James K., *Anatrek SMT Training Guide*, Anatrek, Norfolk, VA, 1987, pg. 31.

6. Ibid, pg. 29.

7. *Holy Bible*, King James Edition, Matthew 7:12.

8. Leibson, Steven H., "EDN's Hands-On SMT Project—Part 5: Automated Testing of SMT PC Boards," *EDN*, July 23, 1987, pg. 76.

9. Leibson, Steven H., "EDN's Hands-On SMT Project—Part 2: Selecting the Surface-Mount Components," *EDN*, June 11, 1987, pg. 165. (See also Part 1, *EDN*, May 30, 1987 issue.)

10. Op. cit., ref. 8, pg. 75.

11. Schmitt, Charles A., "Elements of ESD Control," *Printed Circuit Assembly*, June 1987, pg. 6.

12. Op. cit., ref. 1.

Figure 5 High-speed modem, miniaturized by SMT. *(Courtesy Universal Data Systems.)*

Figure 5, selected for our first edition back in 1989, nicely illustrates the old acorn, "Everything changes, but everything remains the same." Today, modems make great use of very large-scale and ultra-large-scale integration in ICs. Modems that used to require large circuit cards are now easily packed into PCMCIA cards having many times the performance of the 1989 behemoth. Yet circuit designers are still struggling with the integration of through-hole and SMT parts. If anything, the situation has gotten considerably more complex, as fine-pitch technology has generated an additional tier of challenges in harmonizing various technology-specific manufacturing processes.

5 Printed-Wiring Layout Using SMCs

SMT is one of the major design tools that is being applied to the shrinking of computer modems. The high-speed unit shown in Figure 5 is from Universal Data Systems. It is an excellent sample of what we'll be covering in this chapter. How do we integrate tiny SMCs with chunky transformers and other insertion components? How do we orient parts, and how close can we pack them? What do good lands look like? In this chapter, we'll answer questions relating to PWB layout using SMCs.

We've finally completed the preliminaries, and now we're ready for the main event, SMT PWB design. Since this book is intended for working engineers, the textural material on design is contained in this chapter, where it fits naturally as one reads from cover to cover. Drawings of the components and land geometries are collected in Appendix A, for ready reference now and in the future. Note: The six years since publication of our first edition have proven the folly of listing all components. While most of the mainstay packages are listed in Appendix A, no attempt has been made in this edition to provide a complete listing. New packages evolve far too rapidly for such an effort to succeed. Instead, we will caution the reader to consult component suppliers and standards bodies for the most current offerings. We will concentrate here on the pervasive packages, and on how to develop land patterns for the newly arrived.

5.1 Component Orientation

SMT component orientation is determined by both constant and variable rules. Let's first cover the constants, those rules which apply regardless of the application or assembly style. Then we'll discuss rules which are applied only to certain fabrication techniques or situations.

5.1.1 *Steadfast Component Orientation Rules*

As a rule of SMT or any PWB layout, the components should be oriented to keep the copper coverage nearly uniform across the surfaces of the board. Uneven distributions of copper can create problems in the electroplating of boards and can contribute to warpage of completed PWBs.

The most basic rules of good component orientation apply across the board, regardless of the application or manufacturing style. In fact, these rules apply to through-hole as well as SMT assemblies. Thus, they apply to both SMCs and IMCs on mixed-technology boards. While the following rules certainly can be violated, they should stand—except when there is a compelling reason to depart from them.

Orthogonality

Wherever possible, all components should face north/south or east/west. In other words, the sides or longitudinal axes of the components should be either parallel or perpendicular to one another. If the circuit board is generally rectangular or square, component axes should be parallel to the board edges.

Applications may force a modification of this rule to allow some components to be placed with axes every 45°. Where even 45° orientation is too constraining, the ruling factor becomes the placement equipment. If it is capable of resolving very small steps of component rotation, parts may be in any axial position. However, some placement machines can only rotate components in 5° or 1° steps; thus, they limit the axial orientation to only those points of the compass that the machine can visit.

Polarity and Pin-1 Alignment

Consistent component orientation is a boon to hand assembly, repair, and troubleshooting operations in the factory and in the field. The best practice, as shown in Figure 5-1, is to align all number 1 pins, all SOT single leads, and all cathodes the same. Actually, we should extend this rule to cover any component with a clear physical or electrical characteristic.

Population Distribution

For SMT assemblies, good urban development planning is a world apart from the work of your city planning commission. SMT designs should avoid open areas like city parks. Instead of downtown congestion balanced by suburban open streets and grassy areas, we shoot for urban sprawl. The objective is to have a balanced distribution of heat-sink mass across all areas of the board. While this is critical in reflow soldering applications (see Chapter 3, Section 3.3.5, for a further discussion of heat-sink distribution), it is good practice for any design to promote solderability and reflow repair techniques.

Figure 5-1 Electrical and physical alignment of components.

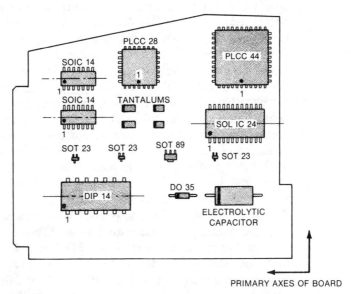

PRIMARY AXES OF BOARD

NOTES:
1. Component axes orthogonal and aligned with the primary axes of PC boards.
2. Consistent orientation of physical or electrical component features assists in assembly, troubleshooting, and repair.

Accessibility Clearances

Good design practices squeeze components no tighter together than necessary. One millimeter of extra space between each component can be of great benefit to visual inspection and repair/rework operations. More details on "design for inspection" procedures are provided in Chapter 6.

5.1.2 *Situation-Dependent Component Orientation Rules*

Many of the rules governing the best component orientation for SMT are invoked only by certain applications or assembly processes. We'll present these situation-dependent rules, and discuss the circumstances that mandate their practice.

Probe-Ability

Probe-ability is a coined word describing the layout of PWBs so as to place probe points on easily accessible grids for test purposes. The 100-mil (2.54-mm) grid of through-hole assembly fame is an example of probe-ability determined design. The board is laid out such that all component leads are centered on an imaginary 100-mil grid on the board assembly. Thus, we ensure easy test access using a "bed of nails" fixture, with probe pins on 100-mil centers.

Probe-ability is a process-determined design rule. Many simple boards, and even some complex memory boards, do not require probe-ability. They are either tested by a "hotshot" mock-up or are fully bus-addressable for I/O-connector test access. Probe-ability design rules are invoked by the need to probe, for testing, those circuit nodes not routed to external I/O.

When probe-ability is required for test, SMT poses new challenges for PWB layout. We can see why by comparing surface-mount probing with through-hole mounting. Placing IMCs on a 2.54-mm (0.100-inch) grid is simple enough, because standard IMCs are designed with leads on 100-mil centers, or multiples thereof. SMCs, which are on 50-, 40-, 25-, 20-, and 10-mil centers, or variables in between, make probe-ability design a more interesting task. We may bring traces out from fine-pitch SMC lands to coarse-pitch 100-mil grid vias or test points. These test points may be designed as addressable from the non-component side. Thus, an SMT board may be designed for a 100-mil-grid bed of nails. This will probably halve the bare and loaded board-testing cost compared to the same number of test points on 1.27 mm (0.050 inch) centers. But the real estate required by these probe points, associated traces, and single-side assembly rules, conflicts with the miniaturization goals of surface mounting. Fine-pitch bed of nails fixtures are available. With these, probe points are located on more space-efficient grids, generally of 50-mil but, increasingly, at 0.635 mm (0.025 inch) centers. The trade-off in specifying fine-pitch probes is the cost per probe point, and the fixture durability. Probes are available down to pitches of 10-mils or below, but costs rise inversely and exponentially with decreasing center distance.

Another factor differentiating SMT and IMC probe-ability is that we do not probe on SMC leads, as we do on IMCs. This is because probe pressure may force an open lead to contact a solder land and falsely pass a defective solder connection. Also, probe-to-lead misalignment can damage fragile SMT leads and fine-pitch probes. Since we don't probe SMC leads, we have a far larger degree of freedom in locating components for SMT probe-ability than we enjoyed with insertion-mount devices. In theory, the surface-mount component could be anywhere on the board, as long as we routed traces from its lands to probe-able test points on our probe grid.

However, there is little to be gained in locating components as close as paving stones, only to give up the real estate saved by running traces out to test pads which must be on a grid. We would compromise visual inspection and repair operations while gaining little or nothing in miniaturization. Therefore, where probe-ability is an issue, it is common practice to center components on the test grid and thus minimize the track length from component lands to test points.

For instance, 1206 passives are workhorse R&C components of SMT industry in the United States. They are nominally 0.126 inch long by 0.0625 inch wide (3.2 x 1.6 mm), and can be placed on 2.54 mm (0.100 inch) centers, whereas the larger 0.126 inch by 0.098 inch (3.2 x 2.5 mm) 1210 devices must be placed on 3.81 mm (0.150 inch) centers to fall on a 1.27 mm (0.050 inch) test grid, and 5.08 mm (0.200 inch) to fit a 2.54 mm (0.100 inch) grid. The smaller 0805 and 0504 parts are gaining some market share, driven by stringent real estate demands. However, test grids still dictate a placement on 2.54 mm (0.100 inch) centers. Therefore, gridded designs profit little from the assembly effort expended in using ultra-small parts except where a tight grid spacing, well below 50-mil, is used.

Assembly Clearance

The intercomponent clearance is application dependent, being influenced by the method of, and the equipment used in, board assembly. When assembling boards by hand, components could conceivably be laid side by side, with only enough clearance to accommodate the maximum dimensional tolerances of each part. (In practice, more clearance is desirable to simplify inspection and rework.) Automated assembly introduces some additional clearance requirements. The placement equipment may be very accurate, but all machines fall short of absolute perfection. Chapter 3 (Section 3.3.3 and Figure 3-16) covers the calculation of required clearances needed to ensure collision-free automated placement.

Assembly Flow Through the Solder Furnace

Solder defects may be minimized by proper component orientation in relation to workpiece flow through the soldering operation. Just what constitutes "proper component orientation" depends on the soldering method in question.

The chief gremlin in reflow soldering, assuming the proper solder-paste deposits are present, is tombstoning. This defect is produced by a combination of the wetting and surface-tension forces in molten solder. It occurs when forces at one end of a component are not reasonably balanced by forces at the other end. The imbalance may be produced by land-geometry variations, component-termination variations, or time-of-reflow variations. Whatever the source, when the solder-force imbalance is great enough, tombstoning is the predictable result. Figure 5-2 illustrates how uneven reflow can be a factor in tombstoning.

Figure 5-2 Diagram showing how uneven reflow can contribute to tombstone defects.

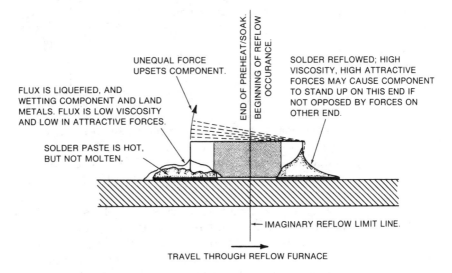

We'll consider imbalances from land geometries or component terminations in separate sections. Here, we're concerned with time-of-reflow variations, as illustrated in Figure 5-2. This phenomenon may occur when an assembly migrates from the homogenous environment of batch soldering to the moving reflow band of the linear furnace. We can imagine a reflow limit line exactly where assemblies reach reflow as they travel through the oven. In reality, this limit line is a moving target pushed back and forth in the furnace by heat-sink loading and furnace temperature stability. We'll take up its movement in a moment, but for now, let's think of the reflow limit line as stationary.

As an assembly moves through the reflow limit line, solder melts (reflows) precisely as it passes through the line. Some components on our imaginary board assembly are oriented so that a line through their two terminations is parallel to the reflow limit line across the furnace. Others are situated with their termination-intersecting line perpendicular to the reflow limit line. It follows that components so oriented that their terminations hit the line simultaneously will not experience the unequal force diagram shown in Figure 5-2. However, parts oriented so that one termination contacts the line first will be subject to the unequal solder forces we've identified as contributing to tombstoning. Figure 5-3 shows the preferred component orientation needed to minimize tombstoning and component-misalignment defects for standard components in linear reflow soldering.

Figure 5-3 Diagram of preferred component orientation for linear reflow soldering.

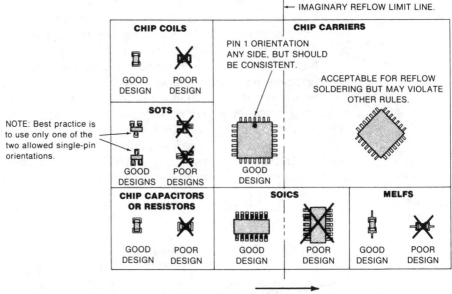

Occasionally, an application may demand that parts be reflowed without regard to the preceding preferred orientation. If tombstoning plagues such a reflow operation, you may resort to batch reflow to avoid the moving reflow limit-line effect, or you may glue or fixture the offending components to prevent their movement during reflow.

Lest the preceding rules seem too simple, remember that we discounted the effects of unequal thermal-mass distribution for our discussion of component orientation. Inequalities in thermal-mass distribution cause the reflow limit line to move within the reflow furnace, and to waver and bend across the width of the board, much like isotherms on a weather map.

Figure 5-4 Special orientation may lessen the solder defects caused by thermal-mass peaks.

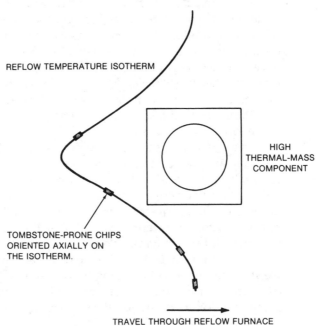

Therefore, thermal-mass rule number 1 is to orient components to distribute the thermal mass as evenly as possible. This rule is of particular importance for linear reflow soldering, but gross departures from it will also produce solder problems in batch systems. At the extreme, localized heat-sink masses can make a board impossible to reflow solder without causing overheating and consequent damage to low thermal-mass areas. Where one component has a large thermal mass, leave some vacant area around it to minimize the heat-sink congregation in its neighborhood. In particular, components that are likely to pop wheelies should be kept at a distance from large heat sinks. Where they must be located in close proximity to the heat sinks, experimentation is suggested in order to determine the ideal orientation for the part which is prone to tombstoning. Figure 5-4 suggests thermal-mass-dictated departures from the rules of Figure 5-3.

Detailed isotherm mapping of boards may be done with multichannel recorders and heavily thermocouple-monitored samples. Such maps serve to guide any special orientation. In the absence of equipment necessary to produce an isotherm chart of the assembly, the "cut and try" approach will have to do.

Flow soldering invokes a completely different set of component orientation rules. Here, one of the driving considerations is in not creating eddy currents that can cause solder skips by blocking the solder from some terminations as the assembly moves through the molten solder. The other primary concern is to position the parts to avoid bridging. Since one defect involves too little solder deposited while the other is due to too much solder, the solution to one problem may work against the other. The rules presented below have been empirically determined as the best compromise for maximum flow-soldering yields.

Component orientation for solder-skip (or open) avoidance is pictured in Figure 5-5. The preferred anti-open rules should be rigidly enforced if solder opens are the prevalent solder defect in your flow-soldering operation.

The rules for solder-bridging (or short) reduction are illustrated in Figure 5-6. Bridging is often a pernicious problem in the flow soldering of SMCs. Therefore, anti-short rules should be followed for all designs.

Certain flow-soldering systems are particularly prone to bridging problems. With these, bridging may be reduced on SO and SOL ICs by deliberate violation of the rule illustrated in Figure 5-5. If orienting the SOIC with its long axis perpendicular to the direction of flow will reduce bridging defects enough to offset the increase in opens, do it. Experimentation will tell. You can easily try it by rotating a sample board, or section of board, 90° and passing it through the flow-solder system. If the experiment indicates a need, reorient only the SO and SOL ICs on the board.

Thermal Management

Thermal management impacts the component orientation for high-dissipation circuits. Where significant heat will be generated in the circuit operation, spread the high-dissipation devices evenly across the circuit board, or as otherwise dictated by cooling strategies. Thermal transfer through the substrate will be improved by leaving as much copper as practical on each layer of the board.

Double-Sided Assembly

Placing components on both sides of SMT boards doubles the real-estate savings. But, placing components on both sides may more than double the headaches for design and manufacturing. The best component-orientation rule is to place the first side for low density, if possible; with moderate density, if needed. Then, place the second-side components to follow the rules given above. For two-pass reflow soldering, locate all high-mass components on the side that will be up in the second pass through the furnace.

Figure 5-5
Component orientation for minimizing flow-soldering opens.

Note: Today's SMC wave solder machines may obviate the need for some of these rules. However, the production environment, state of repair of equipment, operator training, and assembly vendor selection throughout the entire life of a product are much less under the control of the circuit designer than is the layout board. Once these layout rules are well understood, their cost is minimal, and need be paid only once per design. The expense of violating them can be substantial and will be exacted many times per day over the life of the manufacturing cycle.

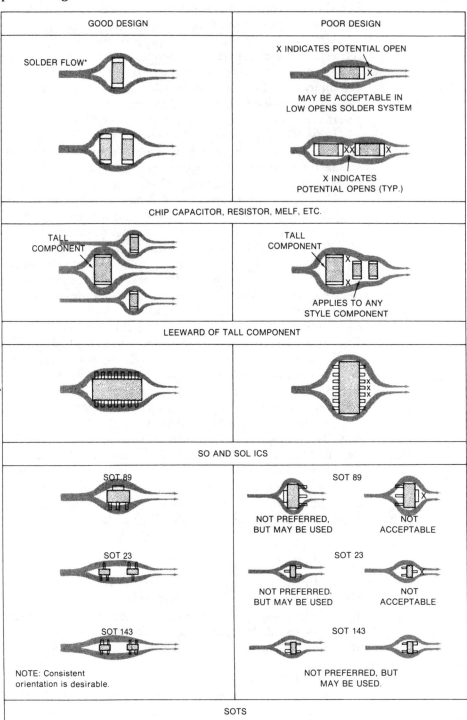

*Solder flow arrows indicate flow relative to component; typical for all devices illustrated.

Figure 5-6 Component orientation for minimizing flow-soldering shorts.

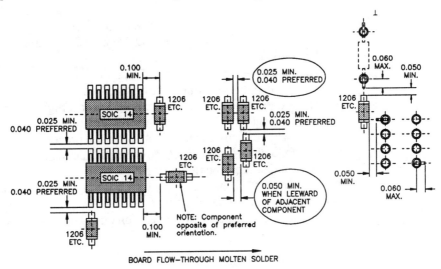

NOTES:
1. 4-sided I/O devices (PLCCs and LCCs) are not recommended for flow soldering.
2. These recommendations are for flow-soldering lands (refer to Appendix A, Section 1.3).
3. Dimensions are in inches.

Additional concerns are the need to place second-side components clear of all test points that are to be accessed from the bottom side. Conversely, place bottom-side components so that any top-side test points will be accessible from the top. Be particularly mindful of heat-sink mass distribution. Double-sided assemblies double your opportunities to create soldering nightmares with lumped-up thermal masses. Likewise, in designing to manage high-dissipation problems, double-sided assembly complicates the task. Clearly, we wouldn't intentionally locate high-dissipation parts one under the other. But, it is an easy thing to do if you're only concentrating on the population of the side on which you're currently working.

All the challenges of double-sided assembly design, test, and rework are indeed formidable. But the real-estate rewards are great enough that it would be pointless to argue for a single-side-only blanket rule. Don't go double-side if you can help it. And if you must use both sides, allow extra time to do the needed design homework.

Partitioning
Partitioning refers to the division of circuitry into blocks. Electrical and/or mechanical considerations may influence the component orientation for the sake of circuit partitioning.

Mechanically, partitioning may be done to facilitate assembly processing. I/O-intensive sections of componentry might be located near the edge connectors. Also, sensitive components may be partitioned into a protected area to shield them from hazards in the environment. High-dissipation parts may be located to suit heat-removal strategies.

Electrically, circuits are commonly partitioned:

+ To separate analog and digital sections.
+ To separate high-frequency areas from medium- and low-frequency areas.
+ To divide the audio and digital sections.
+ To position test points for functional tuning.
+ For the management of power dissipation.
+ To control noise and crosstalk. For instance, high-speed switching areas are often separated from low-level, high-grain amplifiers because the fast-switching noise can be isolated in the process.
+ To modularize the assembly to promote manufacturability.
+ To separate functional blocks for testing.
+ To reduce trace parasitics or cluster components of particular impedances on multiple-impedance boards.
+ To place matching/tracking parts on one card in a system.[1]

5.2 Clearances Around Components and Test Pads

Several issues must be considered in deciding the clearance around components and test points. We'll discuss these next, and develop rules for standardized spacing. Such rules should be built directly into CAD systems wherever possible.

Figure 5-7 Component spacing rules for visual inspection and repair.

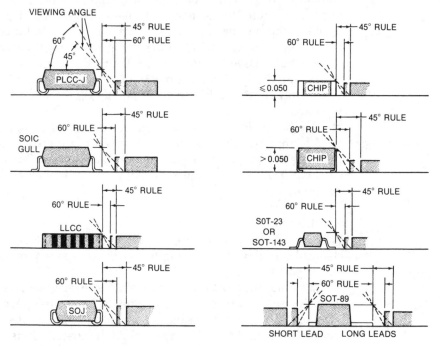

NOTE: All clearances are calculated from nominal JEDEC component dimensions and are rounded to the nearest 0.005 inch.

5.2.1 *Component Clearances for Visual Inspection and Rework*

Component spacing must be, at a minimum, adequate for an error-free assembly operation. In Chapter 3, Section 3.3.3 (and Figure 3-18) presents a formula for calculating this minimum assembly clearance. But any reasonably accurate placement system will safely place parts far closer than desirable for visual inspection and repair.

Figure 5-7 and Table 5-1 give the clearances recommended between typical components for inspection/repair. Where real estate permits, the minimum spacings should be established by the 45° rules. If required, the tighter 60° rules will allow a higher density at the price of complicating visual inspection and repair operations. To use Table 5-1, select, for two adjacent components, the larger clearance required by the viewing angle rule in force.

Table 5-1. Dimensions for Typical Component Spacing Rules

Component	Clearance Recommendation[*]	
	45° HIGH VISIBILITY	*60° RED. VISIBILITY*
Chip component (≤0.05 inch Hi)	1.27/0.050	0.76/0.030
Chip component (>0.05 inch Hi)	Component Height	0.6 × Height
SOT-23/SOT-143	1.02/0.040	0.64/0.025
SOT-89 (Both sides)	1.52/0.060	0.89/0.035
SOIC (Gull-wing lead)	1.65/0.065	0.89/0.035
SOLIC (Gull-wing)	2.54/0.100	1.40/0.055
SOJIC	3.56/0.140	2.03/0.080
PLCC (50-mil centers)	4.83/0.190	2.79/0.110
LLCC (50-mil typ. centers)	1.78/0.070	1.02/0.040

[*] All dimensions are mm/inch; inch dimensions control.

Note that there are plentiful examples of violations of these clearance rules in production today. Some of these are quite manufacturable, others are not. Since many successful designs ignore the clearance rules, they are only guides, not laws. Where very-high-density boards are manufacturable despite close-order spacing, clearance violations have been carefully considered in design. If density demands that the visibility rules be ignored, consider this a warning flag for the team to spend extra effort on manufacturing, inspection, and repair concerns, beginning at the earliest possible phase of design and continuing through production start-up.

5.2.2 *Spacing Minimums for Reflow Soldering*

In Section 5.1 above, we presented spacing rules to minimize bridging in flow soldering. In reflow soldering, we also must observe minimum spacings to prevent bridging. Reflow bridging may occur on closely spaced lands for several reasons.

Solder Application Misregistration
Misregistration can leave solder paste printed between separate lands. If the lands are adequately spaced, such bridges will disappear during the reflow operation. However, lands that are spaced too closely may allow solder shorts to remain intact, producing a solder defect. This is particularly true under leadless compo-

nents, such as chip carriers. Boards with leadless components having fine-pitch leads warrant special attention to solder print registration.

Solder Smearing

Solder smeared across several lands can act just like misregistered printing to form bridging defects. Solder smears also are a potential source of solder balls and should always be scrupulously avoided. This phenomenon, like misregistration bridging, is most prevalent under leadless components that have closely spaced terminations.

The Swimming Effect

The swimming effect refers to a tendency of SMCs to move during the reflow soldering operation. This occurs because most surface-mount components displace more than their own weight in molten solder, and literally do swim or float. Components floating in solder move due to the influence of several interacting forces.

As solder paste goes into a molten state during reflow, solder climbs up the sides of component terminations and spreads across exposed circuit-board metallizations, drawn by wetting forces. Gravitational force acts in opposition to the upward movement of solder on device terminations. The result is the familiar meniscus formation. Finally, surface tension forces the liquid solder to seek its smallest possible surface area. The balance of these three forces determines where the swimming component goes.

To see how swimming might affect assembly, let's consider a simple 1206 chip placed on standard lands. Assuming all the metallizations are clean and easily wetted, the solder can achieve its smallest surface area when the component is symmetrically aligned within the land pattern. Thus, under ideal conditions, swimming works to correct the component alignment in reflow. But what would happen if only one side of the terminations were wettable? That one wettable area of the termination would move to the center of the lands. By the same token, uneven wettability of land metallizations destroys the centering operation of the swimming effect.

Figure 5-8 Clearances for bridge-free reflow soldering.

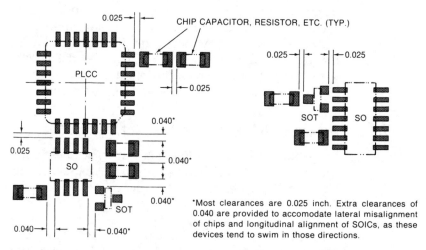

*Most clearances are 0.025 inch. Extra clearances of 0.040 are provided to accomodate lateral misalignment of chips and longitudinal alignment of SOICs, as these devices tend to swim in those directions.

While it is true that reflow soldering generally aids component alignment through the swimming effect, we cannot rely on a perfectly even wettability of all device metallizations and lands. Therefore, we provide clearance so that components that misalign during reflow will not cross over to adjacent lands and produce shorts. Figure 5-8 shows minimum spacings needed to assure that swimming doesn't cause solder shorts.

5.2.3 *Spacings Around Tall Components*

Usually tall components create several concerns for component spacing. They block vision in a conical area around themselves, and they may interfere with placement head travel. To really simplify design and manufacturing, the rule with tall components (greater than 6.35 mm/0.25 inch high) is, don't have them. Where this can't be followed, simple Rule 2 is to add them after all other things are said and done. If neither of these actions are practical, follow the guidelines below to ensure adequate clearance around tall components.

The Cone of Visibility

The size of the clearance core required around a tall component depends on the height of the part and the visibility level needed to accomodate inspection and rework. For very tall components, the rules of Section 5.2.1 would require an excessive clear area. For instance, a metal-can capacitor that is only 1.27 cm (1/2 inch) in diameter and 3.81 cm (1-1/2 inch) tall sits on only 4.84 sq. cm (0.196 sq. inch) of circuit board. Yet, using rigidly enforced 45° clearance rules, it would need a clear area 8.9 cm (3-1/2 inch) in diameter, or nearly 50 times its mounting area. Of course, this is ridiculous. We would relax the 45° visibility rule since an inspector could get a good view by a glance around that capacitor. But, significant visibility clearance would still be required. SMT boards must be inspected from the component side. This is quite a contrast from just flipping a through-hole board over and looking at the solder joints on its IMC leads.

Clearance for Placement Apparatus

In Chapter 3, we considered component clearance to assure a collision-free automated assembly. Tall components complicate this picture. Most placement end effectors are shaped somewhat like a cone with its focal point on the component. Thus, there is a cone around tall components, determined by placement-head form factor, which must be free from obstruction during the placement operation. This clearance cone is graphically explained in Figure 5-9A.

Figure 5-9 Clearances around tall components.

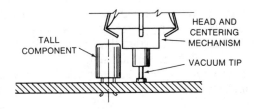 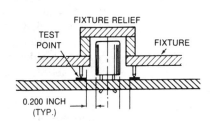

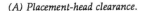

(A) Placement-head clearance.   (B) Text fixture clearance.

Test Points Near Tall Components

Tall components also interfere mechanically with test probe fixtures. A tall component may require relieving the fixture, and this will dictate a 5.08 mm-wide (0.200 inch) annular ring free from test points, around the component in question.[2] Figure 5-9B depicts this clearance.

5.2.4 *Spacing Around Test Points*

Some clearance is required around each test pad, even when adjacent components are less than (or equal to) 6.35 mm (1/4 inch) tall. We must ensure that probes do not land on components for two reasons. First, the spring pressure in probes can cause an open connection to contact a land and falsely test as a good joint. Secondly, a glancing contact between a probe and component can damage either or both. And lastly, we don't want false failures caused by test probes landing on the wrong conductor. By allowing a 0.475 mm (0.018 inch) clear annular ring around each test point, we avoid these concerns.[3]

5.2.5 *Spacing Around Fine-Pitch Components*

When FPT is mixed with 50-mil pitch on a single board, high-yield soldering often becomes a challenge. SMCs with a 50-mil pitch need plenty of solder (7- to 10-mils of wet paste is typical) to avoid defects such as insufficient solder and opens. FPT components require much thinner solder deposits (3- to 6-mils is common) in order to minimize bridging between their closely spaced leads. These disparate requirements may be dealt with by use of selective solder plating on the board, zippered printing, thin stencils, and overprint of the 50-mil pitch areas, or stepped stencils. The first three approaches do not impact board layout, but the use of stepped stencils may impose large increases in minimum space from the FPT lands to adjacent parts. Exact spacing requirements must be determined by process engineers, and are a function of the step variation, the hardness of the squeegee, and printer dynamics.

5.3 Traces and Spaces

SMT lead spacings have forced us to shrink tracks and spaces to suit component dimensions. Tracks and spaces of 0.305 mm (0.012 inch) are typical for IMC PWBs. SMT, on the other hand, rarely uses such generous rules since there would not be room for a conductor between lands of a 1.27 mm (50-mil) pitch device or 50-mil centered vias.

One of the early decisions for the design team is the minimum space and trace dimensions to be used for a project. Below, we'll discuss the issues influencing this decision, and will present five rule sets for standard trace and space selections. These rules, too, are guidelines and not laws. Once understood, they are meant for guidance in selecting spacing and feature size rules.

The cost curve for fine geometries turns up at around 0.2 mm (8-mil) line-and-space rules. One PWB fabricator, commenting on yields in his shop, explained why. This shop was able to consistently get yields of between 85% and 90% on boards with 0.15 mm (6-mil) line/space, but the yields dropped to 60% for 0.1 mm (4-mil) line/space.[4]

A small percentage of contract PWB shops specialize in fine-line high-density work. They can produce 0.05 to 0.076 mm (2- to 3-mil) lines. In such shops, 0.127 mm (5-mil) lines, while not cheap, are not a problem. Use of 0.127 mm (5-mil) line-and-space rules allow designers to run two tracks between 1.27 mm center (50-mil) SO or PLCC lands. But, 0.127 mm (5-mil) rules will significantly reduce the number of board vendors available for a project. Figures 5-10 through 5-14 show trace and space rules for varying densities. The team should pick, from these or a hybrid of them, the least dense rules that will satisfy the needs of the project at hand.

Fine-line board manufacturability can be improved in many ways. In the PWB etching process, some undercutting of copper occurs. The thicker the copper foil on the laminate, the more troublesome undercutting will be. Fine-line boards, particularly multilayer designs, are generally specified as ≤ 1/2 oz. copper (0.018 mm or 0.0007 inch thick), instead of the 1 oz. or 2 oz. (0.036 mm / 0.0014 inch to 0.071 mm / 0.028 inch) material commonly used in through-hole technology. By reducing undercutting, thin copper-foil laminates promote manufacturability of fine-line multilayer boards.

Fine-line manufacturability is also enhanced by photoplotting lines slightly oversized and spaces slightly undersized of the final desired product. For instance, if the design calls for 0.20-mm (8-mil) lines and spaces, plotting 0.23-mm (9-mil) lines with 0.18-mm (7-mil) spaces will allow for line shrinkage during board fabrication, and will produce a closer approximation of the desired result.

Some CAD systems produce excessive and unnecessary jogs in lines as part of their design optimization process. Where such features interfere with an even distribution of metal on a layer, the result will likely be uneven plating due to unequal current distribution. For fine-line designs, metal distribution should be kept even across the surface of each layer. For multilayer designs, the "homogenous distribution" rule applies in the Z axis as well as the X and Y. Designs with one disproportionately heavy routing layer are more prone to warpage. Where heavy power and ground layers are required, they should be placed in symmetrical balance close to the center of the laminate.

By laser plotting the artwork, it is possible to avoid long runs of necked-down traces between via pads. Instead, full-size via pads may be shaved to provide clearance, where needed, for feed-throughs. Of course, minimum annular ring rules must still be followed.

Accurate phototools are a must for fine geometry work. Photo- or laser-plotted artwork and CAD-generated drill tapes should be used. Polyester film, where used, should be a minimum of 0.18 mm (7 mils) thick to avoid excessive dimensional influences from environmental changes. Avoid exposing films to temperature and humidity extremes. Work requiring very-high accuracy may be plotted on glass or special ultrastable film.

Figure 5-10
Trace and space rules—Level 1 density.

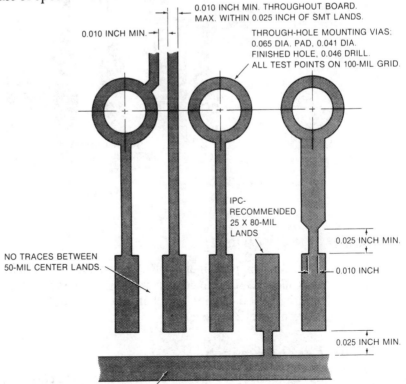

0.010 INCH MIN. THROUGHOUT BOARD.
MAX. WITHIN 0.025 INCH OF SMT LANDS.

0.010 INCH MIN.

THROUGH-HOLE MOUNTING VIAS:
0.065 DIA. PAD, 0.041 DIA.
FINISHED HOLE, 0.046 DRILL.
ALL TEST POINTS ON 100-MIL GRID.

IPC-RECOMMENDED 25 X 80-MIL LANDS

NO TRACES BETWEEN 50-MIL CENTER LANDS.

0.025 INCH MIN.

0.010 INCH

0.025 INCH MIN.

OVERSIZED TRACES MUST NOT ENTER SMT LANDS.

5.3.1 *Level 1 Density*

Level 1 density, per Figure 5-10, is recommended wherever it meets density requirements without forcing use of an undue number of layers. Level 1 rules make for very manufacturable PW boards, and a multitude of suitable PWB vendors. Level 1 trace and space rules, with Level 3 via rules, are a good combination for pad cap layers of MLBs. Level 1 rules will only be suitable for routing layers on very low-density boards, however.

5.3.2 *Level 2 Density*

Level 2 rules, per Figure 5-11, are common for routing layers in SMT. These rules allow reasonable densities, and yet are very manufacturable. Most quality PWB vendors can handle such dimensions in stride.

Figure 5-11
Trace and space
rules—Level 2
density.

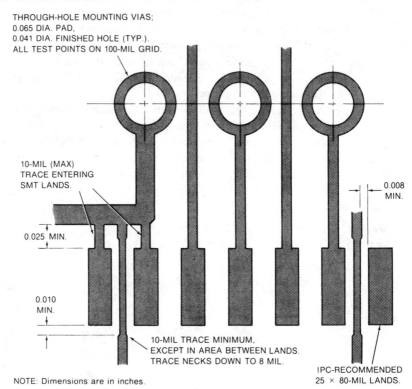

THROUGH-HOLE MOUNTING VIAS;
0.065 DIA. PAD,
0.041 DIA. FINISHED HOLE (TYP.).
ALL TEST POINTS ON 100-MIL GRID.

10-MIL (MAX)
TRACE ENTERING
SMT LANDS.

0.008
MIN.

0.025 MIN.

0.010
MIN.

10-MIL TRACE MINIMUM,
EXCEPT IN AREA BETWEEN LANDS.
TRACE NECKS DOWN TO 8 MIL.

IPC-RECOMMENDED
25 × 80-MIL LANDS.

NOTE: Dimensions are in inches.

5.3.3 *Level 3 Density*

Level 3 rules, per Figure 5-12, are widely used for SMT routing layers. These rules allow reasonable densities and simple test probing. PWB shops can generally furnish Level 3 density, and there should be little or no cost premium in stepping up from Level 2 to Level 3.

5.3.4 *Level 4 Density*

Level 4 rules, per Figure 5-13, may come at a cost premium. However, they allow high densities, and may not cost as much as the added layers that Level 3 rules would impose on a crowded design. Savings may be substantial where increased density translates into double-sided vs. multilayer, or one board instead of two.

Figure 5-12
Trace and space rules—Level 3 density.

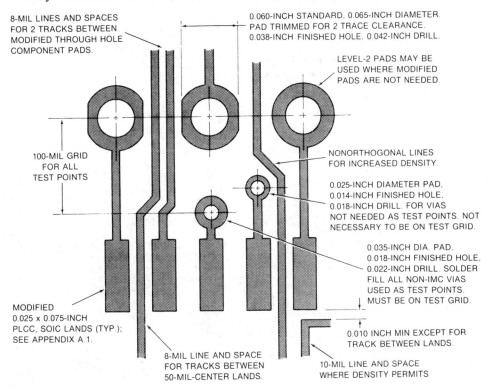

8-MIL LINES AND SPACES FOR 2 TRACKS BETWEEN MODIFIED THROUGH HOLE COMPONENT PADS.

0.060-INCH STANDARD, 0.065-INCH DIAMETER. PAD TRIMMED FOR 2 TRACE CLEARANCE. 0.038-INCH FINISHED HOLE. 0.042-INCH DRILL.

LEVEL-2 PADS MAY BE USED WHERE MODIFIED PADS ARE NOT NEEDED.

100-MIL GRID FOR ALL TEST POINTS

NONORTHOGONAL LINES FOR INCREASED DENSITY.

0.025-INCH DIAMETER PAD, 0.014-INCH FINISHED HOLE, 0.018-INCH DRILL. FOR VIAS NOT NEEDED AS TEST POINTS. NOT NECESSARY TO BE ON TEST GRID.

0.035-INCH DIA. PAD, 0.018-INCH FINISHED HOLE, 0.022-INCH DRILL. SOLDER FILL ALL NON-IMC VIAS USED AS TEST POINTS. MUST BE ON TEST GRID.

MODIFIED 0.025 x 0.075-INCH PLCC, SOIC LANDS (TYP.); SEE APPENDIX A.1.

8-MIL LINE AND SPACE FOR TRACKS BETWEEN 50-MIL-CENTER LANDS.

0.010 INCH MIN EXCEPT FOR TRACK BETWEEN LANDS.

10-MIL LINE AND SPACE WHERE DENSITY PERMITS

Figure 5-13
Trace and space rules—Level 4 density.

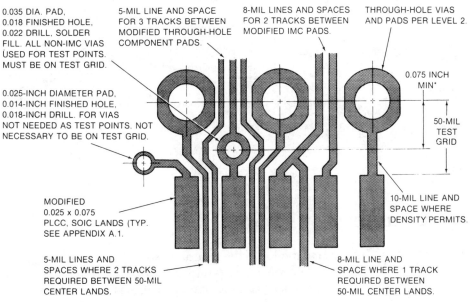

0.035 DIA. PAD, 0.018 FINISHED HOLE, 0.022 DRILL, SOLDER FILL. ALL NON-IMC VIAS USED FOR TEST POINTS. MUST BE ON TEST GRID.

5-MIL LINE AND SPACE FOR 3 TRACKS BETWEEN MODIFIED THROUGH-HOLE COMPONENT PADS.

8-MIL LINES AND SPACES FOR 2 TRACKS BETWEEN MODIFIED IMC PADS.

THROUGH-HOLE VIAS AND PADS PER LEVEL 2.

0.075 INCH MIN*

0.025-INCH DIAMETER PAD, 0.014-INCH FINISHED HOLE, 0.018-INCH DRILL. FOR VIAS NOT NEEDED AS TEST POINTS. NOT NECESSARY TO BE ON TEST GRID.

50-MIL TEST GRID

MODIFIED 0.025 x 0.075 PLCC, SOIC LANDS (TYP. SEE APPENDIX A.1.

10-MIL LINE AND SPACE WHERE DENSITY PERMITS.

5-MIL LINES AND SPACES WHERE 2 TRACKS REQUIRED BETWEEN 50-MIL CENTER LANDS.

8-MIL LINE AND SPACE WHERE 1 TRACK REQUIRED BETWEEN 50-MIL CENTER LANDS.

NOTES:
* Assuming short clenched leads, auto insert and clench. This dimension may be reduced to 0.045 with solder-cut solder processing.
1. Reflow soldering only of high-density side.
2. SMOBC required for trace-bearing surface layers. Must be photo-imaged solder mask.

5.3.5 *Level 5 Density*

Level 5 rules, per Figure 5-14, also may carry a price premium. They should be invoked only when needed. As of this second edition, there are still a limited number of PWB fabricators capable of Level 5 manufacturing, but a growing number can handle even finer geometries. For standard PWB processes, the leading edge is around 0.025 mm (1-mil) line and space as of this writing. Finer geometries are made with thin films (see Chapter 9).

Figure 5-14 Trace and space rules—Level 5 density.

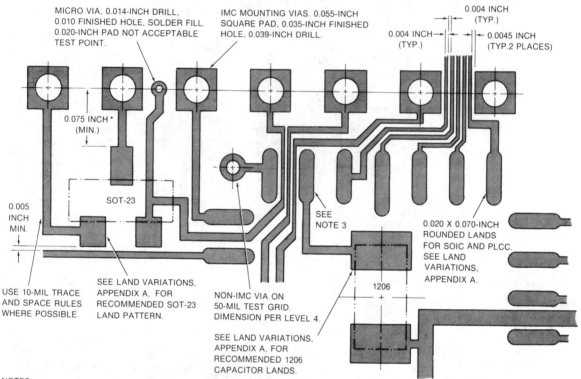

NOTES:
* For short-clenched auto insert and lead clench. May be 0.0375 inch for solder-cut solder.
1. Reflow soldering only of high-density side.
2. SMOBC required for trace-bearing surface layers. Must be photo-imaged solder mask.
3. 4-mil line and space with 5-mil space between line and land.
   Use only where required for 3 traces between special 50-mil-center lands.

5.4 Component Lands

As we mentioned in the introduction to this chapter, land geometry drawings for specific components are collected for easy reference in Appendix A. Next, we'll cover the generic rules for land geometries and the traces entering them.

5.4.1 *Traces Entering Lands*

Where traces enter lands, special rules apply to the trace size and geometry. These rules cover influences on component swimming, control of thermal paths, and solder migration along traces.

Influences of Traces on Component Swimming

In Section 5.2.2, we discussed the swimming effect. We mentioned that one of the determining factors in the direction of swimming is the geometry of metallized area wetted by solder. Where a trace enters a land, therefore, it must either be solder masked to prevent its wetting or it must be included in the swimming equation. Since screened-on masks must be set back from lands, traces entering lands from screened-on masks must be considered as unmasked traces. Figure 5-15 illustrates some rules for exposed traces entering lands.

Figure 5-15
Rules for exposed traces entering lands.

Note: The use of photo-imaged SMOBC has become so prevalent with the second edition of this book that the rules shown here are seldom required. There are applications, however, such as pad-cap boards and boards having no solder mask, where these rules still come into play.

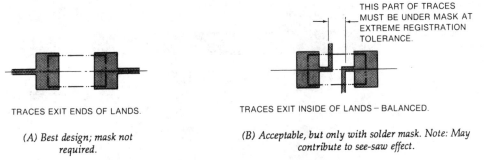

TRACES EXIT ENDS OF LANDS.

(A) Best design; mask not required.

THIS PART OF TRACES MUST BE UNDER MASK AT EXTREME REGISTRATION TOLERANCE.

TRACES EXIT INSIDE OF LANDS – BALANCED.

(B) Acceptable, but only with solder mask. Note: May contribute to see-saw effect.

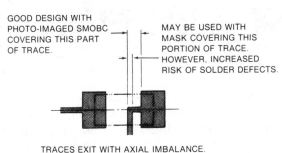

GOOD DESIGN WITH PHOTO-IMAGED SMOBC COVERING THIS PART OF TRACE.

MAY BE USED WITH MASK COVERING THIS PORTION OF TRACE. HOWEVER, INCREASED RISK OF SOLDER DEFECTS.

TRACES EXIT WITH AXIAL IMBALANCE.

(C) Acceptable, but only with solder mask and under conditions listed on sketch. Note: May contribute to see-saw effect.

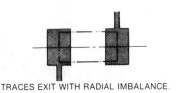

TRACES EXIT WITH RADIAL IMBALANCE.

(D) Unacceptable, except where entire trace is covered with photo-imaged SMOBC.

For ICs, traces may enter the lands from either end, but avoid those designs where a preponderance of soldering forces are congregated on one axis. As long as the forces exerted by solder wetting remain equalized about the component axes, good alignment should result.

Control of Trace Thermal Paths

In reflow soldering, even small differences in thermal mass between the lands of an SMC will influence the time of reflow of those lands. If one land is late in reflowing, the result may be tombstoning or device misalignment. To keep reflow times uniform, avoid excessive thermal coupling with the lands. Do not connect traces more than 0.25 mm (0.010 inch) wide to solder reflow lands. Where a wider trace must connect to a land (i.e., power and ground buses), reduce the trace to a 0.25 mm (0.010 inch) width for at least 0.635 mm (0.025 inch) before entry, as shown in Figure 5-16.

Avoid Solder Starving

Tin/lead or tin plating on traces under a solder mask melts and forms a capillary during reflow soldering. Solder from SMC lands can migrate along such capillaries, leaving solder starved joints and open circuits behind. Therefore, solder masking over bare copper is recommended for non-pad cap SMT boards. Specifying bare copper traces on pad cap designs also eliminates solder migration along traces and into adjacent vias. For multilayer boards, the two additional layers required for pad caps often come at no premium over solder-mask costs for surface-routing layers. And pad cap designs may yield valuable economy on the manufacturing floor.

Figure 5-16
Reduce large traces before entry to reflow lands.

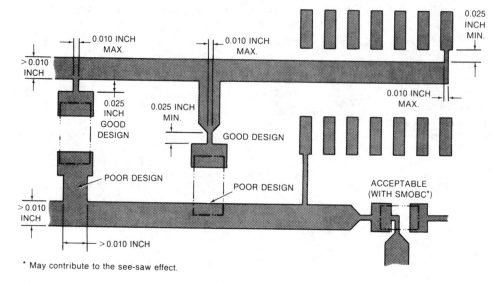

* May contribute to the see-saw effect.

High-Shear Stress Resistance

Thermal excursions, shock, or other environmental conditions imposing significant shearing stresses on solder joints will dictate the use of special rules for trace entry to lands. Abrupt 90° corners, where traces enter lands, can set up stress concentration points in solder joints. By rounding ends of the lands and adding a fillet radius to each trace at its entry point, stress risers are eliminated.

5.4.2 Land Specifications

Bare Copper Traces and Solder Coated Lands

Bare copper trace boards, with tin/lead coating on the lands only, are produced by hot-air solder leveling over the solder mask, by selective tin/lead plating, or by selective tin/lead stripping. Since the hot-air solder leveling process requires a solder mask, it is not generally applied to pad cap boards. Solder-surface finish and height varies by process and vendor, and should be critical in the process selection. Select only those processes which your PWB supplier can control sufficiently to guarantee you flat, solderable lands, and a smooth surface on the PWB.

Bare Copper Boards

Boards are also fabricated without any tin/lead plate on traces or land areas. Several coating processes are available to protect the exposed copper areas from oxidation. In some instances, where just-in-time delivery cycles can be tightly enforced, no coating is used. Finished boards are moved immediately from board manufacture, where copper is squeaky clean, to board assembly.

Selective Plating Near Lands

Lands located in areas which will be dipped into selective gold plating baths must be masked in order to prevent gold contamination. Gold forms gold/tin intermetallics during the soldering process. This intermetallic is extremely brittle, destroying solder-joint reliability. Wherever possible, locating lands above the selective plating areas, per Figure 5-17, saves masking.

Figure 5-17
Locate lands above the selective plating areas.

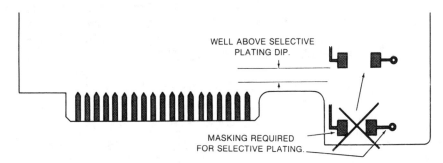

5.5 VIAS

In surface mounting, greater freedom of design is afforded by the fact that vias need not be directly associated with each component lead, but can be anywhere in a trace connecting that lead. In return for this freedom, the designer has to worry about where to locate each and every via required for layer connection. In considering via specifications, we'll deal with via size and via location.

5.5.1 *Via Size*

Via size is much more of an issue for debate in SMT than in IMC design. Real estate savings are a primary goal of the technology, and vias no longer need to be sized to accept a through-hole lead. Therefore, the question becomes, "How small can my vias be?" Limits to via size reduction come from two areas.

Relationship of Via Size to Via Cost
First, the cost of drilling via holes begins to increase in typical shops as drill diameters go below 0.635 mm (0.025 inch). The cost curve becomes non-linear in most shops at around 0.39 mm (0.015 inch), and thereafter, drilling costs may climb sharply with decreasing hole size. A 0.457 mm (0.018 inch) drilled hole is a comfortable number for high-quality vendors, with 0.254 mm (0.010-mils) being a typical lower limit for moderate cost boards.

Pad Size for Vias
Second, tolerance buildups generally require that pads be a minimum of 0.89 mm (0.035 inch) in diameter for test probing of bare and loaded boards. If a 0.46 mm (0.018 inch) hole is drilled and a through-plating buildup of 0.05 mm (0.002 inch) thickness is specified, the finished hole will be 0.36 mm (0.014 inch) in diameter. Allowing an annular ring of 0.25 mm (10-mils) gives a pad diameter of ~0.86 mm (0.035 inch). Where such dimensioning does not interfere with necessary testing of inner layers before lamination, inner layer pads may have an annulus dimension of 0.13 mm (0.005 inch) for a pad diameter of 0.64 mm (0.025 inch). Tear-dropping is used when needed to prevent breakout of trace junctions. This is illustrated in Figure 5-18. For any holes less than 0.5 mm (0.020 inch) in diameter, a minimum plating buildup of 0.038 mm (0.0015 inch) should be specified.

The filling of test-point vias prevents conical probe points from becoming lodged in holes during testing. Where vias are not filled, crown point probes are used. Filled vias improve the sealing on vacuum hold-down fixtures. However, solder-filled vias may bail due to on-board or environmental thermal cycling because the thermal coefficient of solder and that of copper are not well matched. Vias may be copper filled by plating shut small holes (0.33 mm/13-mil) or with solder by the solder fill and reflow method.

Figure 5-18 Via miniaturization for moderate cost boards.

Note: Tear-drops, or the less attractive alternative of key-holing, is used to reduce the risk of breakout of a mis-registered frill where annular ring allow-ances must be small. Some CAD systems, numerous Gerber post-processors, and many PWB front-end CAM systems are able to add tear-drops or key-holed via flashes.

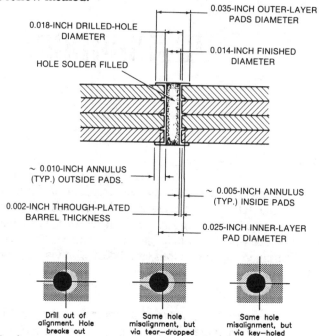

Via Aspect Ratio

Another issue influencing hole-size specification for vias in thick boards is the *aspect ratio*. Aspect ratio refers to the board thickness over the via diameter. High aspect ratio holes may be subject to cracking in through-plated via barrels because of the disparate TCEs of board and via barrel material. Unless special precautions are taken, the aspect ration should not go above 6:1, and 3.5:1 is preferred for high reliability. All high aspect hole plating should be specified as high-ductility copper. Where aspect ratios will go above 4:1, vias should be copper filled in the bare board. An alternative precaution against barrel cracking is the use of substrates with restrained Z-axis expansion approximating that of the copper barrel.

Note, however, that low expansion boards may be unrestrained in the Z-axis. In fact, polymer boards restrained in X-Y expansion by Kevlar™ fibers actually are worse in Z-axis TCE than unrestrained laminates of the same polymer. This is because Kevlar's apparent negative thermal coefficient of expansion results from the Kevlar fibers becoming shorter and fatter as they are heated. Since the fibers are oriented along the X and Y-axes of the board, their contraction reduces the expansion of the board in the X and Y-axes. But the fattening of the fibers adds to the substantial expansion of the polymer in the Z-axis.

5.5.2 *Via Location*

As we mentioned earlier, SMT via locations are not predetermined by component lead penetrations as most through-hole vias are. The surface mount via could potentially be located anywhere in the PWB. In practice, however, relationships to components, other features, test pads, and lands limit the choices somewhat. We tend to mitigate these limits by specifying filled vias. An unfilled via located in a reflow solder land can cause defects in several ways.

1. Solder is pulled away from the joint and through the via by capillary attraction.
2. Uneven topography of the lands due to unfilled vias contributes to component swimming.

Figure 5-19 illustrates via filling further.

Figure 5-20 details the relationships between via location and the location of lands, components, and features.

Figure 5-19 Vias located in reflow solder lands.

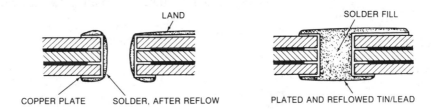

NOTES:
1. Vias filled by either of these methods available from some PWB vendors.
2. Open vias in reflow lands can cause solder defects.

Via Location in Relation to Lands

For reflow soldering, vias should not be located in or connected intimately to solder lands unless special procedures are followed to prevent solder migration down the via. As Figures 5-19 and 5-20 show, unfilled vias for reflow solder processes are normally isolated from the lands by at least a short trace.

This rule reverses for flow soldering. Open vias are encouraged near flow-solder lands because they provide an outlet through which any entrapped gasses can escape during soldering. Where vias can't be close to the land, it's permissible to locate them directly in the land. However, since a component lead may block the gas escape in a land, the "connected" or "close" via is preferable to the "in land" design.

Via Location for Test Pads

Unless custom probes are to be used, test pad vias must be located on the test fixture gird. Keeping track of which vias are so placed requires either CAD/CAT integration or a manual mark-up strategy of some sort. The wrong way to handle this challenge is to go back and locate the test points as an afterthought. Also, clearance must be provided between adjacent test points and the components so that no probes miss the intended site or hit the wrong target. For typical 1.27 mm (50-mil) probe cards, a 0.381 mm (15-mil) separation between test points and a 0.457 mm (18-mil) annular ring of land-free area around the test points accomplishes this.[5]

Vias Beneath Components

Before locating a via beneath an SMC, several questions need to be answered. Will the component block test-probing of a needed test node? If so, locate the via elsewhere, place a separate test point somewhere else in the trace, or consider costly double-side probing. Will the board be wave soldered? If so, is there any possibility of solder or flux bubbling up through the hole and creating a short circuit under the device? If soldering defects could occur, either relocating the hole, using a dry-film mask to tent the via, or filling the vias with solder during board manufacturing will eliminate solder "bubble up" concerns.

Vias Near Gold Plating

Don't locate vias where they will be discolored by the selective plating of gold contact fingers. As shown in Figure 5-17, vias that are less than 2.54 mm (100 mils) from the fingers are affected. The defect is strictly cosmetic, so you are welcome to just grin and bear it if holes must be located in close proximity to the fingers. Or, for a price, you can have the holes tented with plating resist and thus avoid discoloration.[6]

Figure 5-20
Relationships influencing via location.

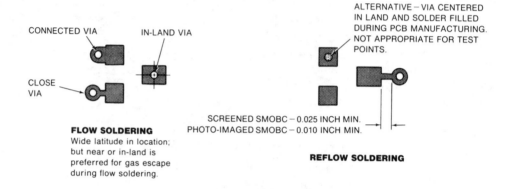

(A) Via location for soldering.

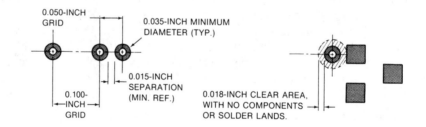

(B) Via location for test points.

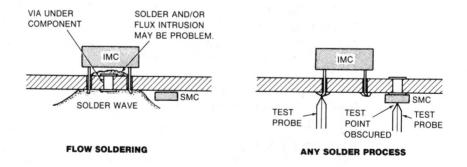

(C) Vias beneath components.

5.6 Substrates and Materials

Very early in the design process, the team should consider alternatives and specify substrate and material. Typical SMT substrate properties are shown in Table 5-2. Table 5-3 shows common applications of various substrate materials.

Table 5-2. Substrate Material Properties[7] (Typical Candidates)[***]

Substrate Materials	Glass Transition Temperature (°C)	TCE X & Y Axis (ppm/°C)	TCE Z Axis (ppm/°C)	Thermal Conductance (W/M°C @25°C)	Flexural Strength (K psi @25°C)	Dielectric Constant (@1 MHz and 25°C)	Volume Resistance (Ω-cm)	Surface Resistance (Ω)	Moisture Absorption (% of Weight)
Glass/Epoxy[††]	125	15	48	0.16	45–50	4.8	10^{12}	10^{13}	0.10
Glass/Polyimide	270	15	57.9	0.35	97	4.4	10^{13}	10^{12}	0.32
Kevlar/Epoxy	125	6.5	~50	0.16	40	4.1	10^{12}	10^{13}	0.10
Kevlar/Polyimide	225	5.0	~60	0.35	50	3.6	10^{14}	10^{12}	1.80
Alumina Ceramic	N/A	6.5	6.5	2.1	44	8	10^{14}	10^{14}	0.00
Beryllia Ceramic	N/A	8.4	8.4	14.1	50	6.9	10^{15}	10^{15}	0.00
Porcelain Coated Steel	N/A	10	13.3	0.06/—[**]	†	6.3–6.6	10^{11}	10^{13}	0.00
Porcelain Coated CCI[**]	N/A	7	†	0.06/57[*]	†	6.8	10^{11}	10^{13}	0.00
Polyimide/CCI core	270	6.5	†	0.35/57[*]	†		10^{12}	10^{12}	0.35
Epoxy/Aluminum core	125	15	†	0.16/203[*]			10^{11}	10^{13}	0.10
Epoxy/Graphite	125	7	~48	~0.16			10^{12}	10^{13}	0.10
Polyimide/Graphite	250	6.5	~50	1.5		6	10^{14}	10^{12}	0.35
Glass/Teflon	75	55				2.2	10^{14}	10^{14}	0.00
Glass/Triazine	220	12				4.5	10^{13}	10^{12}	0.015
Glass/Polysulfone	185	30			14	3.5	10^{15}	10^{13}	0.029
Quartz/Epoxy	125	6.5	48	~0.16		3.4	10^{12}	10^{13}	0.10
Quartz/Polyimide	270	9	50	~0.35	95	3.4	10^{13}	10^{12}	0.40

†† Advances in modified epoxies are occurring very rapidly. Two areas are seen as primary objectives. One is the development of high -T_g materials to withstand the high temperatures in SMT assembly and rework. The other is the search for materials having a low dielectric constant suitable for high-frequency boards, but inexpensive and similar to epoxy in all other properties.
*Surface coating/core material.
**Copper-clad Invar 20/60/20.

***Properties shown are averages for comparison purposes only. For exact engineering calculations, consult the substrate supplier.
† Depends on core-to-surface ratio.

For usual applications, solder coatings on substrates should be approximately 63% tin and 37% lead (the tin/lead eutectic). Most wave solder and reflow operations are optimized for eutectic solder-coated boards. Where another solder formulation will be used for assembly, special matching solder coatings should be specified. When the environment dictates conformal coatings, the selection of coating materials should be guided by matching coating requirements and coating properties. Table 5-4 lists some coating-material properties.

Table 5-3. Typical Applications of Various Substrate Materials[8]

Material	Potential Application
Kapton Film	Flexible circuits with flexural segments and/or rigidized populated areas. Circuits conforming to irregular areas, such as inside cameras.
FR-4 and Other Epoxy/Glass	Standard commercial and industrial SMT circuits.
Alumina	Relatively small circuits requiring moderately high heat-sinking capacity or using leadless ceramic components. High-frequency circuits.
Beryllia	Relatively small circuits requiring very high heat-sinking capacity. High-frequency circuits.
Porcelainized Steel	High heat-sink capacity and the ability to form a structural element of system, such as in auto instrumentation panels.
Polyimide/Kevlar	Matching TCE of ceramic components by balancing the mix of polyimide and Kevlar.
Copper/Invar/Copper or, Copper/Molybdenum/Copper	Very high heat-sink capacity and rigidity. Core material for matching TCE of ceramic components.
Teflon	High-frequency boards.
Glass	Displays and other see-through applications.
Polyimide/Glass	High glass transition (high operating or processing temperatures). Readily available, light weight, and low cost.

Table 5-4. Guide to Conformal Coating-Material Selection[9]

Characteristic	Acrylic	Urethane	Epoxy	Silicone	Paraxylylene
ELECTRICAL @ 23°C					
Dielectric stress (V/mil)	3500	3500	2200	2000	to 7000
Vol. resistance (Ω/cm) 50% RH	10^{15}	10^{14}	10^{14}	10^{15}	to 10^{17}
Dielectric constant 60 Hz	3.0–4.0	5.3—7.8	3.5–5.0	2.7–3.1	2.7–3.2
1 kHz	2.5–3.5	5.4–7.6	3.5-4.5	2.6–2.7	2.7–3.1
1 MHz	2.2–3.2	4.2–5.2	3.3-4.0	2.6–2.7	2.7–3.0
THERMAL					
Continuous heat resistivity	120 °C	120 °C	120 °C	200 °C	150 °C
TCE in PPM/°C	50–90	100–200	40–80	220–290	35–70
Thermal conductivity in 10^{-4} Calories/(sec)(cm^2)(°C) (cm)	4 to 5	4 to 5	4 to 5	3.5 to 8	3
CHEMICAL EFFECTS					
Of common solvents	Softens, dissolves	Generally resistant	Generally resistant	HydroCrb soften	None
Weak acids	None	Minimal	None	Minimal	None
Weak bases	None	Minimal	None	Minimal	None

5.7 Handling Thermal Coefficient of Expansion Mismatches

Two major strategies have been applied to management of TCE mismatch. Either the substrate may be tailored to the TCE of components or some form of compliant member may be introduced to absorb stresses between components and the substrate. We covered compliant leaded-component approaches in Chapter 2, Section 2.2.5. We'll devote additional attention to substrates in Chapter 6, Section 6.4. The discussions from Chapters 2 and 6, together with the following material, give engineering guidelines for applications where TCE mismatches produce unrelieved solder-threatening stress levels.

Figure 5-21 details the thermal properties of common electronic construction materials. This data is used to predict stress levels from a given environment.

Figure 5-21
Thermal properties of electronics materials.[10,11]

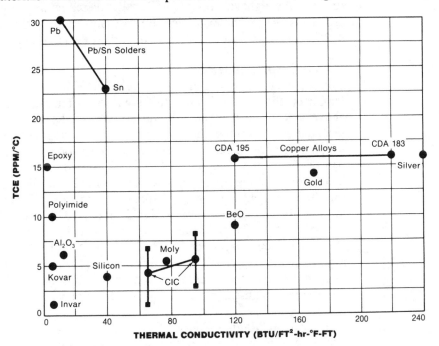

5.7.1 *Restrained Expansion Boards*

Matching Board and Component Materials
The most obvious way to tailor the substrate's TCE to that of the component is to use the same material for both. Since most CLLCCs are made of 96% alumina (polycrystalline Al_2O_3), which is also a common hybrid substrate material, this seems like an elegant approach. Indeed, alumina is used for SMT boards for just this application. Unfortunately, alumina has several drawbacks which prevent it from becoming the substrate of choice in mainstream SMT design. Ceramic boards are limited in size. Very few are produced in the 6-inch or 15-cm-square range, and almost none are produced in larger sizes. Alumina boards are relatively expensive.

The material is about three times the cost per square inch of FR-4 epoxy/glass. Ceramic boards are very brittle, and this is a problem in environments where shock loads might break them. The ceramic boards are also costly to drill.

Where the above concerns dictate a nonceramic approach to TCE matching, common alternatives are the use of restraining layers or restraining reinforcement fibers to reduce the TCE of resin boards.

Metal-Core Boards

Low TCE restraining-layer materials, such as copper/Invar™/copper (CIC) and copper/molybdenum/copper, used as power and ground layers, form functional units in the board and limit thermal excursions to match the components. These materials also add substantial heat-sink capacity, useful where cavity-up CLLCCs will dissipate heat into the substrate. Cost, weight, and limited board vendors are the negatives of this approach. However, as Table 1-2 of Chapter 1 indicates, the substantial weight savings of SMT easily overcome the weight penalty of restraining layers. All of the examples shown used metal-restraining layers, and the net-weight savings ran from 2/1 to 5/1 over their through-hole counterparts.

Copper/Invar/Copper Cores

Copper/Invar/copper is generally supplied in 20% copper, 60% Invar, 20% copper laminated sheet stock. To be effective as a restraining layer, the CIC core (in the above percentage laminate) should be 25% of the thickness of the board. Note: Drill diameters through the metal cores must be kept above 0.5 mm (~0.020 inch) to minimize drill breakage. The surface-layer copper is generally 1/2 oz. (0.0007 inch).[12] To comply with the MIL-STD-275 standard for rigid PWBs, where boards must be tested to MIL-P-55110, a minimum dielectric thickness of 0.089 mm (0.0035 inch) is required. General practice, even for commercial products, is to specify a 0.114 mm (0.0045 inch) dielectric. However, by using MIL-STD-2118 for rigid/flexible PWBs, which will be tested to MIL-P-50884, flexible dielectric layers of 0.038 mm (0.0015 inch) thickness may be used. For high-frequency applications, the resulting reduction in dielectric-layer thickness markedly lowers the impedance and propagation delays.[13] For a more complete discussion of high-frequency design, see Section 5.10 in this chapter.

Copper/Molybdenum/Copper Cores

Copper-clad molybdenum is supplied in 13% copper, 74% molybdenum, and 13% copper roll-bonded laminates. Standard thicknesses are 0.25, 0.76, and 1.52 mm (10-, 30-, and 60-mils). Strength for strength, copper-clad moly is lighter than copper-clad Invar.

Restraining Fibers

Filaments of Kevlar™, quartz, e-glass, or carbon may be used to control the X- and Y-axis expansion of epoxy or polyimide resin boards. Note that Kevlar filaments exhibit a negative coefficient of thermal expansion in the long axis. As mentioned earlier in our discussion of barrel cracking in vias, this is because the filaments shorten and grow fatter with increasing temperature. Thus, while Kevlar restrains the X- and Y-axis expansion, it adds to the Z-axis thermal expansion. For reliability concerns introduced by Z-axis expansion, and cures for these concerns, see Chapter 7, Section 2, and Section 5.5 of this chapter.

5.7.2 *Compliant Leads to Absorb Thermal-Expansion Stresses*

The substrate matching method works well with moderate I/O devices. As chip carriers exceed 68 pins, however, compliance between component and board is necessary. Internal device heat can cause a great enough differential temperature between the board and device to over-stress solder joints. You may have to turn to compliance between component and board to manage the thermal stress.

Compliant Lead Components
One method for providing compliance between the board and the component is the use of leaded components. Where leaded chip carriers are not available, post-lead attachment may provide the necessary compliant members. Post-leaded chip carriers are discussed in detail in Chapter 2, Section 2.2. Sockets may also be used to provide compliance between leadless chip carriers and a board.

Compliant Surface Layers
Compliance may also be built into the surface layer of the substrate. 3M has developed a copper-clad polyimide compliant layer which is laminated to the surface of a board using B-stage adhesive. The board is then completed using ordinary subtractive processing. The 0.127-mm (5-mil) polyimide provides compliance to absorb stresses generated by TCE mismatches between the ceramic components and organic PWB.[14]

5.8 Thermal Management

On the positive side of the thermal equation, most surface-mount ICs are constructed on high-thermal-conductivity copper-bearing lead frames instead of the less conductive Kovar and Alloy 42, which is commonly used in DIPs. For the same silicon, most SMCs exhibit a θJA lower than their IMC counterparts. But they are surprisingly close.

However, surface-mount components have far less heat-sink mass than their through-hole counterparts. SMC lead frames are generally smaller, and the individual leads provide less of a thermal path than IMCs. Also, the high densities inherent with surface mounting allow us to pack far more dissipation into a given square area. The net result is that surface-mount technology dictates a close attention to thermal management.

Next, we'll look briefly at sources for thermal engineering data. Since this book is intended for working engineers, we've made no attempt to cover the basic engineering principles which apply to thermal calculation for either IMC or SMC design. These principles are well explored in a number of engineering texts for through-hole design. Instead, we will cover thermal management strategies that are substantially unique to surface mounting.

5.8.1 *Interpreting Thermal Data from Component Specifications*

Where data is available on θJA, θJC, θCA, and P_D, the task is a relatively simple one of interpolating data to application conditions in the same way that IMC thermal design is handled. Where package data is unavailable for a given SMC, thermal design may take a bit more time. Data from another device may be used, provided the other device is in the same style package, with the same lead frame and encapsulant materials, and same die size. In the likely event that no such analogous component data is found, testing will be necessary to develop thermal data for the device. For additional material and tables on thermal design, see Reference 15.

5.8.2 Heat-Sink Layers

Very early in the design process, the basic heat removal strategy must be selected. Will convective and radiant cooling alone remove heat from the system, or will forced cooling be required? Will heat from devices flow primarily into the board, or into the surrounding air? These are the principle questions influencing strategy. Where high-dissipation circuits must be cooled without forced air, thermal layers in the board are often an answer.

For very high wattages per square inch, corrugated inner layers filled with a circulating coolant may be used. For less stringent cooling demands, a more common approach is the use of an outer or inner layer(s) of high-thermal-conductivity material. Thermal layer materials, such as copper/Invar/copper or copper/molybdenum/copper, may serve several useful purposes as part of an SMT laminate. Do not underestimate the value of copper. SMT boards can be constructed using one ounce clad plated up to two ounces for outer, inner, or both layers. The high thermal conductivity and mass of copper makes it a good place to dump heat. The low TCE of Invar and molybdenum makes them useful in restraining the TCE of the polymer portion of the laminate. As discussed above, proper proportions of bimetal to polymer allow a close matching of the substrate and the ceramic chip-carrier TCEs. And, the thermal layers double as ground/power planes, stiffen boards, and provide RFI shielding.

Single-side boards may also be sandwiched to a metal heat sink of a high-thermal-conductivity material. Where this is done, symmetry of the full assembly is important. If only one board were bonded to a metal core, the dissimilar TCEs of the board and core would produce significant warpage during thermal cycling. Likewise, asymmetrical circuits on two sides of a metal core can cause warpage if the dissipation is concentrated in different locales on opposing sides.

The thermal properties of bimetal layers and metal cores are dependent on layers above the core, and also, the polymer thermal properties. Current data on specific core materials is available from thermal-material vendors and printed-wiring fabricator who specialize in metal-core boards.

5.8.3 Thermal Vias

In designs where heat will be removed primarily into the circuit board, particularly where metal-core layers are part of the thermal management strategy, thermal vias afford substantial improvements in θJC for closely mounted components. Figure 5-22 shows several thermal via designs, and discusses their relative merits for heat removal.

Additional heat may be carried to thermal layers by attaching unassigned or blank pins in order to direct vias to the ground plane/metal core. For this purpose, use blind vias and solder fill them during board manufacturing. Since thermal vias are generally solder filled for conductivity's sake, it is always good practice when using thermal vias, to solder fill all the vias that are not required to be open for lead insertion. Again, solder-filled vias are not as prone to electrical failure due to through-plated-hole barrel cracking, and they provide better vacuum sealing in test fixtures.

Figure 5-22
Thermal vias carry heat to a metal core.

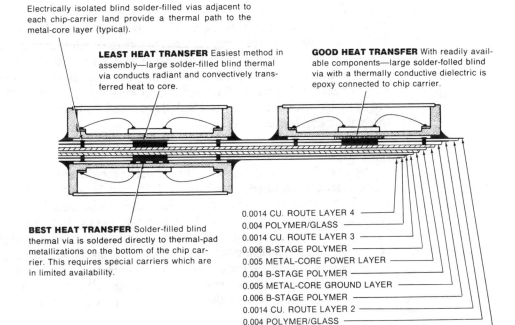

Electrically isolated blind solder-filled vias adjacent to each chip-carrier land provide a thermal path to the metal-core layer (typical).

LEAST HEAT TRANSFER Easiest method in assembly—large solder-filled blind thermal via conducts radiant and convectively transferred heat to core.

GOOD HEAT TRANSFER With readily available components—large solder-folled blind via with a thermally conductive dielectric is epoxy connected to chip carrier.

BEST HEAT TRANSFER Solder-filled blind thermal via is soldered directly to thermal-pad metallizations on the bottom of the chip carrier. This requires special carriers which are in limited availability.

0.0014 CU. ROUTE LAYER 4
0.004 POLYMER/GLASS
0.0014 CU. ROUTE LAYER 3
0.006 B-STAGE POLYMER
0.005 METAL-CORE POWER LAYER
0.004 B-STAGE POLYMER
0.005 METAL-CORE GROUND LAYER
0.006 B-STAGE POLYMER
0.0014 CU. ROUTE LAYER 2
0.004 POLYMER/GLASS
0.0014 CU. ROUTE LAYER 1

5.8.4 Component Package Styles

Chip Carriers

A wide variety of chip-carrier package types are available. Each has its own peculiar thermal characteristics.

Leadless chip carriers dissipate heat to the board far better than do leaded carriers. In selecting a leadless chip carrier, we must choose a package style appropriate for both the intended thermal management strategy and the assembly process. Types A and D CLLCCs are designed for die up/lid down mounting. This places their heat-dissipating surface away from the board and makes them suitable for forced air cooling. Type A carriers are designed for socket mounting only, while Type Ds are meant for reflow soldering. Types B and C are designed for die down/lid up mounting, placing their dissipating surface close to the board for heat transfer to it. Type C is the mainstay of the CLLCC market today. Some Type C carriers are provided with electrically isolated metallizations, arranged in a matrix across the bottom of the part. Soldering these to thermal vias or dummy lands boosts the package dissipation.

Leaded ceramic carriers may be made by post-lead attachment to Type B, C, and D CLLCCs. For dissipation primarily into the air in forced air cooling, Type D should be used.

For plastic-leaded chip carriers, lead-frame material is important to heat dissipation. High-copper-content lead-frame materials, such as beryllium copper, are far more thermally conductive than Kovar or Alloy 42. Figure 5-21 shows the thermal comparisons for commonly used electronic materials.

Smart Power Devices

Special SMC packaging is being developed to suit high-power applications. PLCCs are available with integral heat spreaders which reduce thermal impedance to the substrate. SGS Semiconductor provides such a package in a 44-pin PLCC capable of handling 2 to 3 watts. SGS was working, as of this writing, on packages capable of dissipating 5 watts.[16] Smart power devices, in SOICs and PLCCs, are available with electrically isolated, large metal leads at the corners, providing a thermal conduit to the board.

SOT-89, DPAK, etc.

High-power SMC packages are also available for the smaller semiconductor devices, resistors, rectifiers, etc. The SOT-89 is a large-die-cavity SMC package. Even larger dies and higher dissipations can be handled in the DPAK. High-wattage resistors are also available. Years ago, we worked on a project, building 25-amp solid-state relays by surface mounting on a ceramic substrate. While there are limits on the practicality of surface mount technology for power handling, the technology definitely exceeds that of logic level today. If a high-power circuit would benefit from surface mount conversion, a quick component availability review should be made through the distributors of power components.

5.8.5 Process Considerations

Some power components, such as the SOT-89, rely on a good solder bond between their heat-sink tab (also the die flag) and a land on the board for a large percentage of their heat dissipation. Since we can't depend on flow soldering to provide a continuous solder board under the tab, safe design requires that we derate the SOT-89 to 110% of its free-air rating if it is wave soldered. In practice, this speaks against using flow soldering for boards with high-dissipation SOT-89s.

5.9 Solder Masks

SMT densities and fine-line spacings beg for environmental protection and insurance against dendritic growth. This may be provided by a properly designed pad-cap-only board, or by solder masking, or conformal coating. Under conformal coatings, solder masks are useful for controlling soldering results in non-pad-cap designs.

5.9.1 Mask Technology Selection

Screened Masks

Screened masks are well understood, are available from most PWB suppliers, and are relatively low in cost. However, for dense SMT boards, misregistration or smearing of a screened mask can foul lands, leading to excessive solder defects. Also, a screened mask over tin/lead-coated traces tend to wander during reflow soldering. This phenomenon is evidenced by the mask taking on an orange peel appearance over the traces. When this mask rippling occurs, capillary traction may allow solder to migrate away from solder joints along the traces, producing solder defects. Excessive "orange peel" may upset the components and produce solder defects, or can actually part the mask, leaving unwanted exposure of the traces.

Liquid Photo-Imageable Masks

Liquid photo-imageable masks offer several unique advantages as a solder mask. Application speed is better than that of screened or dry-film masks. Coverage is excellent. There is no tendency to "tent," leaving openings around raised features, as shown in Figure 5-23. And, these masks provide the best of the inherent image accuracy of the photo-imaging process The disadvantages include lengthy exposure times costs, environmental concerns of some materials, and an inability to conveniently tent over large holes.

Figure 5-23 The "tenting" definition illustrated.

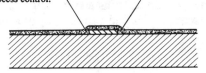

Improper processing of dry-film solder-mask application can leave "tented" openings where traces exit the mask. These openings trap contaminants during processing and pose reliability and manufacturing yield concerns. Therefore, they must be avoided by a dry-film lamination process control.

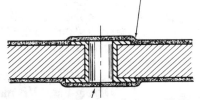

Properly applied, dry film penetrates fully into the corners.

Dry film allows a deliberate "tenting" over of holes to prevent solder migration through vias and/or to protect holes from subsequent board-processing operations. However, sealed tenting such as this entraps atmosphere that may expand and rupture the tent during soldering operations. Ruptured tents can entrap cleaning agents and contaminants which may lead to corrosive failure of hole plating.

Dry-Film Photo-Imageable Masks

Dry-film photo-imageable masks offer the photographic process resolution near that of liquid photo-imageable masking. Dry films also provide the ability to tent over holes, and present no concern about unwanted penetration into holes. However, dry films are more costly than screened masks in many PWB shops. Also, dry-films are generally too thick for use on SMT boards, where maintenance of a planar surface, including both masked areas and lands, is necessary in order to maintain acceptable soldering-process yields. And, as shown in Figure 5-23, poorly applied dry films may leave openings for contaminants to enter around the traces. Good board-manufacturing process control will strictly control tenting, however, and will leave no reason for lamination quality concerns when specifying dry film. With dry films, deliberate tenting may be used to mask tooling holes during through-plating, thus preserving their drilled tolerance. Tenting also allows for protecting holes from selective-plating processes.

5.9.2 *Mask Design Concerns*

Screened Solder Masks

Screened solder masks may be applied over bare copper, controlling the orange peel defect. But the inherent inaccuracy of screened masks limits their use for SMT. Where screened masks are selected, the mask should stop 0.25 mm (10-mils) from all solder lands to ensure misregistration of mask does not cause a fouling of the solder-land areas. The mask should also be omitted under passive components, because bubbling and orange peeling of the mask during soldering can contribute to solder opens and defects.[17] These design rules are illustrated in Figure 5-24.

Figure 5-24
Design rules for
screened solder
masks.

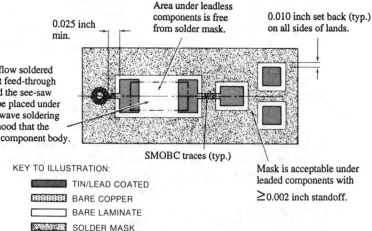

Mask may be place under reflow soldered
parts where needed to protect feed-through
traces. Observe rules to avoid the see-saw
effect. Mask should always be placed under
bottom-side components for wave soldering
n order to increase the likelihood that the
adhesive dot will contact the component body.

KEY TO ILLUSTRATION:

◼ TIN/LEAD COATED
▦ BARE COPPER
☐ BARE LAMINATE
▨ SOLDER MASK

Some Cad systems generate mask artwork from SMT lands and do not allow
the options of editing the setback from lands and placing the clear areas under passives. With such CAD systems, use a pad cap design, or a photo-imageable mask
instead of a screened variety.

Photo-Imageable Masks

During the PWB manufacturing process, photo-imageable masks are registered in
the same manner, and use the same features, as solder lands. Therefore, they may
be produced in very close dimensional relation to the features that are important in
SMT assembly. No setback from lands is required for photo-imaged masks. The
masks are generated from a reverse image of the lands or from solder-screen
artwork.

Dry-Film Masks

Dry films are typically available in 0.025 mm to 0.102 mm (1-4-mil) thicknesses.
Using a 0.76 or 1.02 mm (3- or 4-mil) film, vias up to 1.27 mm (50-mils) in finished diameter may easily be tented by the solder mask.[18]

One design concern raised by dry-film masks is the *see-saw effect* illustrated
in Figure 5-25. The see-saw effect can be minimized by using thin dry films, using
1/2 ounce copper in lieu of 1 ounce, or, better yet, excluding the mask and traces
under small components where they might contribute to the tombstoning problem.
Thick dry films (>1-mil) generally aggravate the see-saw effect. Wet films in thin
applications will reduce the see-saw effect.

Where none of these solutions are possible, tombstoning can be minimized by
running two or more tracks under any problem component, instead of just one
track. Run a track as close as design rules will allow to each land, creating a balanced geometry instead of a fulcrum point. If you don't have two traces available
to go under a problem component, substitute a short dummy trace. The dummy
trace should extend just to the edges of the component. This set of features may be
incorporated in the component library for all reflow-soldered chip components. A
separate component-library pattern is used for the part if it is placed on the bottom
side for wave soldering.

Pad Caps in Lieu of Masks

For multilayer boards, it is often possible to add two more layers containing all the
solder lands connected to routing layers by closely coupled interconnect vias.
Such designs are called Pad Cap boards, and do not need solder masks. In many

cases, the two extra layers of a pad cap board will cost no more than the appropriate solder masks.

Figure 5-25 The See-Saw Effect.

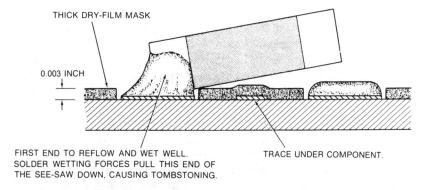

THICK DRY-FILM MASK

0.003 INCH

FIRST END TO REFLOW AND WET WELL.
SOLDER WETTING FORCES PULL THIS END OF
THE SEE-SAW DOWN, CAUSING TOMBSTONING.

TRACE UNDER COMPONENT.

For reflow, best design procedure excludes mask and traces from
under small leadless parts which are prone to tombstoning.

Failing to Plan Is Planning to Fail

In our consulting work, we've been surprised by the number of clients who don't know what kind of solder mask they'll get on their board, or if they'll have any mask at all. Such important decisions should not be left to a board vendor who has no idea of the intended application or assembly processes for your product. For the high density and fine lines typical in SMT, solder-mask defects may account for more than half of all board rejects.[19]

5.10 High-Frequency Design

As digital-circuit speeds increase, and corresponding rise times decrease, circuit-board traces begin to act as transmission lines. Transmission lines affect the circuit operation just as components do. Therefore, great attention must be paid to the detail of trace design for high-speed circuits.

It is beyond the scope of this book to offer an extensive treatise on high-frequency design. We will cover the general topics of particular concern for surface-mount technology, and encourage engineers venturing into microwave frequencies for the first time to research the references listed in Section 5.11 at the end of this chapter.

5.10.1 *Defining High Speed*

As a first step in our treatise on SMT design for fast circuitry, let's define when signals must be considered high-frequency. Designers sometimes speak of circuits with rise times below 4 nanoseconds as fast. But how much below? And should the cutting edge be at 4 nanoseconds or some other number? The answer is, it depends.

Clock rates will yield even less specific insight to the high-speed boundary. Some circuits require transmission-line design at 50 MHz and others run at over 1 GHz, using ordinary interconnect design rules.

These disclaimers are offered not to confuse the issue, but to demonstrate that a simplistic answer is not adequate. Lee Ritchley, president of Shared Resources,

San Jose, CA, has offered a more useful gauge to high-speed design-rule requirements. He explains that circuits should be considered "...high speed when the rise-time transition is completed before the signal has made a round trip along the conductor." When rise time is less than round-trip time, signal reflections and crosstalk are likely to degrade signal accuracy, unless design rules are applied to control these factors.[20]

Ritchley applies the rule of thumb that signals travel on FR-4 or polyimide at between 5 and 6 inches per nanosecond. Thus, if a device with a 1-nanosecond rise time is at the end of a 3-inch signal trace, the 6-inch round trip may not occur before the device transition, and you're in the danger zone. A device with a 6-nanosecond rise at the end of an 18-inch line would similarly be a fast circuit.[21]

Of course, in both cases, if we could reduce that signal trace to 2 inches, you would be back out of the circuit designer's twilight zone. Since SMT is such a champion at shrinking trace lengths on PWBs, surface mounting obviously has a great deal to offer designers of high-performance assemblies.

From the preceding definition, we can readily see that relatively slow (8 to 12 MHz) digital boards may have areas where the gate rise times qualify them as fast. Don't be one of the designers who, familiar with digital logic, is caught unawares by the technological advances of device edge speeds. Don't violate high-speed design considerations. The result may be a board that works in one slot of a backplane, but not in the next. Or, your board may work with one IC, but fail when you trade it for another identical part number.

5.10.2 *Transmission-Line Design*

Types of Transmission Line
When circuit miniaturization won't get the job done, and Ritchley's rule indicates that circuit traces must be treated as transmission lines, several design alternatives are commonly used. Figure 5-26 illustrates four of these approaches.

Figure 5-26
Classification of transmission lines illustrated.

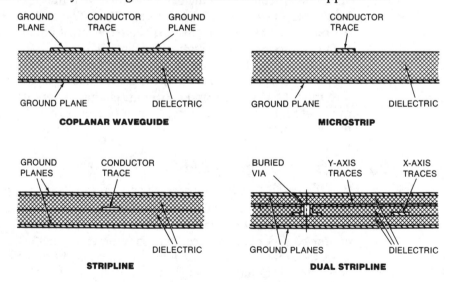

5.10.3 *Controlled Impedance*

For fast devices such as ECL, terminated transmission lines must be designed to produce controlled impedances. Higher impedances limit current flow, and, therefore, consume less power. But the impedance chosen must match the input impedance of the devices used on the board. Where multiple-device families are required, several layers or board areas with individual impedances may be necessary.

Impedance is influenced by the style of transmission line used, the dielectric constant of the board material, the dielectric separation of the line from its ground plane, and the width and thickness of the trace.

Effects of Transmission-Line Style

For a given dielectric separation and trace dimension, microstrip transmission lines inherently exhibit less propagation delay and higher impedances than stripline or dual stripline. However, both stripline styles offer far better EMI shielding of fast switching lines. And, because of the shielding effects of the dual ground planes, stripline styles allow conductors to be placed closer together without crosstalk problems. Thus, stripline styles allow higher densities of circuit traces than does microstrip.

Dielectric Constant of the Board

Propagation speed along a transmission line is in inverse relationship to the dielectric constant (K) of the insulative material around the conductor trace.

Dielectric constant is a measure of a material's ability to store an electric charge when acting as a dielectric in a capacitor. K is the ratio of the charge that would be stored, with air as the dielectric, to that stored, with the material in question as the dielectric. For this computation, air is assigned a K value of 1. Table 5-5 shows the K factors for some typical substrate materials.

Table 5-5. Dielectric Constants of Typical Substrate Materials

Material	K @ 1 MHz	Material	K @ 1 MHz
Alumina Ceramic	8.0	Polyimide/Kevlar™	3.6
Epoxy/Glass	4.8	Polyimide/Quartz	3.4
Epoxy/Kevlar™	4.1	PTFE/Glass	2.2
Polyimide/Glass	4.5	RO2800™	2.9

We mentioned earlier that microstrip transmission lines exhibit smaller propagation delays than stripline types. This is because of the low dielectric constant of the air bounding one side of microstrip structures. Inner-layer microstrips are bounded on both sides by the substrate dielectric material. Thus, they do not display the speed of surface-layer traces.

Dielectric Separation from the Ground Plane

Dielectric thickness between a transmission line and its ground plane(s) has a direct influence on impedance. By decreasing the separation, we decrease both impedance and propagation delays. We can also reduce impedance and propagation delays by using a dielectric with a lower K factor. But, as the impedance formulas indicate, dielectric thickness has a more direct relation. Formulas for the exact relationships between impedance, propagation delays, separation, dielectric material, and transmission-line style are presented next, in the discussion of "transmission-line formulas."

Figure 5-27
Conductor shape affects high-frequency reflections.[22]

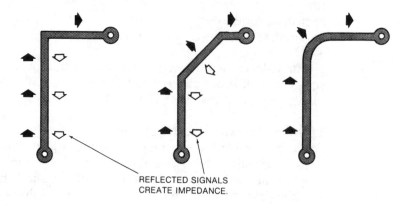

REFLECTED SIGNALS
CREATE IMPEDANCE.

Figure 5-28
Ground-plane separation continuity.

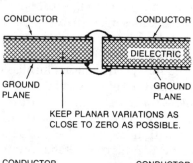

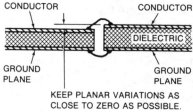

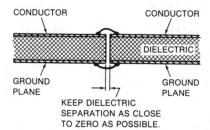

Conductor Form Factors

Conductor width, and to a lesser degree, conductor thickness, determines the impedance of transmission lines. Formulas for calculating conductor impedance for a given size are presented next. Conductor geometry is also important in high-frequency design. Abrupt changes in line direction or any separation from the ground plane will cause reflections—creating unnecessary impedance, damaging system performance, and potentially degrading the reliability. Figure 5-27 shows preferred line geometries for reducing reflections, while Figure 5-28 details some ground-plane treatments.

Transmission-Line Formulas

As frequencies move above 3 MHz, signal propagation along a conductor occurs increasingly at the surface. Figure 5-29 shows this phenomenon, called the skin effect, and formulas to calculate it, while Figure 5-30 shows the relationship between frequency and skin depth.

Figure 5-29 Skin effect in high-frequency signal propagation.[23]

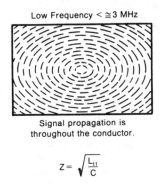

Low Frequency < ≅ 3 MHz

Signal propagation is throughout the conductor.

$$Z = \sqrt{\frac{L_{11}}{C}}$$

(A) Low frequency.

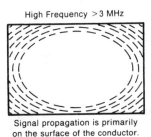

High Frequency > 3 MHz

Signal propagation is primarily on the surface of the conductor.

$$Z = \left(\frac{\sqrt{R^2 + 4n^2f^2L2_{11}}}{\sqrt{G^2 + 4n^2f^2C^2}} \right)^{1/2}$$

(B) High frequency.

Figure 5-30 Skin depth vs. frequency.[24]

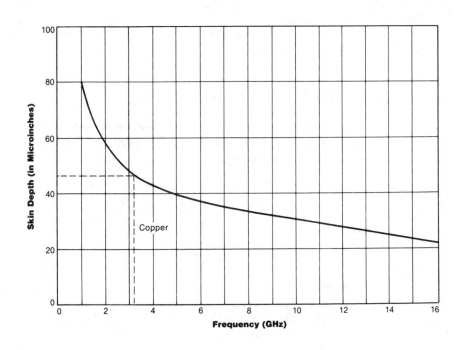

Low-frequency signal propagation can be calculated using the equation[*]

$$Z = \sqrt{\frac{L_{t1}}{C}}$$

(Eq. 5-1)

where,
Z is the characteristic impedance in ohms,
L_{t1} is the inductance in henrys/unit length,
C is the shunt capacitance in farads/unit length.

while high-frequency signal propagation is calculated using[*]

$$Z = \left(\frac{\sqrt{R^2 + 4n^2 f^2 L2_{t1}}}{\sqrt{G^2 + 4n^2 f^2 C^2}}\right)^{1/2}$$

(Eq. 5-2)

where,
Z is the characteristic impedance in ohms,
$L2_{t1}$ is the inductance in henrys/unit length,
C is the shunt capacitance in farads/unit length,
R is the resistance in ohms/unit length,
f is the frequency,
G is the shunt conductance in ohms/unit length.

We can calculate impedances of microstrip and stripline transmission lines using the form factors shown in Figure 5-31 and the following formulas.

Figure 5-31
Dimensional factors for transmission lines.

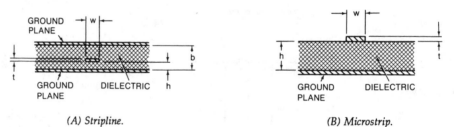

(A) Stripline.   *(B) Microstrip.*

To calculate the propagation delays of microstrip and stripline, use the following formulas for transmission-line impedance. For microstrip, the characteristic impedance is calculated:

$$Z_o = \frac{87}{\sqrt{K_r + 1.41}} \ln\left(\frac{5.98h}{0.8w + t}\right)$$

(Eq. 5-3)

and

$$t_{pd} = 1.017\sqrt{0.475 K_r + 0.67} \ \text{nS/ft.}$$

(Eq. 5-4)

where,
Z_o = The microstrip characteristic impedance,
K_r = The relative dielectric constant of the substrate material,
t_{pd} = The microstrip propagation delay.

---

[*] Per MIL-HDBK-176.

Figure 5-32
Capacitance of
microstrip trans-
mission lines.
*(Reprinted by
permission of
Motorola, Inc.)*

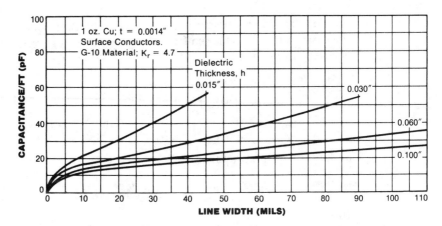

Figure 5-33
Capacitance of
stripline trans-
mission lines.
*(Reprinted by
permission of
Motorola, Inc.)*

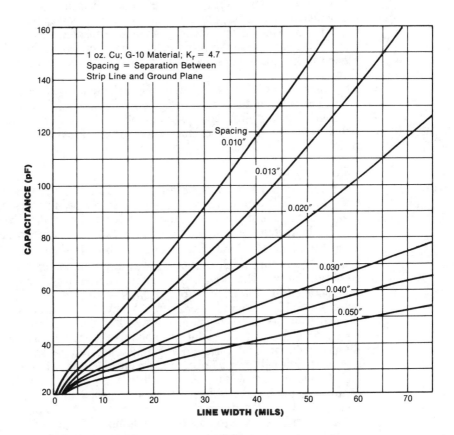

The inductance per foot equals $L_o = Z_o^2 C_o$, where the capacitance, C_o, is taken from the graph in Figure 5-32.

For stripline propagation delays, the inductance per foot is $L_o = Z_o^2 C_o$, where the capacitance, C_o, is taken from the graph in Figure 5-33, and the stripline characteristic impedance Z_o is:

$$Z_O = \frac{60}{\sqrt{K_r}} \ln\left(\frac{4b}{0.67\,\pi\,w\,\left(0.8 + \dfrac{t}{w}\right)}\right) \qquad \text{(Eq. 5-5)}$$

and, the stripline propagation delay is:

$$t_{pd} = 1.017\sqrt{K_r}\ \text{nS/ft.} \qquad \text{(Eq. 5-6)}$$

High-Density Design

As we've established above, reducing the dielectric layer thickness reduces both the impedance and propagation delays of transmission lines. The impedance needed is determined by matching the input impedance with the logic family on the board. Therefore, thin dielectric layers may drop impedances low enough to allow reducing the line widths to gain the needed impedance, thus increasing circuit density.

Complex high-frequency design is a challenging chess game built around the above variables and a host of design considerations (discussed in the references in the following high-frequency bibliography). We have been restricted by space to only a cursory discussion of microwave design. Please refer to the following listed sources for more detail.

Layout Rules

SMT follows the same generic rules for high-frequency board layout common in the through-hole world. However, we see so many instances of total disregard or plain ignorance of these rules that a brief discussion of them herein seems in order.

For EMI immunity, a good goal is to design to tolerate the H field generated by a $\Delta V\Delta T$ or 10^{10} Volts per second. On board relays switching inductive loads can easily produce such events, as can any number of environmental events in close proximity to the circuit. Decoupling, shortening of traces, addition of protection diodes, filtration of input/output lines -- all these strategies may be needed to achieve the 10^{10} immunity goal. Fortunately, SMT simplifies or at least reduces the real estate requirements needed for each of these approaches.

Ground integrity is vital to minimize EMI/RFI emissions from a high-frequency design, and to ensure its proper operation. A very minimal impedance in a ground path will present absolutely no problem to a low-frequency circuit. To a high-frequency signal, that same path will appear to be an open circuit. Ground conductors must be as robust as circuit dimensions will permit. Ground routing must provide for the orderly flow of currents from the circuit to ground. The topology must be chosen to absolutely minimize voltage differentials from one point to another across the ground. Ground planes greatly simplify adherence to good topology rules, and, when correctly placed, reduce H field coupling by several orders of magnitude.

EMI/RFI emissions can easily be aggravated if allowed to pass from the board to the cabling and I/O connected to the circuit. The use of selective-pass filters or ferrite beads can help in this. Where size is at a premium, ferrites and filters are now available in surface mount packaging. Decoupling is generally required for signal lines entering ICs. The closer the capacitor can be placed to the input it protects, the more efficacious it will be. SMT is great for decoupling for two reasons. Small SMT capacitors are easy to pack in adjacent to an IC. The fact that they can be placed on the secondary side of the board, opposite the side where the ICs reside, means that they can often be used with minimal impact to routing tracks approaching a busy IC. Still, in some cases, through-hole parts provide an easier avenue for trace routing. The designer should feel equally free to select either packaging format dependent on the needs at hand. Placement of decoupling is critical to success of EMI damping. Follow guidelines listed in the references at the end of this chapter. Note that, when chaining SMT decoupling together, entry of the ground lines to the lands should be at the two sides of the land, not a single entry from the end with a T connection to the ground track.

Termination networks can be kept to a minimum of board real estate by selecting between IMC or SMT networks or SMT discrete resistor chips. Ultraminiature chips are sometimes used for terminators where dissipations will be low. For dense boards having large numbers of resistors with a close value range, buried resistor techniques may solve thorny miniaturization problems. (see Figure 6-8)

Coupling of adjacent traces to fast-transition paths is a problem requiring all the vigilance the layout engineer can muster. The obvious rules of the road are just a starting point. Keep all runs on a given layer in essentially the same axis (N-S or E-W). Route adjacent layers in perpendicular axes. If two traces where crosstalk could be a problem must run parallel to each other, place a ground between them or route them on separate layers. Where layer-to-layer crosstalk might be a problem, increase the separation of the offending traces, or place them on opposing sides of a plane. Beyond these basics, it is a good idea to identify all the fast-transition paths on a composite check-plot, and make cross-talk possibilities the subject of frequent design reviews.

5.11 Additional References on High-Frequency Design

Interested readers are directed to the following resources on high-frequency for additional information on high-frequency circuit design.

1. *Motorola MECL System Design Handbook*, 1986, Motorola Semiconductor Products, Inc., P.O. Box 20912, Phoenix, AZ, telephone: (602) 944-6561.
2. Motchenbacher, C.D. and Connelly, J.A., *Low Noise Electronic System Design*, John Wiley & Sons, New York, NY, 1993.
3. Harper, Charles A., Ed in Chief, *Electronic Packaging and Interconnection Handbook*, McGraw-Hill, Inc., New York, NY, 1991.
4. Seraphim, Donald P., Lasky, Ronald, and Che-Yu, Li, *Principles of Electronic Packaging*, McGraw-Hill, Inc., 1989.
5. *FAST Applications Handbook*, 1987, National Semiconductor, Digitial, 333 Western Ave., South Portland, ME, telephone: (207) 775-8700.
6. *Applications and Performance of GigaBit 40 I/O Chip Carriers*, 1986, Gigabit Logic, Inc., 1908 Oak Terrace Lane, Newbury Park, CA 91320, telephone: (805) 499-0610.
7. *Guidelines for the Use of Digital GaAs Ics*, Gigabit Logic, Inc., 1908 Oak Terrace Lane, Newbury Park, CA 91320, 1987.
8. *Thermal Management of PicoLogic and NanoRAM GaAs Digital IC Families*, Gigabit Logic, Inc., 1908 Oak Terrace Lane, Newbury Park, CA 91320, 1988.
9. Ramu, S., and Whinnery, T. Van Duzer, *Fields and Waves in Communications Electronics*, 1984, John Wiley & Sons, 605 3rd Ave., New York, NY 10158, telephone (212) 850-6000.
10. Davidson, C.W., *Transmission Lines for Communication*, John Wiley & Sons, 605 3rd Ave., New York, NY 10158, 1978.

5.12 References

1. Hollomon, James K. Jr., *Focus on SMT Design*, Anatrek, Norfolk, VA, 1987.
2. "Testability Guidelines," SMTA Edina, MN, 1987.
3. Ibid.
4. Patterson, Brian T., and Landolt, Roger H., "Fully Additive Plating with Permanent Additive Resist," *PC Fab*, June 1987, pg. 54.

5. Op. cit., ref 2.
6. Leibson, Steven H., "EDN's Hands-On Project -- Part 4: Assembling the Surface Mount Project Board," *EDN*, July 9, 1987, pg. 75.
7. Op. cit., ref. 1.
8. Ibid.
9. Conner, Margery S., "Technology Update -- The Conformal Coating of Your PC Boards Will Enhance Their Environmental Resistance," *EDN*, June 11, 1987, pg. 89.
10. Op. cit., ref. 1.
11. Op. cit., Chapter 1, Reference 12.
12. Knodle, John M., "Manufacturing PCBs with Invar Cores," *PC Fab* , June 1987, pg. 58.
13. Cantwell, Dennis J., "Designing SMT PWBs for Manufacturability," *PC Fab*, June 1987, pg. 69.
14. Fronek, Daniel R., and Grasse, Dr. Peter B., "Compensating for TCE Mismatch Between LCCCs and PWB Substrates," *Surface Mount Technology*, October 1987, pg. 38.
15. *SMD Thermal Data*, Signetics Corp., Sunnyvale, CA, 1987.
16. Bindra, Ashok, "For Surface Mounting, It's Full Speed Ahead," *Electronic Engineering Times*, October 19, 1987, pg. T-24.
17. Leibson, Steven H., "EDN's Hands-On Project -- Part 3: CAD and Surface Mount Technology," *EDN*, June 25, 1987, pg. 212.
18. Goromdy, Emery J., "The Continuous Flow Dry Film Solder Mask Process," *PC Fab*, October 1987, pg. 59.
19. Esposito, Donna, "The Liquid Photoimageables," *PC Fab*, October 1987, pg. 15.
20. Markstein, Howard W. "Packaging for High Speed Logic," *Electronic Packaging and Production*, September 1987, pg. 48.
21. Ibid.
22. Op. cit., ref. 13.
23. Ibid.
24. Ibid.

Figure 6 High-density telecommunications control module assembly requires engineering forethought. *(Courtesy Rockwell International.)*

6 Assembly-Level Packaging and Interconnections

In this chapter, we'll turn our attention from the layout of the board to the layout of the system. Elegant packaging, like that seen in Figure 6, is no accident. It is the result of a well directed and careful system-engineering job carried out by a sharp SMT team at Rockwell International's Communications Group in Dallas, TX. Some SMT problems resolve much easier at the system packaging level than at the board level. For instance, repetitive circuit functions are implemented on modular daughter cards in Figure 6. Thus, the designers and manufacturing people didn't have to constantly reinvent. We'll consider this and other assembly packaging issues in the following chapter text.

In Chapters 1 through 3, we laid the foundation for SMT design decision making. Chapters 4 and 5 developed, on that foundation, the framework for system and detail SMT design. Now we are ready to sheath that framework with tools for fine tuning our design. In this chapter, we'll see how SMT cards are interconnected into working systems.

6.1 SMT Strategies to Reduce System Size

We've stressed team engineering as the key to surface-mount success. Here's how one SMT contractor helps clients apply teamwork in reducing package size. According to Gregory Horton, president of the Chatsworth, California, SMT military contract house *SMTEK*, teamwork begins at the conceptual stage of a project. He recommends,

"The team should develop the approximate component counts, the required board area for each function, power dissipations, I/O needs, etc. Then, they should partition the design into as few segments as possible to keep the board count down. This simplifies I/O, may improve electrical performance, and simplifies board-level interconnects and backplane design."

Horton states that this approach "...boosts reliability and cuts cost. The limiting factor is the practical board-size limit." SMTEK also recommends the standardization of board and panel sizes where more than one module is required.[1]

The drive to save board real estate is certainly not newly arrived with SMT. IMC packaging engineers have been in hot pursuit of miniaturization for many years. But some of the tools that SMT makes available for this quest are unique, and bear discussion herein.

6.1.1 *Common Ways to Gain Real Estate*

Let's look now at some real-estate-saving moves that work in many varied applications. Because these methods are broadly applied, they are worth consideration whenever a design might profit from a space or weight reduction.

Figure 6-1 SMT keeps ZT 8816 NEC V50 Computer to one card. *(Courtesy Ziatech Corp.)*

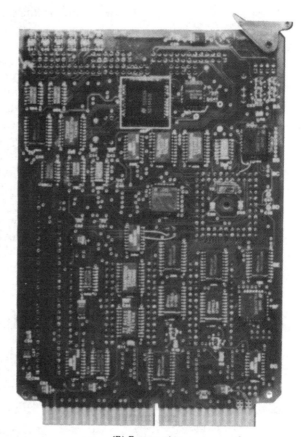

(A) Top view.

(B) Bottom view.

Multiple Boards into One Board

In changing an existing through-hole product to surface mount, it is often possible to crunch the functions of two or more IMC boards onto one SMT card. Where this can be done without undue increases in test and repair costs, the savings are substantial. The raw board is cheaper because the single SMT assembly uses about half the PWB material of its multiple IMC predecessors. Cabling and connectors are saved because there are no board-level interconnects required. And, the elimination of the board-level interconnects (a common failure point) improves system reliability. Figure 6-1A shows the use of SMT in a single-board computer, with dense, mixed technology. This computer is equal in performance to a 6-MHz 80286 device and would certainly have required 2 or 3 cards using IMC technology. Figure 6-1B shows the moderate density, all SMT, backside of the same board. This side, alone, would have required a separate board, using IMC.

SIP and DIP Modules

SIP and DIP modules, like those shown in Figure 6-2, serve miniaturization and economy needs in several ways. As we'll see later, modules can satisfy standard circuit requirements or can simplify the conversion of just a part of a circuit to SMT.

Figure 6-2 SMT SIP and DIP modules.

Where certain circuits, such as amplifier stages, comparators, etc., are used repetitively in designs, converting them to SMT modules can save considerable space and money. Instead of implementing the standard circuit with discrete IMCs on each assembly part number where the function is needed, the SIP or DIP module may be integrated easily into existing and future through-hole boards. And the module may also be used as an inserted module on Type 2 and 3 SMT designs.

For designs which must remain partially through-hole, SMT modules may simplify conversion of those areas that would most benefit from SMT. The SMT module can be built with cost-effective Type 1 technology. Its lead frame allows inserting it in an otherwise through-hole board, and permits wave soldering the assembly on a standard IMC production line.

Double-Sided Population

In certain manufacturing circles, suggesting the double-side population of surface-mount assemblies is tantamount to heresy. Lest we be thought unfaithful to sound manufacturing principles, we'd best preface this discussion with a caution that double-sided assembly should be a last-resort miniaturization method. Nonetheless, many two-sided SMT products are succeeding and earning their manufacturers good incomes today. Products like the ultradense memory board shown in Figure 6-3 just wouldn't exist without the real-estate benefits of SMT. Figure 6-3 is a perfect example of the on-target application of double-sided SMT.

Figure 6-3 High density achieved with double-sided board population. *(Courtesy XeTel Corporation, Austin, TX.)*

TOP VIEW

MIRROR VIEW OF BOTTOM

6.1.2 *Application-Determined Strategies*

Here, we'll look at real-estate-savings moves that are peculiar to certain applications. These methods should be considered whenever a design of a type covered in the following discussion needs a space or weight reduction. Some of the lessons of these application-specific miniaturization strategies will benefit circuits other than pure memory arrays, or circuits with a large number of decoupling capacitors, or any airborne or spacecraft electronics. Thus, they bear study by all engineers interested in ways to shave inches and pounds from their products.

Memory Arrays

For memory boards, several SMT component alternatives should be explored. PLCC or SOJ memory packages save considerable real estate, compared to DIP equivalents. However, they do not offer real-estate savings comparable to SIMM™ memory modules or their SIP clones. SIMM memories are tiny daughter boards carrying several PLCCs, which form a complete memory array, as illustrated in Figure 6-4. While they cost more than an equivalent amount of memory in discrete packages, they offer several advantages. They occupy substantially less real estate, as shown in Table 6-1. SIP memories greatly simplify the routing of memory-array boards. And, they offer socketability at reasonable cost and high density. Socketing PLCC or SOJ discretes is expensive and wasteful of space. Figure 6-4 also shows the use of an angled socket for space-efficient mounting of the SIMM memory module.

Figure 6-4 Typical 256K × 8-bit plus parity SIP memory module. *(Courtesy Texas Instruments Incorporated.)*

Table 6-1. Memory Packaging Compared[2]

Package Type	PWB Area (Sq. Cm./In.)	Number Of Holes Required	Ease Of Socketing
DIP	167.7/26.0	1296	Moderate
PLCC	116.1/18.0	432	Poor
SIP	48.4/7.5	240	Good

Note: Packaging estimates for 2-Mbyte memory arrays (8 bits plus parity), using 256K-bit DRAMs.

Decoupling Capacitors

Circuits with a significant number of PLCCs that require decoupling can be miniaturized by placing the decoupling under the body of the PLCC. Memory modules like the one shown in Figure 6-4 are produced in great volume by Texas Instruments using this technique. TI has employed both vapor phase and infrared soldering methods for these assemblies, so IR shouldn't be discounted as a viable process because of shadowing.

Figure 6-5 Design rules for decoupling capacitors under PLCCs.

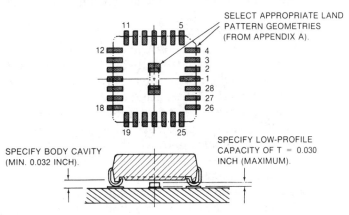

The shadow created by the piggyback PLCC will not necessarily defeat IR reflow of a decoupling capacitor. IR soldering involves heat transfer by radiation, convection, and conduction. In IR systems with any substantial convective heating component, convective heating and conductive transfer through the PWB will reflow the shadowed capacitor.

Figure 6-5 shows suggested design rules for hidden decoupling of PLCCs. Note that, since hidden decoupling capacitors are not visually inspectable, probe access to the chip carrier's V_{ss} and V_{DD} pins is a must.

6.1.3   *Additional Technologies for Real-Estate Improvement*

Several real-estate miser methods are developing which, while they hold significant promise, are not in the mainstream manufacturing practice as of this writing. Because they are not in widespread use, no industry-wide design standards have yet been adopted. Where they are in use, the technology is proprietary to the manufacturers that have spent development time on it. They merit mention, however, as they may well develop into standard approaches in the life of this work. Even where standardization is not forthcoming, these methods are workable, and may be independently developed or acquired by licensing, where needed.

Silicon-on-Silicon

One emerging variation of SMT, silicon-on-silicon (SOS), promises ultrahigh densities rivaling those obtained by elusive methods, such as wafer-scale integration. Silicon-on-silicon SMT is flip-chip and discrete-element placement on a hybrid circuit that is fabricated on a silicon substrate. By using a silicon substrate, many passive and active elements may be fabricated in situ on the circuit board. Where it is impractical to build up a particular component on the substrate, it is added as an SMT discrete.[3] Since this method is really a special variation of a hybrid circuit, design rules for SOS may be drawn from Chapter 9, SMT hybrids. We mention SOS here because of the enormous miniaturization potential it offers.

The advantages of silicon-on-silicon are an extremely high density, excellent heat dissipation, and, in very high volume, a low cost per functional module. Distinct from wafer-scale integration, SOS offers the advantage of discrete-device repairability. The limitations are sources for flip-chip components and technology, high front-end costs, and relatively long product-development cycles.

Polymer Thick-Film Components

Special conductive and resistive inks compatible with polymer substrates may be used to print resistors directly on circuit boards. Combinations of conductive ink layers covered by dielectric ink layers allow printing of low-value capacitors as well. Conductive ink or standard circuit-board metallizations may be formed in spirals to make small inductances.[4] For design guidance in these techniques, see Reference 4. Also, gather specific data on polymer ink behavior and properties from the material suppliers.

Integral Components

Resistive/conductive laminate layers, or resistive traces of polymer thick-film ink, may be built into inner layers of multilayer boards. Capacitive and inductive components may also be formed in situ as described earlier, and buried in a MLB.

However, serious manufacturing, test, and repair-access problems must be addressed when using buried-circuit elements. Because of these concerns, the use of buried elements is currently limited primarily to resistive components.

Capacitors are more easily fabricated by multilayer techniques, putting one plate on each layer and terminating the plates at alternating vias at the ends of the plates, per Figure 6-6.

Figure 6-6 In situ capacitors in MLBs.[5]

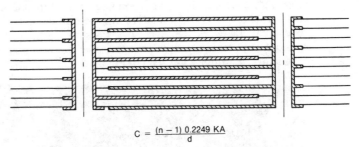

$$C = \frac{(n-1)\ 0.2249\ KA}{d}$$

where,
n = Number of conductive layers,
C = Capacitance in pF,
K = Dielectric constant of insulating material (see Chapter 5, Table 5-2),
A = Area of electrode in square inches (minimum),
d = Thickness of dielectric between layers.

Integral inductors are generally formed as spiral coils on the board's surface. While this same approach may be used on inner layers, use of the surface layer ensures test access and repairability. Figure 6-7 shows multiple turn loops for a spiral coil, modified circles, and a square coil.

Figure 6-7 In situ inductor forms in PWBs.[6]

SPIRAL COIL   MODIFIED CIRCLES   SQUARE COIL

The inductance, L, of a single-turn inductor loop is calculated as follows:

$$L = 0.002\ A(\log B - 2.451) \qquad \text{(Eq. 6-1)}$$

where,
A = The mean circumference of the loop,
B = 2A/(w + t) for rectangular conductors of width w and thickness t.

The *spiral-coil* inductance is:

$$L = 0.0215\ a\ N^{5/3} \log \frac{8a}{c} \qquad \text{(Eq. 6-2)}$$

where,
a = The mean radius ($\frac{1}{2}(R_1 + R_2)$),
N = The number of turns,
c = $R_2 - R_1$ (the radial depth of winding).

(Note: To maximize Q, thus minimizing conductor length, use c/a = 0.8.)

Modified circles approximate the spiral-coil inductance and are easier to lay out.

Square spirals are easily laid out and maximize the inductance per square area of PWB. A square spiral of side A will provide 12% greater conductance (L) than a spiral coil of outside diameter A.

Buried resistors of 10% tolerance can be fabricated for lamination within a PWB. Typical applications are boards with large populations of low-tolerance resistors, such as pull-up and pull-downs, and termination networks. Buried resistors are made using special resistive-layer laminate or polymer thick-film printing techniques.

The resistive laminate method involves electroplating a nickel/phosphorus resistive layer onto a copper foil, then laminating the bimetal foil, resistive layer down, to a polymer sheet. PWB fabricators photoetch the exposed copper to leave only the conductive traces as required. The resistive layer is then etched to form individual resistors, by leaving only the resistive connections desired between the copper conductors.[7] Examples are seen in Figure 6-8.

Figure 6-8 The buried resistor laminate process. *(Samples from Ohmega Tech., Inc.; Photo by AMTI.)*

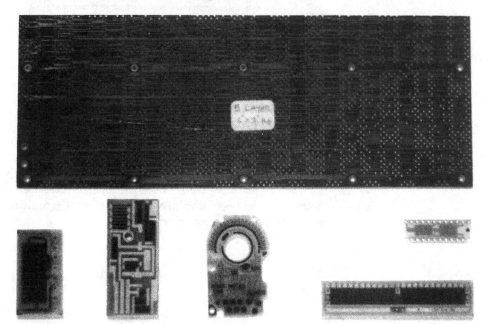

Ultra-Large-Scale Integration

While not by any means unique to SMT, ULSI devices, often having very high pin counts, are very often available only in SMCs or in a choice of SMC packaging or PGA. The PGA is often too expensive and too difficult to deal with in board assembly and repair. Thus, SMT, TAB, flip-chip technology, and MCMs play large roles in facilitating the move to ever higher levels of silicon integration and the resultant enormous real-estate savings.

Multi-Chip Modules

It is beyond the scope of this book to offer a treatise on MCM design. There are numerous works dealing with MCM developments, and those developments are proceeding at such a pace that any book covering them would soon be out of date. We will deal with the rudiments of MCMs in chapter 9. Suffice to say here that they provide the system engineer with a powerful set of miniaturization tools.

6.1.4 SMT Patches on Through-Hole Boards

Isolated SMCs

Occasionally, a solid IMC design gets sabotaged by one or two components which are available only in SMT packaging. To the SMT purist, this smacks of heresy, but there are sockets that adapt SMCs to 2.54-mm-center (100-mil) IMC mounting. They may be useful in downsizing an IMC board by replacing a handful of chips with custom silicon. Often, high-pin-count custom silicon is available only in SMC packaging.

Figure 6-9 shows just such a solution used by VLSI Technology, Inc., of Phoenix, Arizona, in implementing an IBM PC AT™ motherboard using a highly integrated chip set. This 5-chip set cuts the PC AT motherboard count from 110 ICs to 16 (excluding memory).

Figure 6-9 Sockets provide through-hole compatibility for SMCs. *(Courtesy VLSI Technology.)*

For circuits where timing is not critical, sockets may provide a convenient breadboarding tool for SMT implementations, as well. Sockets on 2.54-mm (0.100-inch) centers adapt SMCs to 100-mil proto boards. However, the additional parasitics introduced make this approach useless in high-speed and timing-sensitive designs.

High-Performance Sections

Surface-mount performance is particularly attractive in designing high-speed circuitry. Where only a portion of a board will run at high speeds, an SIP or DIP SMT module allows taking advantage of surface-mount performance where it's needed, while sticking primarily to through-hole technology.

Figure 6-10 SMT DIP telecommunications module. *(Courtesy Mitel Semiconductor.)*

SMT Daughter Boards

Some SMT modules with SIP and DIP lead frames were shown in Figure 6-2. These circuits were used as standardized functional modules in otherwise through-hole board manufacturing. The same approach may be used to solve other design challenges. Figure 6-10 shows a DIP module built on a ceramic substrate. The ceramic substrate matches the TCE of the chip carrier, eliminating thermal cycling concerns. The ceramic board also allows high-performance resistors to be printed directly on the circuit and laser trimmed to close tolerances. The circuit is a telecommunications module, made by Mitel Semiconductor, Ottawa, Ontario, Canada.

6.2 Board-Level Interconnects

We covered packaging for miniaturization, reduction of board count, and convenience in dealing with manufacturing and design challenges above. Now, let's turn to the interconnection of SMT boards to one another and, also, to the outside world. We will look at those areas of board and system interconnects that are unique to SMT. We will not attempt to cover through-hole connectors and interconnect strategies, as these topics are well documented in other texts. However, we'll start our discussion with a note to not rule out through-hole connectors because of surface mounting.

6.2.1 *SMT and IMC Connectors Compared*

One of the first design decisions regarding interconnections is answering the IMC vs. SMC question. SMT processing requires that connectors which are surface mounted must differ from the IMC versions in several ways. We'll explore these variations now. They are instructive in both the SMC/IMC selection process and the design of SMT board interconnects.

Soldering Approach

Flow soldering IMC connectors is no particular problem. While the connector solder tails pass through molten solder on the board's solder side, the delicate portion of the connector is safely out of harms way on the benign component side of the board. Not so in surface mount. SMT wave soldering involves passing components directly through the solder wave. As previously mentioned, many components cannot survive such treatment. Should a connector need to be on the solder side, it would fall among such components. Therefore, SMT connectors are usually designed for reflow attachment. Or, through-hole connectors are used on the component side of flow-soldered SMT assemblies.

Reflow soldering subjects connectors to temperatures beyond the safe limits for many housing dielectric materials. Therefore, connector manufacturers have turned to high-temperature plastics. Be cautioned, however, that some high-temperature plastics are not as mechanically rugged as standard plastic housing materials. Also, connector-material choices will have a direct bearing on the choice of various reflow-soldering techniques and temperature profiles.

Mechanical Integrity

IMC-connector solder tails are attached firmly to boards by solder fillets on both sides of through-plated holes. The solder joints of SMT connectors afford significantly less holding power. Estimates vary with the style of solder connection, but IMC connectors typically provide 2.5 to 20 times more holding power per pin.[8]

Even so, SMT-connector solder joints have plenty of holding power for a few cycles of mating and unmating. However, where connectors will be cycled often, some additional hold-down is generally in order when surface mounting connectors. Hold-down methods include bolting or riveting, heat stake rivets, snap-in detent rivets, and press-fit bosses. All involve special assembly considerations and accurate hole alignment in the board for the mechanical attachment devices.

Zero- and low-insertion-force connectors are quite popular in SMT because they keep insertion forces well within the range surface-mount solder joints can accommodate.

Figure 6-11
Reflow-mountable through-hole connectors.
(Courtesy 3M, Electronic Products Div., Austin, TX.)

(A) 26-pin vertical plug.

(B) 12-pin male header.

Another solution is to use connectors that are provided with solder preforms shaped like donuts on their solder tails. This allows through-the-hole soldering to be accomplished in a reflow operation. A variety of such connectors is shown in the two photos of Figure 6-11.

6.2.2 SMT or IMC Connectors?

As we mentioned above, a major board-level interconnect decision is whether or not to use surface-mount connectors. This must be reconciled early in the design, as certain connector styles may rule out specific manufacturing styles and process selections, which, in turn, will determine the design rules. There are a growing number and variety of SMT connectors available, and some provide very dense I/O. However, many are not compatible with standard placement equipment. Some require dedicated robots or hand assembly. Also, many SMT connectors actually take more board real estate than their through-hole counterparts. And some SMT connectors carry a price premium. Thus, SMT connectors may be a poor choice on a board that will require flow soldering for other IMCs. However, if selecting an SMT connector will allow an exclusive reflow soldering of the assembly, the benefit in process simplification probably will offset these negatives.

6.3 Board Flexure

Flexure of boards produces stresses on solder joints, land-to-board bonds, device terminations, board lamination bonds, device bodies, etc. In short, it stresses the entire assembly. Flexure may be produced by gravity, shock loads, uneven heating, uneven thermal expansion, board handling, and other mechanically imposed loads.

Board flexure is not exclusively an SMT issue. IMC boards give concern, also, in severe environments or situations where they must be plugged in and unplugged often. But the relatively weak SMC solder joint, coupled with the lack of any mechanical attachment of SMCs, makes flexural integrity particularly important to SMT assembly reliability. In the following, we'll discuss methods of improving the flexural resistance of SMT boards.

Flexure is an acute disease when it produces solder-joint stresses that exceed the safe limits of the joints on an assembly. The acute form is easily recognized; components pop off the board like tiddledywinks as the board bends. The more common chronic form of the disease slowly destroys joints by work hardening and crystallizing the solder in them. Eventually, such joints fail, just as a coat hanger fails and breaks when we flex it back and forth to make a tool for opening a locked car.

Where calculated bending stress levels approach limits of concern for a given solder material and desired B-10 life, the following directions should be employed to ensure solder-joint integrity.

6.3.1 Board Stiffeners

Let's start with the simple and direct approach of stiffening the PW board. When a completed design proves to be too flimsy for its intended task, this may be the only avenue open—short of scrapping the design and starting over.

External to the Substrate

Stiffening hardware may be added directly to the board as shown in Figure 6-12, or the card carrier may be stiffened to resist board flexure. If the flexure concern arises from card insertion and removal, the board-stiffener route should be selected, as its protection travels with the assembly. Before reinforcement design may be undertaken, the electrical and circuit-board layout engineers must give the mechanical engineer an accurate description of the board material, its thickness, the weight distribution of components, and the handling stresses envisioned. Given this data, the modeling of external stiffeners becomes a straightforward mechanical engineering task.[9] Reference 9 lists stress formulas for stiffener members acting as simple beams, cantilevers, etc.

Figure 6-12
Military avionics card stiffened for safe handling.
(Courtesy Martin Marietta.)

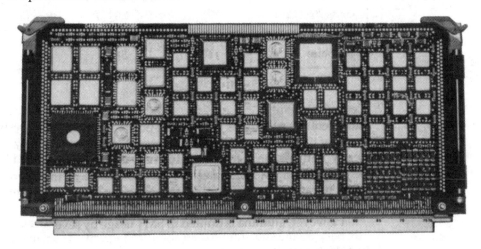

Within the Substrate

External stiffening is perhaps the easiest approach to beefing up boards for flexural stress. But what if there's no room for external ribs? We may still be able to save an existing design by adding a metal core or cores to replace the polymer ground/power layers, or to simply stiffen and restrain the board without any electrical function. Copper/Moly/Copper is particularly effective in strength-to-weight ratio, but any metal core will add considerable structural integrity, and will also markedly improve the thermal characteristics.

There is a cost premium associated with metal-core boards, and PWB vendors are limited. However, the alternative of redesigning an assembly because of flexural concerns may be a larger negative than the cost and availability of the PWBs. Where this is the case, as with low-manufacturing-volume projects, metal cores may well satisfy the need for board stiffness.

6.3.2 *Substrate Materials*

When a flexural problem is identified prior to completion of design, more avenues are open to build structural integrity into an assembly. One method is the use of a high-rigidity substrate. Table 6-2 presents the mechanical data for some common substrate materials.

In some instances, trading polymer-based devices for more rigid substrates may satisfy flexure concerns in existing designs, as well. However, selection of relatively rigid substrate materials will often have a significant impact on design rules. A circuit designed for FR-4 may not translate directly to rigid substrate materials.

Table 6-2. Mechanical Properties of Substrate Materials

| Material | Mechanical Properties | | | |
	Flexural Strength in PSI $\times 10^3$	Modulus Elasticity in PSI $\times 10^6$	Tensile Strength (ft-lb/in)	Izod Impact Notch Test (PSI $\times 10^6$)
Epoxy/Glass	G-10 45–50	2.5–3.0	5.5–7.5	2.5
Alumina	45–50	44	Brittle	25–30
Beryllia	39–49	50	Brittle	22
Porcelain/Steel		5–8P./25–30S.		
Glass 7059 Corning	1	9.8	Brittle	1
Quartz Fuse	7	10.5	Brittle	7
Sapphire	60	51	Brittle	60

Ceramics

Most ceramic substrates are of high-purity alumina (polycrystalline Al_2O_3, with traces of metal glasses to tailor properties). Alumina is widely available, provides nearly 15 times the thermal conductivity of epoxy/glass, and is very resistant to bending. Alumina substrates are the workhorse of the hybrid-circuit industry. Their refractory nature permits use of stable high-firing-temperature thick-film inks critical to many hybrid applications.

However, alumina is limited in SMT usage by several factors. Alumina generally restricts the substrate size to about 230 sq. cm. (36 sq. in.), and the per-square-inch costs begin to climb with substrates above 100 sq. cm. (16 sq. in.). Size for size, small alumina substrates are about 3 times the price of FR-4 epoxy/glass. Alumina is brittle, must be handled with reasonable care, and must be protected from high impact forces while in service. And the dielectric constant of alumina is roughly twice that of epoxy/glass, making alumina unsuitable for some high-frequency applications.

Beryllia ceramics (polycrystalline BeO, with traces of metal glasses to tailor properties) share many of the positive properties of alumina. They are chosen over alumina for applications where their high thermal conductivity (nearly 100 times greater than epoxy/glass) is needed to manage thermal dissipation. The lower dielectric constant of beryllia coupled with its thermal conductivity makes it well suited for microwave circuits with high dissipations.

Beryllia substrates also share alumina's negatives. They are even more costly than alumina. And, unlike alumina, beryllia is highly toxic when inhaled in dust or fume form. Therefore, beryllia is used primarily where its unique properties are called for.

Ceramic-Coated Metal Core

Coated metal-core substrates offer a number of attractive features. They may be fabricated in any size and virtually any shape. Standard metalworking operations, such as bending, forming, and punching may be done prior to coating. Thus, not only stiffening ribs, but circuit and housing mechanical features may be built directly into the substrate. In some cases, metal-core boards form both substrate and housing for an electronic system.

Metal-core boards are shock resistant, strong, moderate in cost, and provide a built-in ground plane. By selecting a metal core with a TCE close to that of ceramic chip carriers, thermal-expansion mismatches between substrate and component may be minimized.

Where in situ components will be added by hybrid thick-film processes, inks must be of the low-firing-temperature varieties. Tooling costs make metal-core boards unattractive for short run and low-volume products.

Glass

Glass substrates are generally selected for applications such as displays, where their transparency and optical clarity are needed. Glass is highly available, can be produced in any size, is low in cost, and rigid. Glass can be produced with very smooth surface finishes, low camber, and minimal warpage, and in qualities useful in thick- or thin-film hybrid processing.

However, the low thermal conductivity of glass (alumina is 30 to 50 times more thermally conductive) rules out the use of glass for high-dissipation circuits. Like coated metal-core boards, glass requires low-firing-temperature inks for in situ fabrication of thick-film components. And, repeatable, reliable, thick-film processes are a challenge on glass.[10]

Quartz

Quartz substrates are primarily made by heating silica to form fused silica quartz. The material is readily available, and may be fabricated into boards of large sizes. The low K factor of quartz makes it of interest in high-frequency applications.

The limitations of quartz as a substrate restrict its use, however, even in the microwave area. The thermal conductivity of quartz, while higher than glass, is still just one-twentieth that of alumina and less than one-hundredth that of beryllia. Thus, quartz is limited to low-power-dissipation applications. Quartz is also highly brittle. Care must be used in its handling, and it must be protected from shock and impact while in service. Single crystal quartz, as opposed to fused quartz, is used in *standing acoustic wave (SAW)* devices, where its ordered crystalline structure justifies its high cost.[11]

Sapphire

Sapphire (monocrystalline Al_2O_3) is produced in boules similar to silicon for electronic applications. Substrates sliced from a boule are limited in size by the boule diameter. An alternative process, *edge-defined film-fed growth (EFG™)* allows the fabrication of substrates of about 10 cm × 15 cm (4 inch × 6 inch) as of this writing, and larger sizes are planned. Sapphire is used where its radiation hardness is needed, and in hybrid microwave devices where alumina's camber and surface finish are inadequate for element fabrication. Applications also include hybrids, using silicon-on-sapphire devices. Sapphire's relatively low thermal conductivity (about one-tenth that of alumina), high K factor, and very high cost restricts its usage.

6.3.3 *Lead Compliance*

In many cases, with attention to device package selection, designers can accommodate anticipated board flexure with compliant leads on the components. This strategy is particularly viable where flexural excursions are small (not over 3.18

mm/0.125 inch per 25.4-cm/10-inch linear board surface, bidirectional), and components are not large (not larger than 44-pin chip carriers). Other methods, discussed herein, are called for where more severe bending will be encountered, or in moderate flexure applications where reflowed components are particularly large (high-pin-count devices and long connectors, for example).

6.3.4 Long Connectors

Long connectors, reaching across more than 38.1 mm (1.5 inches) of the circuit board, are a particular challenge in high-flexure applications. Several manufacturers offer floating-lead designs. Floating leads supply extra compliance to absorb flexure and thermal-expansion stresses without placing undue strain on the connector solder joints.

6.4 Environmental Concerns and Packaging

SMT brings with it a heightened packaging concern about environmental stresses. We will cover the ways that SMT differs from conventional circuit approaches herein. We will not review any standard engineering practices that apply equally to surface mount and inserted mounting. This SMT bias does not, however, exempt surface mounting from basic electrical engineering practices regarding packaging for survival in the intended environment.

SMT calls for special environmental-engineering attention to the problems of thermal cycling, high vibration, G forces and shock loads, connector cycling, frequent card insertion/withdrawal, radiation resistance, and operation in threatening ambient conditions.

6.4.1 Frequent Temperature Cycling

We discussed temperature cycling and engineering solutions to the stresses it produces in Chapter 5, Section 5.7. At this point, let's review why surface-mount technology differs from through-hole technology in thermal cycling, and develop a yardstick to determine when thermal cycling is likely to be a problem. Then, we'll briefly review engineering solutions covered in detail in the previous chapter.

Differences in IMC and SMT Thermal-Cycling Resistance
The entrance of surface-mounting technology brings, along with its many benefits, concerns about long-term reliability in thermal cycling. These concerns arise from two significant differences between SMT and IMC.

First, SMCs generally have no mechanical connection to the board outside of the solder joint. The SMT solder joint must furnish both mechanical rigidity and electrical contact with the circuit. IMCs have an inherent mechanical attachment, due to leads being inserted through holes. Clenched leads provide an even more robust connection for through-hole components.

Second, SMT differs in the compliance furnished by leads. The long DIP lead is quite compliant in X- and Y-axis motion relative to the board. The shorter PLCC and SOIC leads generally provide less of a flexural element than DIP leads. In the worst cases, the leadless SMCs, there is no stress-relieving flexural element at all.

For these two reasons, designs which were quite reliable with through-hole components may come a cropper when we just switch the IMCs for SMC parts without considering TCE engineering.

A Yardstick for Estimating Thermal-Cycling Concerns

This raises the question, "When should we become concerned about thermal cycling?" Several rules of thumb can help provide a yardstick by which to measure designs for potential TCE mismatch concerns.

Table 6-3. A Yardstick to Gauge TCE Concern

Environment Severity Rating	Application Characteristics		
	Component	*Substrate*	*Concern Level*
Moderate	Leadless Ceramic, L* = ≤0.500 inch. SO and SOLIC, PLCC, <84 Pins. Leaded Carriers (all types).	Polymer	Very Low
	Leadless Carriers, L = ≤1.250 inch.	Match TCE	Very Low
	Leadless Carriers, L = ≥1.250 inch.	Any	Add leads to high-dissipation component. Match substrate TCE for low dissipation.
Rigorous	Leadless Ceramic, L* = ≤0.500 inch. SO and SOLIC, PLCC, ≤68 Pins. Leaded Carriers (L = ≤1.000 inch).	Polymer	Low
	Leadless Carriers, L = ≤1.000 inch.	Match TCE	Low
	Leadless Carriers, L = ≥1.000 inch.	Any	Add leads to high-dissipation component. Match substrate TCE for low dissipation.
Harsh	Leadless Ceramic, L* = ≤0.375 inch. SO and SOLIC, PLCC, ≤44 Pins. Leaded Carriers (L = ≤0.750 inch).	Polymer	Concern rises as component size increases.
	Leadless Carriers, L = ≤0.750 inch.	Match TCE	Add leads to high-dissipation component.
	Leadless Carriers, L = >0.750 inch.	Any	Add leads to high-dissipation component. Match substrate TCE for low dissipation.
Extreme	Leadless Ceramic, L* = ≤0.250 inch. SO and SOLIC, PLCC, ≤28 Pins. Leaded Carriers (L = ≤0.500 inch).	Polymer	Concern rises as component size increases.
	Leadless Carriers, L = ≤0.500 inch.	Match TCE	Add leads to high-dissipation component.
	Leadless Carriers, L = ≥0.500 inch.	Any	Add leads to high-dissipation component. Match substrate TCE for low dissipation.

* L = The longest component dimension, length; typical throughout the table.

We can start by drawing lines between various environments, and labeling each by degree of temperature-cycling severity. Office products, consumer appliances, and some other electronics enjoy the luxury of a "moderate thermal environment," by the right of coexisting with temperature-sensitive humans. High-reliability industrial products are generally designed to survive 400 cycles from 0 °C to

+100 °C. While this is less than brutal in temperature swings, it wouldn't make for a cozy office, so we'll call this a "rigorous thermal environment." Military products complying with MIL-STD-202, Method 107F Condition B, and industrial products intended for harsh temperatures, are generally designed to endure 400 cycles from -65 °C to +125 °C. We'll call this a "harsh thermal environment." For long-term survival in truly punishing environments, SMT boards may be designed to survive 1000 cycles from -65 °C to +125 °C. We'll call this an "extreme thermal environment." Using these environmental severity definitions, Table 6-3 shows several concern levels, as determined by application variables in a given thermal-cycling severity. These data are not intended as a categorical statement that there will or won't be difficulties, but they should serve as a guide to flag engineering concern. As such, the statements of Table 6-3 tend toward the conservative.

Where thermal cycling, either from the environment or dissipation within a device, produces thermal expansion differentials between the substrate and component, stresses may damage the solder joints. Later, we'll review the ways covered in Chapters 2 and 5 for bringing stresses back within safe limits. Exactly what is a safe limit is determined by the solder alloy used. Consult your solder vendor for data.

Compliant Leaded Components

Compliant leads on components flex to absorb thermal-expansion-induced stresses, protecting the solder joints (see the cautionary notes on compliancy of J leads under components in Chapter 8, Section 8.2.1). For leadless components, it is possible to attach leads. There are services available for postlead attach, or it may be done internally. Where post leading is not desirable, sockets may add compliance. Special flexural columns are also available for solder attachment between the board and leadless component. Properly engineered, any of these methods can accommodate typically encountered TCE mismatch stress levels.

Matching Component and Substrate TCEs

Thermally generated solder stresses come from two sources. Mismatches in TCE of the substrate and component materials produce stresses when temperature changes from the environment or circuit dissipation act on the mismatched pair. Due to differential temperatures when under power, circuit dissipation also produces solder stress, even in situations where the component and board TCE match. While tailoring substrate TCE to that of components will not eliminate internal temperature-differential stresses, these forces are low enough to be of no concern for moderate-sized IC packages and typical dissipation levels. As long as very large packages and/or very high internal dissipations are not involved, matching or approximating the component and substrate TCEs will keep stresses within safe limits.

6.4.2   Vibration

SMCs are generally more vibration tolerant than equivalent IMCs. At first blush, this is somewhat surprising. After all, surface mounting provides no mechanical attachment to the PWB except through the solder joints. And the SMC's leads are narrower and thinner than the DIP's. But, a 16-pin SOIC weighs only 130 milligrams, while a 16-pin DIP weighs 1200. The differential in mass translates to far less vibration-induced stress in the leads and solder joints for the lighter surface-mount device. Table 6-4 shows weights of typical SMC packages for vibrational and shock force analysis.

Table 6-4. Approximate Weights of Various SMT Packages

Component	Weight (mg)	Component	Weight (mg)
SOIC-14	100	PLCC-68	4280
SOIC-16	130	CLLCC-20	470
SOLIC-20	480	CLLCC-28	750
SOLIC-28	530	CLLCC-44	1750
PLCC-28	900	CLLCC-52	2290
PLCC-44	1850	CLLCC-68	3400
PLCC-52	2500	PQFP-132	4100

Figure 6-13 Standard SMC lead dimensions.

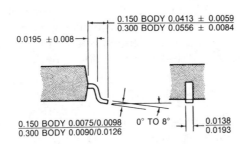

(A) SO and SOL IC (JEDEC).

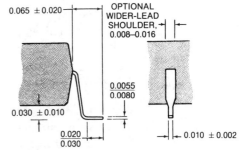

(B) PQFP, 25-mil centers (JEDEC MO069).

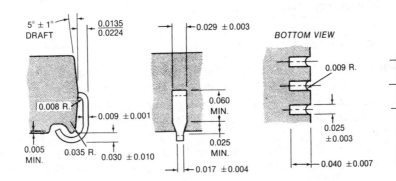

(C) PLCC, 50-mil centers (JEDEC).

(D) CLLCC, 50-mil centers (JEDEC MS004).

Putting lighter weight to a practical test, one major off-road equipment manufacturer was astonished when a destructive vibration test appeared to shear the leads of a 40-pin DIP just above the solder fillets while an adjacent PLCC-44 remained unscathed. Mechanical modeling showed that the DIP leads were actually deforming into a waving "S" curve under vibration. The natural frequency of the heavy DIP was near resonant with the vibrations produced by a large reciprocating engine. The DIP's leads soon work hardened, and broke. The low mass and short protected leads of the PLCC left it on board and rattling away unharmed. Figure 6-13 shows dimensions of standard SMC leads for use in stress analysis calculation. Note that the lead material may vary according to vendor. Consult component manufacturers for verification of the physical properties of the lead-frame material. With the dimensions of Figure 6-13, the component weight from Table 6-4, and data on solder material properties, calculation of solder-joint strength is allowed. See Reference 12 for formulas for various joint configurations, and application factors affecting the calculations.

Natural Frequency of the Assembly

The natural frequency of an assembly may be tuned away from resonance or harmonics with environmental vibration. This is most often done by adding weight to lower the assembly's resonant point. The addition of a metal core can serve this purpose while simultaneously adding stiffness and improving the thermal properties of the assembly.

Mechanical Integrity Considerations

Since the solder joints of SMCs must supply the total connection to the PWB, soldering quality and repeatability are of particular concern in mechanically threatening environments. Special consideration should be given to process controls, nondestructive testing, and destructive sampling of the solder bonds.

6.4.3   High G Forces and Shock Loads

We've just discussed, above, how their low mass and low profile make SMCs more vibration resistant than equivalent-type IMCs. These same properties work to SMT's benefit in high G-force and shock-loading environments. And, the same engineering solutions work to ensure survival in high acceleration or impact loadings. Table 6-4 and Figure 6-13 provide data that can be used to calculate shock and G-force resistance.

6.4.4   Connector Cycling

In Section 6.2, we mentioned the frequent cycling of connectors as a major factor in connector selection. The repeated operation of connectors can easily break SMT solder connections, where the stronger through-hole bonds of inserted connectors would comfortably survive. SMT connectors are particularly at risk when they are large, have a high pin count, are without mechanical attachment, or are not of low- or zero-insertion-force design. See Section 6.2 and Chapter 2, Section 3, for details on engineering solutions to maintaining connector integrity in high cycling environments.

6.4.5   Frequent Card Handling

The frequent insertion and removal of SMT cards raises concerns about connector-bonding and handling-induced bending of the card. Threats to the SMT assembly increase as the component and card size increases, and also as the card stiffness decreases. See Section 6.3 for a full discussion of card flexure.

6.4.6   Radiation Resistance

Harsh radiation environments are encountered in space, in nuclear reactor design, and in radiation equipment. While we pray the feature is never needed, we design military electronics to survive such radioactivity, as well.

It is well-known and understood that radiation can destroy or impair the function of semiconductors. Devices that depend on junction isolation rather than on a dielectric, or the silicon-on-insulator isolation of on-chip elements, are particularly at risk.[13] So Step One in engineering for high-radiation environments is the selection of components with rad-hard silicon inside. Figure 6-14 shows the radiation resistance of various silicon semiconductors, as compared to carbon units.

Figure 6-14 Radiation resistance of various technologies.[14, 15]

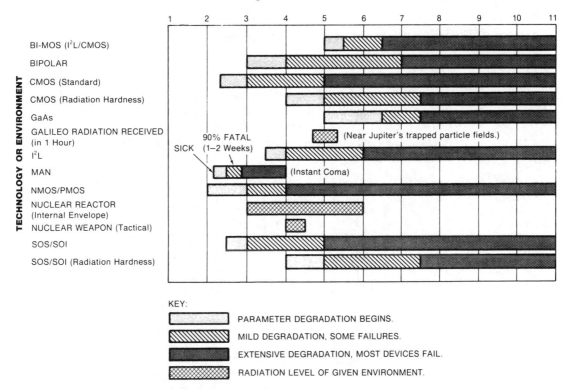

KEY:

☐	PARAMETER DEGRADATION BEGINS.
▨	MILD DEGRADATION, SOME FAILURES.
■	EXTENSIVE DEGRADATION, MOST DEVICES FAIL.
▦	RADIATION LEVEL OF GIVEN ENVIRONMENT.

While selection of rad-hard silicon is vital, it is not the totality of designing for radioactivity survival. Shielding, grounding, and decoupling must all be designed to accommodate the worst-case rad dosages that the system may see. The system should be designed with error correction. And, in cases where periodic high-intensity radiation is a possibility, the design should provide for a controlled shutdown of the system to protect sensitive areas.[16]

Radiation in the form of high-energy cosmic rays, or electrons traveling in the Earth's magnetic field, is typical of the space environment. Such radioactive particles can induce ionizing radiation in circuits, momentarily imparting enough energy to change the state of a bistable element. Unless special rad-hard technology is used, CMOS devices are particularly sensitive to this effect, known as a "single event upset error." Such events may result in a momentary or a permanent failure.[17]

Radioactive-particle bombardment can also displace the atomic structure of semiconductor and electronic-material latices. "Displacement damage," as this defect is called, can temporarily or permanently alter semiconductor performance.[18]

Large dosages of X rays produce electromagnetic pulses (EMPs) in system transmission lines and interconnections. Unchecked, EMP can fuse semiconductor metallizations and can permanently damage or destroy device operation.[19]

Solar ultraviolet radiation is also capable of degrading electronic assemblies. An example is found in the electric power meter which resides (in a glass dome) on the back of your house. Power-meter circuit designers derate the solder tensile strength of the circuits by 50% for the long-term effects of ultraviolet radiation.

For additional material on "designing for radiation resistance," please see References 13 through 19.

6.4.7 *High Operating Temperatures*

A high-temperature operating environment complicates thermal management. Elevated ambient temperatures, coupled with temperature rises from device dissipation, can also significantly weaken solder joints. Figure 6-15 shows the fall off in strength of 50/50 tin/lead solder compared to the higher-melting pure tin. Data on other alloys are not widely available, but solder-paste manufacturers suggest any near eutectic or eutectic tin/lead alloy would closely parallel the 50/50 tin/lead curve.

Figure 6-15
Solder strength versus operating temperature.[20]

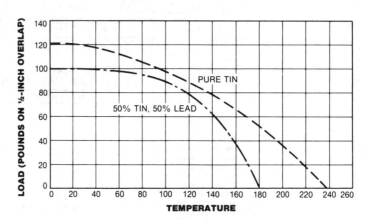

Tall components are of concern where they might impede the flow of cooling air around SMCs mounted close to the board. Locate high-dissipation components where they will receive a full flow of cooling air.

6.4.8 *Humidity*

High humidity raises concerns about surface conductance between the close leads of SMCs, and dendritic growth, corrosion, and penetration of moisture into devices. Use of pads-only surface layers reduces the PWB concerns, but it does not eliminate them. And pads-only layers provide no protection above the board. For severe conditions, conformal coating is recommended. Table 5-4 in Chapter 5 lists the performance characteristics for various coating materials.

When designing for humid environments, particular care should be given to metals combined in the assembly. Metals with a significant difference in electromotive potential between themselves, or the solder joining them, are prone to corrosion. The electromotive potential of common electronic materials is shown in Table 6-5.

Table 6-5. Properties of Common Electronics Metals

Metal	Property			
	Electromotive Potential (V)	*Resistivity (@ 20 °C μΩ/cm)*	*TCE PPM/°C*	*Tinsel Mod. (lb/in² × 10⁶)*
Aluminum	+1.67	11.5	25.74	10.0
Antimony	-0.10	39.0*	11.34	11.3
Beryllium	+1.70	106.8*	12.42	4.6
Bismuth	-0.20	119.0*	13.32	4.6
Cadmium	+0.40	3.43*	29.88	3.0
Copper	-0.34	1.726§	16.38	16.0
Gold	-1.68	2.19*	14.40	12.0
Indium	+0.38	8.37*	32.40	11.0
Lead	+0.13	20.65	28.80	2.6
Nickel	+0.25	6.84	13.68	30.0
Palladium	-0.82	10.8	11.88	17.0
Platinum	-0.86	9.83*	7.74	21.0
Silver	-0.80	1.59	18.90	11.0
Stainless Steel	-0.09	74.0	16.56	29.0
Steel	-0.58	18.0	12.06	Dependent on Alloy
Tin	+0.14	11.5	23.40	6.0

* At 0 °C.
§ At 23 °C.

6.4.9 Reactive Atmospheres

Some materials present in atmospheric environments are particularly threatening to electronic assemblies. Threats may range from the muddy, salt-spray-filled air around shipboard and military gear to the highly acid atmospheres in some chemical processing plants. Such atmospheres are particularly difficult to control when high humidity and high temperatures are also part of the potential environment. Atmospheric contaminants are dealt with by the choice of materials, conformal coatings, and protective hermetic housings.

6.5 References

1. Ball, Michael, "SMT: A Demanding Master," *EDN News*, July 16, 1987, pg. 62.
2. Texas Instruments Incorporated, "SIP Memory Module Application Notes."
3. Weiler, Peter M., "How SMT Copes with High Lead Counts," *Electronic Engineering Times*, October 19, 1987, pg. T20.

4. *Hybrid Microcircuit Design Guide*, ISHM-1402/IPC-H-855, ISHM, Reston, VA, and IPC, Evanston, IL, 1982.

5. Harper, Charles A., Editor, *Handbook of Wiring, Cabling, and Interconnecting for Electronics*, McGraw-Hill Book Co., New York, NY, 1972, pg. 8-18.

6. Ibid.

7. Mahler, Bruce, and Schroeder, Paul, "Planar Resistor Technology for High-Speed Multilayer Boards," *Electronic Packaging and Production*, January 1986.

8. Mosley, J. D., "Surface-Mount Connectors," *EDN*, October 1, 1987, pg. 143.

9. Oberg, Erik, et al, Ed. Henry H. Ryffel, *Machinery's Handbook, 22nd Edition*, Industrial Press, New York, NY, 1984.

10. Op. cit., ref. 4, pg. 55.

11. Ibid.

12. Manko, H. H., *Solders and Soldering*, McGraw-Hill Book Co., New York, NY, 1979.

13. *Rad-Hard/Hi-Rel CICD Data Book*, Harris Custom Integrated Circuit Div., Palm Bay, FL, 1987, pg. 1-8.

14. Ibid.

15. Gauthier, Michael K., and Dantas, Armando Roberto V., "Radiation-Hard Analog-to-Digital Converters for Space and Strategic Applications," *JPL Publication 85-84*, NASA Jet Propulsion Laboratory, Pasadena, CA, 1985, pg. 5-3.

16. Pearson, Bob, "Designing Low-Power, Rad-Hardened Satellite Systems," *EDN*, August 21, 1986.

17. Andrews, J. L., Schroeder, J. E., Gingerich, B. L., Kolanski, W. A., Koga, R., and Diehl, S. E., "Single Event Upset Error Immune CMOS RAM," *IEEE Transactions on Nuclear Science*, Vol. NS-29, December 1982, pgs. 2040–2043.

18. Srour, J., "Basic Mechanisms of Radiation Effects on Electronic Materials, Devices, and Integrated Circuits," *Proceedings of the IEEE Annual Conference on Nuclear and Space Radiation Effects*, 1982.

19. Ibid.

20. Op. cit., ref. 9, pg. 88.

Figure 7 SMT puts new emphasis on engineering and controlling processes for quality, reliability, and yield. *(Photographs courtesy of Texas Instruments Incorporated.)*

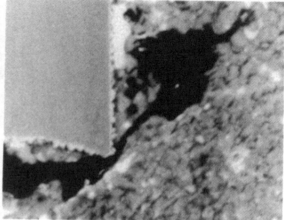

(A) Crack through a void at corner of a leadless chip carrier (100X).

(B) Partially fractured joint of an I-leaded part.

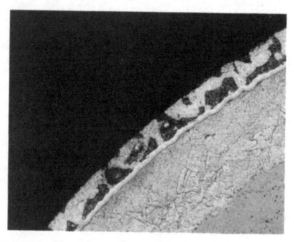

(C) Duplex layer of Cu₃Sn Epsilon phase intermetallic on a copper-bearing base metal with Cu₆Sn₅ above (1000X).

7 Quality Assurance in SMT

Surface mounting is not just another way to slap parts into assemblies. This is true in all facets of SMT, but nowhere is it more glaringly obvious than in quality control and quality assurance (QC/QA). What has always been an interesting slogan, "Design for Test," has become a way of thinking for companies trying to achieve good SMT manufacturing yields. As a result, phenomena we seldom gave any thought to ten years ago are now of grave importance. For instance, solder-joint integrity and intermetallic growth, as indicated in Figure 7, are now being heavily researched. These photos of processes that are poorly engineered or controlled will serve to launch this chapter's topic, *Quality Assurance for SMT*.

Figure 7A shows a crack propagated through a void at the corner of a leadless chip carrier. Poor solderability is also evident in the dewetting seen under the carrier. (Photo taken at 100X power.)

Figure 7B illustrates a partially fractured joint of an I-leaded part. The micro-section has been etched to reveal an intermetallic layer. The crack is along the boundary of intermetallic and through voids spawned by the poor solderability it caused.

Taken at 1000X power, Figure 7C shows a duplex layer of Cu_3Sn Epsilon phase intermetallic on a copper-bearing base metal, with Cu_6Sn_5 placed above. This combination is damaging to the solderability of solder-coated copper-bearing leads and to the reliability of solder joints. The two-layer geometry tends to passivate, becoming very brittle.

One of the great prophets of *statistical quality control (SQC)*, W. Edwards Demming, has noted that both Japanese and American companies implementing SQC have repeatedly proven its value as follows:

- a tenfold to hundredfold (or more) reduction in defects,
- a tenfold improvement in manufacturing cycle time,
- an inventory reduced to one half the previous levels,
- manufacturing floor space is cut by one half.[1]

It is not our intention to offer a course on SQC herein. We do want to note, however, how vital SQC is to surface-mount manufacturing. Many of the tricks ("Band-Aid" cures) that worked to cover sloppy process controls in through-hole assembly will result in enormous disasters on an SMT line. Successful SMT usage demands that you have your SMT processes under control.

In the following discussion, we'll cover areas of specific interest to SMT quality and reliability. The SMT team should include "SQC for surface mount" as part of the assignment to integrate surface-mount technology into a manufacturing organization.

7.1 Component Reliability and Quality Assurance

How can we assure SMT-component quality and reliability. First, we must understand and agree on what these terms mean. Such agreement doesn't come without effort. To illustrate, let's look at a perfect example where a confusion with terms caused trouble. A major corporation called in a high-power QC consultant to find why its quality program wasn't yielding the desired results. The consultant found, when he asked various people "What is Quality?", that he got a different answer from each individual surveyed. One thought it was conformance to specifications. Another saw it as defect-free manufacturing. Some believed it was long-term operation without a failure. No wonder the quality program was disappointing when the people implementing the program hadn't come to a basic agreement on the meaning of its most critical word.[2]

Being forewarned, let's begin by stating what this text means by quality and what's meant when we say reliability. For our purposes, a quality electronic product is one which, immediately after manufacture, works per specifications when it's turned on. If it keeps on working per specifications for as long as intended, then it's reliable.

Given these definitions, let's look now at how SMCs stand up in quality and reliability when compared to through-hole devices.

7.1.1 *How SMCs Compare with DIP Quality*

The drive to push component quality defect-density into the low PPM range was fueled, in part, by SMT developments, such as tape-reel packaging. End users who had been content to buy 1100 ICs to get 1000 live ones could no longer afford such wasteful practices. Components in tape can't be cycled through a receiving test conveniently. And the cost of correcting failures caused by 10% defective components would be enormous. Typically, costs of correcting defects escalate about tenfold, as shown in Table 7-1, with each successive step in the manufacturing process.

To see how the suppliers have met this challenge, we telephone surveyed the major semiconductor manufacturers regarding variances in device quality in DIP vs. SMC packages. The consensus, as of this writing, is that, for electrical quality of the major logic and memory families, there is no statistically significant difference between the outgoing quality levels of PLCC or SOIC and DIP packages. Note that since SMC visual/mechanical standards must be higher than DIPs for coplanarity's sake, we considered only electrical quality, where the standards for acceptance are the same for both styles of packaging. If visual/mechanical qualities are included, there is a slight bias in favor of DIPs.

Table 7-1. Cost of Correcting Component Defect Vs. Detection Point

COST IN DOLLARS	COMPONENT MANUFACTURING (Final Test)	END USER			
		Incoming Inspection	Assembly Line	Final Test	Field Failure
1000					$1000.00
900					
800					
700					
600					
500					
400					
300					
200				$100.00	
100	$0.10	$1.00	$10.00		

7.1.2 *How SMCs Compare with DIP Reliability*

Since SMCs are so much smaller than DIPs, one of the first questions most engineers raise is *reliability*. Doesn't it make sense that a major reduction in plastic encapsulant material spells less protection for the semiconductor die? The answer, as simple logic would lead us to guess, is a qualified yes. Early SMC package reliability bears this out. SMT process-induced micro-cracking of plastic around the lead-frame entry to the package led to infant mortality. Thermal characteristics of packages also made them poor candidates to replace DIPs.

Recognizing SMC package limitations, semiconductor houses turned to their laboratories to develop materials suitable for the new packaging assignment. How well did they succeed? Table 7-2 shows the results of testing conducted by Texas Instruments Incorporated on PLCC, SOIC, and DIP packages housing both bipolar and linear ICs. These tests, selected from MIL-STD-883 and industry standards, are of particular interest because they are directed at package integrity.

Table 7-2. Comparison of Package Reliability[3]

Test	Failure Rates			Units
	SOIC	*PLCC*	*DIP*	
Life Test, 125 °C	2	2	5	Fits*–60% UCL
85 °C/85% R.H.	0.6	0.4	0.5	Percentage/1000 Hours
Autoclave	0.3	0.0	0.5	Percentage/240 Hours
T/C, -65/+150 °C	0.03	0.2	0.5	Percentage/1000 Cycles

* Derated to 55 °C, assuming eV activation energy.

At first glance, the data in Table 7-2 are somewhat amazing. How could SMCs be more reliable than DIPs? In answer to the need for high-integrity small-outline packaging, semiconductor manufacturers introduced new encapsulant and new lead-frame materials for SMCs. However, the most dramatic reliability improvements have been in the semiconductor die passivation (protective glass covering) itself. Since the same dice are used in SMC and DIP packages, the question, "Why are SMCs more reliable?", has merit.

The answer is probably our old friend, TCE. Epoxy encapsulants expand with heating at about 3 times the rate of silicon. And the lead frame (beryllium copper) expands at near the rate of epoxy. TCE mismatches between materials in IC packages set up internal stresses in the package. Although the SMC and DIP TCE differentials are similar, the SMC has less internal stress because of its smaller absolute size.[4]

Similar tests from other semiconductor manufacturers add credibility to this theory. Table 7-3 shows 74S and 74LS logic product-reliability test data gathered by Signetics between 1983 and 1985. No confidence levels or extrapolations have been applied to these results. While reliability has improved since, this early data confirm the trend of SMT's advantage.

Table 7-3. Signetics Reliability Test Results, 1983–1985[5]

Test	Failure Rates SMCs	Failure Rates DIPs	Units
Dynamic/Static High Temperature Life	0.05	0.11	Percentage/1000 hours
85 °C/85% Relative Humidity	0.04	0.13	Percentage/per 1000 hours
Pressure Potential	0.02	0.05	Percentage/100 hours
Thermal Shock, Liquid/Liquid	0.025	0.005	Percentage/100 cycles
Thermal Cycle, Air/Air	0.12	0.08	Percentage/1000 cycles

7.1.3 *Assurance of Component Quality*

Receiving Inspection Electrical Test

In Section 7.1.1, we established that surface-mount components are available in qualities matching that of through-hole devices. We mentioned that tape reeling of SMCs was a factor forcing quality improvements. It is inefficient to 100% inspect devices supplied in tape and reel. For incoming tests, relatively robust packages, such as 1.27-mm (50-mil) PLCCs and SOICs, may be handled stick to stick like DIPs. If such handling is intended, Purchasing should be sure to obtain components in plastic tubes, not tape reels. This minimizes both the handling and purchase costs of components.

For more fragile components, Aetrium, Inc. of Saint Paul, Minnesota, provides a special handler (shown in Figure 7-1) for contacting devices in protective shipping carriers. While carriers are more expensive than tape reel, and may afford little more in shipping protection, they allow test contacting of a fully protected device. The considerable risk of lead damage in the receiving-inspection handling of dereeled parts, and the high cost of coplanarity loss, may well outweigh the cost premium for carriers over tape.

Clearly, where practical, the most cost-effective strategy is to use no incoming verification. Simply purchase high-quality parts sealed in protective antistatic tape, and review the vendor's quality procedures, not the parts. This strategy is made all the more effective when vendor SPC data are available via modem. The handling involved in just passing 150 PPM PLCCs through a good test handler may add the same number of failures as it weeds out.[6]

However, it is comforting to know that there is a full complement of receiving inspection tools available to monitor SMC incoming quality where needed. All the major test-handler manufacturers have systems tailored for SMT.

Figure 7-1 Test handler for fragile SMCs in carriers. *(Courtesy Aetrium, Inc.)*

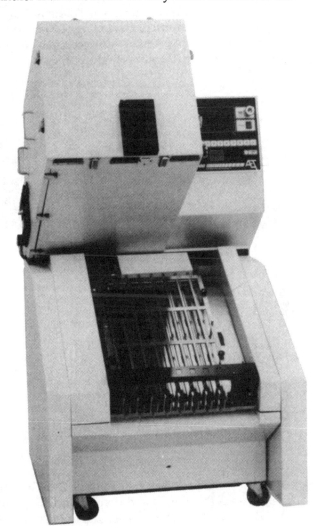

Figure 7-2 A simple immersion wetting test, using various types of components.

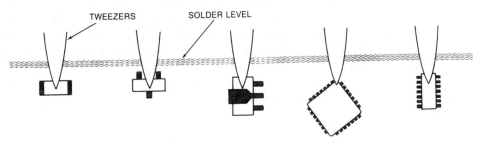

TWEEZERS   SOLDER LEVEL

NOTE: Shaded areas are points for inspection.

Subjective Solderability Testing

A quick, and simple to perform, solderability test is the dip/visual inspection. This test requires nothing more exotic than a solder pot filled with 63/37 solder (or equal), a pair of tweezers of nonsolderable material, and a stereoscopic microscope having between 10- and 20-power magnification. The SMCs are dipped as shown in Figure 7-2.[7]

After a 10-second immersion in solder at 240 °C, the parts are observed under magnification. Any visual dewetting is a concern. A greater than 5% nonwetted area of termination is certainly unacceptable. Figure 7-3 compares three leads and shows the relative degrees of nonwetting.

Figure 7-3 Visual keys to the solderability immersion test. *(Courtesy Signetics Corp.)*

(A) Good wetting.

(B) Cause for concern.

(C) Clearly unacceptable.

Good wetting is shown in Figure 7-3A. The tinning is nearly defect free. There are no areas of congregated defects. It is 100% wetted. In contrast, Figure 7-3B shows a cause for concern. There is visible dewetting. This may be acceptable, as dewetting affects less than 5% of the termination area. Figure 7-3C illustrates a condition that is clearly unacceptable. Over 5% of the termination area is dewetted.

Solder surface tension and cohesive forces can mask considerable areas of poor wettability. Figure 7-4 shows solder coverage over poor solderability lands and component terminations. The board looks well soldered. However, after passing the molten solder through a high-pressure air knife, large areas of dewetting are exposed on the same sample in Figure 7-5. Therefore, for high yields, any dewetting on chip components or boards must be a cause for concern.

Figure 7-4 Solder coverage immediately after immersion soldering. *(Courtesy Hollis Automation.)*

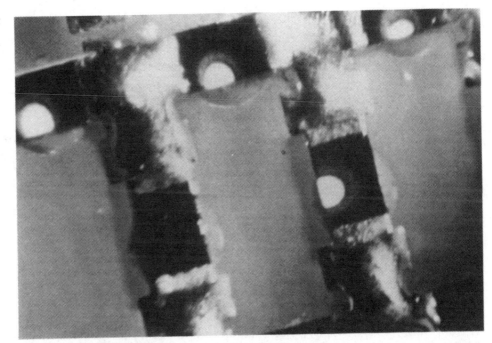

Figure 7-5 Dewetting exposed by air knife cleanup of sample. *(Courtesy Hollis Automation.)*

Dewetting of the tips of SO and gull-wing carrier leads is not of great concern. Dewetting of lead tips may indicate no more than the exposed raw lead-frame material that was left untreated for solderability after lead-frame trimming. Other areas are of varying significance. Figure 7-6 diagrams the areas of a gull-wing lead, and indicates the level of concern for solderability of each distinct region.

Figure 7-6 Wetting concern varies by regions—Gull-Wing Leads.[8]

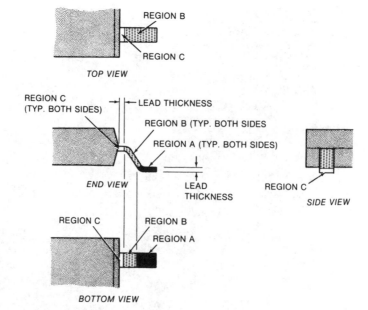

Region A, in the bottom and side views of Figure 7-6, includes the side faces of the lead to within a lead thickness of the body, and the bottom of the lead upward around the bend to a height equal to the lead thickness (the heel). This area is critical to solder-joint strength. It must be fully wetted with a smooth, shiny coating of solder. A small amount of scattered imperfections in the smooth surface may be accepted. Imperfections concentrated in one area are a cause for rejection.

Region B covers the upper sides of the lead, the bottom of the lead above the heel, and top of the lead—all to within the lead thickness of the body. Region B is not so critical to solder-joint strength as is Region A. However, Region B provides a large exposed and visible area for use in the observation of lead solderability. Any significant dewetting of this area may indicate a solderability problem, and should be investigated. Dewetting may be produced by random contamination, fingerprints, overly oxidized lead finishes, or improperly finished leads. A physical and chemical analysis laboratory can determine the nature of soils causing solderability problems, and such information will often point to adjustments in the handling and processing procedures, which bear great fruit for process yield improvement.

Region C, as shown in Figure 7-6, includes the noncoated cut end of the lead and the area near the body, where the lead may be contaminated with mold flash. Solder wetting is not required in these areas.

Objective Solderability Testing

MIL-STD-202, Method 208, and MIL-STD-883, Method 2022, are the commonly applied tests of solderability for DIPs. However, these tests are not appropriate for leaded and leadless SMCs. Martin Marietta has developed a method called the Martin Marietta Orlando Aerospace Solderability Test (MMOAST), which defines procedures for proven wetting balance testing of SMDs.

Purely objective solderability testing, in lieu of the simple but more subjective immersion test, is needed where documented results must be available for after-the-fact inspection. The MMOAST test is a solderability test using a modified wetting balance (Meniscograph), a vibration-free environment, and precision tooling to measure solder-pot insertion and withdrawal force and solder-pot weight gain/loss.[9]

7.1.4 *Assurance of Component Reliability*

Earlier, we considered the problem of assuring component quality. To build SMT products that work the first time, we must have components that work the first time, or we face enormous troubleshooting and rework costs. A closely related, but more challenging problem is the assurance that the components will continue to work properly for the intended life of the product. Fortunately, the tried and true IMC QA technologies can be adapted to surface mount. Next, we will investigate the adaptations and look at strategies for SMT component-reliability assurance.

Burn-In Testing
An electrical test of components operating in elevated ambient conditions, *burn-in testing* will weed out most of the infant mortalities in incoming components. Burn-in sockets, as shown in Figure 7-7, are available for all major SMC package families. Testing may be performed using standard burn-in equipment with SMC sockets. Dependent on the burn-in board and the placement equipment available, burn-in board loading may be automated, using assembly pick-and-place systems.

Destructive Physical Analysis
Destructive physical analysis (DPA) is a powerful tool for quality and reliability assurance. A sampling of components can be sectioned. Photomicrographs, SEM pictures, spectrographic material analysis, etc., can show potential problems and point to their source long before systems start failing in the field.

Electrostatic Discharge (ESD)
In Chapter 4, Section 9, we covered the need for ESD protection and the potential savings that a well planned static protection program may provide. But which components really need protection? Table 7-4 lists the electrical over-stress sensitivity of typical families of devices. Of course, if any components are ESD stress susceptible, all should be treated as sensitive, in order to develop proper handling as a way of life. It is false economy to save a few pennies on a resistor by buying it in unprotected tape. While the resistor may be quite static tolerant, the cover tape being dereeled from it may zap a $35.00 ASIC next door in the assembly operation.

Table 7-4. Semiconductor Static Susceptibility

Type Of Semiconductor	Danger Level/Volt
CMOS	250/2000
Diode, Schottky	300/2500
ECL Hybrid (PWB level)	~500
EPROMs	100/300
GaAs FET	100/300
MOSFET	100/200
Op Amps	190/2500
SCR	680/1000
Transistors, Bipolar	380/7000
TTL Schottky, LS and ALS	100/200

Figure 7-7 Typical burn-in sockets for SMCs. *(Photographs courtesy of Loranger International Corp.)*

(A) PLCC sockets.

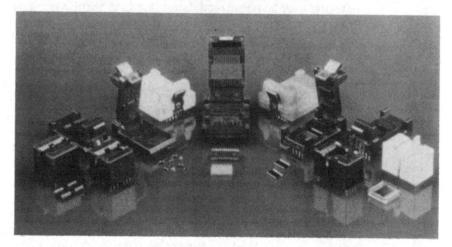

(B) SOT and SOIC sockets.

(C) Chip capacitor/resistor sockets.

Compare the data from Table 7-4 with the electrical potentials generated by typical sources of electrostatic charge, as shown in Figure 7-8, and you can see why quality experts raise so much static about ESD. Data are available from many journals, ESD protection suppliers, and consultants for static-control program help.

Figure 7-8 Static potentials developed by various ESD sources.[10]

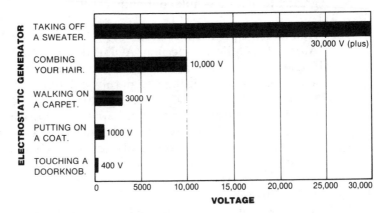

7.2 Printed-Wiring Board Quality and Reliability

We've just covered the quality and reliability of electronic components. Ensuring the quality and reliability of printed-wiring boards is equally important. Bare boards are normally tested for three factors. These are isolation (detecting the fault, shorts), continuity (detecting the fault, opens), and contamination. We'll now discuss quality and reliability assurance for bare boards.

7.2.1 Printed-Wiring Board Quality

Bare-board testing is generally required today for products of any significant complexity, and so is recommended for SMT boards. Bare-board fault distribution is, in order of occurrence, as follows:

1. Shorts (most often the result of slivers).
2. Opens (often from overetch).
3. Cuts and holes.
4. Excessive leakage (features too close).
5. Contamination.[11]

A good testing program should be designed to intercept all these flaws.

Bed of Nails Testing
Bed of nails testing may be performed at low voltages to find shorts and full opens. The test will detect defects on all layers of multilayer boards, and the test can readily be performed on boards with test-pad grids of 1.27 mm (50 mils) and greater. However, test costs are not insignificant for 2.54-mm (100-mil) grid

boards, and costs are in inverse relation to grid size. For grids below 1.27 mm (50 mils), test costs are high and there are a limited number of vendors capable of supplying test services.

High-stimulus voltages and high isolation-resistance resolutions allow bed of nails detection of all five of the faults listed above, but the cost is significantly greater than for a simple continuity check.

Automated Optical Inspection

Automated Optical Inspection (AOI) is useful in finding opens, and some shorts. While it is not as certain as bed of nails probing for these purposes, AOI carries the added benefit of detecting design rule violations, such as traces too close to features or holes out-of-concentricity-specification with annular rings. For multilayer boards, the PWB manufacturer may apply AOI to each layer before lamination and do a continuity check on a final bed of nails probe to find shorts and gross opens only. This can reduce bed of nails test costs while delivering very-high-quality boards.

X-Ray Inspection

X-ray inspection is well adapted to verifying the internal alignment of layers in multilayer boards. Such inspection forms a powerful tool when combined with AOI and bare-board continuity/isolation testing.

Visual Inspection

Visual inspection under low-power magnification can be used to detect PWB flaws, such as delamination, measling, etc. However, visual inspection alone is far from perfect at catching other flaws, such as opens and shorts. For SMT boards, visual testing of board coplanarity should be specified.

Cleanliness

The cleanliness of PWBs should be monitored throughout the manufacturing process. The final testing for contamination is generally a 100-megohm-resolution isolation-resistance test.[12]

Destructive Testing

The forcing of board failure by stress may detect latent defects before they become field problems. DPA also allows a board manufacturer to monitor the processes and ensure that they stay in control.

The IPC provides standard test patterns for single-sided, double-sided, and multilayer boards. By testing standard test patterns as well as the actual product, test results may be compared to a large bank of statistical data, and variations between individual PWB suppliers may be determined.

7.2.2 *Printed-Wiring Board Reliability*

Printed-wiring board reliability is assured primarily by following proven design practices and using materials suited for the product's operating and storage environment. Short of monitoring a product over its lifetime, environmental testing is the most accurate method of proving whether a given design achieves the mark. Where environmental contaminants or dendritic growth under power are found to interfere with PWB reliability, conformal coatings can add the necessary protection to assure reliability.

7.3 Quality and Reliability of Materials and Supplies

Many materials enter into the electronic assembly reliability equation. Space will simply not permit a full discussion herein. Also, individual variations between commodities are clouded by marketing claims and confused by constant product innovations. Therefore, we'll stick to items of a generic and established nature.

7.3.1 *Solder-Mask Quality and Reliability*

Solder-mask reliability is specified in IPC-SM-840A for Class 1 (Consumer), Class 2 (Industrial and Computer), and Class 3 (Military and Life Support) boards. Table 7-5 compares Class 2 and 3 requirements.

Table 7-5. IPC-SM-840A Specifications, Class 2 and 3 Solder Masks[13]

Test	Class 2 Requirements	Class 3 Requirements
Adhesion Over Copper	≤5% Loss.	No loss allowed.
Over Laminate	≤5% Loss.	No loss allowed.
Over Tin/Lead, Tin	≤50% Loss.	≤10% Loss.
Hydrolytic Stability	No degradation.	No degradation.
Insulation Resistance	1×10^8 ohms minimum.	5×10^8 ohms minimum.
Humidity and Insulation Resistance	1×10^8 ohms minimum.	5×10^8 ohms minimum.
Electromigration	Not applicable.	None allowed.
Dielectric Strength	500-V DC peak/mil.	500-V DC peak/mil.
Flammability	94V-1 or better.	94V-1 or better.
Chemical Resistance	No degradation.	No degradation.
Thermal Shock	Not applicable.	No blistering, measling, crazing, or delamination.

As we have previously stated, photoimagable solder masks are preferred for all but the coarsest SMT densities. Both dry-film and wet-film photoimagables are available. Properly applied, there does not appear to be a reliability bias for either wet or dry. However, wet films are said to be suitable for finer geometries than are dry films. To be reliable, there must be no openings around conductors produced by dry-film tenting around the raised conductor topography. Wet films must achieve 100% coverage of the raised features specified to be masked.

7.3.2 *Solder Quality and Reliability*

The most commonly used solders in SMT are tin/lead combinations near the tin/lead eutectic. Figure 7-9 shows the tin/lead phase diagram. This is a chart showing the relationship of temperature to the alloy composition in the movement from solid through plastic to the liquid phases for tin/lead. The plastic range is a point at which lead-rich crystals form. At the eutectic point (63% tin/37% lead), the melting point is lowest and there is no plastic range. On either side of the eutectic point, the melting point increases toward a peak at pure metal. The plastic range variation is also shown for noneutectic alloys. The secondary phase transformations occurring at the lower temperatures at the end of the chart are of so little consequence to material properties that they may be ignored.

Figure 7-9 Tin/ lead phase diagram.[14]

NOTE: Environmental regulators have expressed growing concern about the use of lead in electronic soldering. While there is little evidence of environmental impact stemming from the manufacturing of lead bearing electronics, their argument is that the manufacturers do not (and probably cannot) control disposal of their products. They are concerned that improper disposal may result in lead contamination of soils and ground water. As a consequence, great effort has been focused on finding a lead-free alloy that meets the needs of electronics assembly. Any candidate must exhibit good wetting properties on electronic assembly alloys without the requirement for aggressive fluxes. The material must have a eutectic temperature around that of tin/lead, and must be near tin/lead in ductility. If the material were more creep resistance than tin/lead, that would be a great plus. As of this writing, no true drop-in replacement for tin/lead has been identified, but the time may soon come when we will be forced to change regardless of the industry impact that change may entail.

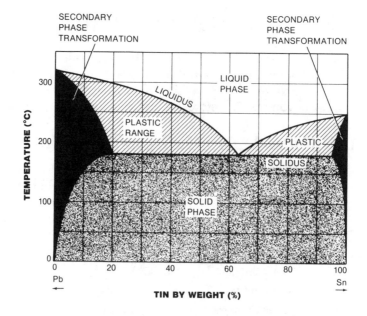

High-tin-content solders are used for high-temperature work, and sometimes for soldering parts on the first side of a double-side populated board. Indium-based, or other low-melting solders, may conversely be used for a second low-temperature soldering operation where eutectic tin/lead is used in the first soldering pass. Indium and gold-bearing solders are often used for gold land and conductor protection as well. In the following sections, we'll cover the reliability of typical solder materials.

Solder Bar Stock

Quality assurance for solder bar, or slab, stock for flow soldering can be approached several ways.

By a simple wetting test on clean metal, we can determine the wettability of newly procured solder. The wetting test involves reflow of a small chunk of the solder material on a test coupon of clean metal. Provided the wetting test is preformed with controlled time and temperature on a perfectly clean smooth coupon of specified base metal, no flux is used. Thus, the flux activity level does not need to be considered. We look for a full even spread of solder on the coupon, and observe the dihedral angle of the sample with the substrate after cooling. If total wetting occurred, the angle, as shown in Figure 7-10, should approach 0°. If the angle is greater than 90°, little wetting took place. An angle of 180° (which could only be observed by sectioning the sample to see under the droplet) would indicate no wetting.

Interpreting the wetting test requires a foreknowledge of solder and coupon materials. With clean CDA copper finished to a reflective surface, the dihedral angle should be well under 90°. A commonly set limit is no greater than 75°, but experience in the given production environment should be the determining factor. A much lower limit, perhaps 15°, might be called for in reflow soldering.

Materials analysis is used to detect potentially harmful contaminants in solder and also to flag incorrect alloys. Two particularly harmful contaminants in tin/lead solders are sulphur, which should be limited to no more than 0.001%, and phosphorus, which should not exceed 0.01%.[16]

The reliability assurance of solder materials is accomplished by a destructive analysis of actual solder joints. Such testing comes as a by-product of thermal cycling, vibration, shake/rattle/roll, and board-flexure tests which stress the whole assembly system.

Figure 7-10
Wetting-test
interpretation.[15]

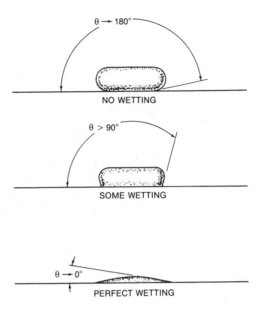

Solder Pastes

In addition to the wetting test described above, solder-paste quality assurance includes inspecting for flux activity, viscosity and rheology (critical to screening and dispensing), and shelf life.

The ceramic coupon test is a quick check for solder paste that bolsters the wetting test described above. A small dab of fresh paste is placed on a clean, smooth, unmetallized ceramic coupon and reflowed. Reflow should be at a correct time-temperature profile for the given paste. With good paste, virtually all the material should coalesce into one shiny, nearly round ball, as shown in Figure 7-11. The flux should form an even and well-spread coating around the solder droplet, and there should be no solder balls (small, detached, solder particles) visible in the flux. Figure 7-12 shows a partially contaminated paste. Here, the flux was not active enough to reduce the oxides in the less-than-fresh paste, and some solder balls are visible as a result. Figures 7-13 and 7-14 show contamination increasing, and Figure 7-15 shows highly oxidized paste. Very little coalescence is evidenced. The sample appears to be mostly detached solder balls lumped into a mud pie. Only samples passing the ceramic coupon test, like that shown in Figure 7-11, should be used in SMT. Any significant solder balling of newly received paste is unacceptable. The samples shown in Figures 7-11 through 7-15 are courtesy of Dr. Barbara Roos-Kossel of Heraeus-Cermalloy. The photography is by AMTI.

Viscosity is tested automatically using a viscometer, a device resembling a cake mixer, which measures the resistance to stirring the paste. Viscosity must be monitored throughout the use of the paste, since it is subject to change due to the evaporation of volatiles in the paste.

The rheology of the paste is a function of its viscosity, thixatropic nature, and solder particle size. Proper rheology is necessary for screen, stencil, or dispenser operation. The simplest of tests for determining screenability is to deposit paste samples in the assembly pattern, using the assembly application method. Results should show uniform depths of deposit, with clean-edge definition of the patterns and full coverage of the print area. Unfortunately, this test, while simple, is highly subjective.

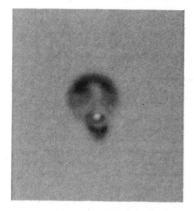

Figure 7-11 High-quality paste passes the solder ball test.

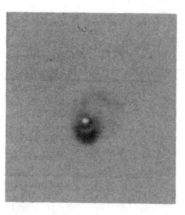

Figure 7-12 Beginning evidence of solder balling.

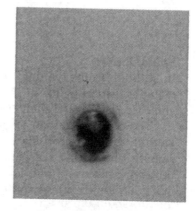

Figure 7-13 Moderate evidence of solder balling.

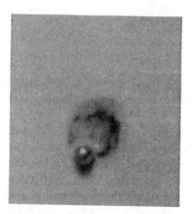

Figure 7-14 Advanced solder balling.

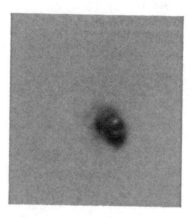

Figure 7-15 Material suitable only for mud pies.

For more quantifiable testing, a uniform pattern may be generated. The pattern in Figure 7-16 serves to numerically specify stencil print quality. The *print quality number (PQN)* of a paste sample is defined by the average of the three smallest features reproduced with good quality. By good quality, we mean uniform depths of deposit equal to the stencil thickness, clean-edge definition of pattern, no bleeding between patterns, and full coverage of the feature area. Inspection under 10X magnification will show where this is achieved. The results of this test may be related to the print quality required for a given assembly job, and a minimum number can be set for inspection to verify paste printability.

Paste shelf life is a function of material properties and solder particle size. Spherical particles of nearly uniform size minimize the surface area for a given metals content and a specified particle size. Elongated particles, or excessive particle-size variation resulting in a large number of "fines," will both reduce screen printing quality and increase the surface area, which can increase the rate of paste oxidation. However, some pastes prove to print better with the inclusion of a few "long" and a moderate number of "fines" to increase the angle of repose of the paste particles. High-power microscopic, or SEM, inspection will allow particle inspection, as shown in Figure 7-17.

Figure 7-16
Stencil-print test pattern.

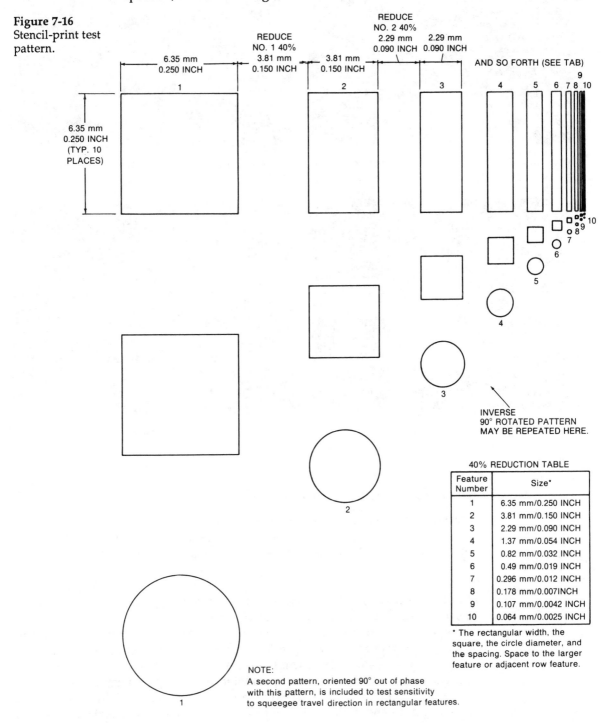

NOTE:
A second pattern, oriented 90° out of phase with this pattern, is included to test sensitivity to squeegee travel direction in rectangular features.

40% REDUCTION TABLE

Feature Number	Size*
1	6.35 mm/0.250 INCH
2	3.81 mm/0.150 INCH
3	2.29 mm/0.090 INCH
4	1.37 mm/0.054 INCH
5	0.82 mm/0.032 INCH
6	0.49 mm/0.019 INCH
7	0.296 mm/0.012 INCH
8	0.178 mm/0.007INCH
9	0.107 mm/0.0042 INCH
10	0.064 mm/0.0025 INCH

* The rectangular width, the square, the circle diameter, and the spacing. Space to the larger feature or adjacent row feature.

INVERSE 90° ROTATED PATTERN MAY BE REPEATED HERE.

Figure 7-17 SEM photographs showing various particle morphologies. *(Courtesy Heraeus-Cermalloy.)*

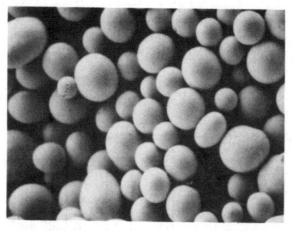

(A) Acceptable sample—uniform-sized spherical particles.

(B) Excessive particle-size distribution—too many fines.

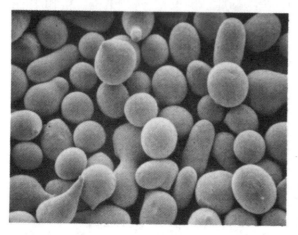

(C) Elongated and nonspherical particles.

(D) Excessive particle-size distribution.

7.3.3 *Adhesive Quality and Reliability*

Adhesives are used for a wide range of electronics assembly applications. Most are covered sufficiently in texts on adhesive technology. However, the unique application of adhesives to Type 3 SMT does bear some discussion.

Single Part
With a single part adhesive system, our primary concerns are its dispensing qualities, its curing ability, and its properties once cured.

An adhesive's dispensing quality is influenced by material viscosity and rheology, and these properties are tested using the methods described in Section 7.3.2 for solder paste. Also, adhesives must be protected from whatever it is that initiates their curing process if they are to retain their dispensing properties over time.

How thoroughly a material polymerizes can be assessed by dispensing a measured drop and curing it in the production curing process. The test is most valid when conducted with worst-case production parameters. In other words, if a U-V curing epoxy must cure with only one third (nominal) of the dot exposed beyond the component, try curing it with one fifth dot exposure (see Chapter 3, Figure 3-17). After curing, component pull testing and sectioning will yield data on how thoroughly the material polymerized.

The strength of adhesives can be tested at room temperature with tensile and shear testers. We also must verify the high-temperature strength of the adhesive. The most critical function of the material will be to hold a component on board during a trip through molten solder. After it has succeeded in that, the adhesive could just as well go away. In fact, engineers and chemists have been working on the development of an adhesive system that would do just that in the cleaning process. The reason is that the adhesive residues after assembly present both a reliability and rework concern. Reliability issues center around the adhesive outgassing, or remaining partially tacky, and trapping contaminants on the board. Rework and repairability demand that the adhesive not have such a tenacious grip that, at reflow temperature, it prohibits component removal or damages the board during component removal.

Two Part
In addition to all the issues presented above for single-part systems, receiving must ensure that the right two parts are on hand for a two-part system. In an application with assembly line equipment, does the material catalyze and cure properly?

Thermoset
With thermally activated adhesives, quality control includes a determination that the specified time/temperature profile will fully cure the material. For reliability, storage and manufacturing environmental temperatures are an issue.

U-V Activated
In addition to all of the preceding, U-V activated adhesives must be protected from environmental U-V radiation during storage and use. The comments regarding environmental temperatures for thermoset adhesives apply also to U-V systems, since their curing process is dramatically accelerated at elevated temperatures.

7.3.4 *Socket Quality and Reliability*

Where sockets are required for production boards, both leaded and LLCC sockets of proven reliability are available. Methode's 84-pin leaded chip-carrier socket has passed MIL-STD-1344A, Method 2005.1 vibration testing and Method 1003.1 thermal cycling from -50 °C to +125 °C. Amp's 84-pin LLCC socket has proven itself in ±20G force testing from 10 to 2000 Hz. The socket has also passed MIL-STD-883, Method 1011 thermal cycling from -50 °C to +125 °C.[17]

7.4 Test and Process Control in Manufacturing

Since we are dealing primarily with design in this book, we can't delve too deeply into process controls and other areas of assembly and manufacturing. It is useful, however, for each team member to have a basic idea of how the manufacturing processes are monitored, and know what test methods are available to ensure the processes are in control.

7.4.1 *Inspection of Materials at Receiving*

Sections 7.1, 7.2, and 7.3, above, lay out the rules for quality assurance for incoming components, boards, and materials. Specific tooling in receiving inspection equipment has been tailored to suit SMT demands, but the basic inspection methods used in IMC manufacturing apply to surface-mount processes as well.

The difficulty of testing tape-reeled SMT components is perhaps the greatest difference actually impacting the philosophy of surface-mount receiving inspection. This, plus a drive for Just In Time (JIT) manufacturing and the Japanese successes with supplier partnerships in quality, is combining to push component-quality responsibility to the component manufacturer, and out of the assembly front end.

7.4.2 *Storage of Materials*

In Chapter 6, Section 4, we covered environmental concerns and materials. We further discussed ESD concerns, earlier, in Section 7.1.4. ESD sensitivity of SMCs has not been shown to be significantly different from that of IMCs. However, several storage and handling differences with surface-mount devices are worth noting. These are covered below.

Figure 7-18
Coplanarity
tester detects
damaged leads.
*(Courtesy MCT
Synerception, Inc.,
Santa Cruz, CA.)*

Coplanarity

DIPs were forgiving bugs in the hands of warehouse personnel. Their leads can be slightly bent during handling and still work without a glitch on the factory floor. Even if the leads were bent beyond allowable limits, the parts are readily salvageable with a lead straightener. Not so with SMCs. New habits must be learned by all who touch or move fragile surface-mount ICs. Bending the leads no more than 0.05 mm (0.002 inch), a feat which may be accomplished by just nudging a reel of components with a stray elbow, will cause disaster when the parts hit production soldering. And there's not any good way to correct damaged leads.

The basic answer to lead coplanarity is to buy good parts and then teach all the people who might ever handle them (including engineers who rummage through the stock room) how fragile they are and how much to respect them. Pay particular notice to the handling of fine-pitch parts, as their tiny lead wires can almost be bent by your breath. Semiconductor manufacturers and component users alike employ scanning equipment, like the coplanarity tester shown in Figure 1-18, to weed out nonplanar parts.

Lead fragility has pushed some designers to reject PROM burning in SMC sockets for fear of losing coplanarity and yields. On-board PROM programming, as illustrated in Figure 7-19, may be selected as an alternative, in order to ensure lead integrity.

Figure 7-19 On-board PROM burning avoids socket damage to leads. *(Courtesy DATA I/O, Inc.)*

Solderability

Solderability of component leads has always been a concern in electronics manufacturing. But DIPs with poor solderability often yielded to aggressive fluxes, high solder temperatures, and multiple passes through flow soldering. None of these "Band-Aid" procedures fit surface-mount products. Parts must be solderable, or they can't be used on a surface-mount line. This places a limit on how long SMCs should be stored, and suggests storage in a cool low-humidity environment. It also implies that solderability testing be undertaken for the surface-mount process, if it was not already in place for IMC.

Moisture absorption by plastic component bodies presents another storage concern. Entrapped moisture can vaporize and cause components to rupture in a soldering process. Low humidity or desiccant storage is a preventative. Once wet, baking is a cure. It is human nature to want to open sealed bags when parts are received from vendors. Unless this is actually required for part inspection, this urge should be resisted, and all people who touch and process incoming material should learn the reasons why. SMT parts, with their great need for coplanarity and solderability, are often shipped in desiccated bags with ESD protection and mechanical packaging to prevent lead damage. Opening the seal on the bag negates the effect of the desiccant in preserving lead solderability. Taking the parts out of the covering destroys the other two levels of protection offered by the shipping container. Unless there is a compelling reason to do otherwise, leave SMCs alone until you are ready to use them.

Kitting

Organizing the parts to build a dozen through-hole boards was a minor chore. Discrete components were available in bulk and ICs came about twenty to a stick. If you didn't need one full stick of ICs, you could take the required number out and stuff their pins down into the conductive foam. Voila, a kit. With the neat reel packaging common to surface mount, kitting can be more of a problem. Unused components in reels must be restocked. And the stockroom will want to know how many are still in the reels when they come back. Since some may have been wasted during the assembly operations, just deducting the "bill of materials" requirement is not necessarily going to answer that question. Equipment like the pocket counter shown in Figure 7-20 counts and respools partial reels.

Figure 7-20 Tape pocket counter. *(Courtesy General Production Devices, Inc., Grand Junction, CO.)*

7.4.3 *Solder Paste and Adhesive Application*

The basic issues governing control and inspection of attachment media are the same whether the media is adhesive or solder paste. Therefore, we'll cover both solder paste and adhesives together, issue by issue.

Material Concerns

Whatever the method of media application, the attachment-material properties must be within limits or the application process will be out of control. QC and QA testing of attachment materials is described in Sections 7.3.2 and 7.3.3.

Dispenser Control

All dispensing operations share one element in control. Dispensing must occur within a predictable distance from the board surface. If we were dispensing on a granite surface plate, this would be simple. It's a little more difficult on a warped and bouncy printed-wiring board. We must either put the board into a known planar location or sense the true relation of the board to the dispensing mechanism.

Other areas of dispenser control are specific to dispenser style. Dispensers may generically be divided into two categories, the time/pressure type and the positive displacement type.

The time/pressure type is controlled by monitoring the pressure applied to the media, and how long that pressure is applied. Some systems simply apply pressure to a syringe to dispense, and then shut off the pressure to end the dispense cycle. Others use a brief reverse suction to reduce stringing. Still others time the dispense operation with a pinch valve near the nozzle. In any case, time/pressure dispensing is not a closed loop system. The desired action, dispensing of a specified volume of material, is inferred from the fact that a given pressure for a given time dispensed that amount in previous trials with the same equipment and nozzle size. Obviously, any change in material viscosity or rheology, or reduction of nozzle opening (clogging), will invalidate this assumption. Therefore, after-the-fact verification of time/pressure dispensing is wise. (See the discussion on "Inspecting Attachment Media Application" given at the end of this section for details on verification methods.) This is particularly true for the time/pressure dispensing of solder pastes, because solder pastes are prone to a change of material properties under dispensing pressures. Pastes may (as noted in Chapter 3, Section 3.1) separate and become increasingly high in metals content. When this condition occurs, clogging of the dispensing nozzle will result.

The positive displacement method of dispensing uses low pressures and large line sizes to feed media to a piston pump which dispenses metered volumes of material on demand. Provided the stroke of the pump is monitored, this method approaches closed-loop operation. We say "approaches" because it is still possible that a glitch might occur in the feed line to the piston pump. It is also remotely possible that the dispensed material might escape from a rupture in the nozzle or pump assembly and not really be going onto the board. Experience teaches that verification of media supply to the pump is generally adequate to verify high-yield dispensing. Where this is provided in dispensing equipment, after-the-fact verification of dispensing may be relegated to visual spot checks rather than an automated scanning of each operation.

Transfer Printing Control
Transfer printing involves the laying out of a metered depth of media on a smooth surface, and then picking up a dot of media on a transfer pin and offset printing it from the pin to a desired location. Accurate transfer printing is established by a complex balance of material rheology, affinity of the media for the metered layer surface, pin surface, and printing site surface, and the printing speed and pressure. Regarding printing pressure, remember our earlier comments about the nonplanar and unpredictable surface topography on which we are printing. The metered depth of media is usually established by a doctor blade being swept across the surface of the media before each print. As long as an adequate flood of proper rheology material remains before the blade, the results should not vary.

Screening and Stenciling Control
Screen and stencil printing of media are done on the same piece of equipment. Only the setup and pattern mask are varied. Thus, many of the process controls are identical. We'll discuss the two topics together, noting where they differ.

Snap-off height refers to the distance between the relaxed mask and the material to be printed, as illustrated in Figure 7-21. For screen printing, this distance usually is between 0.25 mm (0.010 inch) and 0.635 mm (0.025 inch), and is specified by the printer manufacturer. It can be measured by a feeler. Because the stencil mask will not easily deflect, stencil printing is generally done with a zero snap-off, or "on contact." However, some shops mount stencils on open-center, highly tensioned, polyester screens to allow a minimal snap-off.

Figure 7-21
Snap-off height
defined.

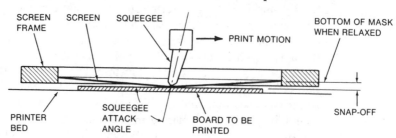

Squeegee material also has an impact on printing quality. The printer manufacturer can suggest durometers which work well for specific applications. Ranges from 30 to 90 durometer are common. Fortunately, the effect of durometer changes is minimal and does not usually need monitoring. The point is to make the proper choice in the beginning. The blade tip of the squeegee will wear during printing. The rate of wear is influenced by printing pressure, blade hardness, tip shape, and mask material. Worn blades should be periodically restored as a part of routine maintenance.

Squeegee attack angle refers to the angle of the squeegee to the screen, as shown in Figure 7-21. This angle is adjustable on most printers, and should be set between 45° and 90°, based upon the manufacturer's recommendations. Attack angle does not vary unless the adjustment is changed, so it is not normally monitored as part of the process control.

Squeegee speed across the print area is important to the print quality, and may be monitored as part of the process control. Process control might also include verification of a full sweep of the squeegee and flood bar strokes across the mask. (The flood bar is a secondary squeegee which wipes paste across the screen between each print without applying printing pressures.)

Squeegee pressure against the screen is adjusted on the screener, and should be periodically checked due to changes caused by blade wear.

Paste rehology has a major influence on printing quality. Environmental temperature and humidity variation must be controlled to keep rehology variations predictable.

The screen mesh size must be selected to suit the paste used and the minimum feature size to be screened. Clearly, the openings in the screen mesh must be larger than the solder particles in the paste. In fact, to avoid any chance of clogging, the screen openings should be large enough for at least three average-size paste particles (the 3X rule) to pass through simultaneously, as shown in Figure 7-22. Where more than one mesh size fits this rule, the finer wire mesh will generally reproduce finer features, and the coarser wire screen will last longer. Table 7-6 shows the dimensions for typical paste particles and screen meshes, and pairs them to accommodate the 3X rule.

Figure 7-22
Defining the 3:1
particle-to-mesh
rule.

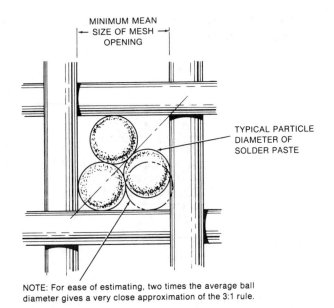

MINIMUM MEAN
SIZE OF MESH
OPENING

TYPICAL PARTICLE
DIAMETER OF
SOLDER PASTE

NOTE: For ease of estimating, two times the average ball
diameter gives a very close approximation of the 3:1 rule.

Table 7-6. Paste Particle and Screen Mesh Dimensions*

Paste Specifications		304 Stainless-Steel Screen Specifications				
Mesh	Ball Diameter**	Mesh	Mean Size	Open Area (%)	Mesh Thickness	Minimum Print†
100	150 micron (0.0059 inch)	80	269.2 micron (0.0106 inch)	70.6	106.7 micron (0.0042 inch)	76.2 micron (0.0030 inch)
200	75 micron (0.0029 inch)	80	223.5 micron (0.0088 inch)	49.6	203.2 micron (0.0080 inch)	101.6 micron (0.0040 inch)
325	45 micron (0.0018 inch)	105	165.1 micron (0.0065 inch)	46.9	177.8 micron (0.0070 inch)	83.8 micron (0.0033 inch)
400	38 micron (0.0015 inch)	150	104.1 micron (0.0041 inch)	37.2	157.5 micron (0.0062 inch)	58.4 micron (0.0023 inch)

* Screen mesh data courtesy IRI Division of BMC, Gardena, CA.
** Less than 1% balls greater, 90% this to next smaller size.[18]
† Print thickness with no emulsion extension past mesh.

Screen mesh to frame orientation is important, particularly where small feature sizes must be printed. As Figure 7-23 illustrates, mesh wires are oriented on the same axis as the screen frame and features may partially obscure fine-print openings. Mounting the screen mesh at 45° to the frame avoids any chance of this happening.

Emulsion thickness is the main factor in determining print thickness for screened solder paste. The print thickness of paste is controlled for several reasons. First, too little paste will cause an increase in opens from reflow soldering. Second, too much paste increases risks of bridging, and tombstoning defects. Also, overly stiff joints caused by too much solder are more prone to stress-related failures in thermal cycling, vibration, etc. For standard-pitch SMCs, wet-paste thickness is generally controlled between 0.15 mm (0.006 inch) to 0.25 mm (0.010 inch). The lower number is preferred for all leadless assemblies, while the thicker paste deposit reduces opens when using leaded parts with coplanarity problems. For fine-pitch components, paste deposits generally range between 0.05 mm (0.002 inch) to 0.15 mm (0.006 inch).

Figure 7-23
Screen mesh orientation.

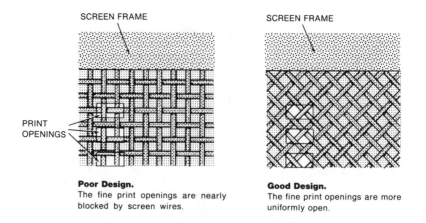

SCREEN FRAME

PRINT OPENINGS

SCREEN FRAME

Poor Design.
The fine print openings are nearly blocked by screen wires.

Good Design.
The fine print openings are more uniformly open.

Texas Instruments suggests the following formula to determine the emulsion thickness required for a desired wet-print thickness.[19]

$$Et = Pt - (Mt \cdot Ao)$$

where,
  Et is the emulsion thickness (beyond mesh),
  Pt is the desired print thickness (wet paste),
  Mt is the mesh thickness ($>2 \times$ mesh filament diameter),
  Ao is the proportion of open area in the screen.

Using this formula, if a print thickness of 0.20 mm (0.008 inch) is desired, and we are using a standard 80-mesh screen with 0.09398-mm (0.0037-inch) diameter filaments, a mesh woven thickness of 0.2032 mm (0.0080 inch), and the screen has 49.6% open area, we get, in millimeters,

$$\begin{aligned} Et &= 0.20\,\text{mm} - (0.2032 \times 0.496) \\ &= 0.09931\,\text{mm (about 4 mils)} \end{aligned}$$

To avoid having to calculate print thickness with the formula, we have included theoretical minimum wet-print thicknesses for the screen meshes listed in Table 7-6. By simply adding the emulsion extension thickness past the mesh to the figures given under "Minimum Print," you derive the actual print thickness for a given emulsion extension.

In practice, some emulsion extension below the bottom (print side) of the screen is needed by some screen manufacturers to smooth the rough surface of the weave. Typical rules of thumb call for 12 to 25 microns (0.0005 to 0.001 inch) of emulsion past the bottom surface. However, Frank Greenway, a research scientist for Advance International of Chicago, reports good results down to 9 microns (0.0004 inch). And, IRI reports a process yielding "zero" emulsion extension.

The mesh characteristics, as defined in the preceding formula, plus the minimum required bottom-side emulsion extension, determine the minimum wet thickness that a given screen material will accurately serve to print. Within limits, an emulsion thickness past the screen-mesh outline adds wet-paste thickness at a 1:1 ratio, as the formula indicates.

(A) The AOI System used to find and photograph these defects.

(B) Excess paste.

(C) Bridging between L. H. lands.

(D) Insufficient paste, third land from the right.

Stencil thickness is a bit easier to deduce. If the desired wet-paste thickness is 200 microns (0.008 inch), use shim stock of that thickness to construct the stencil. Print thickness should follow a 1:1 relationship with stencil thickness.

Inspecting the Attachment-Media Application

Where semiautomated printers are applying attachment media, a visual inspection by the printer operator is often the only on-line verification used. Wet-print height is checked periodically using a height gauge available from various paste manufacturers. On-line inspection may be implemented using *automated optical inspection (AOI)* equipment, as shown in Figure 7-24. Here, both paste coverage and thickness are automatically inspected after each screener cycle. The individual parts of a system are identified in Figure 7-25.

Figure 7-25 The elements of an Integrated Optical Inspection System.

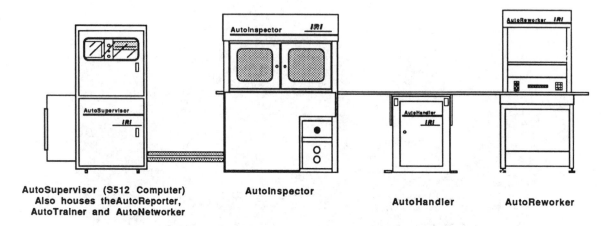

AutoSupervisor (S512 Computer)
Also houses theAutoReporter,
AutoTrainer and AutoNetworker

AutoInspector

AutoHandler

AutoReworker

7.4.4   *Component Placement and Insertion*

Component placement and component insertion actions are monitored and inspected as stated in the following discussions.

Board Registration

Boards are registered for automated assembly operations using mechanical pins, flat surfaces against board edges, or a combination of mechanical and optical schemes. Pins usually register the boards by engaging tooling holes in the board's periphery. However, for cases such as ceramic thick-film boards, two reference edges abutting the artwork zero point for the board are oriented against stop pins. Typical registration methods are as shown in Figure 7-26.

Feature Registration

In insertion machines, pin registration of the tooling holes has long been the standard. For surface-mount technology, however, tooling hole registration alone is only accurate enough for very coarse assembly tasks. Placing chip carriers (particularly high-pin-count and fine-pitch parts) requires a more accurate location of the features on the board. Tooling holes and pins, or reference edges, may be used to establish a rough position of the board; then, AOI equipment is used to accurately locate optical targets on the board and interpolate component land orientations. Sophisticated placement systems can automatically adjust their placement routine to suit routine manufacturing tolerance variations detected by such AOI equipment.

For the extreme accuracy required to place high-pin-count, fine-pitch parts, AOI is used to inspect both the component leads and the component lands, and then actively target one to the other. Such process controls can also be used to flag damaged leads and incorrect form factor components and lands.

Figure 7-26 Typical mechanical and optical registration methods.

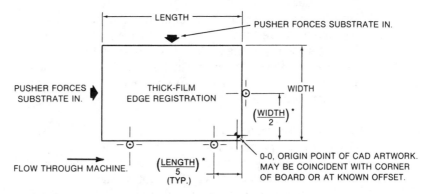

* Recommended design. These dimensions must be same for thick-film print, solder print, and assembly operations.

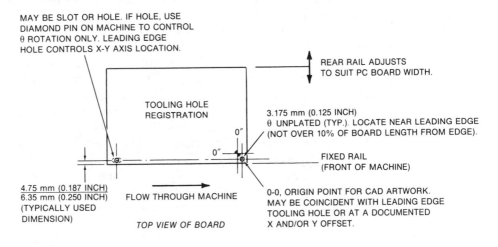

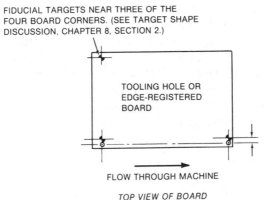

A special application of board-registration methods is required to handle very small boards, flexible substrates, and irregular-shaped substrates. In these instances, one or more substrates are mounted on a tooling plate. Pin or edge registration is used to locate the tooling plate, and pin or edge registration is also used to position the substrate(s) on the tooling plate. AOI may still be used to orient optical targets after their rough position is established mechanically.

Component Inspection

Some insertion and placement equipment incorporates a limited component-testing capability, performing a quick test before the assembly operation. For DIPs, this test is usually a simple short or open test between the V_{ss} and V_{DD} pins of an IC. Both SMT and IMC passives may be verified by RLC measurement. The purpose is to guard against backward ICs and diodes, and to detect setup errors where an incorrect-value discrete has been loaded on the machine.

Placement-Action Monitoring

Surface-mount placement equipment may monitor important parameters during operation to verify its own behavior, check the validity of the setup, and ensure certain types of conformance of assembly parts.

Sensors may detect the arrival, and verify the proper orientation of a board. Bar-code readers may even verify that the correct board part number has arrived, or provide information on the assembly part number to automatically govern the machine assembly actions.

Sensors can also measure the Z-axis excursion during the placement stroke and the Z-axis placement pressure applied. Thus, overly warped boards, failures to contact the board, and collisions with unexpected obstructions can be flagged.

Smart placement heads can provide feedback from the centering jaws, verifying the presence and size of a component. Failure to pick up a component can initiate a recovery sequence. Repeated failures can trigger an empty feeder alarm, cause the machine to switch to an alternate feeder location, or initiate machine shut-down. Component size data can be compared to a lookup table to see if it matches expectations for the desired component.

Optical inspection may verify the features of a picked component and can compare these with the data, stored in memory, on a known good sample. Bar-code scanners can verify the location of a feeder and the markings on tape reels during machine loading.

Placement Verification

After the assembly operation, AOI systems are available to verify placement operations. At the low end, these systems perform a simple check to see that all programmed sites are populated. More sophisticated equipment may measure component sizes and placement offsets, flag out-of-tolerance conditions in both areas, find extraneous components and debris on the assembly, and even provide some confidence that the right component is in each location. AOI systems may also provide valuable data to guide the repair of errors they detect.

7.4.5   *Curing of Media*

Thermal Processes

Solder-paste and thermoset-adhesive curing is generally done in a linear-belt IR furnace, where both the belt speed and temperature are monitored. Temperature measurements are usually made by means of a thermocouple located in close proximity to the travel path of the workpieces. Note that this does not measure workpiece temperature. For instance, a small furnace could be set at 250 °C, but boards passing through it at 100 meters per minute would come out cool to the touch. Boards going through at $1/10$ meter per minute might well be overheated. What we need to monitor is the time/temperature profile that the workpiece encounters.

To actually measure the time/temperature profile, thermocouples may be inserted in a range of points across the board, as shown in Figure 7-27. Then the board is processed at a given set of parameters to characterize the furnace for the workpiece. This is not practical for an on-line measurement, however. Process control, instead, consists of monitoring the furnace settings and the workpiece characteristics to see that no significant changes occur. Loosely controlled time/temperature profiles are sufficient for solder-paste curing. Closer control may be necessary for certain adhesives. Curing parameters will be indicated by the material supplier.

Figure 7-27 A PW board with thermocouples installed. *(Courtesy Transmet Engineering and Texas Instruments Incorporated.)*

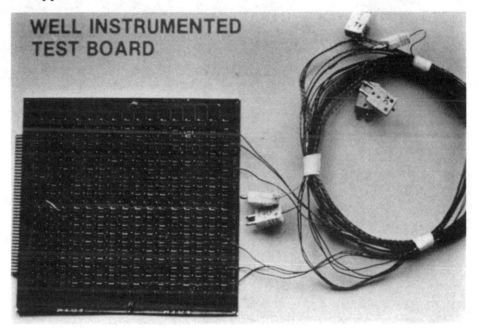

WELL INSTRUMENTED TEST BOARD

U-V Curing Processes
The curing of ultraviolet-activated adhesives is generally monitored by spot checking the curing operation results. Often, elevated temperatures are used to accelerate U-V curing. Where temperature is part of the equation, it is controlled as described in the preceding paragraph.

7.4.6 *Reflow and Flow Soldering*

Reflow Soldering
Reflow oven processes are primarily monitored as described earlier for thermoset adhesives and paste curing. Thermocouple-monitored boards are sent through the

furnace to characterize the setup. Special recording instruments, like the one shown in Figure 7-28, can ride through the furnace with the test board, simplifying the handling of thermocouple wires in linear ovens. After characterization, the temperature of each zone is automatically controlled from measurements of a permanent thermocouple in the zone, and the belt speed is monitored to ensure the desired time/temperature profile.

Figure 7-28 Recording instrument travels with PWB thermocouples. *(Courtesy Electronic Controls Design, Mulind, OR.)*

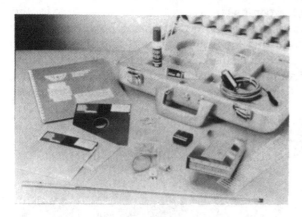

(A) Mole hardware set.

(B) Mole following board passing over a solder wave.

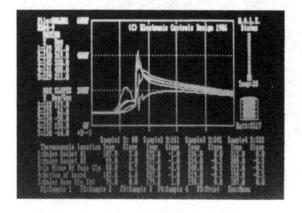

(C) Multichannel measurement graph as displayed by a personal computer.

Advancing sensor technology is being focused on the task of on-line measurement of the actual workpiece temperature in the reflow furnace. Both optical pyrometers and noncontact convective heat-flow sensors have been applied experimentally to the real-time process control of PWB temperature in IR furnaces.[20]

Additional process parameters, such as rate of gas entry and burn-off of combustible atmospheres, must be controlled when special atmospheres are used in reflow furnaces.

Vapor-phase controls ensure that enough heat is introduced to maintain an adequate vapor blanket for the work load, and they monitor the time that the work is in the vapor blanket. For batch systems, the elevator descent and ascent are programmed, and its operation is verified to ensure proper timing. For linear systems, belt speed controls the time at temperature. Additional controls may monitor contaminants in the system and cycle the machine through its filtration process periodically. The control of water flow in the cooling coils may be used, particularly in batch systems, to restrict primary and secondary vapor-blanket height and also prevent vapor escape.

Flow Soldering

Flow soldering of surface-mount devices usually involves passing boards across a high-velocity or agitated wave, or a jet of solder, to ensure application of solder to all the leads. In many cases, this operation is followed by a second pass across a wave having a smooth low-velocity laminar flow. The second wave remelts and removes the excess solder left by the high-velocity/high-pressure wave.

System controls include the liquid fluxer temperature and operation, the preheat panel or emitter temperature, the solder temperature, the height and velocity of solder wave(s) or jets, board travel speed, and the airflow to the exhaust. As shown in Figure 7-29, some machines also are equipped with an air knife to remove any solder bridges not cleared by the laminar wave. Air-knife air pressure, air temperature, flow rate, distance to work, and attack angle are all controlled to maximize this effect.

Figure 7-29 Soldering system equipped with an air-knife debridging unit. *(Courtesy Hollis Automation.)*

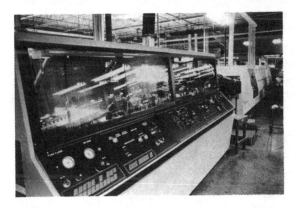

(A) The wave-soldering system.

(B) Close-up view of the dual wave for SMC soldering.

(C) Diagram showing how the air knife works.

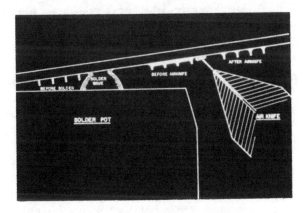

7.4.7 *Cleaning*

Batch Solvent Cleaning

Batch-style solvent cleaners, or vapor degreasers, are not widely used in SMT because of their slow cleaning action under low standoff SMCs.

The controls of degreasers include vapor-blanket condensers, and spray-wand pressure. Some systems also provide a hot fluid bath with controlled temperature. Fluid baths may be ultrasonically agitated to improve tight-space cleaning speeds dramatically. Where ultrasonics are incorporated, both frequency and amplitude should be tightly monitored to ensure the safety of the ICs. Frequencies near an IC's resonance, or excessive amplitudes, can damage or destroy IC wire bonds within the devices.

Linear Solvent Cleaning

Linear solvent systems transport assemblies through either one or a series of vapor blankets, spray zones, and heated fluid baths. Like batch systems, immersion sections may be ultrasonically agitated to enhance cleaning under tightly spaced SMCs. Conveyor belt speed is controlled to produce the correct length of cleaning cycle.

Vapor-section controls include both vapor-blanket height condensers and vapor-generation heater temperature.

Spray controls monitor the pressure on both the top- and bottom-side spray heads. (Top-side pressures are set slightly higher than the bottom side to prevent blowing boards off the belt.) Spray nozzles set the type of spray that each generates. The nozzles may be individually angled in some systems to maximize their effectiveness. The fluid temperature to the nozzles is also controlled.

Batch Aqueous Cleaning

Batch aqueous cleaners use a combination of saponified water-spray/-immersion and deionized water rinses to clean assemblies. Controls monitor the wash and rinse water temperatures, the spray pressures, and the saponifier mix.

Linear Aqueous Cleaning

Linear aqueous cleaners use immersions and sprays of saponified water washes and deionized water rinses to clean. Controls include the wash, spray, and rinse temperatures, the spray pressures, the saponifier levels, and the conveyor belt speed.

7.5 Testing Finished SMT Assemblies

We have touched on "test" previously, both in our discussion of the manufacturing process in Chapter 3, and again in the intervening chapters on design. Developing testable designs should be one of the SMT team's top priorities, and so the topic has been seeded throughout our study of SMT design. Here, in our QC and QA chapter, testing will get our full attention. Below, we've categorized tests used in surface-mount (and IMC) manufacturing, and we present data on each. Some of the following material is a review of Chapter 3, but the importance of SMT testability convinced us that we should cover the topic both under manufacturing and under QC/QA.

7.5.1 *Functional Test*

Functional testing looks at the circuit, or a portion of the circuit, as a complete circuit. A portion of the circuit may include only one, but often includes many, components. Functional testers access the circuit using assembly I/O, such as edge

Figure 7-30 Test fixture provides two-sided access.

(A) Top and bottom access fixture with 1.27-mm (0.050-inch) center probes. (Courtesy Teradyne, Inc.)

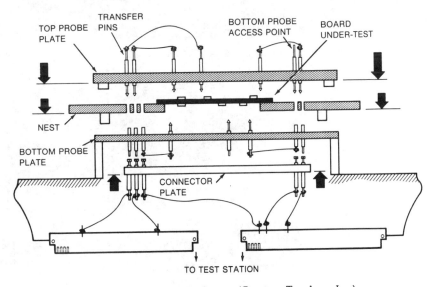

(B) Full access to device under test. (Courtesy Teradyne, Inc.)

connectors or cabling, and the probing of circuit nodes on a bed of nails. Probe access to circuit nodes through one of these means is required if a functional test is to isolate a fault and furnish detailed troubleshooting guidance. If probe access must be limited, active partitioning and built-in test may reduce the costs of troubleshooting on the bench.

Functional test is often coupled with other testing. Efficiency may come from using functional test only for those areas of circuit operation not verifiable in other tests. Less-programming-intensive test methods may then be applied to detect manufacturing defects, weed out component failures, and provide troubleshooting guidance.

Figure 7-30 shows an SMT board with a double-sided bed of nails being used for test access. Full access to the device under test is a boon in both in-circuit and functional test. The design in Figure 7-30B minimizes wiring runs.

Figure 7-31 The in-circuit test of a board can use a clam-shell fixture to access the second side of the board.

(A) The Zehntel 850F/I™ Functional In-Circuit Test System. (Courtesy Zehntel, Inc.)

(B) The Zehntel Access 2™ Test Fixture for dual-sided SMD testing. (Courtesy Zehntel, Inc.)

7.5.2 *In-Circuit Tests*

Figure 7-31A shows a board ready for a single-side in-circuit test on a bed of nails fixture. For application where some nodes cannot be accessed from a single side, the clam-shell fixture shown in Figure 7-31B contacts both the top and bottom nodes when closed on the board. The usual target of an in-circuit test is to individually isolate and parametrically test each and every circuit element. To do this the I/O of every circuit element must be fully addressable for the probes in a bed of nails fixture. And, every circuit element must be completely in-circuit testable.

SMT often works against both these requirements as follows. The close I/O pin spacings of SMCs place difficult demands on probe card design. One solution is to route the traces to test points on 2.54-mm (0.100-inch) centers, but this wastes valuable real estate. High-density designs, with small component clearances, tight track spacings, and hidden/buried vias, further complicate test. And, bottom-side populations interfere with access to the necessary circuit nodes. Finally, SMT is often chosen because it facilitates the very high pin counts needed for VLSI custom ICs. But, these complex devices often can't be fully tested in-circuit.

The major benefit of in-circuit testing is in finding all of the component, as well as the manufacturing, defects. Where it can do this, in-circuit test provides guidance to the repair of defects, without expensive troubleshooting by technicians. However, this benefit is limited to those designs allowing in-circuit analysis of a major portion of the circuit's elements. The value of the test data decreases as the number of elements which can't be in-circuit tested increases.

7.5.3 *The Functional/In-Circuit Combinational Test*

As you may have noted in Figure 7-31, some testers can combine functional and in-circuit modes. The system shown in Figure 7-32 is such an integrated tester. The strategy here is to provide one-shot fault isolation via an in-circuit test, and then add in the advantages of functional test in the detecting of timing-critical faults, the testing complex VLSIs, etc.

Figure 7-32
Integrated Functional/In-Circuit Tester checking a board assembly.
(Courtesy GenRad.)

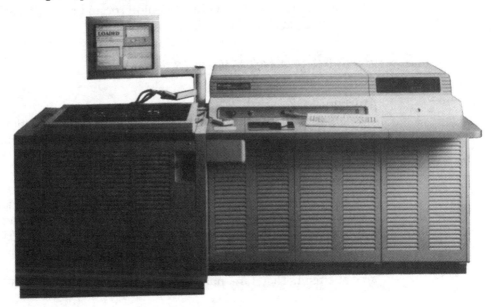

7.5.4 *Manufacturing Defects Analysis*

Manufacturing Defects Analyzers (MDAs) provide a quick test for shorts, opens, and reversed polarity components. Boards under test are probed on a bed of nails fixture. MDAs give rapid feedback on solder opens and bridges, components loaded backwards, and missing components. But, MDAs do not confirm that the correct components were used or that those components are operational. Therefore, MDA testing is usually used to quickly flag certain production difficulties. It is often coupled with other testing to fully qualify the assemblies.

7.5.5 *Burn-In Test*

Burn-in testing means different things in different circles. Some companies call a simple on-line operation of a system, without elevated temperature, a burn-in. If it doesn't begin to smoke during that period, and keeps performing its specified function(s), it passes burn-in. For our purposes here, burn-in means somewhat more than that. Burn-in "involves the monitored operation of the assembly or system at elevated temperatures for a defined period of time. Sophisticated burn-in tests can spot time-of-failure and provide significant detail regarding the failure mode when out-of-specification operation occurs." The exact level of detail that can be extracted from a burn-in test is, like the other test methods mentioned earlier, dependent on access to all the necessary circuit nodes.

By stressing circuits under operation, assembly and system burn-in tests catch infant mortalities. Infant mortalities are circuits that work immediately after manufacture, but which contain a defect that will cause failure early in the circuit's life. Once such infant mortalities are weeded from the circuit population, the remaining circuits predictably survive to a ripe old age. Therefore, eliminating infant mortalities by operation under temperature stress greatly enhances circuit reliability. Since some such failures occur only intermittently under stress, constant monitoring during burn-in is necessary for high-confidence-level testing. Full board or system burn-in will also catch failures instigated by heat build-up. Without such testing, these failures wouldn't occur until the system was in the field under power, at elevated ambient temperatures.

7.5.6 *Assembly Integrity Tests*

Mockup Tests

Mockup testing provides a quick verification of circuit operation to some level of its specified requirements. Mockup tests may be a simple powering up of a system to see if it "comes on." More comprehensive mockup tests may be preformed by plugging the system into a black box, as illustrated in Figure 7-33. The black box may be designed to speed up testing and/or capture more information than what just operating the system would yield. Mockup testing does not require a high capital investment, nor an extensive test-programming development time. It provides a low-cost, fast-response method to find dead-on-arrival circuits.

Most gross manufacturing errors, such as a wrong component, a backward component, a missing component, and shorts and opens, will mean a dead-on-arrival circuit. Therefore, mockup testing provides a fast and low-cost way of monitoring the manufacturing process. However, mockup testing is best applied to testing systems with only a few operating modes and a limited number of response patterns. Complex systems, with many permutations of responses for various stimuli, would take an enormous amount of mockup test time to fully qualify. Desktop computer users can well imagine that their machine might power-up without fault, but fail to properly access a certain disk or respond to a certain command. Trying out everything the machine is specified to do would take days. Mockup testing also produces very little in troubleshooting guidance.

Figure 7-33
Black box traps
test data quickly
in this mockup
test. *(Courtesy
Formation, Inc.,
Mt. Laurel, NJ.)*

X-Ray Analysis
X-ray systems may be used to spot hidden defects, such as voids in solder, delaminations in PWB material, cracked components, and broken internal wire bonds in IC packages. Figure 7-34 shows an SMT X-ray system and the typical defects it detects.

Laser Solder Inspection
Laser thermometry uses thermal signature analysis as an inspection method for solder joints. The system directs infrared laser radiation at an individual solder joint, rapidly heating the joint. The laser is then turned off, and the inspection system monitors radiated heat from the joint. Good solder joints provide a solid thermal path into the board, so the built-up temperature quickly dissipates. Defective joints are flagged because they retain heat longer. Like X rays, laser thermometry finds both visible and hidden solder flaws. For instance, the defective solder joint of Figure 7-35B produced a dramatically different signature than the acceptable joint illustrated in Figure 7-35A.

Figure 7-34 An X-ray machine and the defects detected. *(Courtesy Nicolet Instrument, Inc.)*

(A) X-ray inspection equipment.

(B) Solder voids and solder balls under a chip resistor.

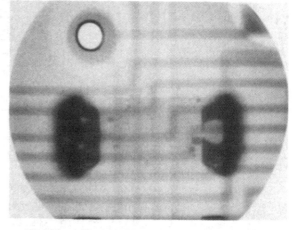

(C) Solder voids and balls in Figure 7-34B viewed from a different Z-axis focal point.

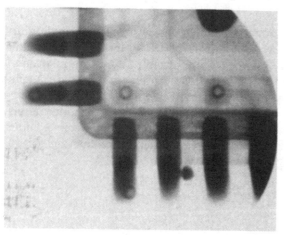

(D) A clearly visible solder ball.

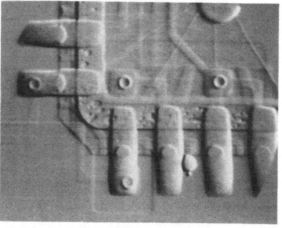

(E) Here, the wire bonds and under-chip solder integrity of the CLLCC in Figure 7-34D are easily viewed.

Figure 7-35
Laser thermometry used to inspect SMT solder joints.

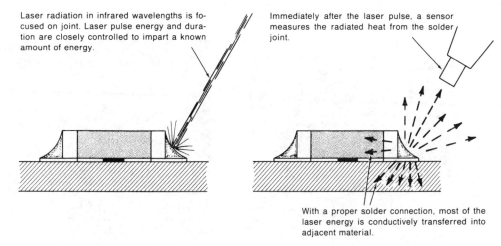

Laser radiation in infrared wavelengths is focused on joint. Laser pulse energy and duration are closely controlled to impart a known amount of energy.

Immediately after the laser pulse, a sensor measures the radiated heat from the solder joint.

With a proper solder connection, most of the laser energy is conductively transferred into adjacent material.

(A) Test procedure.

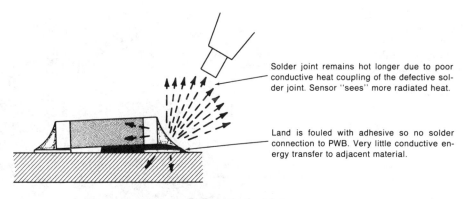

Solder joint remains hot longer due to poor conductive heat coupling of the defective solder joint. Sensor ''sees'' more radiated heat.

Land is fouled with adhesive so no solder connection to PWB. Very little conductive energy transfer to adjacent material.

(B) Example of a defect.

7.5.7 *Nondestructive Physical Analysis*

Thermograhic Analysis

Infrared photographs like the one shown in Figure 7-36 display hot spots on operating circuits and can flag trouble before smoke signals make the trouble all too obvious. Where temperature limits are critical to a board's performance, on-line testing can be done by operators watching television monitors (displaying the results) or by automated inspection techniques.

Cleanliness Testing

Conductometric testing is the standard of IMC PW board cleanliness testing. Cleanliness studies of SMT assemblies indicate that new methods, as discussed earlier in Chapter 3, Section 3.1.5, will be needed in the cleanliness testing of SMT assemblies. One of these methods, a turbidimetric test, is shown in Figure 7-37.

Figure 7-36 Thermographic picture of an operating board. *(Courtesy Hughes Aircraft, Industrial Products Div.)*

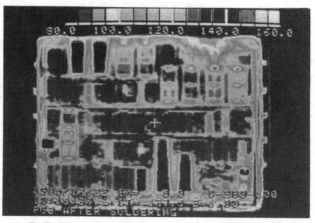

(A) PWB assembly ready for thermal analysis.   *(B) Thermal image of board in Figure 7-36A (under operation).*

Figure 7-37
Nephelometer for
testing of SMT
assembly cleanli-
ness. *(Courtesy
Milton Roy Co.,
Rochester, NY.)*

Vision Testing

Just about every inspection line includes at least a simple eyeball inspection of the boards. A simple operator inspection, with or without magnification, is used to detect missing or grossly misaligned components. Then, microscope inspection may be directed at the solder quality and fine alignment, as shown in Figure 7-38.

Sophisticated high-volume lines may employ AOI or X-ray systems for fast objective testing. Visible-light AOI equipment, with complex accept/reject algorithms, will rapidly detect missing and misaligned components and certain solder defects.

Figure 7-38 Vision inspection with microscope finds SMT defects. *(Courtesy E Light Co., Englewood, CO.)*

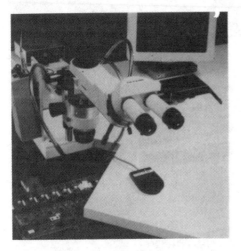

(A) The inspection station with a bar-code reader and a manually positioned cursor.

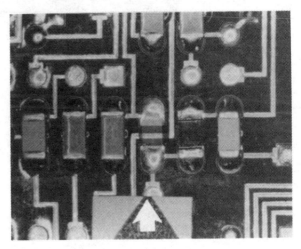

(B) Component missing (4X).

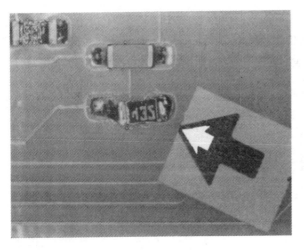

(C) Misalignment producing a solder open (4X).

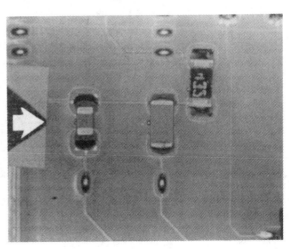

(D) Solder balls (4X).

7.5.8 *Destructive Physical Analysis*

Solder Strength

Shear- and tensile-strength testers are available for the destructive testing of solder joints. Figure 7-39 shows a solder tensile-strength tester that can pull a solder joint apart to record the stress levels required to rupture the joint.

Figure 7-39
Solder shear- and
tensile-strength
tester. (*Courtesy
Quad Group, Santa
Barbara, CA.*)

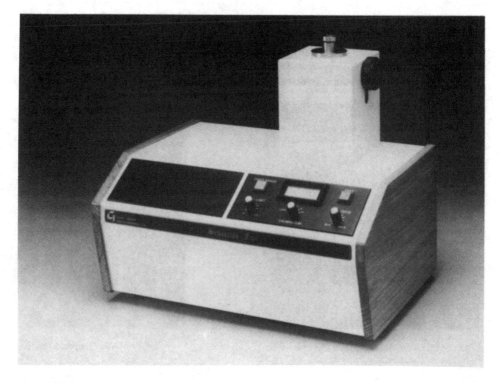

Board Flexure

Flexure tests involve the repeated controlled bending of boards to determine the solder-joint and assembly resistance to recurring mechanical stress. The test may be conducted on home-built equipment, as diagrammed in Figure 7-40.

Figure 7-40 A board flexure test apparatus.

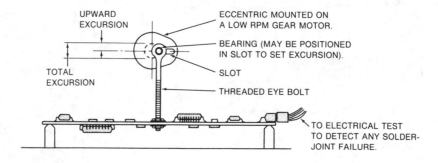

Microsectioning

Sections taken through solder joints, boards, and components can provide valuable information on failure mechanisms. Figure 7-41 shows large voids produced by gas entrapment during the reflow process. The culprit here was probably inadequate paste curing. Undetected, such outwardly good-looking solder joints might inexplicably fail in service.

Environmental Aging

Environmental aging may be produced by putting circuits in high-temperature, high-humidity environments, such as 85/85, by extended burn-in, by radiation, etc. Environmental aging tests are designed to accelerate the real-life environmental impact to circuits in order to verify, within a reasonable time, the survivability of designs for the specified life of the product.

Figure 7-41 Gas-entrapment void revealed by microsectioning.

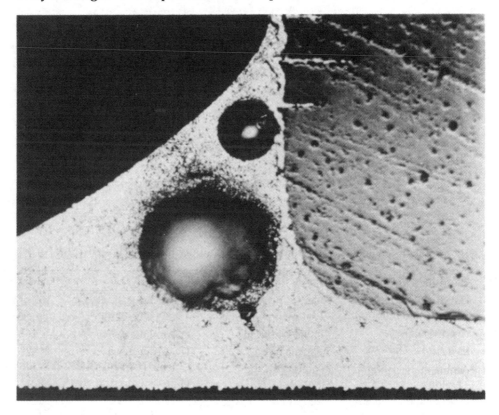

Table 7-7. Environmental Hazard Summary of Shipping and Transportation Stress[21]

Truck Transport	Rail Transport	Air Transport	Sea Transport
Road shock from bumps/potholes.	Rail shock from humping.	In-flight vibration (engine/turbine).	Wave-induced vibration (sinusoidal).
Road vibration (random vibration).	Rail vibration.	Landing shock.	Wave slam shock.
Handling shock (drops/overturn).	Handling shock (drops/overturn).	Handling shock (drops/overturn).	Handling shock (drops/overturn).
High temperature, dry and humid.	High temperature, dry and humid.	Reduced pressure.	High temperature, dry and humid.
Low temperature.	Low temperature.		Low temperature.
Rain/Hail.	Rain/Hail.		Rain/Temporary immersion.
Sand/Dust.	Sand/Dust.		Salt fog.

MIL-STD-810 lists shipping, storage, and operating environments for a number of categories of military products. Environmental ruggedness requirements for military hardware are admittedly stringent, and may constitute overkill for some commercial applications. However, the basic environmental concerns for military equipment engineering are very instructive in guiding any high-reliability design effort. These issues are listed in Tables 7-7 through 7-9. Any considerations relating only to weapons systems will be listed separately in Chapter 10, in our discussion of military SMT design practices.

Table 7-8. Environmental Hazard Summary of Storage and Supply Logistics Stress[21]

Handling/Logistics	Storage In Shelter	Storage In Open
Road shock from bumps and potholes.	High temperature, dry and humid.	High temperature, dry and humid.
Road vibration (random).	Low temperature/ freezing.	Low temperature/ freezing.
Salt fog/Solar radiation.	Salt fog.	Rain/Hail.
Handling shock (drops/over-turn).	Fungus growth.	Sand/Dust.
High temperature, dry/humid.	Chemical attack.	Salt fog.
Reduced pressure.		Solar radiation.
Low temperature/Freezing.		Fungus growth.
Rain/Hail/Sand/Dust.		Chemical attack.

Table 7-9. Environmental Hazard Summary of Operating Conditions Stress[21]

Hand-held Equipment	Mobile Equipment	Shipboard Use	Airborne Use
Handling shock (drops, overturn, slamming).	Road/Offroad vibra-tion (bumps/treads).	Wave-induced vibra-tion (sinusoidal).	Runway vibration (aerodynamic/tur-bine).
Acoustic noise.	Engine vibration.	Wave slam shock.	Maneuver buffeting.
EMI/RFI.	Acoustic noise.	Handling shock (drops/overturn).	Engine vibration.
Atmospheric contami-nation.	Handling shock.	High temperature, dry and humid.	Handling shock.
High temperature, dry and humid.	High temperature, dry and humid.	Acoustic noise.	Aerodynamic environ-ment (high tempera-ture/dry/humid).
Low temperature	Low temperature.	Low temperature.	Acoustic noise.
Rain/Hail/Dust.	Rain/Hail/Dust.	Rain/Hail/Dust.	Low temperature/ Thermal shock.
Solar radiation.	Solar radiation/Salt fog.	Solar radiation/Salt fog.	Rain/Hail/Dust.
TCE storage to use.	Fungus/Chemicals.	Fungus/Chemicals.	Solar radiation/Salt fog.
			Fungus/Chemicals.
			Reduced pressure.

Thermal Cycling

Early research identified the TCE mismatch problem between large leadless ce-ramic components and polymer substrates as a major SMT reliability concern. Thermal cycling of assemblies provides a method of testing solutions to this prob-lem, and has become de rigueur in the engineering development of high-reliability SMT products.

Some engineers have asserted that 1000 cycles from -65 to +150 °C is unrealistic for most product environments, and that, indeed, IMC boards will often fail the same test. We have no hard evidence in hand to show that IMC assemblies typically fail rigorous thermal cycling tests. However, even if they do, the issues of solder-joint integrity and strength are so much more important to SMT than IMC that the argument seems vacuous. Thermal cycling is a valid tool to ensure correctness of engineering solutions to TCE challenges.

7.6 How to Build Testability into Designs

Many of the design guidelines for the testability of SMT boards mirror those for through-hole technology. However, SMT often moves tried-and-true practices from desirable to vitally important. In the following discussion, we'll cover those areas that uniquely relate to surface mounting, and will recap the basic rules made doubly valid by the new demands of SMT.

7.6.1 Basic Rules

The basic rules include:

1. Provide extra gates to control and back-drive clock circuits.
2. Insert extra gates or jumpers in feedback loops and where needed to control critical circuit paths.
3. Tie unused inputs to pull-up or pull-down resistors so that individual devices may be isolated and back-driven by ATE.
4. Provide a simple means of initializing all registers, flip-flops, counters, and state machines in the circuit. (*Note:* Don't rely on built-in test functions on IC chips. They may work well for discrete-device test, but may be too slow for board test applications, or may not even work in-circuit.)
5. Build testability into microprocessor-based boards wherever possible.[22]

Test engineers have been calling for built-in test for decades. In response, design often asks, "Are we really going to gain enough in test cost improvement to justify the design effort and potential cost impact to the boards?" As a Tektronix's project has recently proven, you certainly may.

In answer to increasing competitive price pressure, Tektronix turned to *Built-In Test (BIT)* to reduce the testing cost of their complex scopes. Since test was one of the largest manufacturing-costs components in the product, they reasoned that a test cost-reduction program might be of value. BIT design to the board- and component-level allowed Tektronix to cut test time for their new instruments dramatically. For example, the old-design 7854 scope requires a total of 9 hours test time. The new (and more complex in features) 11400 line, shown in Figure 7-42, is tested to the same confidence level in 45 minutes.

Tom Dye, designer of many of the 11400's BIT programs, notes that Tektronix used functional partitioning to simplify and streamline testing. Otherwise, some of the complex ASICs on their boards would require millions of test vectors to fully verify chip operation.[23]

Figure 7-42 The Tektronix 11400 line of scopes uses BIT to save test time and dollars. (*Courtesy Tektronix, Inc.*)

7.6.2   In-Circuit Test and MDA

Probe-ability design guidelines are important where bare boards will be tested for opens and shorts, or where in-circuit or manufacturing-defects analysis testers will be used on finished assemblies. Because IMC component leads are on 2.45-mm (100-mil) centers, through-hole assemblies naturally adhere to a 2.45-mm (100-mil) grid for test points. Also, inherent in IMC technology, every single circuit node is readily accessible from the solder side of the board. SMT does not automatically impose nodal access and generous probe grids. Therefore, for in-circuit and manufacturing-defects analysis testing of SMT boards, be sure that every circuit node connects to at least one accessible test point. If at all possible, this test point should be probe-able from the back side of the board.

IMC probe points are generally at the lead penetration on the solder side of the board. In SMT, leads may not be probed, and top-side component leads are not available on the bottom side of the board. In order to avoid costly double-side probing for tests, leads are generally connected to on-grid vias leading the board's bottom side. When this is done, some method must be used to identify those vias that will be used as test points, and to record the node to which they are connected. Test engineers will need this listing, and the X-Y offset from tooling holes, in order to develop a test fixture.

Figure 7-43 Board test using a hold-down cover.

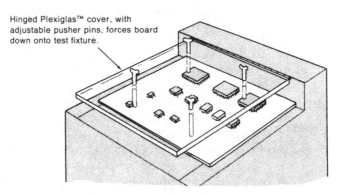

Hinged Plexiglas™ cover, with adjustable pusher pins, forces board down onto test fixture.

Since a 2.45-mm (100-mil) grid is not enforced by component lead spacing, we are free to select a test grid for SMT boards. Be aware that 1.27 mm (50 mil) on-center probe cards may cost two times as much per test point as 2.45-mm (100-mil) cards. That assumes single-side probing; costs escalate for dual-side fixtures. And, 1.27-mm (50-mil) center probes have a service life of about one half to one third that of the 2.45-mm (100-mil) variety. If you must probe assemblies with test grids smaller than 1.27-mm (50-mil), specialized and exotic probe cards and micro manipulators may be required. Use a 2.45-mm (100-mil) grid for test vias wherever possible. And, whenever possible, route all test points to the bottom side of the board, eliminating the need for double-side probing.

Note that device leads or terminations are not acceptable test-probe points. The spring pressure of a test probe might cause an unsoldered or disturbed joint to make good contact for the duration of the test. Thus, a failed board may pass testing. Also, fragile spring-loaded test probes may be bent as they glance off poorly aligned SMCs.

Whenever vacuum fixturing is part of the board test strategy, all via holes should be solder or epoxy filled. See Chapter 5, Section 5, for further discussion of via filling. Figure 7-43 shows a mechanical top cover that is used by Hewlett-Packard's Manufacturing Test Division to replace the vacuum hold-down for an SMT board with unfilled vias. While this gets the job done, it adds cost and slows the production loading of the tester. It also excludes access to the board for functions such as mechanical cycling of switches during test.[24]

7.6.3 *Functional Testing*

With double-side-populated boards, where high densities prevail, it is often impossible to design for full probe-ability for in-circuit testing. In other cases, in-circuit testing may cost more than it saves. Circuits with a bus lend themselves to functional testing, since devices may be individually addressed on the bus. Circuits with even a few devices which can't be in-circuit tested are also candidates for a functional test.

The circuit design team should not consider any one form of testing as a quality-mandated issue. Quality should always be designed in, not tested in. In-circuit testers, functional testers, and manufacturing-defects analyzers are tools which may reduce the test and repair costs involved in producing a quality product. When they don't reduce the costs of producing a quality product, use some other QA strategy that will. Alternatives include a functional test and hot-shot mockup testing. Table 7-10 lists the general advantages and disadvantages of several test strategies.

Table 7-10. Advantages and Disadvantages of Various Test Methods.

Test Issue	Type Of Test			
	Mockup	*MDA*	*Functional*	*In-Circuit*
Typical Uses	Go/No Go	Manufacturing Defect Detection	Performance Verify Verification	Manufacturing and Component Defect Detection
Go/No Go Decision	Very Fast	Slow	Fast	Slow
Fault Sensitivity Level	Manufacturing Defect	Manufacturing Defect	Manufacturing, Component, and Engineering Defect	Manufacturing and Component Defect
Fault Isolation Level	System	Component	Node	Component
Multifault Isolated	No	Yes	Maybe	Yes
Repair Guidance	Poor	Good	Poor	Good
Programming Costs	Low	Low to Moderate	High	Moderately Low
Timing Sensitivity	Critical	Noncritical	Critical	Noncritical
Capital Equipment Cost	Low	Low	High	Medium

In those instances where nearly 100% of the circuit components can be tested, in-circuit testing shines. In such cases, an in-circuit test provides a full-defect analysis for repair. However, most complex VLSI components cannot be fully tested in-circuit, so boards with a large complement of VLSI devices are candidates for a functional test. The timing sensitivity of functional testers allows them to detect timing defects which in-circuit tests will not find.

Where functional test is the method of choice, circuit designers can minimize test and repair costs by providing for the probing of critical circuit nodes. Internal probing allows test engineers to monitor functional test results within the circuit as a signal propagates from device to device. Thus, with complete probe access, it is often possible to isolate specific faults in functional testing. Finding these defects would require considerable technician time if all access were from the edge connector.

Where it is not possible to allow test pads for a PLCC, there are test access caps which fit over the component on the board rather like an inverted socket. The test cap then provides temporary top-side test-probe points.[25] These points allow testing of the mounted component, but may not ensure solder-joint reliability.

Logical Solutions Technology, Inc., of Campbell, California, provides special IC chips which can be added to a board to provide a more detailed fault analysis through functional test. These devices, available in PLCC, SOIC, CLLCC, and DIP packaging, programmably partition the circuitry so that a functional tester can access small areas or even individual ICs, independent of the full assembly. John Torino, president of Logical Solutions, indicates that "Several customers are having other MUX and custom gate circuitry incorporated in our chips, and are using one port for built-in test." Figure 7-44 shows two typical BIT chips from Logical Solutions Technology.

Figure 7-44 BIT chip from Logical Solutions enhances testability. *(Courtesy Logical Solutions Technology, Inc.)*

7.6.4 *Test Strategy and Design*

To simplify application of the preceding testability guidelines, we've condensed the rules into the outline provided in Table 7-11. Once you've selected a test method(s) for your product, this table will highlight pertinent design rules to simplify testing.

7.6.5 *Design for Repairability*

Perhaps the one most important point in design for repair is to always be mindful that the average human (presumably the description of the repair person) can't bend and attenuate their limbs like Gumby™. Discipline yourself for this by recalling the horror of changing a spark plug or oil filter on an option-loaded, big engine Detroit car.

Table 7-11. Test Methods and Design for Testability

Test Strategy	Pertinent Design Guidelines						
	Extra Gates	Tie-down	Initializing	On-Board Test	Probe-ability	Vacuum Sealing	Active Partition
Mock-Up	N/A	CMOS	Maybe*	Yes	No	No	No
MDA	N/A	N/A	N/A	N/A	Yes	Yes	N/A
Functional	Yes	Yes	Yes*	Yes	Yes**	Yes**	Yes
In-Circuit	Yes	Yes	Yes*	Maybe†	Yes	Yes	N/A

Our thanks to Ted Tracy of ATE Labs, Garden Grove, California, for his guidance in the construction of this table.

* Initializing circuits might be faster than forcing the initialization externally. In some cases, ICs can only be initialized by reset lines once they are installed in-circuit.

** To provide access to help partition the board diagnostically. Signals may be forced from the edge connectors, but probe monitoring helps track where the faults occurred.

† Use where on-board or chip testability solves the problem of the components being untestable via in-circuit test.

Figure 7-45
Relationship of package parameters to solder reliability.[26]

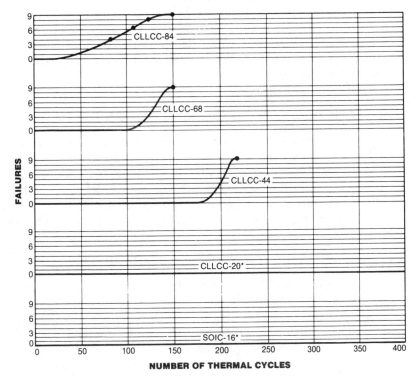

* These parts were tested to 1000 cycles without a failure.

Beyond allowing room for hands and tools, you can also simplify troubleshooting by providing a silkscreen legend identifying the components to the circuit diagram. Component markings help on larger components, but 0805s and SOTs don't provide room for much in descriptive nomenclature. Provided manufacturing will use tape-reeled components, it is worthwhile to specify markings on the resistors and capacitors. However, the law of averages suggest that components fed from bulk, or hand placed, will end up with about 50% of the devices being installed with the markings down. You might as well save the money that markings would cost. Also, locate all nomenclature so that it is readable when the board is on an extender or in the system. Even the best markings are difficult to read while looking into a dentist's mirror. Provide sufficient clearance around the components for test-clip attachment, and for heat shielding from adjacent components during rework soldering.

7.7 Design for Reliability

All the design guidelines herein are aimed at production of reliable assemblies using surface-mount technology. The following suggestions go beyond the norm, in that they call for cost or availability trade-offs. Consider the cost versus the reliability gain.

Leads were originally added to chip carriers to enhance solder-joint reliability by absorbing thermal-expansion and board-bending stresses. As chip carriers got larger, such stresses became more extreme. At some point of increasing size, the PLCCs leads would not provide sufficient resilience to prevent expansion and bending forces from breaking solder joints. As Figure 7-45 shows, there is an inverse relationship between package size and solder-joint reliability in temperature cycling tests. The following presents a solder-stress formula which relates package size and other parameters to solder stress (see also text and Figure 9-9 in Chapter 9).

7.7.1 *Power Cycling Produces Complex Stresses*

The stresses produced by power on/off cycles are complex. TCE mismatches between ceramic components and polymer boards act like a bimetallic strip, producing a bending stress. Differential heating in component and board tends to oppose this bending. Acting on substances of matching TCE, differential heating would produce a bimetallic bending effect in the opposite direction. Power dissipation from a mounted chip produces an out-of-plane displacement.

7.7.2 *The Manson-Coffin Equation Modified*

Expanding on the original work represented in the generalized Manson-Coffin equation, Norris and Landzberg of IBM modified the formula to define and model these complex bending stresses to account for the effects of maximum temperature and frequency of temperature cycling. Shah and Kelly added consideration of the dwell time at high temperature.

Based on his own work, plus research from Hall and his associates, Englemaier of Bell Laboratories modified the original Manson-Coffin equation, producing the following model:

$$N_f = \frac{1}{2}\left(\frac{\Delta\tau}{2\varepsilon'_f}\right)^{1/C}$$

(Eq. 7-1)

where,

N_f = Mean cycles to failure,

ε'_f = Fatique ductility coefficient (0.325),

C = Fatique ductility exponent $(-0.442 (-6)\times10^{-4}\,\overline{T}_s + 1.74\times10^{-2}\,\ln(1+f))$

$\Delta\tau$ = Shear strain range $\left(\frac{L}{\sqrt{2h}}\,\Delta(\alpha\Delta T)_{ss}\times10^{-4}\,\{\text{in percent}\}\right)$

and, where

$\Delta(\alpha\Delta T)_{ss}$ = The in-plane steady-state thermal expansion mismatch $(\alpha_s(T_s\text{-}T_o) - \alpha_c(T_c\text{-}T_o))$,
α_c = TCE of LCC,
α_s = TCE of substrate,
T_c = Temperature of chip,
T_s = Temperature of substrate,
T_o = Temperature in °C (power off, steady state).
$\overline{T}_s$ = Mean cyclic solder-joint temperature (°C),
f = Cyclic frequency ($1 \le f \le 1000$ cycles/day),
L = Chip carrier size,
h = Height of standoff.

Lead form also enters into the reliability equation. The lead form factor is covered in greater detail in Chapter 3.

7.7.3   *Test Methods*

Tests producing the results shown in Figure 7-45 were conducted by Texas Instruments. Nine units of each device type were constructed with internal daisy-chain interconnects, as defined in Figure 7-46. These devices were placed on PWBs with appropriate land patterns—also daisy-chained for continuity testing. Test points were included on the boards to facilitate the location of failures. A failure was defined as an essentially open circuit (whenever the monitored current through the daisy-chain path was interrupted).

Figure 7-46
Daisy-chain
continuity test
interconnects.

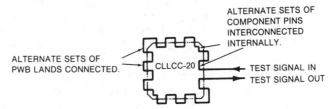

The PWBs used were built up of one 0.508-mm (0.020-inch) G-10 epoxy-glass wiring board, laminated to two circuitless substrates of the same material and thickness. Assemblies were temperature cycled in a relatively benign 0–80 °C excursion for a period of one hour. Cycling was continued until a failure occurred. Upon a failure, the test was interrupted. The failure was diagnosed, recorded, and repaired with a jumper wire. Testing was then resumed, and the procedure repeated until all devices had failed.

7.7.4   *Solder-Joint Reliability*

The same relationship is seen in the comparison of solder-joint reliability in board-bending stress, as shown in Table 7-12. For this test, three 254-mm (10.0-inch) long boards of each design were bent 6.35 mm (0.25 inch) in one direction only, and then allowed to relax. This approximates a smooth radius of 101.6 cm (40.0 inches). The boards were bent and relaxed five times every 2 hours.

Table 7-12.
Solder-Joint
Reliability in
Board-Bending
Stress[27]

Assembly Type	Number Of Cycles	Failures
18-Pin PLCC	1000	None
	6500	None
68-Pin PLCC	1000	15 Opens
14-Pin to 24-Pin SOIC	1000	None
	5000	1 Open

Recognizing that large SMCs place great demands on solder joints, we can enhance surface-mount reliability by using smaller packages. Where devices of greater than 44 pins are required in a design, use 0.635-mm-center (25-mil) PQFP, TapePak, or Micropack devices. A 132-pin PQFP with 0.635-mm (25-mil) centers is the same size as a 68-pin PLCC with 1.27-mm (50-mil) centers. The 0.508-mm (20-mil) TapePak would provide 168 leads in the same space.

7.7.5 *ESD/EOS Resistance*

ESD/EOS hardness follows the same rules whether the board has SMT, IMC, or a mix of the two. We do see numerous examples of products where these rules are ignored, however. Therefore, we will review the rudiments here.

Electrostatic Discharge (ESD) is a common occurrence in any dry environment, and can easily subject parts designed for 3 to 6 volts to momentary surges in the order of 15 to 30,000 volts. When a spark jumps from your hand to some exposed piece of metal on a circuit (a shaft of a control, a board-mounted bezel, etc.), it will follow the path of least resistance to ground. Predicting that path in the absence of deliberately designing it is nearly impossible, since it will be affected by such imponderables as ionization paths in the air, contaminants or humidity on the surface of the board, spacings of parts, clenched leads, and so on. The only way to harden the design against damaging ESD is to eliminate entry points or to arrest the discharge and drain it through a low impedance path to ground before it enters any areas where it might harm circuit elements. To do this, zener diodes can be placed across lines that might sustain a hit. Guard-band traces can be routed around entry points and connected to ground. Spark gaps can be designed into the board to provide a path to ground around I/O. A clever design for a spark gap is shown in Figure 7-47.

Figure 7-47 The pointed conductors lower the voltage required to propagate a spark. Discharges may burn off the copper but carbonized PWB material will continue to serve as a gap.

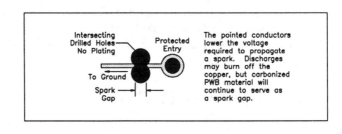

Electronic Overstress (EOS) can occur when a sudden overvoltage, or surge, enters through the AC mains or from the driven load. Line surges occur within household, office, and industrial settings, and may be produced by a number of sources. Disconnecting an inductive load causes an inductive kickback that can be quite substantial in voltage and rapid in rise. Capacitive and resistive loads cause inrush surges in amperage when switched on. Lightning strikes can induce very large voltages on power lines. The designer must evaluate where inrushes may enter the circuit, and provide protection such as spark gaps, MOVs, Sidactors, or other voltage clamping devices.

7.8 References

1. Roome, Diana Reynolds, "Management Techniques & Processes—Cutting Through Complexity: How SQC Can Help?," *Printed Circuit Assembly*, July 1987, pg. 16.

2. Richter, Ed, "Let's Talk Business," *Quality*, September 1986, pg. 31.

3. Hutchins, Dr. Charles, and Ganden, Howard, "Pitting SMT Against Standard Mounting," *Electronic Engineering Times*, October 19, 1987, pg. T8.

4. Ibid.

5. *SMD Reliability Data*, Signetics Corporation, Sunnyvale, CA, 1986, pg. 23.

6. Hollomon, James K. Jr., "The Case for On-Line Test on Placement Systems," *SMART 1985*, New Orleans, LA.

7. *SMD Component and Substrate Solderability*, Signetics Corporation, Sunnyvale, CA, 1986, pg. 2.

8. Ibid.

9. Lish, Earl F., and Weber, John O., "Solderability Testing of Leaded and Leadless SMDs by Means of a Modified Wetting Balance," *11th Annual Electronics Manufacturing Seminar Proceedings*, Naval Weapons Center, China Lake, CA, February 1987, pg. 283.

10. Anderson, Dan C., "The Doorknob Syndrome," *Assembly Engineering*, September 1986, pg. 14.

11. Hroundas, George, "PCB Test Strategies for Manufacturing Yield Improvement," *Proceedings of NEPCON West 1987*, Cahners Exposition Group, Des Plaines, IL, February 1987, pg. 913.

12. Ibid.

13. IPC-SM-840A, *Qualification and Performance of Permanent Polymer Coating (Solder Mask) for Printed Wiring Boards*, Institute for Interconnecting and Packaging Electronic Circuits, Lincolnwood, IL.

14. Manko, Howard H., *Solders and Soldering*, McGraw-Hill Book Company, New York, NY, 1979, pg. 44.

15. Ibid, pg. 7.

16. DOD-STD-2000, *Military Standard—Soldering Technology, High Quality/High Reliability*, June 10, 1986, pg. 10.

17. Bussert, Jim, "Socketing Military VLSI Chips," *Connection Technology*, October 1987, pg. 23.

18. ANSI/IPC-SP-819, *Solder Paste for Surface Mount Applications*, Institute for Interconnecting and Packaging Electronic Circuits, Lincolnwood, IL.

19. Mullen, Jerry, *How to Use Surface Mount Technology*, Texas Instruments Incorporated, Dallas, TX, 1984, pg. 2–8.

20. Rall, Dieter, and Hollomon, James, "Advances in Reflow Soldering Process Control," *Proceedings of NEPCON West 1987*, Cahners Exposition Group, Des Plaines, IL, February 1987, pg. 393.

21. MIL-STD-810, *Military Standard Environmental Test Methods and Engineering Guidelines*, July 19, 1983, pg. 7.

22. Small, Charles H., "Surface-Mount Technology Forces Engineers to Follow Testability Guidelines," *EDN*, May 14, 1987, pg. 93.

23. Conner, Margery S., "Good Engineering Decisions Are Key to Improving US' Competitive Stance," *EDN*, October 1, 1987, pg. 73.

24. Leibson, Steven H., "EDN's Hands-On SMT Project—Part 5: Automated Testing of SMT PC Boards," *EDN*, July 23, 1987, pg. 70.

25. Neal, Thomas N., "Production Testing of Board Mounted PLCCs Made Easy," *Evaluation Engineering*, October 1987, pg. 28.

26. Op. cit., ref. 19.

27. Op. cit., ref. 3.

Figure 8 Products for high-volume markets must be engineered right before production startup. *(Photographs courtesy Compaq Computer Corp.)*

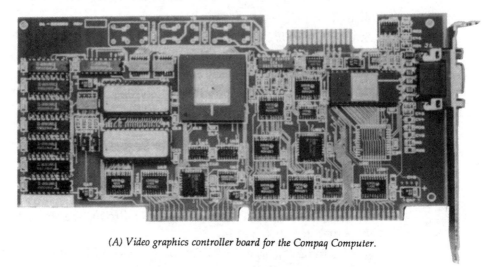

(A) Video graphics controller board for the Compaq Computer.

(B) The Compaq Computer.

8 Reviewing a New SMT Design—Is It Manufacturable?

8.1 Design Constraints and Manufacturing Resources
8.2 Design for Manufacturability
8.3 Nonmanufacturing Issues in Design Reviews

SMT made its U.S. debut in the laboratories of the major electronics manufacturing firms. In a lab environment, the manufacturability and testability of designs were all but irrelevant. But times have definitely changed. Design review, this chapter's topic, is vital, if we are to ensure that a product can be profitably manufactured. For instance, the circuit seen in Figure 8-A is built in high volume. Thus, any manufacturing or test difficulties could have a devastating financial impact on its maker. A volume product demands a careful design review. The board assembly is the video graphics controller for a Compaq® computer (shown in Figure 8-B). Fortunately, a PC is one product which might help in the review process. The software for design analysis, thermal modeling, and high-frequency performance analysis, etc., is transforming the PC into a major tool directed at design review.

Chapter 1 introduced SMT and Chapter 2 covered the components. In Chapter 3, we discussed SMT manufacturing methods, and the impact of manufacturing style on design. With that as a background, Chapters 4 through 6 carried us through the design process. Then, Chapter 7 covered quality assurance. The next step might be called Design Review. This is not to suggest, however, that the design review should be the last step in the R&D process. To be effective, design review must occur frequently from the first phases of conceptual design forward throughout the design effort. If design reviews are delayed, by the time problems are identified in the review process, the cost and schedule impact of corrections will likely be so great that the review will be pointless.

Design review really includes several issues, such as cost analysis, a manufacturability study, resource review, the build or buy decisions, etc. This chapter will cover design review, but we will not recap all the rules we've laid out in the foregoing chapters. And, we will not cover any application-specific SMT hybrid and military rules, which will be the subjects of chapters 9 and 10. What we're presenting here are some additional considerations for design review, points which add to the rules established in our other chapters. So, if design fits all of the requirements established outside this chapter, and passes the following suggested review points, it should be good, manufacturable SMT design.

We will cover the design directions used to promote manufacturing ease, cost effectiveness, testability, repair simplicity, etc. And, we'll touch on the nonmanufacturing issues which make a design either a *win* or a *sin*.

8.1 Design Constraints and Manufacturing Resources

Even the best design may be of little value if you can't build it. And, if nobody can build it, it's of no value at all. Some manufacturing-imposed design constraints are easy to work around, and some others can be solved by subcontractors. Other constraints stand firm until advancing technology solves them. In the following discussion, we'll relate manufacturing capability to the requirements imposed by various SMT designs. We'll sort between limitations, challenges, and the advantages of various process steps.

8.1.1 *Equipment Available*

Available manufacturing equipment does not dictate what can and can't be done in SMT. But it does determine what can successfully be done in-house without capital expenditure. And, creating the need for surprise capital expenditures can dictate the premature end of a career. Thus, manufacturing-resources review deserves to be high on the priority list in analyzing a new design.

Figure 8-1 Hand test fixture for SMCs. *(Courtesy GenRad.)*

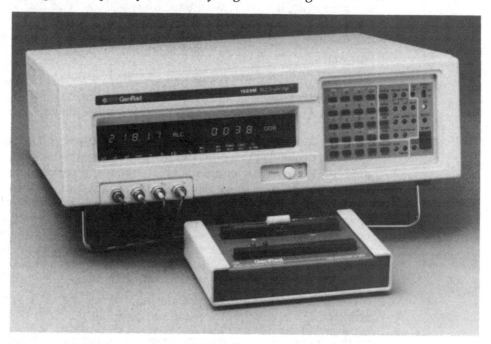

Receiving-Inspection Equipment

Receiving inspection can generally be contracted to vendors for a price. And, under some conditions, with proper purchasing clout, it may be replaced by a *No Incoming Verification (NIV)* program. NIV substitutes source inspection and review of vendor-quality data for receiving inspection. However, the operative word in effective NIV is usually "purchasing clout." Without this, projects must be guided in component selection by a knowledge of what the QC department can inspect, or inspection must be farmed out. Below, we'll look at the special resources required for SMT inspection.

Component Test

Hand test, in a simple fixture such as shown in Figure 8-1, will suffice for the test of a small number of parts. SMC handlers are needed, however, for anything more than the sample testing of incoming components. Figure 8-2 shows a flexible receiving inspection handler for ICs. Designed for receiving-inspection use, flexible handlers have interchangeable accessories, so they can be tooled for the test sorting of SOICs, PLCCs, and CLLCCs. Separate equipment will be needed to handle discrete components.

Figure 8-2 Flexible handler for the receiving inspection of SMT ICs. *(Courtesy Micro Component Technology, Inc.)*

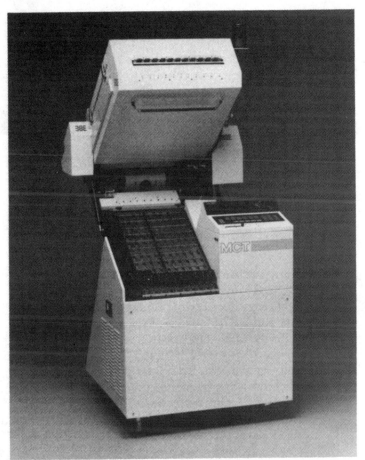

Solder-Paste Test

Solder paste is inspected for particle size using high-power optical microscopes, or SEMs. The paste must be checked for its dispensing or screening quality. Also, the paste is examined for contaminants. In Chapter 7, Section 7.3.2, we presented testing methods for these aspects of solder-paste inspection. Viscosity is also measured at receiving, and periodically throughout the paste's use. Figure 8-3 shows a device that looks like an automatic cake mixer. No ordinary kitchen appliance, this machine measures the resistance to its stirring action, and thus checks the viscosity of a sample of paste.

Figure 8-3 A viscometer for measuring the shearing stress in stirring a sample of solder paste. *(Courtesy Brookfield Labs.)*

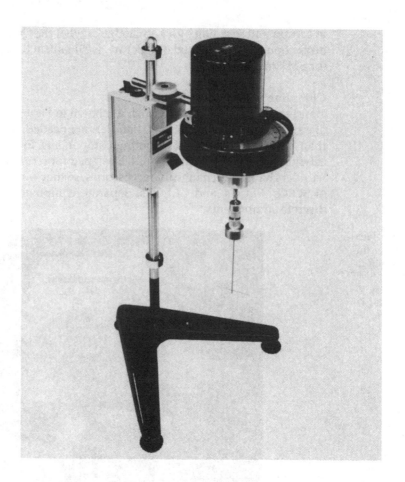

Printed-Wiring Board Test

The need to inspect printed-wiring boards is certainly not unique to surface-mount technology. Bare board probing, as shown in Figure 8-4, is a common link between SMT and IMC. However, SMT does raise new quality issues about the inspection of PWBs. Many SMT boards require fixtures with probe centers closer than the standard 2.54 mm (0.100 inch). Tighter 1.27-mm (0.050-inch) probe cards are available, but at a cost. Even finer pitches can be probed, but below 100 mils, the test point cost is an inverse function of center distance. Figure 8-4 shows some SMT test adaptions. Figure 8-4A shows the test points for accessing one side of a two-sided board. The view is looking down on the fixture shown in Figure 8-4B, which has 50-mil (1.27-mm) center probes. Figure 8-4E shows a fixtureless bare-board tester, which uses a moving probe to solve access and gridless test problems. Other issues are discussed below.

Glass transition temperature, or *Tg* (i.e., the temperature at which a material begins to flow), is important to SMT. Fine features and small-aspect ratio holes are easily damaged when the board Tg is exceeded and the substrate expansion triples. Astute inspection departments have found that some laminate and prepreg suppliers have material Tg in tight control, and others do not. Also, Tg is a function of the level of the PWB supplier's curing of B-stage material. This variation of glass transistion with cure time and temperature is illustrated by the graph of Figure 8-5.

Figure 8-4 Bare board testing adaptations for SMT.

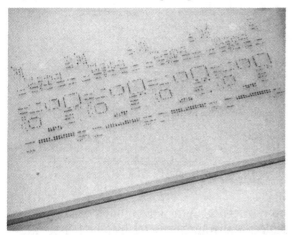

(A) Test points for accessing on one side of a two-sided board.
(Courtesy Trace Instruments.)

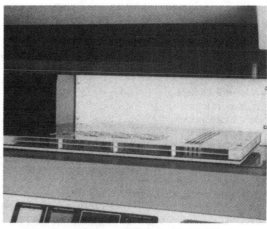

(B) Bed of nails open and ready for a board. (Courtesy
Trace Instruments.)

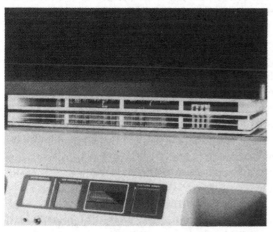

*(C) Double-side access of a board with a bellows-actuated top-
side fixture.* (Courtesy Trace Instruments.)

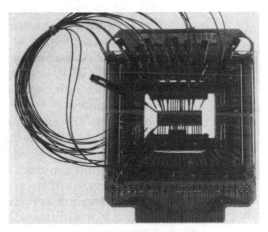

(D) In some cases, special fixtures are needed. (Courtesy
I. C. Probotics.)

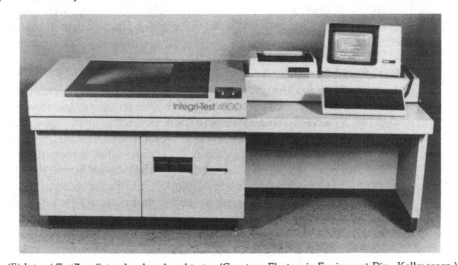

(E) Integri-Test™, a fixtureless bare-board tester. (Courtesy Electronic Equipment Div., Kollmorgen.)

Figure 8-5 Graph of the relationship of cure time to Tg of the PWB material.[1] *(Courtesy DuPont. Reprinted by permission from November 1986 issue of Electri·Onics. ©1986. Lake Publishing Corp., 17730 West Peterson Road, Libertyville, IL 60048 USA.)*

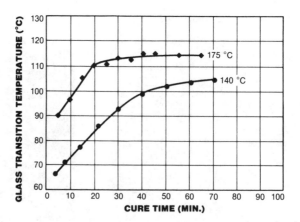

For fine-line and tight-geometry boards, Tg may need to be inspected. *Thermomechanical analysis (TMA)* measures the dimensional effect of temperature changes on materials. *Thermogravimetric analysis (TGA)* measures the weight change due to outgassing, decomposition, or solvent removal under temperature changes. *Differential scanning calorimetry (DSC)* determines the cure state of prepreg resins by measuring the Tg of a sample. *Dynamic mechanical analysis (DMA)* measures the stiffness of prepregs and laminates as a function of heat. Complete PC-based inspection setups are available from several sources for bench-top TMA, TGA, DSC, and DMA testing. Services are also a source of thermal analysis tests for infrequent requirements.

After all the QC data in Chapter 7, and the above additions on QC/QA, one might ask, "Isn't this a bit of overkill?" We'll answer with a military anecdote which aptly illustrates what usually happens when you leave things to chance. As a student in the Navy's torpedo maintenance school, Willard Paul, Jr., was impressed with the instructor's repeated discussions of the importance of built-in safety interlocks. At one point in the training, Mr. Paul asked the instructor if there was an interlock to prevent the opening of the inner door while the torpedo tube was filling with water. The instructor replied haughtily that, "A torpedoman *always* checks his level indicators *first*." He then promptly turned around and opened the breech of the training tube, dumping 3500 gallons of water in the classroom.[2] If anyone's surface-mount program ends up all wet, we'd rather it be because of things that they choose to leave out, instead of something we left out of this text.

PROM Programming

Programming PLDs and PROMS in SMT packages is no problem, if you have a reasonably new device programmer. But older machines generally do not have SO and PLCC sockets. To avoid replacing a perfectly good PROM burner, you can obtain simple and inexpensive sockets which adapt standard DIP tooling on older equipment to surface-mount packages. Such a socket is shown in Figure 8-6.

One word of caution, however. Where conversion from DIP to PLCC required a pin-count change (for instance, a 24-pin DIP PLD converted to a 28-pin PLCC), vendors are at liberty to choose how they will pin out the new device. They may exercise this liberty to achieve best wire-bond routing on their silicon rather than to standardized devices for you, the user. So several sockets, one for each pinout, may be required for a single package size. Emulation Technology offers nine different adapters for the 28-pin PLCC alone.[3]

As mentioned in Chapter 7, Section 7.4.2, on-board PROM programming is an alternative to burning PROMs in sockets. This technology is finding application in SMT to protect the fragile leads from loss of coplanarity in socketing.

Figure 8-6
PROM-burning adapter sockets. *(Courtesy Emulation Technology, Santa Clara, CA.)*

8.1.2 *Size of Board*

For automated assembly, there is some upper limit of board size which can be processed by the manufacturing line. This limit includes maximums of how wide, how long, and how thick a board may be. There will also be a lower limit, past which boards must be carried on tooling plates. If tooling-plate handling can be avoided by making a board just 5 mm (0.197 inch) bigger, the change will save considerable tooling and handling costs. Lower limits may also be avoided by processing several circuits in a panel. More information on panel handling given in Section 8.2.1.

Upper board-size limits cannot be exceeded without risk of serious grief in the manufacturing operation, so they are quite important in design review. The upper and lower limits are imposed by each required piece of process equipment. Even when all the equipment comes from one vendor, it is quite unusual for there to be much agreement on board-size limits, machine to machine. Thus, the line limit is set by the least common denominator of maximum and minimum width, length, and thickness for the individual equipment in the manufacturing system.

Note that size of the boards may have a far greater impact on per-square-inch cost than one might expect. This is because raw materials for PWB manufacturing are typically supplied in standard sizes such as 1 meter square, 36 inches square, or 36 by 48 inches. Let's look at an oversimplified example to illustrate this principle. If you designed a board as 19 by 25 inches, only one circuit could be fabricated out of a 36 by 48 inch sheet of laminate. Of course, the board shop could use the scrap from the laminate to make other, smaller boards. But who will pay for the material? If you guessed, "Both customers," give yourself 100 points. By simply reducing the board dimension to 19 by 24, you would sacrifice 4% of the board area but cut the cost of raw material in half. If you could squeeze the board down to 18 by 24 inches, you would give up around 9% of your 19 by 25 inch board's area, but cut the raw-material cost by 75%! When boards include small projecting areas, elimination of these, or panelization in nested orientation, may produce even more dramatic results. Now, as I said above, this is an oversimplification. In the real world of board shops, raw stock is generally cut into a more manageable size (also called a panel) for circuit manufacturing. If raw stock of 36 by 48 inches is used, panels may be 24 by 36 or 18 by 24 inches. Shops vary. Also, because of manufacturing process requirement in the board shop, some perimeter of the standard panel must be left clear of the imaged circuit. Therefore, to intelligently select a board size for maximum material usage, you must know a good deal about the practices of the vendors you will use for board manufacturing. Good designers will learn these facts. Purchasing departments do well to build a strong partnership with a limited number of quality suppliers rather than paper the world with RFQs searching for today's bargain-basement pricing each time a new requirement arises.

8.1.3 Shape of Board

Irregularly changed boards require special handling methods, just like undersized boards. Odd shapes may be processed multiply in rectangular panels, or loaded on tooling plates as shown in Figure 8-7.

8.1.4 Panelization

If panel handling is part of the manufacturing plan for a product, depaneling methods must be available. For a very small number of boards, breakaway tabs may be cut with tin snips or heavy diagonal pliers, and the rough edges removed by grinding (while wearing appropriate breathing protection to avoid inhaling the dust). For anything beyond prototype volumes, however, automated depaneling equipment should be available, or a depaneling contractor should be secured. Figure 8-8 shows several styles of depaneling machines.

Figure 8-7 Irregularly shaped boards present special handling needs. *(Courtesy Telxon, Inc.)*

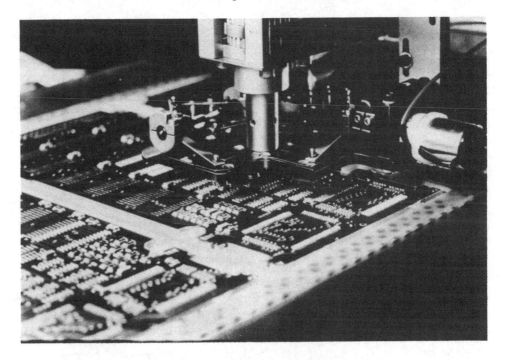

Figure 8-8 Depaneling machines automate removal from panels. *(Courtesy Cencorp.)*

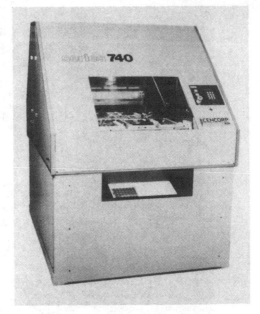

8.1.5 *Clear Areas on Board*

Each automated processing station in the line will impose certain requirements for areas of board clear of components. Such clearances may be required for guide rails, pushers, and locating pins for automated board handling and registration. **Figure 8-8** shows typical clear area requirements. Individual equipments vary, so establish clear-area requirements for your line based on the worst case for all stations in the line.

Figure 8-9
Typical clear-area
requirements for
automated
handling.

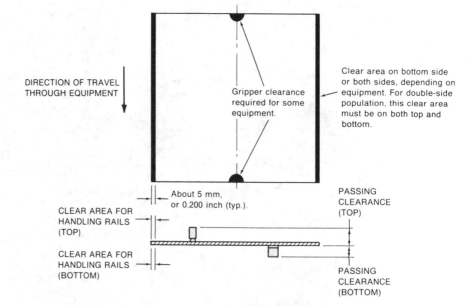

DIRECTION OF TRAVEL
THROUGH EQUIPMENT

Gripper clearance
required for some
equipment.

Clear area on bottom side
or both sides, depending on
equipment. For double-side
population, this clear area
must be on both top and
bottom.

About 5 mm,
or 0.200 inch (typ.).

CLEAR AREA FOR
HANDLING RAILS
(TOP)

CLEAR AREA FOR
HANDLING RAILS
(BOTTOM)

PASSING
CLEARANCE
(TOP)

PASSING
CLEARANCE
(BOTTOM)

Figure 8-10 PW
board assembly
after taking a
dive during wave
soldering. (*Cour-
tesy U.S. Navy,
Electronic Manufac-
turing Productivity
Facility, China Lake,
CA.*)

*NOTE: For panelized
boards, if the panel
is wide, stiffening webs
help minimize warp-
age during thermal
processing such as
wave soldering. For
panels over 200 mm
(8 inches) wide, ex-
ternal stiffeners may
be needed to ensure
that boards do not
submarine as the
wave temperatures
warp them.*

B0004333

8.1.6 *Passing Clearances, Top and Bottom*

Additional clearance considerations are imposed by the proximity of the undercarriage and overcarriage of a piece of equipment to the boards passing through it, as shown in **Figure 8-9**. Remember that the under- and overcarriages include any moving parts mounted thereon, which may pass across the board during processing. Also, for boards requiring components on both sides, remember that they must pass through the line in both orientations. Thus, topside components in the first pass will need to comply with bottom-side clearance requirements in order to clear the second pass. Likewise, if guide rails require clear edges on the bottom, the inverted second pass will dictate that the same clearance is on the top.

8.1.7 *Heavy Components*

Rule 1 for heavy components is the same as Rule 1 for rework. "Don't have any." Unfortunately, both rules have, at times, proved difficult to follow. Where heavy components must be used, add them after the automated assembly operations, if at all possible. If the chunky component absolutely must be on the board for the trip through the SMT line, locate it near the board's edge, where it will have maximum support. Watch out for heavy parts on the board during the wave-soldering operation. They can cause the board to act like a submarine, as illustrated in **Figure 8-10**.

Board carriers may be used to add stiffness. As a last resort, "pop-up braces" may be installed on each machine, to raise up after the board is registered and support the sagging assembly. Of course, this is a costly and time-consuming solution that is to be avoided whenever possible.

8.2 Design for Manufacturability

In the following paragraphs, we'll discuss rules for reviewing and enhancing the manufacturability of SMT assemblies. Enforcing these rules while the assembly can still be changed with an eraser can save many a dollar, and can prevent needless gun fights between manufacturing and engineering.

8.2.1 *Standards*

Great gains in manufacturing efficiency are available in the area of internal standards. The following are some suggestions regarding the areas to standardize for your SMT program, and the ways to derive standards for them.

Databases
Maintain a common database throughout design and manufacturing. Use protocol conversion software, if necessary, to connect nonstandard equipment. Hewlett-Packard's Entry Systems Operation in Cupertino, California, has implemented a common database, and they can shed some light on how it helped in their case.

According to engineering managers, Thomas K. Landgraf and Felix Guerra, Jr., the schematic capture and data checker on the front end, and the file conversion for creating programs for production and test equipment, required the greatest effort. But the result was well worth their trouble.[4] The standardized database permits programming of the entire operation from one post-processed CAD/CAE database. Over the life of a factory, programming costs for line equipment and test can quickly exceed the capital acquisition cost. Therefore, cost-reduction efforts which are focused on programming expenses can yield great savings.

Components

Standardize on as few component part numbers as is humanly possible, while allowing construction of the required circuits. Many designs electrically call for the same-value resistor or capacitor, with several different tolerances. If you specify several tolerances for the parts, each part must be a separate part number, which must have a unique feed location on the manufacturing placement equipment. There are physical limits to the number of positions on an assembly machine. When these limits are exceeded, double-pass placement is required, and this not only drives setup costs through the roof, it can destroy first-pass yields as well. There are also hidden costs in purchasing, inspecting, warehousing, and the handling of five different part numbers where one would suffice. Limiting all designs to 1%-tolerance resistors can cut the resistor part number count by over 50%. The substantial manufacturing cost savings are well worth a slight cost-premium in components.[5]

Many ICs are available in several package styles. The same memory may be offered in SOIC, SOJ, PLCC, CLLCC, and TapePak. Consider the alternative packages, and standardize on one type for all ICs in a given pin count, unless other factors absolutely force a deviation. In all cases, endeavor to have no more than one part number for a given IC.

SO packages are more space efficient in devices with less than 20 pins, while 20- and 24-pin SOICs take slightly more real estate than their PLCC counterparts, but they provide a clearer routing access on board. From 28-pins up, the quad I/O devices have a clear size advantage. For parts requiring extra handling steps, such as PROMs or PLDs, SOICs may be a poor choice due to the fragility of their leads. To avoid lead damage during programming, a PLCC is a better package for such devices.

ANSI/IPC CM-770C calls for coplanarity of ±0.05 mm (±0.002 inch) for leaded IC packages. To IC manufacturers, this total variation of 0.1 mm (0.004 inch) is in serious conflict with packaging cost-reduction efforts. In fact, at least one major U.S. semiconductor manufacturer has attacked the coplanarity problem by forming the J leads of PLCCs tightly over molded-plastic mandrels under the device body. The net effect is the production of a highly coplanar package with virtually no lead compliance. This manufacturer sells compliant-leaded devices to the few users aware of the need to ask for them, and boosts their own yield by making coplanar noncompliant devices for everyone else. Compliant J leads do work to relieve damaging TCE-induced stresses to solder joints. The noncompliant devices produced by this one supplier may well be at the core of perceived problems with solder integrity of J leads.

While the component supplier may feel that the coplanarity specification is too tight, the end user suffering with excessive solder opens sees ±0.004 inch as unconscionably loose. For reasonable manufacturing yield, coplanarity variations should be no greater than ±0.025 mm (±0.001 inch).

Component coplanarity specifications determine the assembly parameters required to achieve high solder yields. **Figure 8-11** shows a model for determining the required wet-paste print thickness and the lead penetration into that wet paste for solder-joint formation with a given coplanarity error. The three parameters—

device coplanarity, wet-print thickness, and lead penetratation—can be monitored. So, given the data from the model shown in **Figure 8-11**, we have the tool for specifying and controlling a high-yield soldering process.

Figure 8-11 The determining processes needed to prevent solder opens.[6]

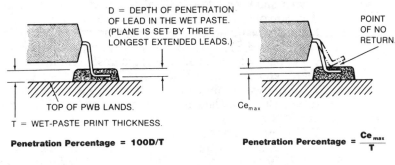

D = DEPTH OF PENETRATION OF LEAD IN THE WET PASTE. (PLANE IS SET BY THREE LONGEST EXTENDED LEADS.)

POINT OF NO RETURN.

TOP OF PWB LANDS.

T = WET-PASTE PRINT THICKNESS.

Penetration Percentage = 100D/T

Ce_{max}

Penetration Percentage = $\dfrac{Ce_{max}}{T}$

NOTE: See text discussion.

The pressure required to have the correct lead penetration can be calculated using the following two equations:

$$\text{Penetration Percentage (\%P)} = 100\ \frac{D}{T} \qquad \text{(Eq. 8-1)}$$

and

$$\text{Penetration Percentage (\%P)} = \frac{\text{Maximum Coplanarity Error } (Ce^{max})}{T} \qquad \text{(Eq. 8-2)}$$

Therefore, in order to apply sufficient placement pressure to yield the proper penetration, and for a wet-print thickness of 0.2 mm (0.008 inch) with standard parts having a coplanarity specification of 0.1 mm (0.004 inch), the device placement system should be set to supply sufficient placement pressure to yield a percent penetration as follows (Equation 8-1):

$$\%P = 100\ \frac{D}{T} = 100 \times \frac{0.004}{0.008} = 100 \times 0.5 = 50\%$$

Note that this is a simplified model which ignores the effects of lead form and which assumes good solderability and good wetting forces of the materials. Also, the joints formed by deformed leads, which barely touch the surface of the paste, will probably not be as strong as joints formed with leads which are not deformed. If joint strength is important in a given application, testing should be used to establish the maximum coplanarity variation which will produce acceptable joints.

Uniform Device Polarity and Orientation
Chip resistors should be oriented with their resistive-element side up. In this position, they are more stable, because the resistive-element side is rougher than the alumina-substrate side. Figure 2-7 in Chapter 2 shows the construction details that explain the variation in flatness. Also, component markings are generally on the resistive-element side. Thus, orienting the device with the element upwards leaves the markings visible after mounting. Taped resistors are supplied with the element side up as the tape is dereeled. Therefore, resistive-element upward orientation comes automatically when using taped components on automated place-

ment engineering. However, bulk resistors must either be optically oriented in special feeder equipment, oriented by hand placement, or randomly placed. When random placement is required, a silkscreen legend, visible after component placement, helps with the troubleshooting, field service, and repair of boards.

ESD Protection

If your shop has a true ESD plan in place, and static charges are controlled to less than 50 volts, you can design freely with all classes of components. Unfortunately, too many manufacturing organizations still peak in terms of being able to "get by with" limited ESD protection. This is only true where a company wishes to "get by with" poor yields, shoddy merchandise, dissatisfied customers, a falling market share, and an eventual demise. Chapter 4, Section 4.9 and Chapter 7, Section 7.1.4 present useful data in determining static control strategies. An engineering review should include a consideration of the ESD protection levels of the devices in the assembly.

PWB Specifications

All bottom side chips should use wave soldering lands, as shown in Figure A-48 and A-49 of Appendix A. One of the first design decisions for a multiboard system is how the board should be panelized. The most common approach is to place each board in its own panel. If the design pushes board tolerance limits or has many layers, board yield may be low. Each panel may have a number of X-outs (rejected individual circuits). In such a case, separate panels is the best strategy. However, where board yields should be high, it may make sense to place one or several sets of boards for the entire system in a panel. By doing this, so long as your testers have sufficient pin capacity, it is often possible to complete both in-circuit and functional tests of the board in the flat. Thus, reworking failures from functional tests is much less difficult. It is sometimes possibl to use axial inserters to "staple" interconnect wiring between circuits in a multiboard panel. After depaneling, the boards are formed to fit the final packaging. This trick provides a very low cost interconnection that is inherently more reliable than a header, and also facilitates the in-the-flat system test we mentioned above.

Giving a manufacturing engineer a fistful of board sizes in small-lot manufacturing is about as likely to win his friendship as giving him a one-way ticket to Siberia. On many pieces of process equipment, each unique board size requires custom tooling. There is also the cost for setup time needed to adjust for board variations. And, with each adjustment, there is the risk of missing the correct mark. When these costs are submitted to top management, the design team may be the ones traveling to the Siberian salt mines. To avoid undesirable relocations while still retaining the right to design the board sizes that an application demands, panelize the designs for batch processing. Most shops can standardize on one form-factor type of panel, which will accommodate at least one of their largest boards. Smaller boards are then built in multiples on the panel. The outer edges of the panel may be standardized with free areas for handling conveyors, tooling holes, fiducial marks, and panel identification markings or bar codes.

In panelizing designs, always put each circuit within the panel on 2.54-mm (0.100-inch) incremental steps, so that probe points are on such centers, board-to-board, within the panel. This factor is often overlooked and causes unnecessary grief when making bare-board test fixtures.

A panel will typically have a perimeter of circuit free material, or selvage. This vacant ground need not be entirely wasted. By clever forethought, the selvage can serve several functions. Most obviously, it provides an unobstructed area for handling on sliding rails, nesting in magazines, moving by grippers, fixturing by tooling holes, and orientation by fiducial marks. But the uses need not end here. If copper thickness is an issue, an extended serpentine pattern in the selvage can be used to measure the copper by doing a quick resistance measurement on each manufacturing lot. Test coupons can be included to allow testing of through-plating in holes, peel resistance of copper, and cleanliness of the board surface. Coupons are defined in IPC-D-330 and MIL-STD-275. Another valuable addition to include in selvage areas is dummy lands. These are lands for mounting SMT parts. As a spot check of soldering adhesions, you can periodically hand-place components on these sites, run the board through the soldering process, then use a shear and/or tensile test to determine how well your soldering process is working. We have even used the break-away tabs to carry tracks from ground planes into the selvage, and back through break-aways of adjacent circuits where we were panelizing a multiboard system in a single panel. This simplified testing of the entire system with high integrity grounds. One word of caution, though. The level of consternation this generates with PCB front-end personnel tells us how uncommon it is. If you don't want to get phone calls, place clear manufacturing notes on the mechanical layer. This won't stop people from calling. After all, who reads the notes first? But it will put you in a position of power with those who call.

Individual circuits within the panel may be partially defined by routing edges, but leave a few bridges of material (called break-away tabs). Each board shop has individual requirements for panelization, and their suggestions should be weighed in the selection of actual panel dimensions.

Generally, four tabs per board are used. Tabs are typically only on two sides, and should not be on sides which must go into card guides. For small boards, two tabs may suffice. Where two tabs are used, boards are less prone to warp out of the plane of the panel if they are located near opposing corners rather than directly across from one another. Round or irregular-shaped boards may use three tabs.

Board stiffness is of concern when using break-aways, particularly where wave soldering will be required. For large panels needing stiffness, the 12.7-mm (0.5-inch) separator strips improve transverse rigidity. Where stiffeners are not required, individual boards should be separated only by a single route line and tabs.

The perimeter, or frame, serves to give the panel stiffness, provides free space for assembly handling, tooling holes, optical targets, and bar-code labels. Circuit board manufacturers also require a frame. If your panel frame meets their minimum requirements, they can save board material by using your frame. Discuss frame and tooling-hole size, and copper-free areas around the optical features, with your PWB vendor early to take advantage of such savings.

The removal of tabs is often approached as an afterthought. Involving manufacturing engineering early in the process can save headaches. A 90-pound assembly operator with a pair of diagonal pliers small enough to fit into the routed slot is a poor, but frequently applied, alternative to forethought on this matter. If no automation is available, a clever solution is to use carbide-blade nibblers. Unfortunately, we don't know of any available from the shelf. You'll probably have to buy steel-bladed nibblers and have a toolmaker replace the blades. This effort produces a tool which self-locates in the slots and removes tabs with a minimum of effort.

Alternatively, individual boards may be depanelized after assembly by specialized depanelizing shears, or by other cutting methods, such as lasers or high-pressure water jets. All boards on a panel should be oriented identically. Panelization within the CAD system and photoplotting of the panelized phototools will minimize any individual feature registration errors.

Additional common methods for panel separation include scribe or notch and break, perforation (a string of closely-spaced unplated drills around each circuit, and punchout-and-reinsert. Scribe or punch techniques are particularly suited to punchable materials such as CEM-1 or CEM-3. Vendors who supply appropriate laminates can give guidelines for depaneling schemes fitting their shop practices.

Whether or not boards are panelized, the tooling holes should be standardized per manufacturing equipment requirements and PWB fabrication considerations. The tooling holes should always be left unplated, to produce closely toleranced diameters. For the sake of accuracy, tooling holes should always be drilled during the first drill operation, prior to through-plating of the holes. To leave the holes unplated, they are tented with the plating resist. Holes larger than 3.1785 mm (0.125 inch) cannot be reliably tented and, therefore, should not be used.

Tooling Holes and Fiducial Marks

Clear areas are generally required around tooling features, such as locating holes and fiducial marks. The amount of clearance required is determined by the equipment used in your line. Equipment vendors can give you these specification for incorporation in your "house standards."

Warpage

Board warpage presents problems to the automated handling equipment, from the front to the back ends of the line. Warpage is a particularly troublesome problem for wave soldering, where extreme cases may cause voids or submarine diving. Any warpage in wave soldering will affect the time that the solder side of the board spends in the wave. Warped boards also may cause problems for adhesive or solder-paste dispensing operations, and for placement equipment. Within board cost constraints, warpage should be kept, by specification, to an absolute minimum.

8.2.2 *Reflow Soldering Yield Enhancement by Design*

Even Heat-Sink Mass Distribution

Keep the heat-sink capacity of the vias nearly even. Where a via makes connection to an internal power or ground layer, neck the connection down to prevent excessive heat sinking through the via, if that via must be soldered or is near a solder land. The recommended thermal-path break to an internal layer is shown in Figure 8-12. This strategy may be built into CAD design rules so that it is always followed, unless there are reasons to override the rule.

Figure 8-12
Breaking the thermal path through a via.

NOTE: Wagon wheels are commonly produced with four spokes where greater electrical conductivity is required.

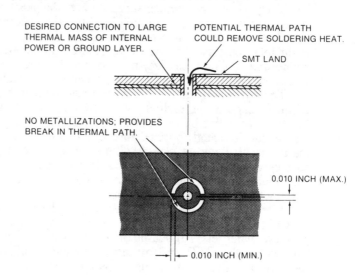

Land Geometries

Certain manufacturing defects relate specifically to land geometries. Lands that are too large or too small will produce a high number of reflow-soldering defects, as illustrated in Figure 8-13. Use these illustrations to help the team understand the results they will see in prototyping.

Large lands bring more solder to the joint, creating an overly stiff structure. Figure 5-20 (in Chapter 5) showed that tin/lead solder has a TCE of 23 to 30 ppm/°C. Ceramic components, such as chip capacitors and resistors, have a TCE of about 8.3 ppm/°C. With the common practice of using a nickel barrier to open up the soldering-process window regarding leaching (see the discussion on chip capacitor terminations in Chapter 2), we introduce a very hard and brittle layer between two objects moving at very different rates. When the solder joint is sufficiently small, the inherent ductility of the tin/lead solder accommodates this mismatch fairly well. However, if too much solder is applied, expect to see cracking of components, the delamination of end terminations, etc. As evidence, look at the TCE-produced stress crack shown in **Figure 8-14**. This chip capacitor has only one end soldered to the board to eliminate any influence of the TCE mismatch between its ceramic body and the board. Nonetheless, it cracked due to the disparate TCEs of the component and the solder. And, the crack is along the nickel/solder boundary.

Figure 8-13
Reflow-soldering land-geometry-related defects.

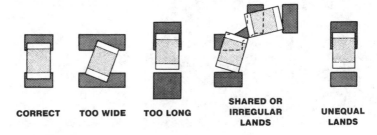

Figure 8-14
Crack in a capacitor induced by solder/ceramic TCE mismatch. *(Courtesy Texas Instruments Incorporated.)*

NOTE: Only one end of this capacitor is soldered to the board, so it is abundantly clear that mismatches between the CTE of the board and component materials had no role in producing this cracking.

When using nickel in a solder termination, be aware of the potential for the formation of tin/nickel intermetallics. There are two intermetallics of concern. Stable Ni_3Sn_4 forms very slowly at room temperature, but its formation speed increases significantly with temperature. Ni_3Sn_4 is a concern where its formation during soldering or rework compromises the solder-joint integrity. A metastable intermetallic, $NiSn_3$, forms at room temperature. Its formation accelerates with increasing temperature up to about 145 °C, and then decreases with any continuing temperature increases. The metastable intermetallic may be a concern for device solderability where devices are stored for long periods.

The stable Ni_3Sn_4 tin/nickel intermetallic is extremely brittle, and this can contribute to cracking along its boundary with adjacent structures, under thermal cycling stress. When microsectioning a failed part with a nickel barrier, and you

Figure 8-15 Tin/nickel intermetallic formation. *(Courtesy Amp, Inc.)*

(A) 75% Sn/25% Pb over nickel, aged 16 days at 75 °C, with solder removed (shown at 1000X magnification. Note that there are virtually no platelets of metastable intermetallic.)

(B) The large platelets seen here (at 1000X magnification) are the metastable SnNi. Sample aged 1.5 years at 50 °C. Photograph shows intermetallic growth in matte Sn, with nickel underplate.

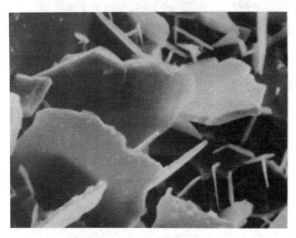

(C) Same sample as seen in Figure 8-17B, shown here at 5000X magnification.

(D) Photograph shows metastable intermetallic growth in pure tin at 1000X magnification. (Sample aged 1.5 years at 50 °C.)

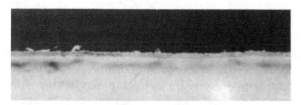

(E) Photograph shows depression of intermetallic growth by 10% lead in a tin/lead coating. (Sample aged 1.5 years at 50 °C, 1000X magnification.)

(F) There is virtually no metastable intermetallic growth in Sn 60%/Pb 40%. (Sample aged 1.5 years at 50 °C, 1000X magnification.)

see a thin intermetallic at the nickel boundary and a tin depletion in the grain structure of the adjacent tin/lead, the Ni_3Sn_4 intermetallic should be a suspect. **Figure 8-15** shows details of both types of intermetallics, plus the depression of metastable $NiSn_3$ growth in the presence of significant amounts of lead. For more information on these phenomena, see Reference 7.

Tombstoning

We've mentioned drawbridging, or popping wheelies. We covered design directions to minimize this defect. But what do you do when your prototype comes out of the reflow oven covered with chips that are standing up like skyscrapers (the "Manhattan effect")? As the formulae in **Figure 8-16** show, component motion is influenced by the aspect ratio of the part and the size of the pad. The formulae provide a mathemathical model for predicting and controlling tombstoning.

Figure 8-16 A model to predict and control tombstoning.[8]

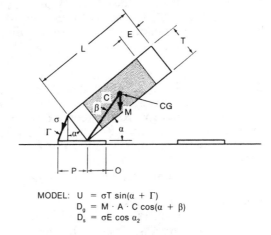

$$\text{MODEL:} \quad U = \sigma T \sin(\alpha + \Gamma)$$
$$D_g = M \cdot A \cdot C \cos(\alpha + \beta)$$
$$D_s = \sigma E \cos \alpha_2$$

Three forces interact to determine chip component motion during reflow soldering. Gravity pulls the chip downward and tends to oppose tombstoning. Surface-attraction forces of solder under the chip also pull it downwards. Opposing these, the surface-attractive forces of the solder fillet (to the left of the chip in the sketch) create a turning moment, acting to lift the chip about its cantilever point.

To predict the motion of the chips, without cumbersome calculations, the following model does not take into account minimal effects of fillet surface tension to the right of the fulcrum point. The model also assumes that the fillet meniscus is a straight line throughout.

The model formulae are:

$$U = \sigma T \sin (\alpha + \Gamma) \qquad \text{(Eq. 8-3)}$$

$$D_g = M \cdot A \cdot C \cos (\alpha + \beta) \qquad \text{(Eq. 8-4)}$$

$$D_S = \sigma E \cos \frac{\alpha}{2} \qquad \text{(Eq. 8-5)}$$

where,

A   = Acceleration due to gravity,
C   = Distance of CG to cantilever,
CG = Center of gravity of part,
D_g = Downward moment from gravity,
D_s = Downward moment of surface tension under the chip,
E   = Length of bottom termination,

L   = Length of component part,
M   = Mass of part per millimeter of width,
O   = Overlap of component over land,
P   = Projection of land past part,
T   = Thickness of component,
U   = Upward moment from fillet surface tension,
α = Angle of bottom of component to substrate,
β = Angle of bottom to dissector of CG; or, arctan (T/L),
Γ = Angle of fillet and chip end; arctan $\dfrac{(P - T \sin \alpha)}{(T \cos \alpha)}$

Tombstoning occurs when the upward exceeds the downward moment, or U > D_g + D_s. From the formulae, we see that a chip of large T/L ratio will have a greater tendency to tombstone. Also, chips with a very short E-termination extension are more prone to the defect. With chips of typical aspect ratios and termination dimensions, the balance between upward and downward moments is tenuous. In such cases, minor influences from vibration, condensing soldering fluids, evaporating gases, etc., can cause a defect.

Reflow: Soldering Component and Trace Orientation
We covered this topic in detail in Chapter 5, Section 5.1. But it is important enough that it bears mention in any discussion of design review. Are the components oriented and spaced for good reflow yields? Have correct trace size and orientation rules been followed? Are the reflow lands sized properly for the process?

Dummy traces, such as shown in Appendix A, Figure A-46, are a necessity where parts must have track routed underneath. Of course, if two tracks are to be routed under the part, the tracks can simply take the place of the dummy tracks.

Materials Specifications
Specify solder pastes based on the manufacturing process to be used to dispense solder on your boards. Table 8-1 shows typical solder paste parameters for the three primary paste application methods. Additional information may be obtained from paste and application equipment suppliers.

Table 8-1. Guide to Solder Paste Specifications

Paste Property	Application Method		
	Screen	*Stencil*	*Dispenser*
Mesh Size	-270 +500	-200 +325	-270 +500
Viscosity KCps	600 900	700 1100	300 600
Percent Metal Weight	86 89	89 92	85 90

8.2.3   *Flow-Soldering Yield Enhancement by Design*

Gas Escape Holes
Gas bubbles can form around components as they pass through molten solder in the flow-soldering process. Bubbles can come from cavitation in the wave,

outgassing of flux volatiles, or outgassing of heated materials. Whatever their source, if there is nowhere for the bubbles to go, you'll think they came from the devil. They will cause solder skips in the flow-soldering operation. The result is that a painstaking and expensive solder touch-up will be required.

By providing escape holes for any entrapped gases, solder yields may be dramatically improved. Some recommended locations for gas escape holes are shown in **Figure 8-17**. Do not place breather holes under a top-side component where flux or solder from the soldering operation might bubble up and hide.

Figure 8-17 Gas escape holes for flow-soldering entrapped gases.

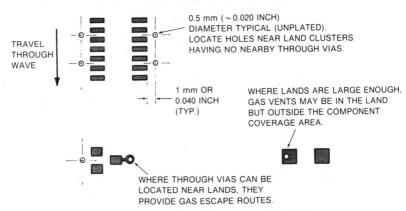

Solder Thieves
In the preceding paragraph, we dealt with a design trick to reduce solder-starving to flow-soldered joints. Curiously, right next door to this problem, we may find solder bridging because of too much solder on the same assembly. *Solder thieves* give us a design trick that can be used to promote a more even distribution of solder. Solder thieves are small extraneous metalized areas, which cause the excess solder to flow away from the lands so it does not form bridges. They work in the same way that plating thieves act to distribute the current densities during board plating. **Figure 8-18** shows some recommendations for solder thieves for flow soldering.

Figure 8-18 Solder thieves soak up excess solder.

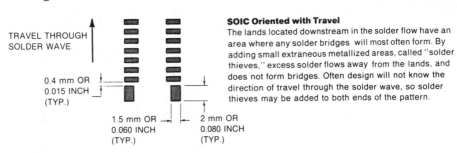

Wave-Soldering Component and Trace Orientation
If component orientation, spacing, and land geometry deserved mention in our previous reflow-soldering discussion, they deserve double attention here. Proper orientation is a critical issue to flow-soldering yields. In your design review, check that all previously developed rules have been rigorously enforced.

8.2.4   *Cleaning Enhancement by Design*

The design team should be aware that, minute by minute after the boards leave the soldering oven, board cleaning gets more difficult. Think of SMT assemblies like dirty dishes. Right after dinner, a quick dip in a little hot soapy water will leave your good china sparkling and squeaky clean. Leave those same dishes on the counter to dry for three days and then washing them becomes a monstrous chore.

Ideally, the product should flow into the cleaner while it's still warm to the touch from soldering. Flux residues polymerize over time, after soldering. Therefore, avoid designs that must be routed to other processes between the soldering and cleaning.

Component standoff has a dramatic impact on how quickly an assembly can be cleaned to a specified level. Where alternatives are available, the package with the greater standoff will enhance the cleaning processes. A side benefit is that a greater standoff will usually spell more compliance to absorb TCE and bending stresses.

8.2.5   *Design for Inspection*

Whether a product will be inspected by operators during hand assembly, or by sophisticated vision robots, automated test equipment (ATE), etc., an adherence to basic inspectability design rules will simplify the job and improve the results.

Visual Inspection

For visual inspection, the clearance between devices must be sufficient to allow a good view of solder fillets on the device leads. Repair operations, as well, demand a clearance, both for safe reflow and for visibility. **Figure 8-19** shows clearance design rules for ensuring visibility during inspection and repair. Table 5-1 in Chapter 5, Section 5.2.1, lists the clearances determined with these 60°/45° rules for typical components. The rules may be used to calculate clearances for any parts not included in the table.

Figure 8-19
Design rules for inspection visibility.

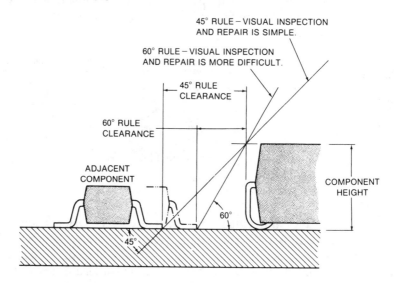

Note that, in spite of numerous articles in the trade press stating the contrary, solder joints on J-lead PLCCs may be given a quick visual inspection easier than can gull-wing SOICs. Where there is insufficient or no solder on a J lead, the immediately discernible point of light reflected by a good fillet is replaced by an obvious black cavern. With the gull-wing lead form, it takes close observation to determine if the reflected light is coming from the lead itself or from a solder fillet. Therefore, where real estate permits, it is good practice to always use the 45° rules or better for gull-wing leads, even on boards using the 60° rules for other components.

In **Figure 8-20** the PLCC solder joints, in view, are all acceptable, as evidenced by the distinct points of light they reflect. Solder-joint integrity of the gull-wing SOICs can't be determined in this photo, even though they are in the foreground.

Figure 8-20
Visual inspection of "J" and gull-wing lead forms.

Automated Vision Systems

The vision systems used in manufacturing can provide valuable process-control data for SPC programs, and an early warning of drift in a process. Early knowledge of process drift lets manufacturing recognize and cure machine wear, misadjustment, etc., before it produces defects at final test.[9] Several design rules may be imposed by vision systems. We'll consider them by category, based on the type of vision inspection in question (Table 8-2).

PWB registration systems use vision to accurately register the surface features of SMT workpieces for a process operation. In IMC assembly, the boards were normally registered by pins engaging tooling holes in the board. This strategy works well for through-hole devices because the holes we are trying to register are in close registration to the tooling holes, having been fabricated in the same close-tolerance drilling operation. However, surface features, such as SMT lands, are not fabricated in the same operation as the tooling holes. Thus, there may be considerable misregistration between the lands and the tooling holes. Board-registration vision automates the orientation of the board's surface features by finding targets (called fiducials) fabricated in the same high-accuracy photographic processes used to construct the lands.

Applications	Design Considerations	
	Clearance Rules by Application	*Requirement for Fiducials*
PW Board Registration	Clearance required by vision system, features to fiducials.	Size, shape, location, and contrast as specified by vision equipment.
Dispensing Inspection	Generally, vision inspection can accept standard design-rule clearances.	Same as PWB registration, where vision inspection is used to register workpiece.
Vision-Guided Auto Placement	Systems aligning to fiducials by component need extra space.	Fiducials may be required near component for targeting.
Post-Placement Inspection	Generally, vision inspection can accept standard design-rule clearances.	Same as PWB registration, where vision inspection is used to register workpiece.
Solder-Joint Inspection	Generally, vision inspection can accept standard design-rule clearances.	Same as PWB registration, where vision inspection is used to register workpiece.

For good results, fiducials should be placed close to three of the four corners of the board. For panelized boards, targets may be on three corners of the panel if individual circuits are closely registered with the panel. However, any error from step-and-repeat operation in panelization may be canceled by placing fiducials on each individual circuit. The three-fiducial approach allows for the determining of separate offset and size corrections for the X, Y, and 0 axes, and for orthogonal errors (perpendicularity errors in the X-Y grid system). Fiducial registration systems assume that all these errors are linear across the board. For 1.27-mm (50-mil) center components on ordinary sized boards, we can mathematically demonstrate that this is a safe assumption. For fine-pitch component handling, systems allowing localized feature recognition and component lead-to-pad correction are a necessity.[10]

Whether fiducials are being registered by the vision equipment or actual SMT lands are used as targets, sufficient clear space (free from circuit features) should be provided around the vision target to preclude the vision system from getting mixed up. Note that the clear area requirement often must be applied to inner as well as outer layers. IR wavelengths are quite distinct from visible light in their propensity to penetrate polymer materials and reflect off inner layers of metallization. Registering from the wrong feature will probably cause undesirable quirks in the assembly process. Consult your vision system vendor for required clearances.

Contrast between the target and board material is important in vision system accuracy. Those of us who don't deal in vision equipment tend to think of the video camera as an eye, and the computer that decides the image as a human brain. Actual experience with vision system capabilities has given me a new appreciation for the wonder of God's creation. Whereas human eyes can distinguish minute density and hue variation, machines routinely see a green PWB mask, a white silkscreen legend, and silvery metallizations as "one continuous field." Human eyes easily adjust to the color temperatures of incandescent or fluorescent light, or sunlight. Machines are miraculously confused by minor changes in light levels or color temperature. The best motto is: "Provide all the contrast possible between target and surrounding board." Be sure that the contrast is evaluated by comparing the reflectiveness of the board surface (mask, etc.) and target material at the wavelength of illuminating light in the vision system. Materials that look very different to us may look very similar to an infrared camera, and vice versa.

Target shape is determined by the vision system used. Better systems allow a wide latitude of feature shapes. Where the choice is up to you, consider some variation of the shapes shown in **Figure 8-21** since they provide visual verification of PWB etch control as well as good optical targets.

Figure 8-21
Recommended
optical target
shapes.

Note: Current standards call for a simple round dot of metallization for fiducials, and most low-end vision systems will not recognize anything else accurately. Still, where vision-system capabilities allow, the feature shown here is much less likely to be confused with other common board artifacts. It provides the additional benefit of a quick visual indication of over or under etch.

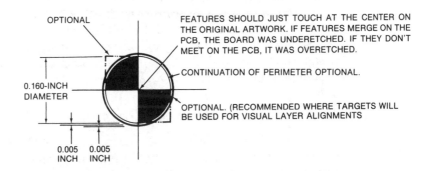

Dispensing inspection systems are used immediately after a dispensing operation to verify the presence and uniformity of the attachment media, such as solder paste or epoxy. Fiducials may be required for the automated registration of the workpiece prior to inspection. For special clearance and contrast rules around the features to be inspected, consult the inspection equipment vendor.

Vision-guided automatic placement refers to the use of integrated vision inspection of the component leads/metallizations and the actual placement land features. Such systems provide correction for both linear and nonlinear PWB error, plus compensation for component errors. Land-to-part systems are often used to achieve the required accuracies for fine-pitch and high I/O-component placement. With fine-pitch parts, the minimal error budget necessary for acceptable placements is generally exceeded by the board and component errors, meaning that the placement system must somehow provide less than 0.0 error. In other words, the placement system must compensate for built-in material errors if reasonable yields are to be achieved.

For design rules regarding clear space around the inspected features, and feature-to-board contrast in land-to-part-guided placement, consult the vision or placement system vendor.

Post-placement inspection involves automated vision verification of the assembly process. Simple systems may do little more than verify that components have been placed in all the required locations. More sophisticated machines will verify the component presence, and will detect the correct orientation of polarity and pin-1 registration features. The most powerful machine can even verify component markings and can do a visual inspection of the components for cosmetic defects.

Such scanners are usually located between the placement and soldering operations, so that any defects may be touched up easily prior to soldering.

Component spacing requirements, compatible components, and contrast rules vary widely between vendors. So, consult your vision equipment supplier for applicable design rules.

Solder-joint inspection is performed by several competing technologies. X-ray systems allow looking inside the solder joints, under components, and even inside boards. The Z-axis focal point of the picture is on-line adjustable, so you can actually dial through the component's wire bonds, the solder joints, and each individual layer of a multilayer PWB. Such systems are very useful for locating hidden defects, like broken wire bonds, gas entrapment voids, blowholes, or hidden contaminants inside solder joints.

X-ray systems are slow inspection tools, and a computer interpretation of their data requires a great deal of custom development work. Therefore, X rays are used more in failure analysis and sample testing than they are in on-line automated test. However, from the designer's viewpoint, an X-ray machine's ability to see through things makes the machine very tolerant of circuit layout variations. Generally, any design complying with the basic spacing and clearance rules given in Chapter 5 should present no problem in X-ray inspection.

8.2.6 *Designing so Rework Is Possible*

"Avoid rework, it's as much fun as a heart attack." SMT veterans have stressed this until it's become a cliche. Actually, SMCs are relatively easy to remove and replace. But the job is quite time-consuming. It requires specialized tools and skills. And, rework expertise is not plentiful as of this writing. There are a host of otherwise high-tech manufacturers who are out of control in their rework process, and they don't have any idea of what reliability their reworked boards may have. The moral of the rework challenge is twofold.

1. Give heavy weight to those design suggestions which improve manufacturing yields. Even infrequently occurring defects should be eliminated by design rather than by rework.

2. Develop controlled rework processes and procedures. Know how much heat adjacent components will see during rework. There should be a limit on rework time in any area of the board, established from observed time at reflow temperatures, before problems develop. Problems can include damage to the PWB, copper dissolution into solder, damage to adjacent components, and the component-termination materials leaching into multiple reflowed joints. Whenever a component is removed, the solder should also be removed. This avoids formation of excessive copper/tin intermetallics, which degrade solder performance.

Documentation
A basic requirement of any repair/rework department is good documentation. But, IMC assemblies are usually easy to second guess when documentation is not available. This is not the case with SMT. The components are small. Many look like identical twins, even though they come from entirely different parents. Often there is little or no nomenclature marked on SMCs to allow us to visually distinguish one from the next. And SMCs may be packed on the circuit like sardines in a can. After several engineering changes, the original circuit designer may not be able to figure out "who's on first" without considerable bench time. Clear and rigorously maintained documentation and revision-level markings are mandatory for SMT work.

8.3 Nonmanufacturing Issues in Design Reviews

This entire section of the design review should be under the tutelage of the marketing and sales representatives of the team. Here is where the designers must somehow get their heart off their sleeve, and look at the new product like the most critical of reviewers—the customers—might view it. Below, we'll develop a few tools to help in this delicate cardiac surgery.

8.3.1 *Material Costs*

First, how much will it have to cost the customer? If it's cheap enough, customers will be a bit forgiving. But all the patriotism in the world won't protect those U.S. manufacturers of consumer goods which cost twice as much, don't last half as long, and have less features than the goods of our foreign neighbors.

8.3.2 *Reliability*

Under this heading, Step one must be the establishment of a reliability target. Most engineers want to build the most reliable product possible. As proof mounts regarding the cost advantages of good SQC, even the stingiest managers are joining the quality-first ranks.

Today's competitive marketplace viciously enforces attention to quality and reliability in the design of products. Customers expect products that:

- Work as they're supposed to work (are quality products)
- Continue to work for a reasonable time (are reliable)
- Require infrequent and inexpensive repair (maintainable)
- Are easy to use (user friendly)
- Are safe to use and don't pollute the environment (safe)
- Are esthetically pleasing (sexy)[11]

But, as pragmatic thinkers, we realize that these goals may come in conflict with material costs. No one doubts the reliability of a Rolls Royce. But drivers who can afford Rolls Royces are few and far between. Within product-cost limits, is it as reliable as it can possibly be? More importantly, is it as reliable as it should be? These are the questions that the team must answer.

One simple way to determine reliability (in review, the product's ability to work according to specifications for an extended period of time) is to put the product in service and see how long it lasts. Unfortunately, this simplistic approach often is too slow in flagging problems. In the process of a design review, the team should be constantly mindful of the uncompromising and often unsympathetic standards that the customer brings to the judgment of their work. To ensure the design will meet and exceed the customer's expectations, the reliability of electronic products can be predicted using methods presented in MIL-HDBK-217D.[12]

8.3.3 *Marketability*

The most basic step of marketability is to see that the product really does what it is intended to do. This sounds so obvious as to not need mention. But, with a little thought, we all can recall irritating examples that prove product reliability is sometimes bypassed in the headlong rush to market new products.

After we've established, by testing breadboards, that the product will do what it is supposed to do, we must ask some disturbing questions. Is this what the customer wants done? How does it stack up against the competitor's products.

Benchmarketing is a management term meaning picking the best in a given category and setting it as a benchmark that must be excelled by your design. By benchmarking every facet of a design, you can assure excellence in any product that meets the tests. Table 8-3 lists areas that generally deserve benchmarks. In some, marketing may target them to be best in a niche, rather than best in the world. Our Rolls Royce example serves to show why.

In any case, benchmarking is a valuable tool when used to defuse design ego and view a product objectively. And we know from personal experience that, after we've lived and breathed and sweated with a design project for 18 months, we may need a powerful tool to separate ego from design. Yet, until this separation occurs, no objective evaluation can be done.

Table 8-3. Guide to Marketing Benchmarks

Item	Competitive Evaluation			
	System A	*System B*	*System C*	*Target*
Credibility	Claimed	Advertised	Customer test	Customer letters
Delivery	30 days	2 to 6 weeks	3 to 4 weeks	<2 weeks
Features	3 standard, 1 new	3 standard	3 standard, 2 new	3 standard, 3 new
Price	$1000.00	$1100.00	$1500.00	$995.00
Reliability	B-10 of 1 year	B-10 of 1 year	B-10 of 2 years	B-10, 32 months
Service	Slow and costly	Few centers	Factory (slow)	Factory (fast)

* Analysis shown is of a hypothetical product and niche.

Credibility
Interpreting Table 8-3, by credibility, we mean how believable will the device's claims be. How easy will it be for customers to understand the product's plus points? Will you be able to win customer support by simply advertising the demonstrable truth about the product?

A major airline once set a milestone in regrettable advertising with a series of commericals showing travelers being herded into the innards of a plane that was filled wall to wall with cattle. The implication was that the other guy's planes had no more space than a cattle car. The visual impact and attention-grabbing value of the advertisement was undeniable. It hit right at the heart of all travelers who felt like they were being herded to the meat market in tiny plane seats. But it didn't achieve sales results for the airline. Why? Because the airline running the commercial had some of the tightest seating in the industry. The clever and catchy advertisement couldn't cover up for a lack of credibility in the basic message.

For top-flight credibility, develop your product with features that are easily understood. Features should be clearly demonstrated to work in solving customer problems and providing customer needs. To beat the benchmark, your product's features should do this better than any competition in the niche.

Delivery
Here, design for manufacturability is the cornerstone of success. Will you be able to get it to the customer before the competition gets him one of theirs? Delivery alone has decided many a sales contest when the customer's need was critical.

Delivery also is important in obtaining the impulse buyer's nod. These two substantial customer segments are virtually hidden from view for the company which lags competitors in delivery.

Features

Benefits are what sell a product. Features are always related to advantages and benefits. For instance, compare a cellular phone to a radio telephone of the past. The cellular phone features automated cell assignment and central-station switching. This means that the phone automatically switches to the nearest receiving station as you move through a city. The advantages are clear crisp sound rivaling fixed-line telephones and enough frequencies to meet the user demand. Older radio telephones suffered from poor reception in some areas, and there just weren't enough frequencies to go around. The benefit to the user is a mobile phone which provides the communications capabilities of an office telephone, regardless of where you take it. The customer likes it because it lets him keep in touch, not because he knows it employs cell-switching technology. In other words, it's the benefit that sells products, not the feature. When viewed by benefits, does your product exceed the market benchmark?

Reliability

Has the product been tested to establish that it will perform as specified over its full life cycle? Can these tests be demonstrated to those customers concerned with reliability and quality for their money? Is the product more reliable than the competitive benchmark?

Service

Even the most reliable hardware occasionally needs a tune-up. Customer enthusiasm can turn to bitterness if a good product costs too much, or if service takes too long. And customers are most unsympathetic when they pay their good money for service, and the thing still doesn't work. Is the product-service strategy, implementation, pricing, speed, and testing the best in your market?

Price

There's nothing arcane about this section of the marketability evaluation. If you can develop a product that meets and exceeds competitive benchmarks on the other five fronts, and it can profitably be sold for less than all the other market entries, you'll have a story customers are likely to find interesting. Advertising and selling such a product is a joy. Sales can look a customer straight in the eye and categorically state, "For the money, we just don't think you can beat us." No sales voodoo or hocus-pocus sells like this sort of honest conviction.

8.3.4 Conclusion

When the team designs a product that fits all the SMT rules of Chapters 1 through 7, meets any required rules in Chapters 9 and 10, and then beats all of the above six market benchmarks, get set for a resounding success. You will have it.

8.4 References

1. Thomas, Leonard C., "Examination of Thermal Analysis for PWB, Electronic Base Material, Part 1," *Electri·Onics*, November 1986.

2. Paul, Willard, Jr., *Reader's Digest*, March 1988, pg. 47.

3. Leibson, Steven H., "EDN's Hands-On Project—Part 2: Selecting the Surface-Mount Components," *EDN*, June 11, 1987, pg. 171.

4. *EDN News*, July 16, 1987, pg. 62 inset.

5. Barton, Martin, "Implementing Surface Mount Technology (SMT) into Telecommications Products," *Proceedings of the 1987 SMTA Technical Symposium*, SMTA, Edina, MN, October 28th, 1987.

6. Boccia, Louis J., "A Simplified Geometrical Model for Lead Planarity," *Printed Circuit Assembly*, April 1988, pg. 12.

7. Haimovich, Joseph, Ph.D., "Intermetallic Compound Growth in Tin and Tin-Lead Platings Over Nickel and Its Effects on Solderability," *12th Annual Electronics Manufacturing Seminar Proceedings*, Naval Weapons Center, China Lake, CA, February 1988, pg. 51.

8. van Buul, J.A.N., and Gnant, Russell S., "Advances in Hybrid Chip Placement," *Semiconductor International*, September 1987.

9. Mead, Dr. Donald C., "Machine Vision in SMT," *Assembly Engineering*, September 1987, pg. 40.

10. Amick, Christopher G., "Machine Accuracy Requirements for Surface Mounted Assembly," *Proceedings of the Society of Manufacturing Engineers E'Ssembly Conference*, Society of Manufacturing Engineers, Ann Arbor, MI, 1986.

11. Pennucci, Nicholas J., "Estimating Reliability Using MIL-HDBK-217D," *Quality*, September 1986, pg. 40.

12. MIL-HDBK-217D, *Reliability Prediction of Electronic Equipment*.

Figure 9 SMT, hybrid, and surface mounting of IMCs combined in a compact DC/DC converter. *(Courtesy Computer Products.)*

9 Hybrid Circuits and Multi-Chip Modules (MCMs)

Together, surface-mount and hybrid technology make a powerful team. In this chapter, we'll focus on how they can be allied to miniaturize and improve electronic circuitry. Figure 9 is a good case in point, and it serves to introduce this discussion. The circuit density of SMT, the power-dissipation ability of a hybrid substrate, and the fact that IMCs can be lead prepped for surface mounting, all combine to make this circuit possible. The assembly is a 3-watt DC/DC converter. Its compact 24-pin DIP packaging takes up just 6.45 square centimeters (1.0 square inch) of board area.

9.1 Overview of Hybrid Technology

9.1.1 The Changing Hybrid Market

The hybrid marketplace, worldwide, is in the midst of a revolution fueled by surface-mount technology. This revolution will drastically alter the technology

[1]NOTE: While we will not take up MCMs until the end of this chapter, all the design rules for ceramic and polymer hybrids, presented in the first 8 sections of the chapter, are germane to that discussion.

and design of circuits. But, more importantly, it will spell enormous growth opportunities for hybrid manufacturers.

The United Kingdom chapter of ISHM conducted a survey of the European hybrid market. Its findings are quite revealing. The overall market growth from 1986 to 1990 will be a robust 75%, or a growth from about $1.75 billion to $2.6 billion. As this growth occurs, the technology mix will shift markedly toward the use of SMCs in small-board modules. Figure 9-1 shows the overall market growth and expansion of surface-mount technology as uncovered by the ISHM survey.[1]

Figure 9-1 European hybrid market performance—1986 to 1990.

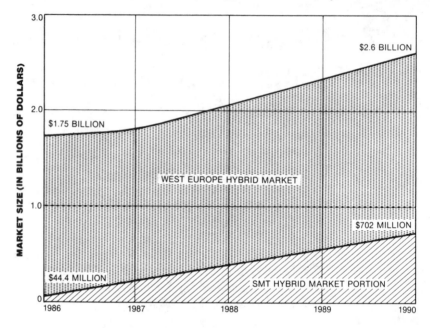

SMT, a 40-Year-Old Hybrid Technology?

The dramatic changes that are underway in hybrid devices, as a result of SMT, should be sufficient evidence to set aside any notion that surface mounting is the same old stuff that hybrid engineers have been doing for the past 40 years. Indeed, components have been surface mounted on hybrids since the late 1950s. The oldest SMT circuit I have seen was a military product designed by Centralabs in World War II. But, recent advances in IC and passive component packaging have brought a host of new SMT benefits to the designer of hybrid circuits. Assemblies like the one shown in Figure 9-2, with a wall-to-wall array or 284-pin TAB devices, would certainly have been news even five years ago, and would have seemed like a Jules Verne fantasy back in 1960. Each of the nine 9.9-mm-square (0.390-inch) chips packs an equivalent of 20,000 2-input NAND gates, and has 238-signal and 46-power/ground pins. The alumina substrate is metallized with copper thick film. We are not talking here about business as usual.

What We Mean by Hybrid

Definitions for "hybrid" are just about as plentiful as hybrid circuits themselves. So let's start our discussion by defining what we mean, in this text, by "hybrid." A hybrid is "any circuit having at least one element which is added to the assembly as a separate discrete component." Note that by this definition, a standard through-hole component FR-4 PWB, with one inductor formed by a spiral trace of printed wiring, is a hybrid. And an alumina ceramic circuit filled with chip-and-wire ICs, but having no substrate integral elements, is not a hybrid. Admittedly, our classification has its humorous challenges. But, for the sake of discussion and study, the line must be drawn somewhere.

Figure 9-2
Screened multi-layer copper thick-film TAB hybrid. *(Courtesy Honeywell Bull.)* The part shown is an early example of a high-density MCM-C.

9.1.2 An Overview of Thick-Film Processing

Thick film refers to the printing method used to fabricate circuit-board features on hybrid circuits named for the process. Thick-film conductors and circuit elements are printed onto circuits using screen or stencil printing techniques. Thick-film printing techniques typically produce film thicknesses from 7.5 microns (~0.0003 inch) to 12.5 microns (~0.0005 inch). Thin films, as described below, are generally from 200 angstroms (0.0000079 inch) to 5 microns (~0.0002 inch).[2]

Figure 9-3 shows a typical process flow which would be used to build a multilayer thick-film circuit. As indicated, these techniques allow the construction of resistive, capacitive, and inductive elements as well as circuit conductors and bonding sites. Special techniques also allow the construction of thick-film opto-electronic and electroluminescent elements.

Figure 9-3 Typical process flow for multilayer thick-film circuits.

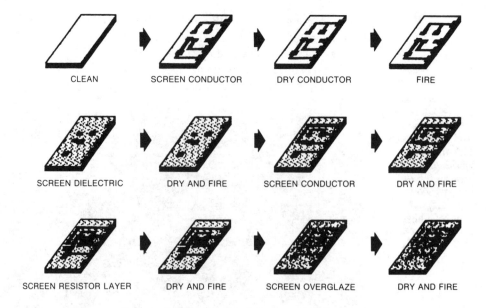

9.1.3   *An Overview of Thin-Film Processing*

Thin films in the thickness range of 200 to 5000 angstroms are produced by vacuum deposition using sputtering or pyrolytic evaporation techniques. Evaporative deposition through a mask may be used to build up additive circuitry. But, subtractive processes are more common. Subtractive circuits are produced by sequential evaporative deposition of thin films over the entire substrate surface, followed by a photolithographic removal of the undesired portions of the film. In the subtractive process, the resistive coating is generally applied first, and then covered by a conductive film. The desired resistors are exposed by conductor etching. A typical thin-film process flow is presented in Figure 9-4.

The preceding discussion is an introduction to what a hybrid is. Since our focus here is on SMT, we'll have to leave the in-depth study of this technology to other texts, a number of which are listed in our bibliography.

Turning our attention to SMT, let's explore the benefits and limitations of surface mounting for hybrid circuits. We will look at the rules for SMT use in hybrid circuits today. And, we will study examples of actual circuits where these rules have been applied.

9.2 What SMT Offers in Hybrid Circuitry

Surface mounting must clearly provide hybrid engineers some important benefits, otherwise the usage growth shown in Figure 9-1 would not be happening. What are these benefits? And conversely, what can hybrids offer to SMT? When would SMT limitations speak against its use in hybrids? Just how do SMT methods compare to chip-and-wire density? We'll answer these questions in the following pages.

Figure 9-4 Typical process flow for thin-film hybrid substrates.

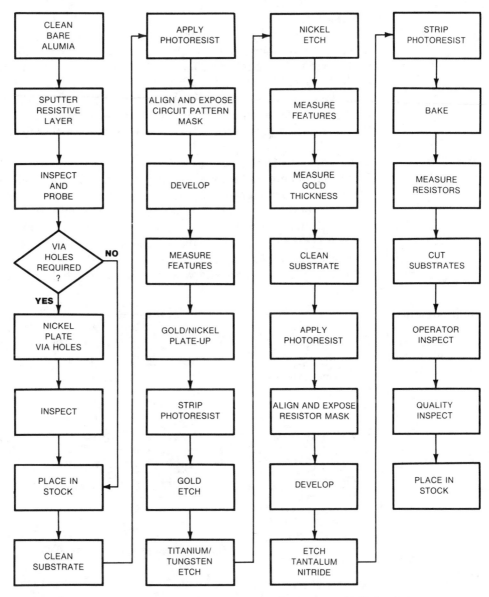

*Thanks to Amplica, Inc., of Newbury Park, California, for technical assistance with this flowchart.

9.2.1 *Benefits of SMT for Hybrids*

SMT benefits for hybrid circuits are somewhat different than the plus points we listed for SMT over IMC. While SMT offers substantial real-estate savings when compared to IMC, it is seldom possible to shrink a hybrid circuit dramatically just by switching from chip-and-wire ICs to SMC packaged devices. But, SMT does provide hybrid engineers with some powerful new tools. In some cases, it may even save board real estate.

Miniaturization

SMT presents a mixed bag in this category. TAB procedures may allow greater miniaturization than chip-and-wire techniques for complex ICs. Some fine-pitch packaged SMCs also compete very well with chip-and-wire density. But 1.27-mm-pitch (0.050-inch) packaged devices generally do not provide the density available with classical hybrid approaches. So miniaturization of existing hybrid circuits probably would involve a combination of custom VLSI and SMT. We'll explore the factors influencing SMT vs. chip-and-wire density further in Section 9.2.4.

Component Protection

Protection of chip-and-wire ICs on a hybrid circuit is a major element of hybrid design. For very benign environments, a simple conformal coating may provide all the protection that's needed. More hostile surroundings often dictate that the entire hybrid be hermetically sealed within a protective package. Such protective packaging can form a major element of the cost of a hybrid circuit. In contrast, SMC packages generally offer all the protection needed for a die. Figure 9-5 compares a hermetically sealed chip-and-wire hybrid with a hybrid using hermetic chip carriers.

Figure 9-5 Comparing approaches for the hermetic protection of a die.

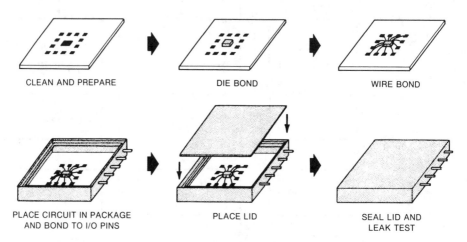

CLEAN AND PREPARE   DIE BOND   WIRE BOND

PLACE CIRCUIT IN PACKAGE   PLACE LID   SEAL LID AND
AND BOND TO I/O PINS   LEAK TEST

(A) Hermetic packaging of chip-and-wire devices.

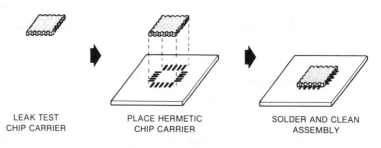

LEAK TEST   PLACE HERMETIC   SOLDER AND CLEAN
CHIP CARRIER   CHIP CARRIER   ASSEMBLY

(B) The prepackaged method.

Component Testability

Bare dies are from difficult to impossible to fully parametrically and environmentally characterize before they are wire bonded to the hybrid circuit. By substituting

SMT ICs, the hybrid manufacturer can fully test devices before assembly. Thus, SMT hybrids tend to provide significantly higher manufacturing first-pass yields. Yield improvements translate directly into lower manufacturing costs.

Ease of Rework

Where chip-and-wire dice are defective, the wires, and eutectic or epoxy die bond, must be removed. The bonding sites must be cleaned and reconditioned. And, a new IC must be die-bonded to the substrate, wire-bonded to the circuit, and tested. This process takes time, costs money, and risks irreparable damage to the circuit assembly. So, as you're doing the repair, you pray the replacement IC passes the first test and doesn't have to go through the whole cycle over again. Using SMCs lets the hybrid manufacturer pretest components and thus reduce the need for rework. And, when rework is needed, it's a simple matter of spot heating the substrate, as shown in Figure 9-6.

Figure 9-6 Spot heating to remove and replace SMCs.

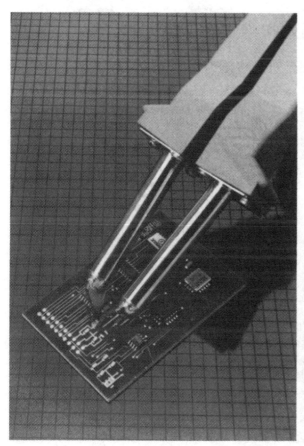

(B) Formed tips for SMT component removal. (Courtesy Hexacon Electric Co.)

(A) Tweezer-style iron. (Courtesy Nu Concept Systems, Inc.)

(C) Special soldering-iron tips for PLCC removal. (Courtesy Hexacon Electric Co.)

Reduction in Screenings

Each successive conductive and dielectric screened layer adds both cost and reliability concerns to a thick-film hybrid. Several resistive screenings may be required for different resistor values. By substituting discrete resistors for odd-value resistances, it may be possible to reduce resistive screenings, thus lowering costs, raising substrate yields, and boosting reliability.

Complex circuits requiring multilayer hybrid interconnect structures are costly. Multilayer hybrids, as of this writing, are limited to about one third the number of layers achievable with polymer MLBs. Using SMT discretes, hybrid designers can replace screened crossovers with discrete components, dramatically lowering the cost of the substrate and the final product. Figure 9-7 compares multilayer hybrid and crossover techniques that are used to achieve the same circuitry.

Figure 9-7 Multilayer hybrid vs. component crossover techniques.

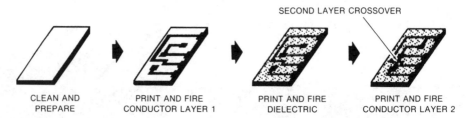

(A) Multilayer crossover.

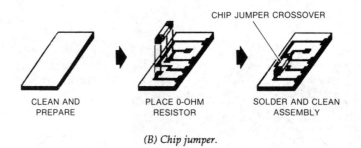

(B) Chip jumper.

Via Fabrication

Vias are costly in all substrate technologies, but particularly so in ceramic boards. In Japan, a component known as the "Zero-ohm resistor" is very popular. It is a jumper used expressly to perform a crossover without resorting to extra layers and associated vias. The one-cent part compares very favorably with the cost of two vias and a trace on a secondary layer.[3] Figure 9-7 shows the use of just such an SMC to avoid layer interconnect vias.

Substrate Limitations

Polymer substrates are restricted, in classic hybrid approaches, by the limitations of polymer-compatible resistive inks, as shown in Table 9-1, and by die and wire-bond site integrity. By allowing use of discrete resistors for high-tolerance requirements, and where TCR is critical, SMT reduces concerns about polymer thick-film (PTF) ink systems. And, packaged SMC semiconductors eliminate concerns about die and wire bonds on polymer-based systems.

Table 9-1. Comparison of Typical Resistive Ink Systems.

Parameter	Thick-Film Technology		Thin-film Technology		
	Ceramic	*Polymer*	*Chromium Si*	*Ni Chrome*	*Ta Nitride*
Conductor Ink (Typical)	Ag, Au, Cu, PdAg, PtAg	Ag, Au, Cu		Al, Au, Cu	Al, Au, Cu
Noise (dB)	–35–+20	0–+35		<–30	<–10
Power Handled (Watts/cm²)	≤15.5	3.875		≤15.5	6.2
Resistance Range (Ohms/Sq.)	1–1000M	10–100K	3.5K–>5K	10–300	5–500
Resistance Stability, in %ÅR. (1000 Hours @ 25 °C/50% RH)	≤ ±0.25	≤ ±2		≤ ±0.1	≤ ±0.1
Resistance Stability, in %ÅR. (Rated Power/1000 Hour/25 °C)	≤ ±0.25	≤ ±3		0.1	≤ ±0.1
Resistance Stability, in %ÅR. (Thermal/150 °C/1000 Hours)	≤ ±0.25	≤ ±2	<0.1	≤ ±0.1	≤ ±0.2
TCR Tracking (PPM)	10	50		1	1
TCR (PPM/°C)	±50–300	±250–1000	–5 to –10	±0 to 50	±40
Trim Stability, in %ÅR.	≤ ±0.25	≤ ±2		≤ ±0.01	≤ ±0.01
Voltage Coefficient Resistor (PPM/V)	0.5–5	25 to 100		0	1

9.2.2 *Benefits of Hybrids for SMT*

The benefit equation is far from a one-way street. Hybrid techniques provided the technological seeds for SMT. And hybrid methods have a great deal to offer SMT engineers. Below, we'll cover some hybrid engineering solutions to surface-mount problems.

Figure 9-8 Comparison of post-leaded and leadless CCCs.

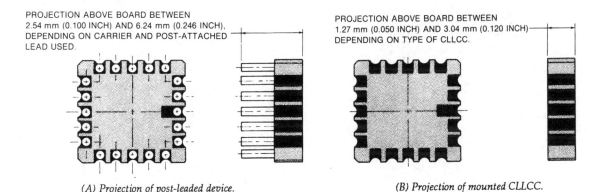

PROJECTION ABOVE BOARD BETWEEN 2.54 mm (0.100 INCH) AND 6.24 mm (0.246 INCH), DEPENDING ON CARRIER AND POST-ATTACHED LEAD USED.

PROJECTION ABOVE BOARD BETWEEN 1.27 mm (0.050 INCH) AND 3.04 mm (0.120 INCH) DEPENDING ON TYPE OF CLLCC.

(A) Projection of post-leaded device.   *(B) Projection of mounted CLLCC.*

Compatible TCEs
When CLLCCs are required for the hermetic protection of ICs, engineers must find a way to eliminate TCE mismatch problems. Adding leads to the chip carriers can solve the problem. But leads add cost and introduce an additional reliability concern by doubling the number of solder joints. Leads also increase the standoff

above the board. For the sake of cleaning, a greater standoff is a benefit, but for close-order stacking of cards, it is a decided disadvantage. These considerations are illustrated in Figure 9-8.

One way to attack the TCE mismatch problem without resorting to leads is to use a substrate matching the TCE of the ceramic chip carrier. Several choices are available. Alumina ceramic is one of these. Alumina is the same material that the chip-carrier package is made from, and also the material of choice for most hybrid substrates.

Using 96% pure alumina substrates, small- to moderate-size CLLCCs may be safely mounted direct to the board, even when high dissipations are involved. Large chip carriers having low internal dissipation may also be direct mounted to compatible substrates. The upward limit is established by the temperature delta between the device under power, and the substrate, and the overall chip-carrier size, as shown by the following formula and the diagram of Figure 9-9.

Figure 9-9 Differential temperature stresses in CLLCCs.[4]

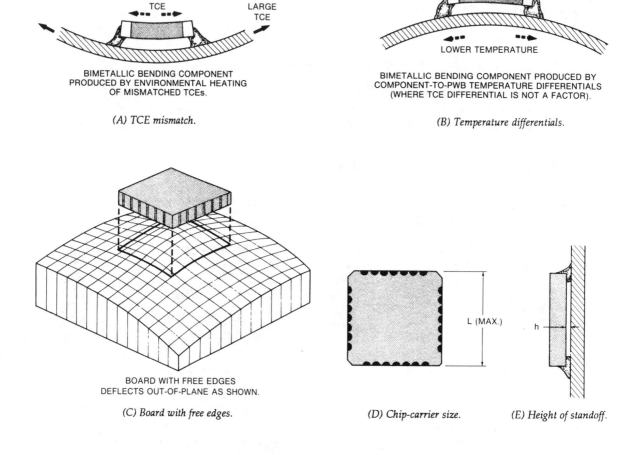

SMALL
TCE

LARGE
TCE

BIMETALLIC BENDING COMPONENT
PRODUCED BY ENVIRONMENTAL HEATING
OF MISMATCHED TCEs.

(A) TCE mismatch.

HIGH TEMPERATURE
DUE TO DISSIPATION

LOWER TEMPERATURE

BIMETALLIC BENDING COMPONENT PRODUCED BY
COMPONENT-TO-PWB TEMPERATURE DIFFERENTIALS
(WHERE TCE DIFFERENTIAL IS NOT A FACTOR).

(B) Temperature differentials.

BOARD WITH FREE EDGES
DEFLECTS OUT-OF-PLANE AS SHOWN.

(C) Board with free edges.

L (MAX.)

(D) Chip-carrier size.

h

(E) Height of standoff.

Power Cycling Produces Complex Stresses

The stresses produced by power on/off cycles are complex, as shown in Figure 9-9. TCE mismatches between ceramic components and polymer boards act like a bimetallic strip, producing a bending stress (Figure 9-9A). Differential heating in component and board tends to oppose this bending. Acting on substances of matching TCE, as shown in Figure 9-9B, differential heating would produce a bimetallic bending effect in the opposite direction. Power dissipation from a mounted chip produces an out-of-plane displacement. This is illustrated in Figure 9-9C.

The Manson-Coffin Equation Modified

Expanding on the original work represented in the generalized Manson-Coffin equation, Norris and Landzberg of IBM modified the formula to define and model the complex bending stresses to account for the effects of maximum temperature and frequency of temperature cycling. Shah and Kelly added consideration of the dwell time at high temperature.

Based on his own work, plus research from Hall and his associates, Englemaier of Bell Laboratories modified the original Manson-Coffin equation, producing the following model:

$$N_f = \frac{1}{2}\left(\frac{\Delta\tau}{2\varepsilon'_f}\right)^{(1/C)} \tag{Eq. 7-1}$$

where,

N_f = Mean cycles to failure,

ε'_f = Fatique ductility coefficient (0.325),

C = Fatique ductility exponent (-0.442 (-6) $\times 10^{-4}\overline{T}_s + 1.74 \times 10^{-2}$ In($1 + f$)),

$\Delta\tau$ = Shear strain range $\left(\frac{L}{\sqrt{2h}}\Delta(\alpha\Delta T)_{ss} \times 10^{-4}$ (in percent)$\right)$,

$\overline{T}_s$

  = Mean cyclic solder-joint temperature (°C),

f = Cyclic frequency ($1 \le f \le 1000$ cycles/day),

L = Chip-carrier size (Figure 9-9D),

h = Height of standoff (Figure 9-9E).

and, $\Delta(\alpha\Delta T)_{ss}$ = The in-plane steady-state thermal expansion mismatch $\alpha_s(T_s - T_o)$ $-\alpha_c(T_c - T_o)$, where

α_c = TCE of LCC,

α_s = TCE of substrate,

T_c = Temperature of chip,

T_s = Temperature of substrate,

T_o = Temperature in °C (power off, steady state).

Printed Components

Another advantage of ceramic substrates is that they lend themselves well to the process of printing components directly on the board. In our discussion of components used as crossovers, we mentioned that SMT often allows the elimination of multiple screenings when printing resistors with a variety of ink-resistive values. Still, hybrid engineers prefer to print components where practical for several reasons. First, by in situ fabrication, solder-bond requirements are reduced and reliability is increased. Also, printed resistors can save space. Resistors as small as 1 mm (~0.040 inch) are simple to produce using thick-film processes. With care, resistive tracks of <0.7 mm (0.030 inch) are producible. And thin-film processes reduce the above limits by more than an order of magnitude.

Small inductances and low-value capacitors may also be fabricated directly on the substrate. Electroluminescent elements and photosensitive arrays can be produced using thick- and thin-film processes. Thick-film carbon inks are also being used to replace selective gold plating for keyboard switch contacts and gold contact fingers.

Thermal Management

The thermal conductivity of alumina ceramic is about 23 times that of organic polyimide film, and 13 times greater than epoxy/glass composites. And beryllia ceramic is about 88 times as thermally conductive as epoxy/glass. High-dissipation circuitry can use the increased thermal conductivity of ceramic substrates to great advantage.

9.2.3   *Limitations of SMT for Hybrids*

Above, we've studied some of the plus points of mixing SMT with hybrid technology. Fortunately for the engineering profession, there are still factors which must be weighed, and trade-offs that must be analyzed before selecting surface-mount technology for a hybrid circuit. Below, we'll look at some hybrid application features which work against SMT's benefits.

Miniaturization

As we've said, some types of SMT will save space on some types of hybrids. The poor miniaturization candidates for SMT are those circuits with few high-pin-count devices. Also, the type of SMC under consideration determines the level of real estate that is required, in comparison to chip-and-wire devices. High-pin-count TAB devices and fine-pitch packages fare well in this comparison. Low-pin-count devices fare very poorly. And, 1.27-mm (50-mil) SMCs are generally on the losing end regardless of pin count. We'll look more closely at the deciding factors in the real-estate wars next, in Section 9.2.4.

Package-Imposed Thermal Limits

Even with cavity-up ceramic chip carriers soldered directly to ceramic boards, there is some standoff height between the carrier and the board. Thus, packaged component thermal resistance is likely to be greater than that of the same die bonded direct to the board. This is particularly true where the primary dissipation path is through the substrate. Thermal-transfer epoxies can help the package-to-board dissipation, but they still don't approach direct chip mounting. Table 9-2 shows the thermal properties of some typical die-bond materials.

Table 9-2. Properties of Die-Bonding Materials.

Material	Property	
	Thermal Conductivity (Watts/°C/cm²)	*TCE (PPM/°C @ 25 °C)*
Epoxy, Conductive	0.0062	15–16
Epoxy Preforms	0.00155	15–16
Gold/Silicon Eutectic	0.8835	
Gold/Tin Solder 80/20	0.6975	
Silicon	0.3565	7.38

Substrate Cost

It is not accurate to make the unqualified statement that thick-film network boards are more expensive than PWBs with the associated resistors needed to equal the hybrid board. It is common practice to build hybrid circuits on FR-4 or even paper phenolic substrates using polymer ink systems. Such substrates may be quite cost-effective compared to PWB and chip-resistor assemblies (comparing apples and apples). But, the limited resistance range and TCR stability of polymer inks often dictates the use of ceramic systems for hybrids. With ceramic thick-film substrates, costs are generally higher than for analogous polymer board assemblies with chip resistors.

Substrate Size Limits

Just as we can't categorically rule in favor of polymer PWBs for cost, we can't give polymer boards the nod in a substrate size contest, either. Hybrid circuits can be made with polymer inks printed on very large polymer- or porcelain-coated metal-core boards. However, the ceramic board, which is the foundation of the hybrid industry, does carry some rather restrictive size limitations. Above 100 mm (~4.0 inch) square, substrate costs per square inch begin to climb sharply. The absolute limits are somewhere between 150 mm (~6.0 inch) and 300 mm (~12.0 inch) square, depending on the processing parameters and substrate thickness.

Frequency Limits of Ceramic Substrates

Ceramic substrates restrict frequencies due to their relatively high dielectric constant. Alumina has a K ranging between 9 and 10. Microwave circuits operating at 1 GHz are routinely built on beryllia, which has a K about 70% that of alumina. But X-Band police radar runs at 24.150 GHz. Such frequencies dictate low-K polymer substrates. Of course, hybrid techniques still may be used on polymer boards.

Higher Capacitance of Ceramics

Ceramic insulation materials spell higher capacitances than polymers typically exhibit. Thus, signal traces of a given geometry will have a lower impedance. For bipolar devices, this lower impedance translates to a higher driving power and to lower termination resistances. Thus, power dissipation will be higher. This equation limits the use of ceramic substrates with power devices.[5]

Via Fabrication on Ceramic Substrates

There are two types of vias in ceramic substrates. One type is formed in the thick-film printing process, and it communicates between layers above the ceramic substrate. The other is a through-feed made by metallizing a hole in the ceramic. This second type provides one method of interconnecting circuitry on opposite sides of a dual-side substrate. However, the feed-through-type via is expensive, and there are a limited number of shops qualified to supply it.

Limits to Number of Layers

Regardless of the substrate material, there are limits to the number of layers that can be reliably produced with the thick- and thin-film processes.

Thin films are usually restricted to single conductive layers. However, Brown-Boveri & Cie, Inc. of Switzerland announced a multilayer process at the International Society for Hybrid Microelectronics (ISHM) International Symposium on Microelectronics (1986). By using a spin-coating of polyimide as a dielectric layer, they reported success with dual conductive layers, and expected to carry the technology beyond two layers.[6] That same year, Honeywell discussed a 4-layer process using thin films of copper and polyimide on ceramic substrates.[7]

Most thick-film circuits have less than ten conductive layers, as of this writing. Toshiba reported a new process based on a polymer silver-conductive ink printed on polycarbonate resin film. Local resin flow into prepunched areas under heat and pressure was used to form vias between layers. Resistors were sputtered from NiCr. The authors stated that they had built a 16-layer prototype, and that the process is not particularly constrained as to the number of layers.[8] None of these technologies, however, have approached the interconnect densities of laminated polymer MLBs. ETA Systems, Inc. says their ETA10™ supercomputer-on-a-board will use a 44-layer board. The board is 406.4 mm (16 inches) wide by 558.8 mm (22 inches) long by 6.35 mm (0.25 inch) thick, has 80,000 drilled holes, and carries 240 custom CMOS SMT chips.[9] Figure 9-10 shows the high density of this supercomputer-on-a-card circuit.

Figure 9-10 High-density PWB puts Supercomputer on a card. *(Courtesy ETA Systems, Inc.)*

(A) The 16 × 22-inch processor board (bottom) and 8 memory modules (top).

(B) One of the 240 chips that make up the CPU board of the ETA10 Supercomputer.

The 41 × 56-centimeter processor board, shown in the lower half of Figure 9-10A, runs in liquid nitrogen for high performance (or air at half speed). Eight of the 16 memory modules (providing 128 Megabytes of memory) are shown in the top half of the photo. Eight additional modules are located on the back side of the assembly.

The chip shown in Figure 9-10B is only one of the 240 chips that make up the CPU board of the ETA10 Supercomputer. Each chip has 284 pins on 0.279-mm (0.011-inch) centers. Each is a custom chip using 1.0-micron CMOS technology. This gate array has about 20,000 gates. Collectively, the 240 chips on the 44-layer CPU contain about 2,700,000 gates.

Technology and Manufacturing Problems
Surface mounting alone is a rather formidable technological hurdle to clear. Hybrid technology, if it isn't already in-house, certainly adds to the learning curve. Even with outside contractors helping, taking on SMT and hybrid manufacturing at one sitting is a challenge. Where outside contractors are not used for hybrid manufacturing, the acquisition of hybrid-process equipment is also a deterrent to mixing hybrid and surface-mount technologies.

9.2.4 *The Density Impact of SMT in Hybrid Designs*

Actual circuit density is heavily influenced by application, IC to passive-component mix, IC technology level, and a host of other factors. Therefore, our comparison is, at best, an approximation based on typical conditions. With this disclaimer stated, Figure 9-11 compares the densities of several interconnect technologies.

Figure 9-11
Densities of interconnect technologies compared.[9]

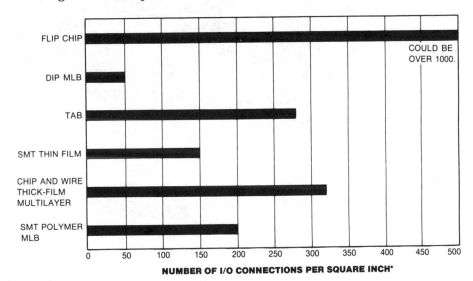

* Estimates, assuming each technology is taken to the edge of its capabilities (as of this writing).

9.3 Actual Examples of SMT Hybrids

With the preceding discussion of the plus and minus points of SMT hybrids fresh in mind, let's look at some actual applications. We can see here how the engineering teams met the challenges listed, and benefited from the mixture of the two technologies.

9.3.1 *Thick-Film Circuits*

We'll start with thick-film circuits. Thick films hold the largest share of the hybrid marketplace. As we'll see below, they are used for a wide range of applications.

Automotive Applications
Figure 9-12 shows a car radio from Delco Electronics Division, a subsidiary of G. M. Hughes Electronics. To the right of the unit, mounted against the finned heat sink,

two power hybrids carry the high-dissipation devices. Just to the left, past the large black electrolytic capacitor, stands a dense hybrid combining chip-and-wire devices, SMT ICs, screened resistors, and numerous discrete passives.

Figure 9-12 Car radio uses IMC, Type 3 SMT, and SMT hybrids. *(Courtesy Delco Electronics Division, G. M. Hughes Electronics.)*

Figure 9-13 Sequential Fuel Injector (SFI) System. *(Courtesy Delco Electronics Division, G. M. Hughes Electronics.)*

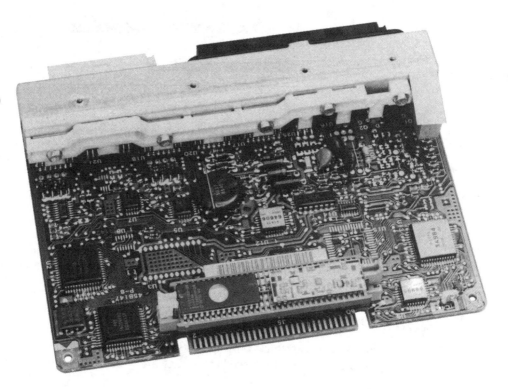

Figure 9-13 shows a Sequential Fuel Injector (SFI) System designed and manufactured by the Delco Electronics Division. It employs surface-mount and multilayered technologies to maximize density. There are over 350 SMCs on this 4-layer board. In the foreground, just to the right of the C8454, there is a small dual-in-line hybrid with one chip-and-wire IC surrounded by surface-mount passives.

Figure 9-14 shows the combination of surface mount with chip-and-wire techniques to produce an automotive controller circuit. The 44-pin ceramic leadless chip carrier in the center of the circuit is a multichip module housing 2 ICs and 4 transistors. The main circuit contains chip resistors, ceramic capacitors, and tantalum capacitors. Actually, this circuit is not a hybrid, since it has no in situ components.

Figure 9-14
Automotive
controller with
multichip carrier.

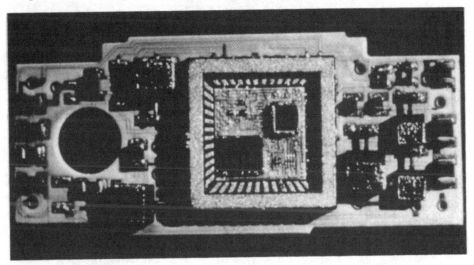

Computers and Peripherals

Cherry Electrical Products used the strengths of SMT and hybrid microcircuits in creating the industry's first DC electroluminescent display panel assembly. Actually, the SMT and hybrid technologies are on separate boards in this assembly. But the application merits our attention because of the unique combination of the two circuit methods which made the DC electroluminescent display possible.

The SMT driver board sits immediately behind the display and makes electrical contact with the glass thick-film hybrid display panel through four elastomeric connectors (Figure 9-15B). The glass display substrate has 640 column electrodes etched in tin oxide. Over the electrodes, a 25-micron phosphor interlayer, and a 1-micron aluminum layer are fabricated. The substrate is then trimmed, rows are scribed by machining, and bridging links are placed to connect the rows to the contact fingers. The result is a display as thin as 1.524 mm (0.060 inch), and weighing less than 0.7 kg (1.5 pounds).

In Figure 9-16, we see a studied mixture of through-hole, Type 2 SMT, SMT hybrid modules, and VLSI to achieve a 1-slot implementation of a bubble memory card, including the associated interface controls.

Figure 9-15 Electroluminescent display panel assembly. *(Courtesy Cherry Electrical Products, Waukegan, IL.)*

(A) SMT controller for the Electroluminescent Display Panel.

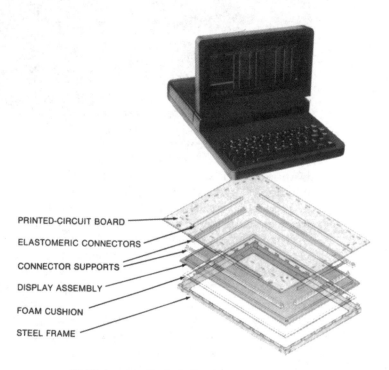

PRINTED-CIRCUIT BOARD

ELASTOMERIC CONNECTORS

CONNECTOR SUPPORTS

DISPLAY ASSEMBLY

FOAM CUSHION

STEEL FRAME

(B) Display assembly, including thick-film glass hybrid.

Consumer, Industrial, and Instrument Uses

Figure 9-17 pictures a hybrid used in a benchtop high-precision multifunction calibrator. The backside of the hybrid carries a network of heater resistors used for temperature compensation. Front- to backside interconnection is provided solely by edge-connector attachment.

Figure 9-16 SMT hybrid helps miniaturize a bubble memory card. *(Courtesy Magnesys, 1605 Wyatt Drive, Santa Clara, CA.)*

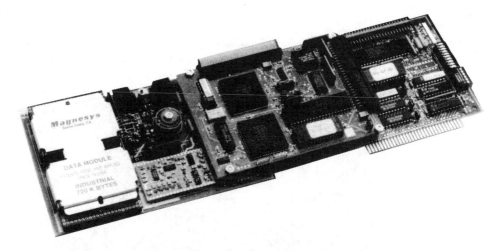

Figure 9-17 Hybrid circuit with temperature compensation. *(Courtesy John Fluke Manufacturing Co., Inc.)*

Figure 9-18 shows a potpourri of unusual components used on consumer product hybrids. The small circuit at upper left is a switch element from a drill. The large beryllium copper tab that is reflow mounted on the substrate is touched and wiped for good electrical contact by the switch plunger.

Figure 9-18
Consumer circuits with unique components.

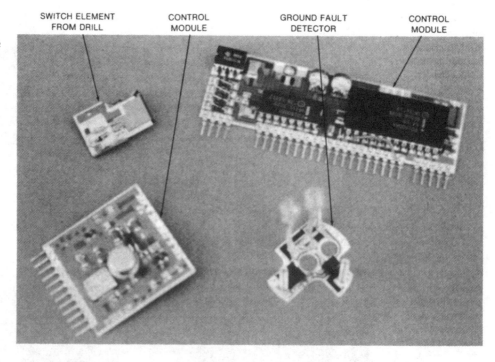

SWITCH ELEMENT FROM DRILL   CONTROL MODULE   GROUND FAULT DETECTOR   CONTROL MODULE

Figure 9-19
Multichip module CLLCCs on a military hybrid.

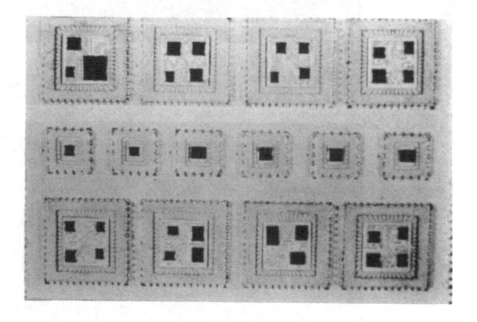

The odd-shaped circuit to the lower right in Figure 9-18 fits inside a small hand-held tester which shows ground faults or out-of-phase conditions when plugged into wall outlets. Note the two small light bulbs standing above the circuit on lead wires. These components are actually mounted through-hole in the ceramic substrate. All components, including the through-hole lights, are reflow soldered. The unusual shape of the board, including the holes for the light leads, is produced by punching green ceramic tape.

The two modules at the upper right and lower left of Figure 9-18 show the use of lead-formed DIPs and a lead-formed metal-can TO package, respectively.

Military Hybrids

High-density military packaging is seen in Figure 9-19. The CLLCCs are de-lidded to show the multichip hybrid modules internally. Here, eight 44-pin multichip modules and six 20-pin CLLCCs pack 36 ICs into a space of about 36 cm² (5.58 in²). This equates to a density of over 5.3 16-pin equivalent DIPs per square inch.

The doughnut-shaped circuit in Figure 9-20 is an impact fuse for an artillery shell. The circuit combines the surface mounting of passives and an impact switch with chip-and-wire mounting of ICs. The bolt-like component shown just above the 9 o'clock position in Figure 9-20A is the impact switch. Directly to the right is a view of one of the bare ICs. This is shown close-up in Figure 9-20B, along with some of the unique feature geometries needed for this specially shaped substrate.

Figure 9-20 Doughnut-shaped hybrid fits in the nose cone of an artillery projectile. *(Courtesy Motorola and Harry Diamond Labs.)*

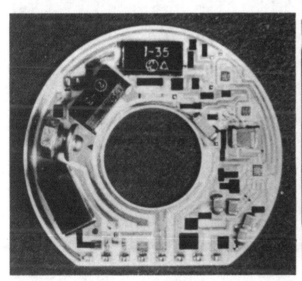

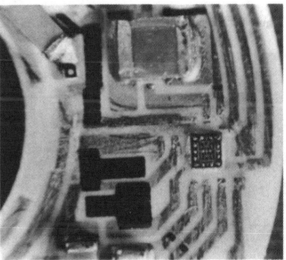

(A) Ordinance fuse. *(B) Close-up of chip-and-wire circuit.*

The super DIP shown in Figure 9-21 is a computer module. The co-fired multilayer ceramic is double-side populated with leadless chip carriers. The populated bottom side, shown on the left of the photo, carries 14 chip carriers, plus decoupling. The top side, shown partially processed on the right side of the photo, will receive nine 24-pin CCCs and one 52-pin device. The 16-pin DIP equivalent density is about 6.8 DIPs per square inch. The circuit is really not a hybrid. There are no integral components on the substrate.

TAB and Flip-Chip Applications

Figure 9-22A shows a TAB tape inner-lead bonded to an IC, while Figure 9-22B shows the high I/O tape before IC mounting. The large square open window in the tape is where the inner lead bonds are made to the IC. The four narrow open windows expose metallizations for outer-lead bonding. They are where the outer lead bonds are made after the device is excised from the tape.

Figure 9-21 A
double-side-
populated mili-
tary computer
module.

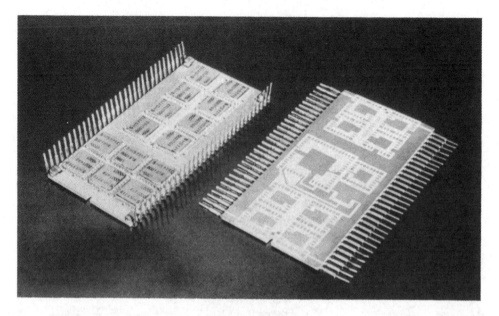

Figure 9-22 Tape automated bonding of ICs on an SMT hybrid. *(Courtesy 3M.)*

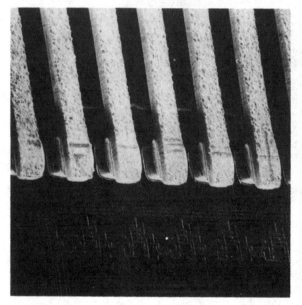

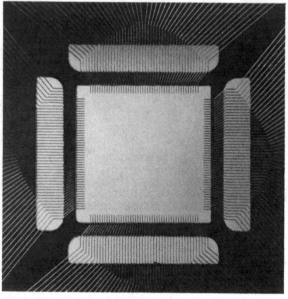

(A) Inner lead bonds on an IC chip.   *(B) High-I/O TAB tape.*

In Figure 9-23, we see how flip-chip technology is used in conjunction with other surface-mount methods in hybrids.

Telecommunications Circuits

In Figure 9-24, we see a telecommunications module, with a ceramic substrate, used to eliminate TCE mismatch concerns with a 24-pin CLLCC. Two other ICs are in lead-formed DIPs, solving SOIC-availability problems at the time this substrate was produced.

Figure 9-23
Diagram of flip-chip process integration in an SMT hybrid assembly.

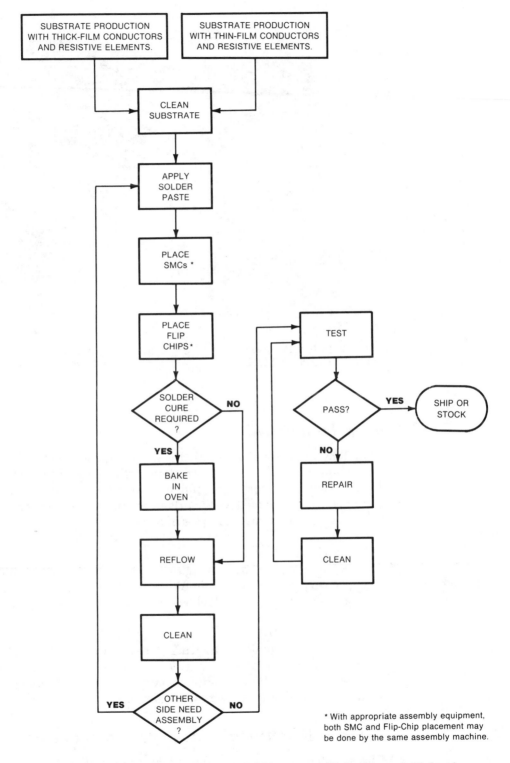

* With appropriate assembly equipment, both SMC and Flip-Chip placement may be done by the same assembly machine.

The same circuit shown in Figure 9-24 is seen again in Figure 9-25, but here, a PLCC replaces the more expensive CLLCC. The land geometry was configured so that the PLCC could be used, if available, and the CLLCC could be substituted, if necessary.

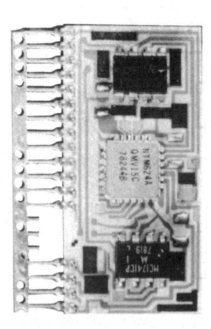

Figure 9-24 Module uses ceramic substrate to match CLLCC TCE.

Figure 9-25 Circuit shown in Figure 9-24 with a PLCC substituted for the higher-cost CLLCC.

9.3.2 Thin-Film Circuits

Thin films are predominantly used where there is a need for the tight tolerances and fine-geometry definition that their photolithographic process allows, or where the very stable, high tolerances of thin-film resistors are required. High-frequency circuits and dense single-layer boards are among the common applications. Below, we'll look at some actual examples of these.

Military Thin Films
Circuits, like the one shown in Figure 9-26A, combine surface mounting with the fine-geometry capability of thin-film technology. The marriage produces very compact, high-frequency assemblies at moderate cost.

Telecommunications
The microwave amplifiers shown in Figure 9-27 are typical of the thin-film circuits used in telecommunications. Figure 9-27A shows the miniaturization achieved by comparing the thin-film amplifier to a dime. Figure 9-27B illustrates typical hermetic packaging for these microwave amplifiers.

9.3.3 SIP and DIP Modules as Super Components

The potential for in situ component fabrication makes hybrids well suited to the construction of dense SIP and DIP modules. By building standard circuit functions into such modules, engineers can quickly combine modules with PCB-mounted circuitry to create new custom assemblies. This both simplifies custom circuit design and reduces the costs of custom circuit manufacturing. A collection of such SIP and DIP modules is shown in Figure 9-28.

Figure 9-26
Thin-film circuits
with SMCs.
*(Courtesy RHG
Electronics Labs,
Deer Park, Long
Island, NY.)*

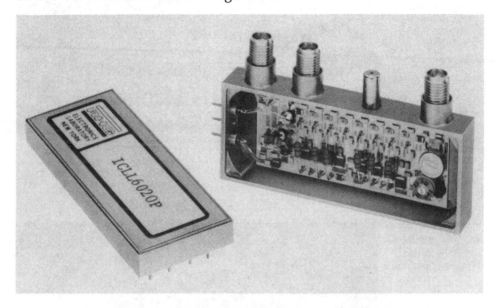

(A) Thin film and SMT are integrated to produce a compact microwave amplifier.

(B) For the SMC placement operation, multiple circuits are loosely registered mechanically. Then optical targets are used to accurately orient them.

Figure 9-27 Microwave amplifiers on thin-film substrates. *(Courtesy Amplica, Inc., Newbury Park, CA.)*

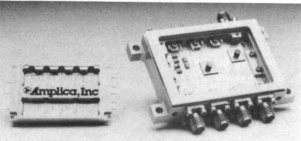

(A) A microwave amplifier implemented with a thin-film hybrid. Chip-and-wire technology is used here.

(B) Typical hermetic-packaging solutions for microwave hybrids. Note use of SMC passives.

Figure 9-28 SIP and DIP modules simplify circuit construction.

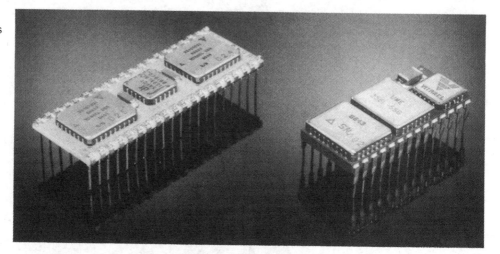

(A) Memory modules.

(B) Telecommunication modules.

9.4 Materials for SMT Hybrids

SMT hybrid materials differ somewhat from those we're familiar with in building SMT assemblies on FR-4 print-and-etch substrates. Let's review the materials, pointing out those areas where differences exist, and those areas where our previous discussion of materials apply to hybrids as well.

9.4.1 *Discrete Components*

We'll start our material review with discrete components, where we'll find a great deal of similarity with our discussion of components covered in Chapter 2.

Resistors

Chip resistors for SMT hybrids are the same as chip resistors for polymer SMT. The major difference that hybrid technology brings to resistor selection is the possibility of fabricating the resistor directly on the substrate, in lieu of placing a resistor chip. In situ components will be discussed below.

Trimming potentiometers are generally added as discrete components, and are specified for hybrids in the same way that they are specified for standard SMT assemblies.

Capacitors

Discrete capacitors are selected in the same way that they are specified for standard FR-4 SMT assemblies. However, the use of palladium silver as a metallization on a ceramic board may relieve the need for diffusion barriers to protect the chip leads from silver leaching. Where silver is present in the substrate metallizations, it is common practice to use a solder containing 2% silver (such as 62/36/2), which substantially slows leaching.

Tantalum or aluminum electrolytic SM capacitors are available. Single-layer silicon capacitors are also used on hybrids.

Capacitors may also be fabricated in the hybrid substrate, but only small capacitances are practical, using thick-film techniques. Therefore, capacitors are usually added as discrete components and are not fabricated in situ.

Trimmer capacitors are specified for hybrids in the same way that they are for standard SMT assemblies.

Transistors

With transistors, we have several options on a hybrid. The familiar SOT family is commonly used. For high-frequency work, the Leadless Inverted Device (LID) and MO types are also available. In addition, direct chip mounting and wire bonding is another alternative that may be considered with hybrid technology.

Inductors

Inductors may be formed in situ, but such devices are limited to very small inductances. Chip inductors like those presented in Chapter 2 are commonly used where higher values are required. Tunable coils are also available, and are selected in the same way they are for nonhybrid SMT assemblies.

Thermistors

Chip thermistors for hybrid circuits are the same as we've studied for other SMT applications, and should be specified for hybrids per the guidelines given in Chapter 2.

Other Components

The full range of components mentioned in Chapter 2 may also be surface mounted on hybrids. To review, this includes connectors, sockets, switches, transformers, relays, fuses, and jumpers (or Zero-ohm resistors).

9.4.2 Integrated Circuits

SOICs

Originally developed for placement on small ceramic circuits for watches, the SOIC has been used in hybrid surface-mount processes since its birth in the late 1960s. Because its dual-in-line I/O allows feed-throughs in one axis under the component, the SOIC is particularly useful for single-layer hybrids. Small SOICs are quite space efficient when compared with quad I/O packages. However, at somewhere between 20 and 28 pins, the real-estate advantage begins to shift toward devices having four-sided I/O.

PLCCs

In 28-pin to 68-pin packages, the PLCC provides a low-cost surface-mountable option to chip-and-wire mounting. However, the 1.27-mm (50-mil) PLCC does not rival chip-and-wire density. Above 68 pins, other packaging alternatives are often chosen for space efficiency.

CCCs

Ceramic leaded and leadless chip carriers provide a hermetic package with a TCE suitable for mounting, within our stated design rules, directly to ceramic hybrids. One caution, however. The stated rules do not include the direct mounting of large carriers where high temperature differentials will exist between the package and substrate due to power dissipation during device cycling.

QPFPs

The fine-pitch QPFP will certainly find a receptive market in SMT hybrids. It is much more space efficient than the 1.27-mm (50-mil) PLCC, coming close to chip-and-wire density. And, the quad flat pack allows a full pretest of components before placement.

QCFPs

QCFPs carry the same space efficiencies and pretest capacities noted above, plus providing a hermetic package. For high-pin-count devices, the leads provide a flexural element, eliminating the concern about temperature deltas between device and board. The lead standoff also makes cleaning under the package much easier than cleaning under CLLCCs. However, the leads add Z-axis clearance requirements, and don't provide the thermal path to the substrate that the leadless carrier boasts. The QCFP will be an interesting package on hybrids, but probably won't replace the CLLCC overnight.

TapePak
Even more space efficient than the QPFP and just as pretestable, the TapePak is expected to become a popular package with hybrid designers.

TAB
TAB rivals and often eclipses chip-and-wire densities. The development of flexible equipment for automatically excising and outer-lead bonding TAB devices will spell an increasing use of this technology, particularly for very high I/O devices.

Flip Chip
Flip Chips are semiconductor dice specially prepared for face-down bonding by the reflow of tiny solder bumps grown on their bonding pads. The technique is diagrammed in Figure 9-29.

Figure 9-29
Diagram of a typical Flip Chip process.[10]

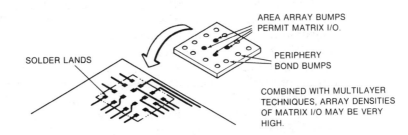

(A) Flip Chip attachment.

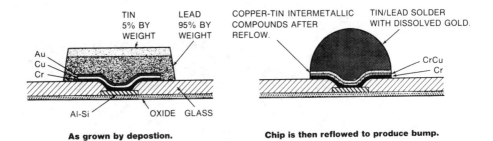

(B) Solder bump production.

9.4.3   *In Situ Component Materials*

Thin-film elements are restricted in value by the materials available and by board-size constraints. Thick-film elements are limited in value both by the conservation of board area and the peculiarities of the screening process. As Figure 9-30 shows, squeegee pressure causes some thinning of the printed image at the center of the screen opening. For small features, this effect is negligible. But larger features show marked height variations from the edge to the center.[11]

Resistors
Thick-film resistor materials are selected to suit the substrate on which they are printed. Ceramic and refractory substrates permit the use of resistor inks with high firing temperatures. These high-firing-temperature inks exhibit better TCR stabil-

ity than do polymer systems, and offer the widest range of resistances of any hybrid resistive materials.

Figure 9-30 Squeegee pressure produces uneven print height.

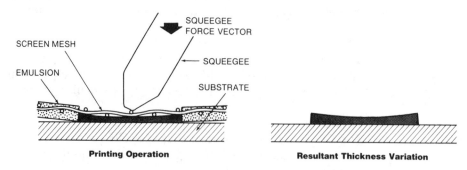

Printing Operation   **Resultant Thickness Variation**

Polymer boards and coated metal-core substrates rule out the use of high firing temperatures for resistor inks. Polymeric binder inks, firing at relatively low temperatures around 220 °C, allow the printing of resistors on nonrefractory board materials, including polyimides and even epoxy-glass.

Sputtered or evaporated coatings of materials, such as nichrome, tantalum nitride, or chromium silicide, are used to form resistors in the thin-film process.

The layout of hybrid resistors will be detailed in Section 9.5.3. Thermal engineering for in situ resistor dissipation also differs from the familiar discrete component models. Reference 12 presents the methods for calculating and modeling dissipation from a thick- or thin-film resistor and discrete components on a hybrid substrate.[12]

Figure 9-31 In situ capacitors for multilayer thick-film hybrids.

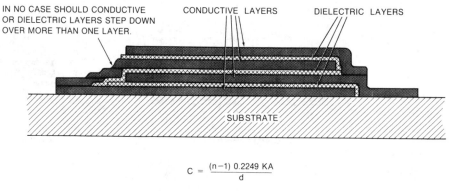

$$C = \frac{(n-1)\ 0.2249\ KA}{d}$$

where,
  n = Number of conductive layers,
  C = Capacitance in pF,
  K = Dielectric constant of insulating material,
  A = Area of electrode in square inches (min.),
  d = Thickness of dielectric between layers.

Capacitors

Capacitors are formed by multilayer thick-film techniques, printing metallized plates that are separated by dielectric layers. Figure 9-31 shows a typical thick-film approach to capacitor fabrication, and restates the formula presented earlier (in Chapter 6) for the calculation of the capacitance. Note, from the formula, that the thin dielectrics and high-dielectric constants inherent in the thick-film process both work in favor of the capacitance that can be achieved with hybrid vs. multilayer polymer boards. Still, there are serious limits on layer construction. And, the costs of multiple print and fire operations also limit the use of in situ capacitors to low values.

Figure 9-32
High-rise hybrid cuts urban sprawl and saves space and dollars. *(Courtesy Kepco, Inc.)*

HYBRID MODULE

(A) Power supply uses two custom hybrids to shrink package.

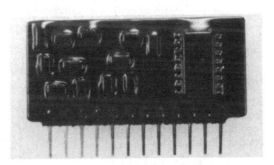

(B) Custom hybrid module after conformal coating.

Inductors

Inductors may be fabricated using thick- or thin-film techniques. Figure 6-7 (in Chapter 6) shows the construction details for designing a required inductance, while the formulae are given in the text. As the formula indicates, only small-value inductors can be practically fabricated in situ. A very large board area would be required to lay out even a moderate inductance. The small line-and-space dimensions which can be manufactured with thin-films allow considerably higher values in a given space than do thick films.

Networks
Resistor, and even RC networks, may be fabricated in situ. In fact, a service industry has grown out of providing resistor and resistor-capacitor hybrid networks, sometimes with SMT active elements, on small SIPs. Such SIP supercomponents can move a whole neighborhood of PWB components from urban sprawl to one high-rise area, as shown in Figure 9-32. The result is space and manufacturing cost economy.

9.4.4 Substrates

Ceramic substrates (mainly alumina and some beryllia) so dominate hybrid manufacturing that the terms, ceramic substrate and hybrid, seem almost synonymous. However, the entire range of materials discussed in Chapter 6, Section 6.3.2, and even more exotic materials, are used for hybrid substrates. Ceramics are popular because of several properties that fit thick-film processing requirements. The relatively camber-free smooth surface of a ceramic substrate allows close-tolerance screen printing. And the refractory nature of the material permits the use of very stable, high-firing-temperature, resistive inks. However, by using special inks and care in printing, the thick-film process has even been applied to paper substrates.

9.4.5 Materials

Hybrid circuits use many of the same coating and attachment materials that we've previously studied for print-and-etch boards. Below, we'll add to and modify our PWB materials discussion.

Dielectric Layers
Insulative layers of various types are used for crossovers, capacitor build-up, and the construction of multilayer hybrids. Dielectrics are chosen based both on the electrical purpose they will serve, and on the substrate material and conductive-ink system on which they will be applied. Table 9-3 lists some commonly used materials and details their physical and electrical properties.

Table 9-3. Dielectric Material Characteristics

Property and Unit of Measurement	Dielectric Type		
	Thick-Film Ceramics	*Thin-Film Ceramics*	*Polymer Film*
Breakdown Voltage (V/mil)	>500	≥100	≤100
Capacitance Range (Typ., in pF)	≤1000	≤50	≤500
Conductive Ink Styles	PdAg/PdAu/Au	Au/Cu	Cu/Ag
Dielectric Constant (@ 1 kHz)	8 to 14		4 to 6
Dissipation Factor Range (%)	<3.5		>5
Hermetic Properties	Excellent	Fair	Poor
Insulation Resistance (Typ., in Ohms)	>10^{11}	>10^{10}	>10^7
TCE Matches with Substrate	~96% Al_2O_3	~96% Al_2O_3	Polyimide
Temperature Coefficient Capacitor (PPM/°C)	<250		
Trimming Methods	Abrasive	No	Abrasive

Masks

Traditionally, hybrid dielectric layers, as described above, function like the familiar solder mask of a printed-wiring board. Even on single-layer circuits with no crossovers or in situ capacitors, a dielectric layer is often applied to protect circuitry and passivate resistors. Wet- or dry-film polymer mask materials may also be used on ceramic hybrids, and they find wide application on polymer-based boards.

Conformal Coatings

The full range of coating materials discussed in Chapter 5, Table 5-4, are useful in protecting SMT hybrid circuits.

Conductive Adhesives

Conductive adhesives sometimes are used in lieu of solder to bond chip passives to circuits where soldering would be difficult because of metallizations or limitations on the processing temperatures. One recently developed adhesive, shown in Figure 9-33, is made selectively conductive when processed. Thus, the material may be applied in a continuous line across lands, and then aligned to conduct in the Z axis over land areas only. Even small separations, such as 0.25-mm (~0.010-inch) spaces between leads, remain fully dielectric. The manufacturer, Uniax, reports dielectric separation is retained down to about 0.03 mm (~0.001 inch). The material offers potential relief to solder yields with fine-pitch components.

Figure 9-33 Z-axis conductive adhesive attaches fine-pitch IC. *(Courtesy Uniax Corp., Troy, MI.)*

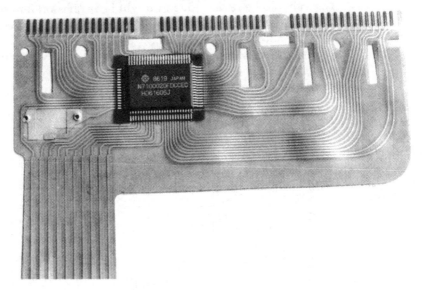

Packages

External packaging is often required for the protection of chip-and-wire hybrid assemblies. The use of packaged SMT ICs may negate the need for such a protective housing. However, Flip Chip and TAB technologies, which are planar mounting methods, offer limited protection for the dies, and may dictate post-packaging of the assembly.

Metal packages offer both hermeticity and a wide range of form factors, but typically, they add considerable cost and weight to the system. Ceramic packages are available in hermetic and nonhermetic designs, with the costs being clearly higher for the hermetic styles. Finally, light and inexpensive plastic packages may be used to shield circuits from impact and environmental contamination. Plastic packages are used where hermeticity is not required.

9.5 Special SMT Design Rules for Thick-Film Hybrids

The design rules we've presented in our first eight chapters, and the special points for military circuitry contained in Chapter 10, are basically sound rules for SMT hybrid design. Below, we'll cover those points where hybrid rules depart from PWB norms.

Please be forewarned that hybrid microelectronics is a major technology. We cannot, in one chapter, begin to catalog the tons of data available and published on the topic. For anything more than a minor use of hybrid techniques in your SMT designs, additional study will be mandatory.

9.5.1 *Component Orientation*

For reflow attachment of components, the general rules for Type 1 SMT circuits, presented in Chapter 5, apply for hybrid as well as standard SMT PWBs. These rules should be reviewed, and might bear revision for conductive epoxy attachment. The adhesive vendor should furnish design-rule guidance, where Chapter 5 rules do not pertain.

In situ resistors should all be oriented for trimming from the same side if possible. This means that their long axes should be parallel and there should be a clear area of substrate (no features) on the side that will be trimmed.

Figure 9-34
Interconnect
technologies
compared.

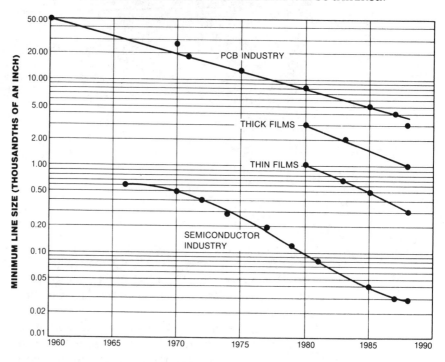

9.5.2 *Interconnect Layout*

Not surprisingly, hybrid thick-film techniques have a dramatic impact on minimum line-and-space rules. Fine geometries at the limit of PWB manufacturing technology are routine in hybrid thick films. Below, we'll analyze the impact of thick-film technology on layout.

Traces and Spaces

Figure 9-34 compares various interconnection methods, and charts the shrinking of feature sizes as technologies have advanced. Minimum line separation is generally equal to minimum feature size, as shown in the chart. So, from Figure 9-34, we see that thick-film techniques can produce 25-micron (1-mil) lines and spaces.

Clearances Around Solder Lands

Clearances around solder lands are determined by the type of mask used to protect the conductive traces from unwanted short circuits. The rules for this are covered in Figure 9-35.

Figure 9-35
Clearance rules for thick-film conductors.

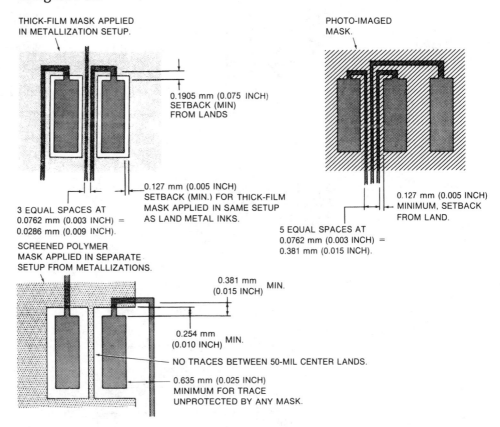

Crossovers

Where only a few crossovers are needed, options are to use Zero-ohm resistors, or print and fire dielectric and conductive layers over the lines that must be crossed. Figure 9-36 illustrates both methods, and presents ISHM guideline dimensions for screened crossovers.[13]

Figure 9-36
Crossover design
rules.

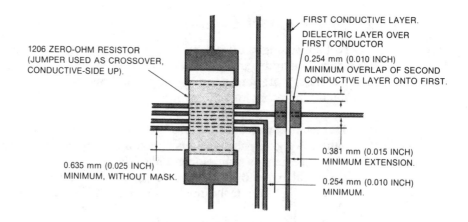

FIRST CONDUCTIVE LAYER.

DIELECTRIC LAYER OVER
FIRST CONDUCTOR

0.254 mm (0.010 INCH)
MINIMUM OVERLAP OF SECOND
CONDUCTIVE LAYER ONTO FIRST.

1206 ZERO-OHM RESISTOR
(JUMPER USED AS CROSSOVER,
CONDUCTIVE-SIDE UP).

0.635 mm (0.025 INCH)
MINIMUM, WITHOUT MASK.

0.381 mm (0.015 INCH)
MINIMUM EXTENSION.

0.254 mm (0.010 INCH)
MINIMUM.

Figure 9-37
Multilayer thick-
film via design
rules.[14]

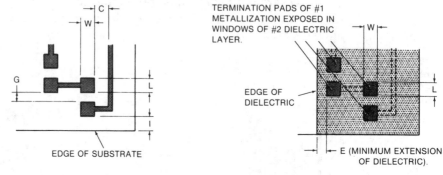

TERMINATION PADS OF #1
METALLIZATION EXPOSED IN
WINDOWS OF #2 DIELECTRIC
LAYER.

EDGE OF
DIELECTRIC

EDGE OF SUBSTRATE

E (MINIMUM EXTENSION
OF DIELECTRIC).

(A) Metallization layer.

*(B) Dielectric layer over metallization layer in
Figure 9-37A.*

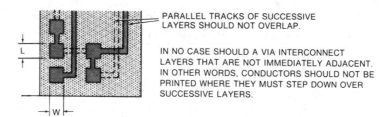

PARALLEL TRACKS OF SUCCESSIVE
LAYERS SHOULD NOT OVERLAP.

IN NO CASE SHOULD A VIA INTERCONNECT
LAYERS THAT ARE NOT IMMEDIATELY ADJACENT.
IN OTHER WORDS, CONDUCTORS SHOULD NOT BE
PRINTED WHERE THEY MUST STEP DOWN OVER
SUCCESSIVE LAYERS.

(C) Second metallization layer.

DIMENSION	MODIFIER	MANUFACTURING SIMPLICITY		
		Simple (mm/inch)	Average (mm/inch)	Difficult (mm/inch)
C	MIN.	0.508 (0.020)	0.381 (0.015)	0.254 (0.010)
E	MIN.	0.508 (0.020)	0.381 (0.015)	0.3048 (0.012)
G	MIN.	0.508 (0.020)	0.381 (0.015)	0.3048 (0.012)
I	MIN.	0.0508 (0.020)	0.381 (0.015)	0.3048 (0.012)
L	MIN.	0.508 (0.020)	0.381 (0.015)	0.3048 (0.012)
W	MIN.	0.508 (0.020)	0.381 (0.015)	0.3048 (0.012)

(D) Table of values.

Vias

Vias may be constructed by multilayer thick-film techniques. Very simple multi-level circuits may require no more than a few crossovers, as discussed above. However, denser circuitry often dictates that several successive layers of traces be stacked up, with insulating layers of dielectric material between. When this is the case, individual thick-film layers may be interconnected using thick-film vias. Multilayer thick-film via design rules are presented in Figure 9-37.

Mixing SMT with hybrid technology presents the designer with a wealth of options for interconnects. Vias, for instance, may be fabricated by thick-film techniques, as discussed above. They may also be constructed using PTHs on double-sided or multilayer PWBs. On ceramic boards, double-sided metallizations may be joined by print through-hole or wraparound methods. The various options for layer interconnects are covered in Figure 9-38. These are often mixed and matched on polymer thick-film boards.

Figure 9-38
Options for layer interconnects in SMT hybrids.

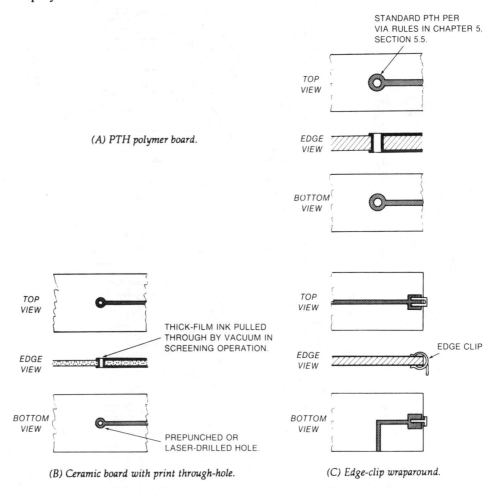

(A) PTH polymer board.

(B) Ceramic board with print through-hole.

(C) Edge-clip wraparound.

9.5.3 *In Situ Component Design*

Hybrids take their name from their mixture of in situ components and the components added to the board as separate elements. Therefore, it is clear that a study of

SMT hybrid technology will include rules for in situ component production. Below, we'll look at some rules for the design of components commonly built as part of hybrid substrates.

Figure 9-39
Resistive values
for standard
resistor pastes.
*(Reprinted by
permission of
ISHM, Reston,
VA, and IPC,
Lincolnwood, IL.)*

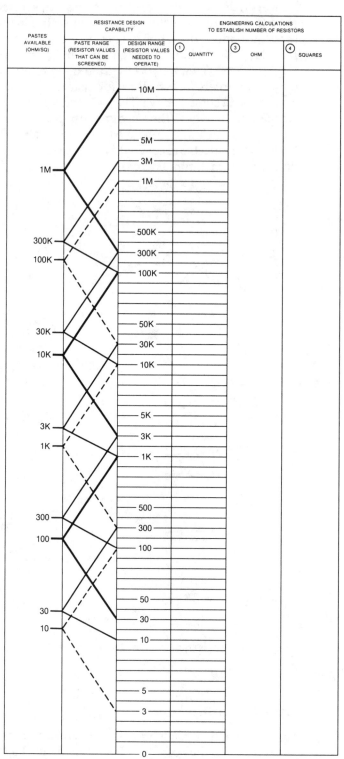

Resistors

Step 1 in resistor design is the selection of the resistive paste or pastes required to produce the necessary resistance values in a given circuit. Figure 9-39 shows the value ranges for standard resistive pastes.

The left-hand column in Figure 9-39, "Pastes Available," lists the standard resistive paste values in Ohms/Square. The brackets extending from each paste value indicate the range of resistors that can be screened using a particular paste, within the design rules on resistor aspect ratios.

Since some values overlap, the designer's job is to choose the minimum number of pastes required for the fabrication of all in situ resistors. Remember, each paste must be applied in a separate print-and-fire cycle.

After selecting the paste to be used, the next step in designing resistors is the calculation of the number of squares required for each individual resistor. This is determined by dividing the resistance required, in ohms, by the paste value in ohms/square. The formula is:

$$S = \frac{R}{P}$$

(Eq. 9-1)

where,

S = Number of squares,
R = Resistance required in ohms,
P = Resistivity of paste in ohms/square.

Note that "ohms/square" is an abstract term. It defines the resistance obtained from a specified thickness of ink (per paste manufacturer's print-thickness specifications) printed in a square pattern of any dimensions. To double that resistance, we would print a resistor two squares long by one square wide. To halve resistance, we could print our resistor one square long by two squares wide. In theory, as long as we could maintain print thickness, it wouldn't matter whether the unit of measure was square angstroms or square miles. The resistance would remain the same. Of course, the maximum power dissipation of the 1 square-mile resistor would be considerably higher than that of the 1 square-angstrom device.

That being the case, "squares," as calculated above, is not a square area. It is, rather, a definition of the aspect ratio required to obtain the desired resistance. We generally select a resistive paste that will not produce a lopsided aspect ratio.

With the number of squares established, the form factor of the resistor must be determined. Form-factor selection is influenced by the area requirements for power dissipation, the grid of the CAD system, and the minimum values for length and width as stated in Figure 9-40. For additional data on dissipation, see Reference 15.

In Figure 9-41, we see all these rules applied to an actual production SMT hybrid. Note the multiply taped resistor to the left of each of the SOTs. Each element of the resistor has been individually laser trimmed between tap leads. Two top-hat-style high-aspect-ratio resistors are seen in the upper right corner of the circuit.

Figure 9-40
Screened resistor
form-factor
rules.[16]

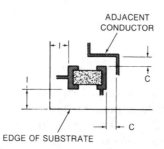

EXTENSION
(TYP.)

EXTENSION
(TYP.)

OVERLAP
(TYP.)

S = NUMBER OF SQUARES
S = L/W AND AS DEFINED BELOW.

A, the minimum area required, is
calculated thus: A = D/50
    where D = the maximum power
    dissipation (in watts).
W, the width of resistor, is calculated
thus: W = $\sqrt{A/S}$
    where S = the number of squares.
L, the length of the resistor, is calculated
thus: L = S • W.

DIMENSION	MODIFIER	MANUFACTURING SIMPLICITY		
		Simple (mm/inch)	Average (mm/inch)	Difficult (mm/inch)
L	MIN.	1.27 (0.050)	1.016 (0.040)	1.016 (0.040)
W	± 1% TOLERANCE MIN.	0.016 (0.040)	0.889 (0.035)	0.889 (0.035)
	± 5% TOLERANCE MIN.	0.762 (0.030)	0.635 (0.025)	0.635 (0.025)
OVERLAP	MIN.	0.254 (0.010)	0.2032 (0.008)	0.2032 (0.008)
EXTENSION	MIN.	0.381 (0.015)	0.254 (0.010)	0.2032 (0.008)
S	MIN.	0.5 SQ.	0.5 SQ.	0.3 SQ.
S	MAX.	5.0 SQ.	10.0 SQ.	10.0 SQ.

(A) Resistor form factors.

ADJACENT
CONDUCTOR

EDGE OF SUBSTRATE

DIMENSION	MODIFIER	MANUFACTURING SIMPLICITY		
		Simple *	Average *	Difficult *
C	MIN.	0.127 (0.005)	0.127 (0.005)	0.127 (0.005)
I	MIN.	0.381 (0.015)	0.254 (0.010)	0.254 (0.010)

* Values are in millimeters and inches.

PROBE PADS ON GRID FOR
TRIMMING OPERATION.
PROBE LOCATED AS CLOSE AS
POSSIBLE TO RESISTOR PADS

WHERE POSSIBLE,
LAYOUT RESISTORS
TO BE TRIMMED
IN UNIFORM DIRECTION.

CLEAR AREA FOR TRIMMING

(B) Resistor near another conductor.

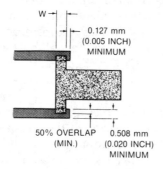

0.127 mm
(0.005 INCH)
MINIMUM

50% OVERLAP
(MIN.)

0.508 mm
(0.020 INCH)
MINIMUM

TOP HAT RULES:
1. Preferred where voltage sensitivity is a concern
2. Noise is reduced by increasing the resistor
   length and/or area.
3. Top hats are not recommended where tolerances
   of less than 1% are required.

(C) Top hat resistor design.

MATCHED RESISTORS RULES:
1. Matched resistor are to be of the same ink.
2. Matched resistors should be in close proximity
   to each other.
3. Matched resistors should be located on a
   common axis.

(D) Matched resistor design.

Figure 9-41
Well-designed
hybrid resistors
on an SMT
circuit. *(Courtesy
John Fluke Manufac-
turing Co., Inc.)*

The chart in Figure 9-39 may be copied and used as a work sheet in selecting the optimum resistive ink(s). Figure 9-42 shows use of the worksheet for a hypothetical circuit, where a wide range of resistance values is required. Where all the values are clustered, one ink may suffice. Widely spread values may require several inks. Since each ink requires a separate print-and-fire operation, it is well to limit in situ resistors to one, or no more than two, inks. If a wide range of values is needed, screenings may be limited by selecting one or two inks which will produce the bulk of the resistors required. Chip resistors may be used to produce the less common values.

To use the resistive ink-value chart as a worksheet in resistor design, make a copy of the chart given in Figure 9-39, and perform the following steps:

1. As shown in Figure 9-42, place a dot on the resistive-value line in Column 4 (the Quantity column) for each resistor in the circuit.

2. Decide which resistors should be fabricated in situ. In the example given in Figure 9-42, twelve 10K, eight 1K, one 100K, and four 5M resistors are needed. The 1K and 10K values, representing 80% of the resistors in the circuit, can be fabricated in a single screening. A second screening of 1M paste will allow the construction of four of the five remaining parts (the four 5Ms), or 96% of the resistors specified.

   As shown in Figure 9-42, both the 1K and 10K values may be produced in a single screening, using a 1K or 3K paste. A second screening with 1M paste would be necessary to fabricate the 5M resistors. And, a third screening would be needed to construct the single 100K resistor.

   The designer should weight the relative value of in situ construction versus the cost of each subsequent screening when deciding which resistors to fabricate using thick-film techniques and which resistors might best be added as chips. If the 100K or 5M parts require a close tolerance or active trimming, the additional screenings might well be justified. It might also be possible to produce both the 100K and 5M parts in a single screening, with a paste blended from 100K and 1M inks.

   After deciding which resistors to screen, circle (on the worksheet) the dots selected for in situ fabrication.

Figure 9-42
Resistive ink-
selection
worksheet. *(Basic
chart courtesy of
ISHM and the IPC.)*

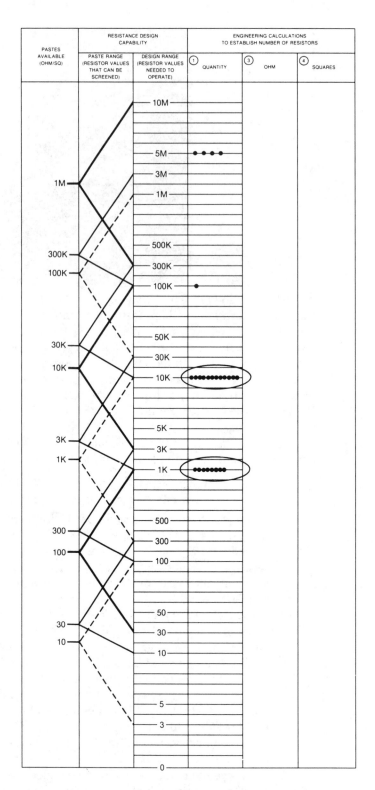

3. Observe, on your work sheet, the clustering of dots that are to be fabricated by thick-film screening (the circled dots). Select as few paste values
as possible to fabricate these resistors. (In our example, one screening of

1K ink will suffice.) Record in Column 5 (the Ohm column) the Ohms/
Square value(s) selected.

4. Compute and record in Column 6 (the Squares column) the number of
squares for each required resistor. (The methods and rules for computing
the required number of squares are discussed in the preceding text.)

Note: For SMT/hybrid cost-effectiveness, it is desirable to limit the screening
to two, and never more than three, inks. The best practice is a single screening,
unless trimming requirements force more. Resistors not screened are added as
SMT discretes.

Where very-close-tolerance resistors (better than ±20%) are required, the resistors may be printed with lower values than required, and then trimmed to the
required value by removing material in a trimming operation. Passive trimming
involves the removal of material while monitoring the resistance. Active trimming
is the removal of resistive material while monitoring some other electrical function
that is influenced by the resistance in question.

Inductors

Inductors are designed in hybrids in the same way that they are designed on
polymer PWBs. Figure 6-7 (Chapter 6) shows some typical geometries and formulae for calculating the inductance of in situ inductors. Figure 9-43 illustrates use of
multilayer thick-film techniques to construct an on-board inductor.

Figure 9-43
Thick-film
construction of
an on-board
inductor.

MULTILAYER THICK-FILM
VIA MAY BE SUBSTITUTED
FOR PTH FEED-THROUGH.

Capacitors

Chapter 6 (Figure 6-6) gave us the following formula for in situ capacitor values:

$$C = \frac{(n-1)\ 0.2249K\ (A)}{d}$$

where,

- C = The capacitance in pF,
- n = The number of conductive plates,
- K = The dielectric constant of the insulative layers,
- A = The area of the electrode plate in square inches,
- d = The thickness of the dielectric layer between each electrode.

We discussed capacitors in Section 9.4. To review, the preceding formula is
valid for both thick-film capacitors and standard polymer PWBs. However, several characteristics of thick films have a bearing on the capacitance values achievable with the technology. There are practical limits on the number of conductive
layers that we can print with thick films. Thus, "n" may be limited to less layers in
thick films than in MLBs. Conversely, the relatively high dielectric constant, "K,"
of glass-frit-type insulative layers, and the thin "d" dimensions producible in thick
films work to the benefit of capacitance values.

9.6 Special SMT Design Rules for Thin-Film Hybrids

The design rules in the first eight chapters, and the military circuitry rules from Chapter 10, plus the preceding thick-film discussion, form the backbone of thin-film SMT hybrid design. Below, we'll cover those points where thin-film technology requires a departure from this.

9.6.1 *Component Orientation*

As with thick-film circuits, the general rules for Type 1 SMT circuits apply for thin films assembled by reflow soldering. The rules may bear revision for conductive epoxy attachment, and adhesive vendors can furnish design-rule guidance as needed, in those instances where thin-film circuits are assembled with conductive epoxy.

9.6.2 *Wiring Layout*

Thin-film technology generally involves the deposition of a conductive film across the full surface of the substrate, followed by masking, photolithographic exposure of the mask, washing away of the unwanted portions of the mask, and etching of the pattern not protected by the remaining mask. Because this process involves the photographic registration of conductive patterns, metallization-feature to mechanical-feature registration is not guaranteed. Therefore, it is a good practice to photolithographically generate a set of orientation features during the conductive pattern-generation process.

Multilayer thin-film techniques are in practice, and may be studied where required. However, currently, very few SMT hybrids are built using thin films with more than one conductive layer. Therefore, we will consider only single-layer thin films herein.

Traces and Spaces
Where space permits, line-and-space dimensions per Level 2 density, as shown in Figure 5-11 of Chapter 5, keep manufacturing costs low. However, on good quality ceramic surfaces, thin-film technology will readily permit the production of 0.025-mm (0.001-inch) lines and spaces. On sapphire or other highly polished substrates, lines as small as 5 microns (0.0002 inch) can be fabricated commercially. So thin-film techniques lend themselves to tight single-layer geometries.

Clearances Around Solder Lands
The clearances around solder lands, as specified in Chapter 5 (Sections 2 and 4), are determined more by circuit-assembly considerations than by metallization techniques. Whenever possible, they should be followed regardless of the substrate manufacturing process.

Crossovers
Crossovers may be accomplished using dielectric covers over small sections of first-layer conductor. Design rules per thick films may be applied to thin-film crossovers.

9.6.3 *In Situ Component Design*

Resistors

Thin-film resistors are inherently stable, exhibit excellent TCR tracking, and extreme trim stability. Resistors may be fabricated to close tolerances using thin-film techniques.

One of the most constraining factors in thin-film resistor design is the general requirement to abide with a single resistive layer. This means that all thin-film resistors must be constructed using only a one-sheet resistivity value.

The thick-film formulae presented earlier apply also to thin-film resistors. The typical geometry limits for thin-film resistors are covered in Figure 9-44. Some resistive thin films are self-passivating and others will require a protective overglaze.

Figure 9-44
Thin-film etched-resistor form-factor rules.[17]

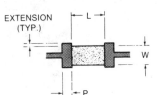

DIMENSION	MODIFIER	MANUFACTURING SIMPLICITY		
		Simple *	Average *	Difficult *
L	MIN.	0.254 (0.010)	0.127 (0.005)	0.127 (0.005)
W	MIN.	0.063 (0.0025)	0.025 (0.001)	0.013 (0.0005)
S *	MIN.	0.2	0.1	0.05
S *	MAX.	100	1000	2000
EXTENSION	MIN.	0.254 (0.010)	0.127 (0.005)	0.076 (0.003)
P	MIN.	0.051 (0.002)	0.025 (0.001)	0.013 (0.0005)

* Values are in millimeters and inches.

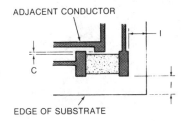

DIMENSION	MODIFIER	MANUFACTURING SIMPLICITY		
		Simple *	Average *	Difficult *
C	MIN.	0.254 (0.010)	0.127 (0.005)	0.051 (0.002)
I	MIN.	0.254 (0.010)	0.254 (0.010)	0.127 (0.005)

* Values are in millimeters and inches.

Inductors

The small fine-line geometries which can be etched using thin-film printing techniques allow the construction of considerably higher inductances than what can be achieved with thick films or PWBs. The formulae for inductors are the same as those presented for PWBs in Figure 6-7 of Chapter 6.

9.7 SMT Hybrid I/O Interconnects and Packaging

Next, let's look at how a circuit interfaces with, and is protected from, the world around it.

9.7.1 *SIP and DIP Lead Frames*

Lead frames are often used to make SIP or DIP components from SMT hybrids. Lead frames on 2.54-mm (0.100-inch) or 1.27-mm (0.050-inch) centers are available. Typical solder land configurations for edge-attached lead frames are shown in Figure 9-45.

Figure 9-45 Solder land dimensions for edge-attached lead frames.

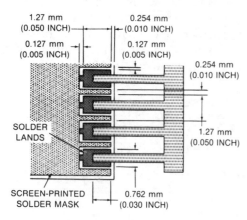

(A) Screened mask—1.27-mm centers.

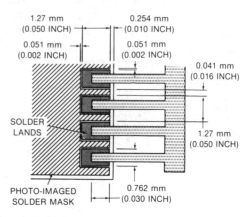

(B) Photoimaged mask—1.27-mm centers.

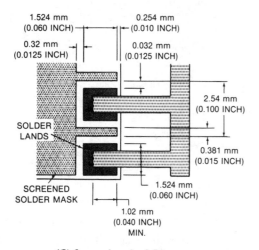

(C) Screened mask—2.54-mm centers.

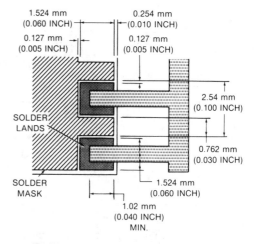

(D) Photoimaged mask—2.54-mm centers.

9.7.2 *Clip Leads for Planar Mounting*

Using the dimensions shown in Figure 9-45 for solder lands, leads may also be attached to form a planar-mounting DIP or quad component. Figure 9-46 shows an ADC-00302 12-bit, 2-MHz, T/H and A/D flatpack SMT hybrid which has planar-mount leads attached for surface mounting to a larger SMT motherboard.

Figure 9-46 SMT
hybrid with
planar-mount
lead frame.
*(Courtesy ILC Data
Device Corp.,
Bohemia, NY.)*

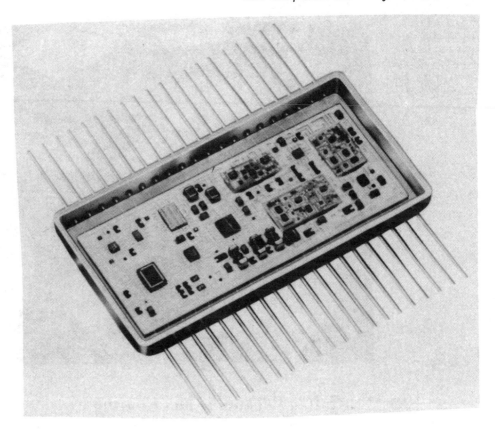

9.7.3   *Connectors and Headers*

Surface-mount connectors and headers may be reflow mounted to ceramic or polymer substrates. However, our previous cautions regarding the weakness in repeated connector cycling may speak against this. Connectors having a mechanical interface to the substrate will provide a far better life under repeated cycling. Mechanical connections generally require holes in the substrate. Thus, widespread use of mechanically fastened connectors may rule against selecting ceramic as a substrate material.

9.7.4   *Package-Mounted I/O*

As mentioned earlier in this chapter, hybrid circuits having exposed semiconductors are often protected with external packaging. Hermetic metal packages, hermetic and nonhermetic ceramic packages, and nonhermetic plastic-molded housings are used for this purpose. But, whatever the packaging direction, system I/O must somehow be routed through the package. An assortment of packages and I/O lead-frame clips is shown in Figure 9-47. Where hermeticity is required, feedthrough pins may be fired into glass in the wall of the package. The pins are usually located at the same height as the top of the substrate, and are connected to the circuit with wire bonds. Packaging vendors can lend assistance in your package and feedthrough design.

Figure 9-47
Assorted packages and lead frames. *(Courtesy Hybrid Systems Div. of Sipex, Billerica, MA.)*

9.7.5   *Protective Packaging and Coating*

Packaging

Where protective packaging must provide a hermetic seal, seal integrity becomes a major quality issue for manufacturing. The protective purpose of the hermetic seal may not be met if there is even a very small leak. During operation, as the circuit heats, gas in the package expands and begins to leak to the outside. This process continues until the internal hot gas reaches equilibrium pressure with the ambient atmosphere, or the device is switched off. After the hot internal pressure has equalized with the ambient pressure, switching off the circuit will cause an internal vacuum with respect to ambient atmosphere, and can draw damp or otherwise contaminated air into the package.

Leak rates are measured in the cubic centimeters of air that would pass through the leak per second at a differential pressure of 1 atmosphere (Atm.cc/S.). Rates greater than 10^{-5} Atm.cc/S. are called gross leaks. Slower leaks are called fine. To graphically depict what is meant by a fine leak, we have shown in Table 9-4 the time required for 1 cc of air to pass through various leaks.

Table 9-4. Comparison of Leak Rates[18]

Leak Rate (Atm.cc/S.)	Time Required
10^{-1}	1 cc of air leaks every 10 seconds.
10^{-3}	1 cc of air leaks every 17 minutes.
10^{-5}	1 cc of air passes through leak in 28 hours.
10^{-8}	1 cc of air passes through leak in 3 years.
10^{-11}	1 cc of air leaks in 3000 years.

Hermetically sealed hybrids are tested for hermetic integrity using several methods. Unfortunately, the methods which will detect gross leaks fail to locate fine leaks, and vice versa. Figure 9-48 shows the leak rate ranges of some common leak-test methods. To ensure all leaks are detected, tests covering the full range of concern, with sufficient overlap, should be selected.

Figure 9-48
Detection ranges of various leak tests.[19]

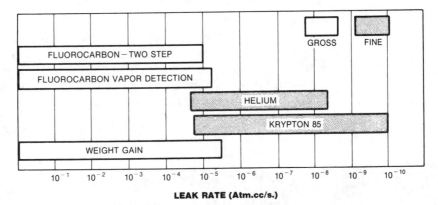

(A) Small cavity tests.

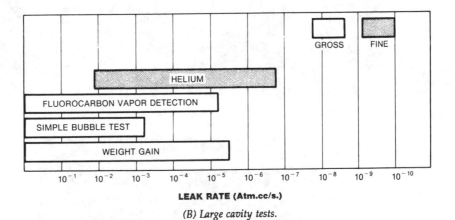

(B) Large cavity tests.

Coatings
Hybrid circuits are generally coated with a protective overglaze. Ceramic thick-film overglazes are usually low melting temperature vitreous glass materials. Overglaze protection is required for many resistive inks. It may also be used to cover conductors and act as a solder resist. Screened or photoimaged solder resists may serve these same functions, and are commonly used on polymer substrates.

9.8 SMT Manufacturing Methods for Hybrids

Below, we'll provide a brief view of SMT hybrid manufacturing methods, and the equipment used in the assembly process. We will not have space to begin to give a complete accounting of various classes of equipment. If you're shopping for a line, please look at more than just this text.

9.8.1 Unique Elements of SMT Manufacturing of Hybrids

Incoming Inspection

Hybrid technology brings new materials to the manufacturing operation. Receiving inspection must be prepared to test or NIV certify resistive, conductive, and dielectric layer materials, substrate materials, and packaging/coating materials for the hybrid process. Figure 9-49 shows some of the procedures QC applies to receiving inspection in a hybrid manufacturing plant.

Figure 9-49 Hybrid QC procedures. *(Courtesy Sfernice International, 199 Bd. de la Madeline, B.P. 17 06021, Nice Cedex, FRANCE.)*

(A) Optical inspection is a major tool of a hybrid QC department.

(B) SEMs provide added analytical and QC vision enhancement.

(C) A wealth of analytical procedures are based on the PC— surface analysis, thickness measurement, etc.

Ceramic Processing

If your SMT hybrid operation will make extensive use of ceramic substrates, you may want to process ceramic boards internally rather than buy them from subcontractors. A ceramic processing line is shown in Figure 9-50.

Figure 9-50 A ceramic-substrate processing line. *(Courtesy Amplica, Newbury Park, CA.)*

Figure 9-51 Firing furnaces in a thick-film deposition area.

Figure 9-52 Thin-film manufacturing of resistor networks. *(Courtesy Sfernice International, 199 Bd. de la Madeline, B.P. 17 06021, Nice Cedex, FRANCE.)*

(A) CAD area—design underway.

(B) Vacuum deposition of thin-film by cathode sputtering.

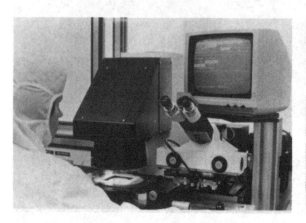

(C) Photolithographic imaging of 1-micron pattern.

(D) Etching bench uses ion-beam machining to precisely remove nonphoto-masked material.

(E) Plasma etching system.

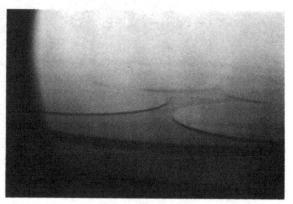

(F) Substrates in plasma etch chamber being passivation-layer etched, with close-up view through porthole of machine shown in Figure 9-52E.

Thick-Film Printing

In Figure 9-51, we see thick-film conductors being screened onto a ceramic substrate. In the background, a firing furnace stands ready to dry and fire the ink on the circuit board. After firing, additional screenings will add resistive and insulating layers.

Thin-Film Printing

Figure 9-52 illustrates the thin-film manufacturing process in a resistor network plant. Unfortunately, the black and white close-up shown in Figure 9-52F fails to fully capture the other-world aura cast by the purple plasma glowing on the operators in their clean-room bunny suits.

Figure 9-53 Laser trimmer in operation. *(Courtesy Electro Scientific Industries, Inc.)*

(A) ESI's Model 44PLUS Laser Trimming System.

(B) Circuit with probes attached undergoes laser trimming. Time-lapse photography shows resistive material being vaporized by an invisible Nd:YAG laser beam.

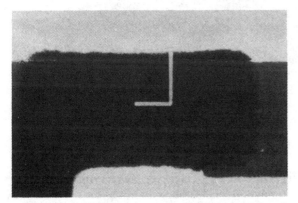

(C) Close-up of a trimmed resistor, showing an L cut. The laser is programmed to turn the cut 90° as the resistor approaches the specified tolerance. This slows the rate of resistance change per length of cut, allowing a more precise trimming.

(D) Extreme close-up of an L cut. Note the uniformity and minimal slag. The cut is in a thick-film resistor.

Figure 9-54
Solder screening
on an edge-
referenced
ceramic board.
(Courtesy De Haart.)

Figure 9-55 SMC
placement on
edge-referenced
substrates.
(Courtesy Zevatech.)

Laser Trimming
Laser trimming is underway in Figure 9-53. Here, laser passive trimming is quickly bringing a network of thick-film resistors to ±1% of specified value.

Attachment-Media Application
The printing of solder paste on hybrids is not very different from solder printing on polymer PWBs. Where substrates are ceramic, they are typically referenced from

two edges, rather than from tooling pins in holes. Figure 9-54 shows a solder-paste screening operation on an edge-referenced ceramic substrate.

Figure 9-56
Linear-conductive soldering of SMT hybrids.
(Courtesy Sikama International, Inc., Santa Barbara, CA.)

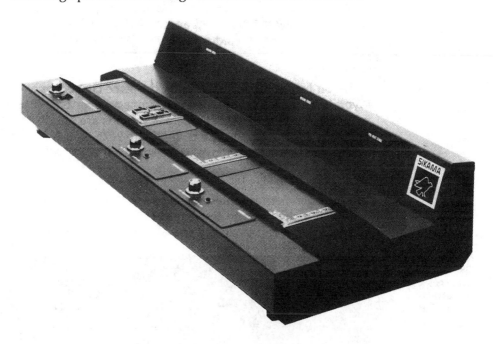

SMC Placement
Like solder application, placement of SMCs on hybrids is similar to standard PWB placement. Again, edge referencing may be required for a given substrate material. In Figure 9-55, we see SMC placement on a series of ceramic hybrid boards that are edge referenced on an indexing mechanism.

Figure 9-57
Linear aqueous cleaning of SMT hybrids. *(Courtesy Hollis Automation.)*

Curing and Soldering

The curing and soldering of SMT hybrids is done in the same way that it is for standard PWBs using SMCs. Few hybrids use through-hole connection techniques, so reflow soldering is the typical approach. Figure 9-56 shows a linear-conduction reflow system used to solder single-side-populated SMT hybrids.

Cleaning

SMCs on hybrid substrates present the same challenges as they pose on PWBs. Hybrids are cleaned in essentially the same manner as PWB assemblies. Extra care may be required to ensure that small hybrids are not displaced on the cleaning system belt by strong spray jets. In Figure 9-57, we see a linear aqueous cleaning system that is used to remove water-soluble flux and soils from hybrid circuits.

Test

The tight geometries achievable in hybrid layouts are a boon to miniaturization, but often a challenge to test engineering. In Figure 9-58, a special probe card provides test access to closely spaced test points on an SMT hybrid.

Figure 9-58
Custom probe
card accesses fine
geometries of
SMT hybrid.
(Courtesy Mepco/
Centralab, Inc.,
Philips Circuit
Assemblies Division.)

Repair and Rework

Repair and rework of SMT hybrids may be done with hot air or with heated non-oxidizing gas, just as is commonly done with SMT PWB assemblies. Single-side-populated hybrids on ceramic boards are also reworked on micro reflow stations. Such a repair station commonly uses a glow-plug-ignited flame on hydrogen gas, which is then directed below the circuit board for reflow of precise spots. Preheat from a resistance element guards against cracking due to thermal shock.

9.8.2 *Typical Process Flows*

Figure 9-59 shows several ways that both through-hole and surface-mount components may be integrated into a hybrid manufacturing process.

Figure 9-59 Two typical flow processes for SMT hybrid production.

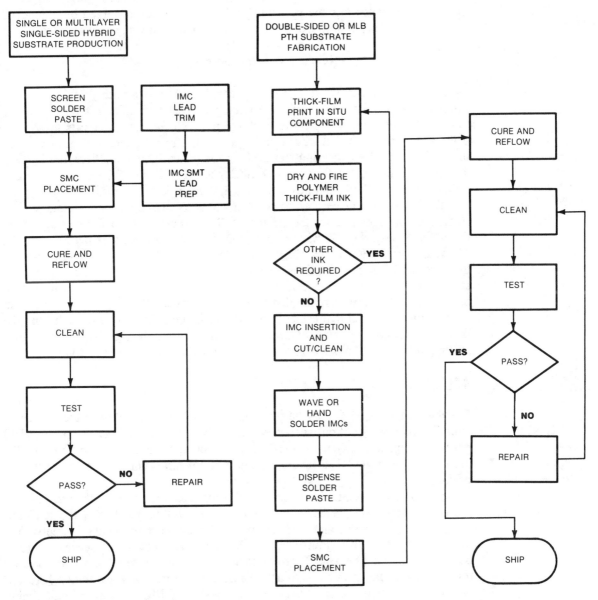

(A) Flow process for Type 1 SMT hybrids.   *(B) Flow process for Type 2 SMT hybrids.*

9.9 Fundamentals of MCM Design

At clock rates of 25 MHz and below, there is very little need for MCMs. As frequencies move to 50 MHz and beyond, the MCM begins to be an attractive packaging solution. They are already widely used in workstations and high-performance computers, cellular telephones, laptop computers, palmtops, and other hand-held electronics. Advanced automotive electronics such as steering control, engine management, suspension controls, collision avoidance, and satellite positioning will provide an expanding market for MCMs for some time to come.

Why are designers so excited about MCMs for high-performance system? The answer is clear when we compare the densities of various packaging techniques. Let us define density as the percentage of substrate area occupied by silicon IC die. This provides a tool to make such a packaging-density comparison. Thus:

$$e_{IC} = (IC\ die\ area/total\ substrate\ area) * 100\%$$

By this measurement, if we completely cover a PC board with PLCCs, we will achieve an e_{IC} of about 6%. MCMs routinely deliver e_{IC} ratings in the range of 12 to 60%. By allowing designers to pack die more closely, interconnects are drastically shortened. For fast-transition signals, shorter interconnects means fewer problems with false triggering from reflected signals, large reductions in interconnect parasitics, and greatly improved management of EMI/RFI. Both coupled and radiated noises are substantially diminished, thus the interest in this technology.

There are three generic types of MCM circuits, based on the technology used to produce the substrate for the circuit. MCM-C substrates are made using hybrid thick-film techniques. The C is drawn from the usual substrate material, ceramic. MCM-D substrates may also be ceramic, but the D refers to deposited, indicating that the circuit is metallized using thin-film deposition. An MCM-L has a substrate made of a polymer laminate, like an ordinary circuit board. We will look at each separately. However, the distinctions between each are not always clear. In the drive for cost and real-estate reduction, techniques from hybrid thick-film and thin-film technology often spill over into laminate circuit boards. The intent in presenting three distinct types is to avoid boundaries to engineering creativity.

9.9.1 MCM-C

The thick-film MCM-C may be made using standard print-and-fire techniques or cofired substrate. As of our second edition, substrates having lines as small as 100 microns have been made using the MCM-C approach. Theoretical limits for multiple layers are around 100 layers, but circuits of more than 10 or 15 layers are rarely seen.

As 9-34 indicated, interconnects in all areas have followed a rather steady progression toward ever smaller feature size. From 1980 to 1990, only the semiconductor industry showed any variance from a linear downward curve. Line widths produced by thick-film, thin-film, and laminate-etch techniques shrunk continuously over the past decade. There is no reason to believe this trend will not continue well into the next millennium. Thus, the feature sizes mentioned here are provided to help designers select between styles of MCMs. For current limits, check with substrate suppliers.

9.9.2 MCM-D

The MCM-D thin-film deposited interconnect structure uses technology born in the semiconductor industry to build the conductive paths on the substrate. Metallizations are made by vapor deposition or ion sputtering. MCM-D substrate, as of

1955, routinely employ interconnects down to 20 microns in width, with similar spacings between adjacent traces. Very high eIC densities are possible. D type MCMs, like Ferraris, are fast and pretty. They are also, like Ferraris, seldom cheap. Both the processes and the materials for the D substrate tend to be costly. Even the dielectric polymers used are precious, costing as much as $200.00 per gram. Dielectric layers are applied by spin coating, so material loss adds to the cost of each successive layer.

9.9.3 MCM-L

The MCM-L is a multi-chip module built on a laminated-polymer substrate. Per square inch, it is the undisputed price leader among MCMs. However, typical laminated-polymer interconnect structures do not come close to ceramics in either interconnect density or the ability to dissipate the heat of a closely packed cluster of fast ICs. Seeing these factors not as limitations, but as opportunities for the low-cost laminate approach, innovative packaging engineers have pushed the density envelope of the MCM-L by adapting MCM-C and MCM-D processes. Thus, while space and trace rules usually start at 100 microns, new techniques have yielded densities down to 25 microns.

9.9.4 *Variants on the Theme*

The MCM-LD is one obvious merger of technology. This packaging approach involves the use of a laminated substrate with a surface layer or layers fabricated using thin-film techniques. The substrate may be a multilayer circuit. Thus, the MCM-LD can provide a very-high-density interconnect structure at a reasonable cost.

Another technology merger, the MCM-LF or laminated-film MCM, rivals the MCM-D in density and adaptability to the needs of high-speed circuitry. It involves fabrication of two or more thin-film-on-polymer circuits which are then laminated onto one high-density interconnect structure. Interstitial connections are formed with microvias as small as ?? Since it avoids the need for spin coated dielectrics, the cost of the MCM-LF is much lower than the typical cost for the MCM-D approach.

Technology advances are pushing the cost of very-high-density MCM substrates downward at a rapid rate. Clearly, the approach will gain rapidly in use. We also expect the substrate fabrication techniques to spread in usage. Why limit high-density interconnect techniques to just a few IC chips? Why not fabricate the entire circuit on a single high-density card? Unfortunately, space does not permit us to give in-depth coverage to the MCM. The subject is complex enough to deserve a book all its own. Let us simply note the importance of the growing technology base known as MCMs, and encourage further study for all who find PWB real estate or performance a critical issue.

9.9 References

1. *Proceedings of Hybrid Microtech '88*, West European Market for Hybrid Circuits, Dr. Nihal Sinnadurai, Chairman, United Kingdom Chapter, International Society for Hybrid Microelectronics (ISHM), London, January 1988.

2. Hamill, A.T., Konsowski, Steven G., et al, *Hybrid Microcircuit Design Guide*, ISHM, Reston, VA, 1982.

3. Balde, John W., Caswell, Greg, et al, *Surface Mount Technology*, International Society for Hybrid Microelectronics, Reston, VA. 1984, pg. 3.
4. Lau, John H., and Rice, Donald W., "Solder Joint Fatigue in Surface Mount Technology, State of the Art," *Solid State Technology*, October 1985, pg. 91.
5. Op. cit., ref. 3.
6. Ackerman, Karl-Peter, Hug, Rolf, and Berner, Gianni, "Multilayer Thin Film Technology," *Proceedings of International Symposium on Microelectronics, Atlanta 1986*, ISHM, Reston, VA, 1986, pg. 519.
7. Kompelien, D., Moravec, T.J., and DeFlumere, M., "A New Hybrid Technology: High Density Thin Film Copper/Polyimide Multilayer System," *Proceedings of International Symposium on Microelectronics, Atlanta 1986*, ISHM, Reston, VA, 1986, pg. 749.
8. Ohdaira, Hiroshi, Saito, Masayuki, and Iida, Atsuko, "A New Polymeric Multiyalers Substrate," *Proceedings of International Symposium on Microelectronics, Minneapolis 1987*, ISHM, Reston, VA, 1987, pg. 515.
9. Smith-Vargo, Linda, "PCB Makers Spiral Into the Future," *Electronic Packaging & Production*, February 1988, pg. 67.
10. Pedder, D.J., "Flip Chip Solder Bonding for Microelectronic Application," *Journal of ISHM---Europe*, No. 15, January 1988, pg. 4.
11. Frecska, Tamas, "Theoretical Model for Multilevel Screen Printing of Thick Film Compositions," *Proceedings of International Symposium on Microelectronics, Minneapolis 1987*, ISHM, Reston, VA, 1987, pg. 314.
12. Op. cit., ref. 2, pg. 74.
13. Ibid.
14. Ibid.
15. De Mey, G., and Van Schoor, L., "Thermal Analysis of Hybrids with Mounted Components," *Journal of ISHM---Europe*, No. 15, January 1988, pg. 28.
16. Op. cit., ref. 2, pg. 77.
17. Ibid.
18. Gillespie, T., "Semiconductor Seal Testing," Journal of ISHM---Europe, No. 15, January 1988, pg. 46.
19. Ibid.

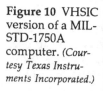

Figure 10 VHSIC version of a MIL-STD-1750A computer. *(Courtesy Texas Instruments Incorporated.)*

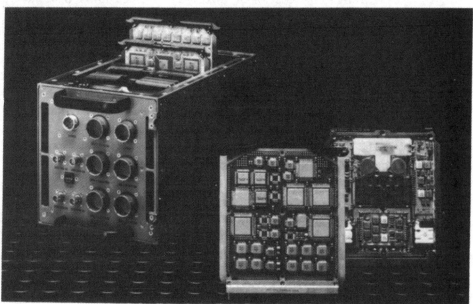

10 Using SMCs in Military and Space-Borne Applications

Surface-mount technology holds important potential benefits for military electronics. The VHSIC 1750A Computer shown in Figure 10 stands as tangible proof of the miniaturization, weight reduction, and performance advances that SMT may deliver. In this chapter, we'll look at state-of-the-art in surface-mounting in the military and aerospace fields.

The miniaturization, performance improvement, and weight reduction potentials of SMT offer great advances for military and space-borne electronics. But, to realize these potentials, SMT design and manufacturing rules must be tailored to the special requirements posed by the military and space environments. Just as with commercial circuits which support human life, reliability must be superlative. In fact, reliability in a miltary circuit can carry much more than one human life in the balance. Our freedom and our very life as a nation could depend on the proper functioning of our military hardware, and on our fighting men being all that they can be.

10.1 The Military's Interest in SMT

The United States military establishment is far too large and too segmented for us to speak of it as a single unit. It is certainly true that the military has shown intense interest in SMT. Table 10-1 shows some of the SMT research underway through Wright-Patterson AFB, Dayton, Ohio. There is, however, an opposing vector resisting any rapid technology change, as shown in Table 10-2. A balanced view is that the military and the space organizations are very interested in technologies that improve performance, reliability, density, and cost. At the same time, they are acutely concerned about obtaining consistently reliable circuitry. SMT circuits present a new and challenging array of process control and inspection requirements. Before the military/space community will accept a vendor's SMT program, they must be convinced that the process controls and inspection procedures which the vendor has in place will yield consistently reliable circuits.

Table 10-1.
Wright-Patterson
AFB Spearheads
SMT Research[1]

Program	Description	Contractor
F33615-82-C-5047	R&D for modeling system for HCCs on PWBs of Kevlar/quartz/polyimide, Kevlar-modified polyimide, and quartz/Teflon.	Westinghouse
F33615-82-C-5071	Evaluation of HCCs on various PWB material, bonded to metal cores.	Texas Instruments, IBM, and Boeing
F33615-84-C-1415	VHSIC low-K PWBs.	Hughes and Norplex
F33615-84-C-5023	Tailorable TCE PWBs for high-density leadless perimeter and grid-array packaging.	General Electric
F33616-84-C-5047	Solder-filler metal development for electronic chip carriers.	Westinghouse

Table 10-2.
Problems Noted
at Twin Bridges
Conference[2]

Problem	Suggested Solution
Inadequate chip solderability.	Test solderability in assembly area. Update leadless component solder specifications.
Ceramic substrate cracking.	Control of baseplate flatness.
Thermal stresses.	Thermal design and match TCE of component and substrate.
Unmarked components.	Laser mark for visual and auto ID.
Low yields due to chip quality.	Improve specifications, diffuse lot bond, package sample test, and lot jeopardy.
No production envolving hermetric copper multilayer system.	Work with ink and furnace manufacturers to define requirements and develop technology.
Copper multilayer furnaces constrict designs.	Work with ink and furnace manufacturers to define requirements and improve furnaces.
Difficult to predict reliability.	Develop guidelines and requirements.

A great deal of progress has been made since the Twin Bridges conference referred to in Table 10-2. The publication of DoD-STD-2000, the surface-mount amendments to MIL-STD-454, the distribution of NAVSEA 5516562, plus IPC and EIA advances in commercial specifications, and the publication of books like this one, all add valuable support to the safe military use of surface-mount technology. In Appendix B, we list current specifications and specifications. Below, we'll present currently acceptable military and aerospace SMT practices. We'll start with a review of SMT benefits and look at how they apply to the needs of high-reliability electronics.

10.1.1 *Miniaturization Using SMT*

The drive for more electronics in less space is not unique to the consumer industry. Today, sophisticated electronic circuitry is fundamental to defense readiness. And the fight for space in a satellite can be a battle royal. Both military and space engineers are vigorously whittling away at circuitry real estate. Increasing integration on the silicon chip was the primary driver for miniaturization until the early 1980s. VLSI, VHSIC, and ASIC implementation is still supplying part of the push. But, second-level interconnects are becoming a major attack point in the battle to shrink supercomputer circuitry into snuffbox-size packages. As we'll see below, SMT is at the front lines in this battle.

Size-Driven Applications
Field-transportable and hand-held electronics are among those applications where SMT is showing its miniaturization value. Designers of avionics controls and flight computers also are using SMT miniaturization to cram more function into the limited space available on a combat aircraft. The AN/ALQ-161 avionics board shown in Figure 10-1 is an electronic-jamming controller/device. It is designed to allow the B-1B bomber to penetrate deep into heavily guarded hostile airspace. Individual boxes communicate over an MIL-STD-1553 hi-rel data bus. The full 100-box system consumes about as much power as 120 microwave ovens operating simultaneously. Clearly, without with high-level miniaturization seen, the system would be unmanageably large for an airborne application.

Figure 10-1 An AN/ALQ-161 electronic countermeasures board for the B-1B bomber. *(Courtesy Eaton Corp., AIL Div. & AMAX Spec. Metals.)*

Cost-Driven Applications

Separating cost- and size-driven applications requires a judgment call. Generally, making a unit appreciably smaller will save on costs of housings, PWBs, etc. When miniaturization goes far enough to reduce a multiboard system to one card, the savings can be substantial. A dividing line might be drawn when making it smaller to reduce the cost shifts over to making it smaller regardless of the cost.

Figure 10-2 shows a use of SMT to dramatically reduce system size to the advantage of system cost. The single-board military specification VAX™ computer, produced by Raytheon under license from Digital Equipment Corp., is tailored for embedded real-time control applications. The entire VAX computer, with a 512K SRAM, 256K ROM, and two RS-423 serial ports, fits on the single card shown, roughly 150 mm × 200 mm (6 × 8 inches). The CMOS system dissipates 20 watts, and weighs less than 1.6 pounds (0.75 kg). The two units in the background are a VAX minicomputer and a superminicomputer.

Figure 10-2 A single-board military VAX. *(Courtesy Raytheon Co., Electronics Div.)*

10.1.2   *Weight Reduction Through SMT*

Airborne electronics designs are typically constrained in both size and weight.

Separating weight reduction from miniaturization also requires a view of design intent more than design direction. In hand-held and personnel-borne gear, weight reduction is often a key design target. An infantry soldier probably doesn't care if his walkie-talkie is the size of a cigarette lighter or a cigar box. But, running for his life across a hot muddy field with bullets burning past his ears, he cares intensely if it is heavy enough to slow him down.

Figure 10-3 shows an incredible reduction in system size from nine cards to one. Weight reduction was one major objective in this IMC to SMT conversion. The final card is one of two that make up the Joint Tactical Information Distribution System (JTIDS) airborne computer. The full system was reduced from 15 through-hole boards to 2 double-side-populated SMT boards. Actually, two 6-layer boards are bonded to a copper/Invar/copper thermal core to form the substrate assembly.

Figure 10-3 A 9:1 reduction in a JTIDS computer project. *(Courtesy Rockwell International, Avionics Group.)*

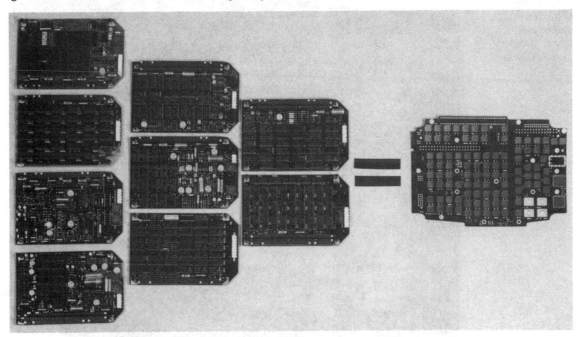

Recalling the JTIDS analysis from Table 1-2 of Chapter 1, the SMT version reduced the system size (volume) by a factor of 8:1 and the weight by 3:1. The system packaging and interconnection costs were reduced correspondingly.

10.1.3 *SMT Design for* Severe *Environments*

In SMT design for rugged military environments, we must remember that some of SMT's attributes are helpful and others detract from the robustness of the board assembly. The lower mass of SMCs is of great advantage in shock situations. Depending on the frequency, SMCs also generally hold the advantage over IMCs in vibration resistance. However, the lack of mechanical attachment furnished by through-hole leads means that SMT solder-joint integrity must be assured at all costs. And, in the few instances where their masses are close to equal, the IMC is likely to be the more resistant component in shock and vibration conditions. With this trade-off in mind, Chapter 6, Section 6.4, lists data for the design of products for high-stress environments.

Where environmental challenges will possibly be difficult, and certainly where the environment has presented problems for through-hole design, a basic research and test program is a must. The board in Figure 10-4 is a test vehicle for General Electric.

Many of the major defense contractors have used such test assemblies to qualify and calibrate design and process variables. Particularly in the area of board processing, much of the data required to assemble high-reliability SMT boards is just not available in the public domain at this time. Often, such data are dependent upon the particular materials and process equipments involved. With the number of permutations possible, it is likely that in-house research and development will be a fact of life for some time yet.

Figure 10-4
Engineering and
manufacturing
process test
board. *(Courtesy
General Electric.)*

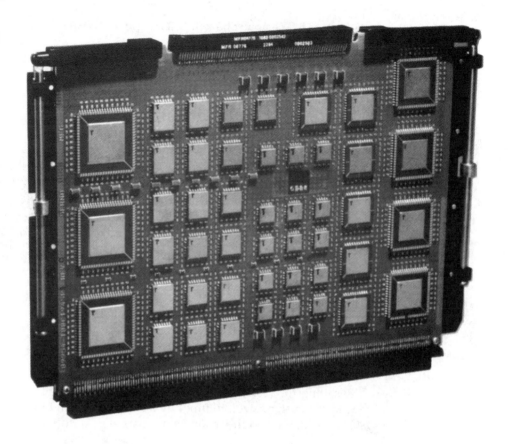

10.2 SMT and the VHSIC Program

Achieving high interconnect density, and the resultant reduction of signal-path lengths, is critical to VHSIC designs. Chip and wire techniques on ceramic substrates are not recommended because of the high dielectric constant of ceramics. TAB on polymer substrates, or on polyimide dielectric layers over ceramic substrates, is one alternative that may be used.

Figure 10-5 shows a high-performance MIL-STD-1750A computer, which carries up to fourteen standard electronic modules (SEMs) in E format. The SEM substrates here are alumina ceramic, chosen for compatible TCE with the CLLCCs. The higher dielectric of the ceramic substrates did slow signal propagation in the system, but the density gain from surface mounting was sufficient to allow its use within system specifications.

The largest carriers on the SEMs are 84-pin devices, on 1.27-mm (0.050-inch) centers. The modules are double-side populated. Each SEM-E module is built up of two alumina SMT cards, which are elastomericly bonded to an aluminum thermal plate. The elastomer prevents stress buildup due to TCE mismatches between the ceramic cards and the metal core.

The combination of VHSIC technology with advanced SMT packaging, high-density power supplies, and modular hardware design reduced the system size and power consumption while simultaneously boosting system performance. The

14-slot three-quarter ATR chassis weighs only 11.34 kg (25 pounds) fully loaded. Yet it provides up to 6.0 MIPS of processing power using DAIS mix. A complete system includes an MIL-STD-1750A processor, an MIL-STD-1553B dual communication interface, 1.7 million words of local memory, two-level system maintenance, built-in test, and a high-density power supply. The computer was developed under a Lockheed contract to produce two demonstration models for the YF-22A Advanced Tactical Fighter.

Figure 10-5
Advanced
VHSIC MIL-STD-
1750A Computer.
(Courtesy Texas Instruments Incorporated.)

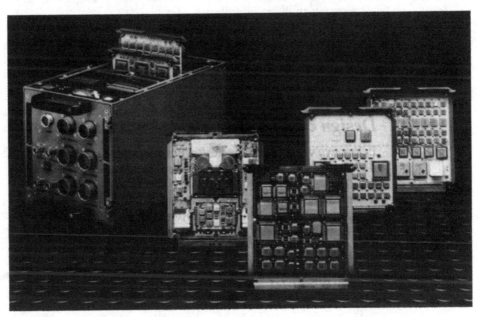

10.3 Adapting SMT to Military Specifications

The rules presented throughout this book are targeted at producing highly reliable circuits. Thus, rules don't change dramatically when we shift to the military and the high-reliability world. Where there are differences, they are covered below.

We need to realize, however, that most current MIL Standards and Specifications were written for insertion technology. In many cases, they are difficult to apply to SMT. Sometimes, they are completely out of phase with the requirements of surface-mount technology. We will discuss cleaning as one out-of-phase example. Inspection method is another area where the needs of surface-mount technology are very different from through-hole technology.

Where there is such a direct conflict between the specifications required by a contract and the needs of SMT, a variance must be obtained. Military contractors using SMT today generally find that exceptions are far from the exception.

Fortunately, industry has not been left to stand alone in the fight for advanced manufacturing technology. The U.S. Navy sponsors a research center called the *Electronic Manufacturing Productivity Facility*, which was designed to work with industry in researching and developing quality materials and manufacturing procedures to advance and improve U.S. electronics manufacturing. The EMPF is not directed at any specific weapons program. Its mission applies equally to all weapons programs, and to private industry as well.

The rationale behind this seeming generous use of tax dollars is pure hard-nosed cost savings. Of the billions of dollars that the Department of Defense spends annually on defense electronics, over 30% goes into the "hidden factory" of finding and correcting errors and defects. By developing and disseminating technology that shrinks the "hidden factory" only 1%, the EMPF can save American taxpayers millions of dollars a year.

The EMPF executive board, composed of senior executives from the nation's leading defense contractors, has targeted SMT as a top priority. The EMPF staff can often help with specification problems. Where they don't have an answer, they can access key technical resources of the companies working in partnership with them. EMPF and its industry partners offer a wealth of information resources.

10.3.1 *Design of Circuits*

Below, we've gathered the design points that are unique to SMT, and the topics that deserve special treatment for military applications.

Components
Military circuits must often survive, and even perform, for extended periods in hostile environments. To ensure that environmental contaminants don't enter components, many high-reliability specifications require the use of hermetically sealed parts. Since hermetic components can be leak tested, it is possible to check the seal integrity. Plastic encapsulants also provide good die protection, but they can't be leak tested. Ceramic leadless chip carriers provide a leak-testable hermetic-packaging alternative that is very popular in military electronics. However, the leadless design introduces concern about the TCE match with substrate, as previously discussed. Where circuits must survive as many as 2000 thermal cycles, some military commands are forbidding the use of leadless components because of TCE concerns.

Since plastic components are generally not acceptable in military projects, there is a significant interest in the leaded-ceramic alternatives. However, leaded parts add standoff height and vibration-resistance concerns that are unacceptable in some applications. Component manufacturers are striving to meet all these confusing and often conflicting needs in one package. In our discussion of components in Chapter 2, we covered hermetic packages and the trade-offs between leadless and leaded varieties. Package outlines are shown in Appendix A. MIL-SPECs for components are covered in Appendix B.

Substrate Considerations
For military boards, substrate materials should have a high enough glass transition temperature to permit rework by semi-skilled personnel. Polymide and high Tg epoxies are a popular alternative to FR-4 epoxy glass. As we'll see later, the thermal conductivity of substrates has a greater emphasis in military design than in many commercial applications.

Since ceramic components are often required for hermeticity, substrate selection may be dictated by TCE considerations. If ceramic ICs will be surface mounted, they should either have flexural leads or be placed on a substrate with a compatible TCE. Rules for the matching or management of TCE were presented in Chapter 5, Sections 5.6 and 5.7, and in Chapter 6, Section 6.4.1.

Land Geometries
The general land dimension discussions of Chapters 5 and 6 pertain to both military and commercial design. Specific land recommendations are given in Appendix A. Note that there are several variants available. You should choose the land-geometry variant needed, based on the manufacturing process and the specification considerations. Some military specifications now make specific mention of land sizes, or reference commercial specifications that cover land size. So, a thorough review of the requirements imposed by the specifications is Step One in selecting land geometries.

Thermal Management
Military hardware is often designed to function in the most trying environments, such as Arctic cold, jungle heat, salt fog, desert sand storms, etc. To avoid drawing contaminants into a system, equipment designed for such environments does not normally use forced-air cooling. Thus, strategies for heat dissipation through the board are attractive for military design.

10.3.2 *Manufacturing Methods Overview*

Type 3 SMT has found very little use in military circles. Type 1 is quite popular because its simple process flow makes for easier process control. The Navy's Deputy Chief of Reliability, Maintenance, and Quality Assurance, Will Willoghby, pointed to process design as opposed to after-the-fact tests as the key to high-reliability electronics. This is particularly true with surface-mount devices, where solder joints are critical to reliability, yet difficult to inspect after assembly.

Cleaning is one area where military and commercial processes are quite different. Many commercial SMT manufacturers use ultrasonics to boost cleaning speed and efficiency. Military specifications prohibit ultrasonic cleaning, and a convincing body of data would be required to obtain an exception. This prohibition is due to concern about potential damage to IC wire bonds. Figure 10-6 shows a damaged wire bond that was caused by improperly controlled ultrasonics. There is work underway at the Navy EMPF to determine if ultrasonics can safely be used for military boards (commercial concerns are capable of controlling the process to build a reliable, clean product). Also, only a few commercial SMT boards are cleaned using aqueous-based systems. In contrast, military projects often require some water cleaning to ensure the removal of polar soils.

10.3.3 *Workmanship Standards*

Workmanship standards have been, until the publication of DoD-STD-2000, either nonexistent or proprietary. DoD-STD-2000 has been of great benefit to those military contractors who are without internally generated documents. Martin Marietta is updating a through-hole workmanship manual which they offer for sale. A company spokesman said they plan to issue a revision that includes SMT after the update is released.

Figure 10-6 Wire bond damage caused by ultrasonic cleaning. *(Courtesy Hi-Rel Laboratories, Monrovia, CA.)*

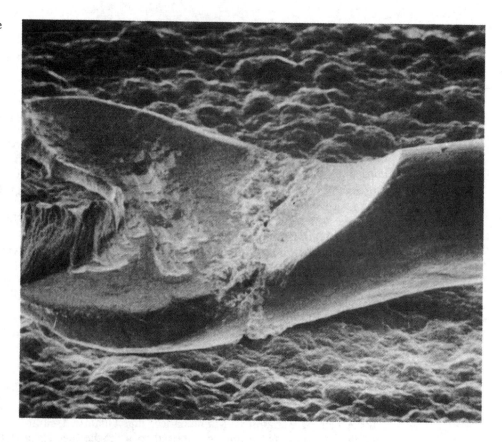

10.3.4 Test and Inspection

It is not by accident that the chapter on SMT quality and reliability assurance (Chapter 7) is the longest chapter in this book. Inspection, testing, and process control are vital to acceptable results with surface mounting. Throughout this text, we've been preaching design for test. SMT, in general, begs for this. But military SMT, driven by MIL-STD-2165, dictates design for test. The specification is deliberately vague on how one tests boards, but it mandates a management approach using design for testability. The Standard requires that the contractor document the steps taken to assure testability. Since we've already dealt with quality and reliability assurance (Q&RA) in some detail, we won't review what Chapter 7 sets forward. Below, we will cover those points unique to military and high-reliability SMT.

Destructive Physical Analysis
Destructive physical analysis (DPA) testing is not unique to military products. But military DPA test requirements are often more rigorous and comprehensive than those applied to commercial circuit assemblies. Testing for thermal cycling, vibration (shake, rattle, and roll), shock, salt spray, radiation, and high-temperature/ humidity conditions, is common to all military and space circuitry. When these tests are required for commercial boards, the test specifications are seldom as punishing. A review of the environments shows why.

MIL-STD-810 lists shipping, storage, and operating environments for various categories of military products. The nonmilitary issues listed here were shown before in Tables 7-7, 7-8, and 7-9, in Chapter 7. The weapons system considerations are given in Tables 10-3 through 10-6. These environmental and operational stress conditions guide the DPA test specification.

Table 10-3. Environmental Hazard Summary of Shipping and Transporation Stress[4]

Truck Transport	Rail Transport	Air Transport	Sea Transport
Road shock from bumps/potholes.	Rail shock from humping.	In-flight vibration (engine/turbine).	Wave-induced vibration (sinusoidal).
Road vibration (random vibration).	Rail vibration.	Landing shock.	Wave slam shock.
Handling shock (drops/overturn).	Handling shock (drops/overturn).	Handling shock (drops/overturn).	Handling shock (drops/overturn).
High temperature, dry and humid.	High temperature, dry and humid.	Reduced pressure.	High temperature, dry and humid.
Low temperature.	Low temperature.	Thermal shock (air dropped).	Low temperature.
Rain/Hail.	Rain/Hail.		Rain/Salt fog.
Sand/Dust.	Sand/Dust.		Temporary immersions.

Table 10-4. Environmental Hazard Summary of Storage and Supply Logistics Stress[4]

Handling/Logistics	Storage in Shelter	Storage in Open
Road shock from bumps and potholes.	High temperature, dry and humid.	High temperature, dry and humid.
Road vibration (random).	Low temperature/Freezing.	Low temperature/Freezing.
Salt fog/Solar radiation.	Salt fog.	Rain/Hail.
Handling shock (drops/overturn).	Fungus growth.	Sand/Dust.
High temperature, dry/humid.	Chemical attack.	Salt fog.
Reduced pressure.		Solar radiation.
Low temperature/Freezing.		Fungus growth.
Rain/Hail/Sand/Dust.		Chemical attack.

Objective Solderability Testing

MIL-STD-202, Method 208, and MIL-STD-883, Method 2022, are the commonly applied tests of solderability for DIPs. However, as we mentioned earlier in Chapter 7, these tests are not appropriate for leaded and leadless SMCs. Martin Marietta has proposed a test called the Martin Marietta Orlando Aerospace Solderability Test (MMOAST) for objective solderability testing of components, including SMCs. The test redefines procedures for using a wetting balance, thus making the equipment appropriate for the testing of surface-mount components.

Table 10-5. Environmental Hazard Summary of Operating Conditions Stress (Equipment)[4]

Hand-held Equipment	Mobile Equipment	Shipboard Use	Airborne Use
Handling shock (drops/overturn/slamming).	Road/Off-road vibration (bumps/treads).	Wave-induced vibration (sinusoidal).	Runway vibration.
Acoustic noise.	Engine vibration.	Wave slam shock.	Aerodynamic turbulence.
EMI/RFI	Acoustic noise.	Handling shock (drops/overturn).	Maneuver buffeting.
Atmospheric contamination.	Handling shock.	High temperature, dry and humid.	Engine vibration.
High temperature (dry and humid)/Fungus.	High temperature, dry and humid.	Acoustic noise.	Aerodynamic/environment (high heat/dry/humid).
Low temperature.	Sand/Mud/Salt fog.	Low temperature.	Acoustic noise.
Rain/Hail/Dust.	Low temperature.	Rain/Hail/Dust.	Low temperature/Thermal shock.
Solar radiation.	Rain/Hail/Dust.	Solar/Salt fog.	Rain/Hail/Dust/Sand impingement.
TCE store to use.	Solar radiation.	Fungus/Chemicals.	Fungus/Chemicals.
Firing and blast shock.	Fungus/Chemicals.	Engine vibration.	Reduced pressure.
Explosive atmosphere.	Road/Off-road shock of holes.	Mine blast.	Gunfire vibration.
Chemical attack.	Land mine blast.	Weapon firing shock.	Solar radiation.
Sand/Mud/Salt fog.	Weapon firing shock/vibration.	Explosive atmosphere.	Salt fog.
Thermal shock.	Explosive atmosphere.	EMI/RFI.	Explosive atmosphere.
	EMI/RFI.	Increased pressure (submarine).	EMI/RFI.
	Thermal shock (storage to use).	Thermal shock (storage to use).	Catapult launch/Arrested landing.
			Air blast (landing/takeoff/maneuvering).

Such solderability testing, in lieu of the simple immersion test, is needed where documented results must be available for after-the-fact inspection. The MMOAST is a solderability test using a modified wetting balance (Meniscograph), a vibration-free environment, and precision tooling, to measure the solder-pot insertion and withdrawal forces, and the solder-pot weight gain/loss.[5]

10.3.5 *SMT Rework and Repair*

Design for repair is always an important part of an SMT program. Its importance is greatly amplified when the system must be maintained by relatively unskilled personnel, either in remote field locations or in shipboards repair shops. Shipboard repair is particularly challenging because aggressive chemicals for conformal-coating removal are not allowed.

Table 10-6. Environmental Hazard Summary of Operating Conditions Stress (Weapons)[4]

Delivery of Projectile	Underwater Delivery	Missle/Rocket Delivery
Firing shock.	Launch acceleration.	Launch/Maneuver acceleration.
Firing acceleration.	Handling/Launch shock.	Acoustic noise.
Handling/Loading shock.	Engine vibration.	Handling/Launch shock.
Acoustic noise.	Acoustic noise.	Engine-induced vibration.
Aerodynamic heating.	Explosive atmosphere.	Aerodynamic turbulence (random vibration).
Explosive atmosphere.	EMI/RFI.	Pyrotechnic shock at booster separation.
EMI/RFI.	Immersion.	Aerodynamic heating.
Thermal shock (storage to use).	Thermal shock.	Explosive atmosphere.
Rain impingement.		EMI/RFI.
Sand/Dust impingement.		Rain impingement.
		Sand/Dust impingement.

The "Design for Reparability" suggestions of Chapter 7, Section 7.6.5, should be followed to the letter. Avoid conformal coatings wherever practical. Where coating is a must, use a material that can be removed with solvents that are available at the repair facility. If possible, allow at least 2.5 mm (~0.100 inch) between adjacent components. Remember, an operation that proves boringly routine in a manufacturer's lab becomes fairly challenging when done on a storm-tossed ship while someone's shooting live ammo at you.

10.4 NASA'S View of SMT

Both NASA's Jet Propulsion Lab and NASA Langley have programs underway to glean available technology from the current commercial and military application of surface mounting. Once, during a tour of JPL, our guide stated categorically that NASA doesn't use SMT. But, when we looked at circuit samples, there sat CLLCCs, flat packs, and the like. In fact, even the IMCs had been lead formed and surface mounted. Actually, not a single component was through-hole mounted on the "non-SMT" assemblies we saw. The engineer went on to explain that JPL "planar mounts" everything. They just don't call it SMT.

One circuit in particular was fascinating. For a spacecraft, it was a relatively large card. And, it was fully planar mounted with single-side population. The back side of the board was bonded to a metallic plate. The circuit was part of the electronics of the Voyager space probe. Many of the components were leaded types which had been custom formed for surface mounting. The system engineering objectives were to meet the environmental challenges within the weight and package size limits set for the spacecraft. To accomplish this, the metallic backside of the substrate was used as an outer skin of the spacecraft, as well as a support and a heat sink for the electronics. Therefore, surface mounting of the components was a must if they were to achieve a smooth hole-free skin for the spacecraft.

10.5 References

1. Markstein, Howard W., "Low TCE Metals and Fibers Prove Viable for SMT Substrates," *Electronic Packaging and Production*, January 1985, pg. 52.

2. Deputy Chief of Naval Materials for Reliability, Maintenance, and Quality Assurance (sponsor), *Proceedings, Leadless Component Conference*, Twin Bridges Marriot, Arlington, VA, 18 April 1985, Naval Material Command, Washington, DC.

3. Hollomon, James K., Jr., "The SMT Panel—Process Before Specification," *Proceedings of NEPCON West 1988*, Cahners Exposition Group, Des Plaines, IL, February 1988, pg. 346.

4. MIL-STD-810, *Military Standard Environmental Test Methods and Engineering Guidelines*, 19 July 1983, pg. 7.

5. Lish, Earl F., and Weber, John O., "Solderability Testing of Leaded and Leadless SMDs by Means of a Modified Wetting Balance," *11 Annual Electronics Manufacturing Seminar Proceedings*, Naval Weapons Center, China Lake, CA, February 1987, pg. 283.

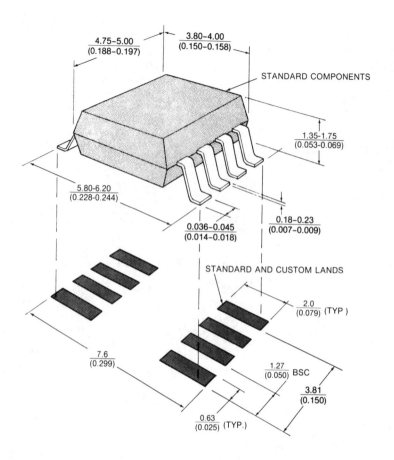

Figure A Working drawing for SMT design.

4.75–5.00
(0.188–0.197)

3.80–4.00
(0.150–0.158)

STANDARD COMPONENTS

1.35–1.75
(0.053–0.069)

5.80–6.20
(0.228–0.244)

0.036–0.045
(0.014–0.018)

0.18–0.23
(0.007–0.009)

STANDARD AND CUSTOM LANDS

2.0
(0.079) (TYP)

7.6
(0.299)

1.27
(0.050) BSC

3.81
(0.150)

0.63
(0.025) (TYP.)

A Component and Land Geometries

A.1 SMT Component Geometries

A.2 ANSI/IPC-SM-782 Land Patterns

A.3 Land-Pattern Variations

In this section, we have compiled the working drawings, such as the one seen in Figure A, which the reader will need for a trip through this book. These same drawings are the tools that the SMT team will need in day-to-day design discussions. Therefore, this appendix might deserve a full-time marker for ready reference.

Also in this section, we will present standard SMT component outlines, standard land geometries for the common SMT components, and recommendations for custom land patterns for specific applications. Except as specifically indicated, referenced EIA and JEDEC specifications are the sources for all component outline and dimensional data given. Unless otherwise noted, metric dimensions control. Inch conversions are direct and are rounded to three places. Where such conversions do not agree with the conversions shown in cited specifications, the correct conversion has been substituted for the specification conversion. Where metric control results in cumbersome inch conversions, round the inch dimensions to the nearest 0.005 inch for inch control. Conversely, where inch control produces inconvenient metric dimensions, the metric numbers should be rounded to the nearest 0.1 mm, unless otherwise noted.

A.1 SMT Component Geometries

The outline drawings for standard SMT components are shown in the following text. This material is meant to be used in conjunction with Chapter 2, in the specification of, and engineering with, SMCs.

A.1.1 *SMT Passives*

We'll start with the components with the simpler form factors, the *passives*. Following are the outlines for capacitors of various types, typical resistors which include trimmers and networks, thermistors, and various coils, transformers, and inductors. All dimensions for passives are from Reference 1, unless otherwise noted.

Ceramic Chip

Ceramic chip capacitors are available in five standard sizes (four preferred, plus one additional). Figure A-1 shows the dimensional information for standard-size chip capacitors.

Figure A-1 Chip-capacitor outline drawing.

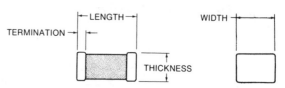

NOTES:
1. For discussion of end terminations, see Chapter 2, Figure 2-4, and related text, plus the discussion here.
2. For land patterns, see Figures A-29, A-46, and A-48. For vias, see Figures A-47 and A-49.
3. For dimensions, see Table A-1.

Table A-1. EIA Chip-Capacitor Standard Sizes

EIA Size Designation	Dimensions (mm/in)*			
	Length	*Width*	*Thickness (Max)*	*Termination*
0805	1.8–2.2 0.070–0.087	1.0–1.4 0.040–0.055	1.3 0.050	0.3–0.6 0.012–0.024
1206	3.0–3.4 0.118–0.134	1.4–1.8 0.055–0.070	1.5 0.060	0.4–0.7 0.016–0.028
1210	3.0–3.4 0.118–0.134	2.3–2.7 0.090–0.106	1.7 0.067	0.4–0.7 0.016–0.028
1812	4.2–4.8 0.166–0.190	3.0–3.4 0.118–0.134	1.7 0.067	0.4–0.7 0.016–0.028
1825**	4.2–4.8 0.166–0.190	6.0–6.8 0.236–0.268	1.7 0.067	0.4–0.7 0.016–0.028

* Inch dimensions control, and metric dimensions are shown rounded to the nearest 0.1 mm.

** Not included in SMTA Component Standards Group recommended sizes.[2]

There is also an infinite variety of nonstandard-dimension parts that are made for special applications. They follow similar form factors, and are rarely larger than 7.62 mm (0.300 inch) in length or width. For sake of handling and mechanical orientation, special sizes should have an aspect ratio that is distinctly apart from 1:1. Cubic parts are particularly troublesome to handle mechanically. In the same way, parts that are nearly the same height as they are in width are troublesome in mechanical sorters.

Chip-capacitor end terminations are generally coated with thick-film mixtures of silver, palladium, and glass frit. Where soldering and/or rework processes will require exposure to typical soldering temperatures (240 °C) for a total of over 10 seconds, a diffusion barrier should be specified.

For a time, diffusion barriers became so common that there was serious talk of deleting the unprotected palladium/silver termination from manufacturers' catalogs. However, several major automotive and telecommunication users objected to the nickel-barrier part, feeling that it was only a solution to poor process control, and might introduce problems of its own. Where nickel terminations contribute to component cracking or solder-joint stress failures, the soldering process should be adjusted to permit the use of plain palladium/silver parts.

Nickel-barrier specifications are covered in the Chapter 2 discussion of chip capacitors.

Figure A-2
Electrolytic "brick" capacitor outlines.

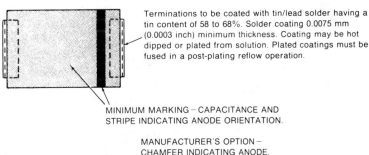

Terminations to be coated with tin/lead solder having a tin content of 58 to 68%. Solder coating 0.0075 mm (0.0003 inch) minimum thickness. Coating may be hot dipped or plated from solution. Plated coatings must be fused in a post-plating reflow operation.

MINIMUM MARKING – CAPACITANCE AND STRIPE INDICATING ANODE ORIENTATION.

MANUFACTURER'S OPTION – CHAMFER INDICATING ANODE.

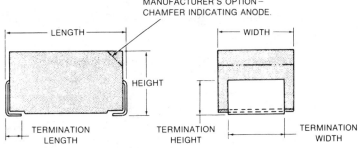

LENGTH
WIDTH
HEIGHT
TERMINATION LENGTH
TERMINATION HEIGHT
TERMINATION WIDTH

NOTES:
1. For dimensions, see Table A-2.
2. For land patterns, see Figure A-30.

Figure A-3
Entended-range electrolytic "brick" capacitors.

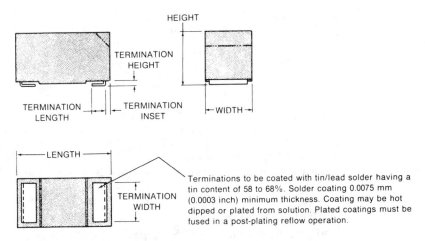

HEIGHT
TERMINATION HEIGHT
WIDTH
TERMINATION LENGTH
TERMINATION INSET

LENGTH
TERMINATION WIDTH

Terminations to be coated with tin/lead solder having a tin content of 58 to 68%. Solder coating 0.0075 mm (0.0003 inch) minimum thickness. Coating may be hot dipped or plated from solution. Plated coatings must be fused in a post-plating reflow operation.

NOTES:
1. For dimensions, see Table A-3.
2. For land patterns, see Figure A-30.

Tantalum and Aluminum Electrolytic Capacitors

Tantalum and other electrolytic capacitors are generally packaged per the outlines shown in Figures A-2 and A-3. Dimensions for these drawings are given in Tables A-2 and A-3, respectively. Note that the inset terminations of the extended-range brick allow some parts to fit the standard brick lands.[3]

Table A-2.
Standard Capacitance Electrolytic "Brick" Capacitors

EIA RS-228 Size	Size Code/Standard Capacitor Range (mm/in)			
	A Case	*B Case*	*C Base*	*D Case*
Length	3.0–3.4 0.118–0.134	3.3–3.7 0.130–0.146	5.7–6.3 0.224–0.248	6.8–7.6 0.268–0.299
Width	1.3–1.8 0.050–0.070	2.6–3.0 0.102–0.118	2.9–3.5 0.114–0.138	4.0–4.6 0.157–0.181
Height	1.4–1.8 0.055–0.070	1.7–2.1 0.067–0.083	2.2–2.8 0.087–0.110	2.5–3.1 0.098–0.122
Termination Length	0.5–1.1 0.020–0.043	0.5–1.1 0.020–0.043	0.5–1.1 0.020–0.043	0.5–1.1 0.020–0.043
Termination Width	1.1–1.3 0.043–0.051	2.1–2.3 0.083–0.090	2.1–2.3 0.083–0.090	2.3–2.5 0.090–0.098
Termination Height	0.7 (min) 0.028	0.7 (min) 0.028	1.0 (min) 0.040	1.0 (min) 0.040

* Inch dimensions control, and metric dimensions are rounded to the nearest 0.1 mm.

Table A-3.
Extended Range Electrolytic "Brick" Capacitors

EIA RS-228 Size	Size Code/Extended-Range Capacitor Range (mm/in)			
	3518 Case	*3527 Case*	*7227 Case*	*7257 Case*
Length	3.3–3.7 0.130–0.146	3.3–3.7 0.130–0.146	6.9–7.5 0.272–0.295	6.9–7.5 0.272–0.295
Width	1.6–2.0 0.063–0.079	2.4–3.0 0.095–0.118	2.4–3.0 0.095–0.118	5.2–6.2 0.205–0.244
Height	1.7–2.1 0.067–0.083	1.7–2.1 0.067–0.083	2.5–3.1 0.098–0.122	3.0–3.7 0.118–0.146
Termination Length	0.6–1.0 0.024–0.040	0.6–1.0 0.024–0.040	0.8–1.2 0.031–0.047	0.8–1.2 0.031–0.047
Termination Width	1.6–1.8 0.063–0.071	2.4–2.6 0.095–0.102	2.4–2.6 0.095–0.102	5.4–5.8 0.213–0.228
Termination Height	0.7 0.028	0.7 0.028	1.0 0.040	1.2 0.047
Termination Inset	0.4–0.6 0.016–0.024	0.4–0.6 0.016–0.024	0.6–0.8 0.024–0.032	0.6–0.8 0.024–0.032

* Inch dimensions control, and metric dimensions are shown rounded to the nearest 0.1 mm.

Variable Capacitors

The current multiplicity of SMT package outlines offered in variable capacitors might lead us to speculate that "variable" applies to the shape of the device rather than to the capacitance value. Work is underway to set standards. However, no standard was available as of this writing. The outlines shown in Figure A-4 illustrate some "typical" packages that are widely available in the commercial market.

Figure A-4
Typical SMT
outlines for
variable capaci-
tors.[4, 5]

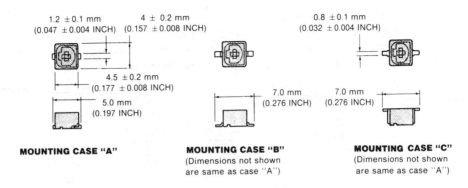

MOUNTING CASE "A"

MOUNTING CASE "B"
(Dimensions not shown
are same as case "A")

MOUNTING CASE "C"
(Dimensions not shown
are same as case "A")

(A) MuRata Erie components available in single plate and multilayer capacitance ranges. Shipping containers are 12-mm reel or stick.

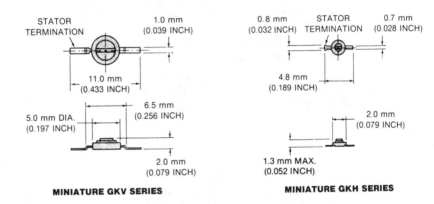

MINIATURE GKV SERIES

MINIATURE GKH SERIES

(B) Two of the many SMT cases available from Spraque. Capacitance ranges from 2.5 to 40.0 pF. These parts are typical of hybrid-style trimming capacitors.

Figure A-5
Standard SMT
chip resistor
outline.

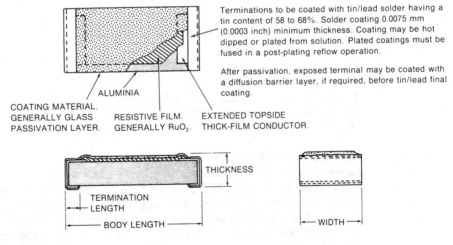

Terminations to be coated with tin/lead solder having a tin content of 58 to 68%. Solder coating 0.0075 mm (0.0003 inch) minimum thickness. Coating may be hot dipped or plated from solution. Plated coatings must be fused in a post-plating reflow operation.

After passivation, exposed terminal may be coated with a diffusion barrier layer, if required, before tin/lead final coating.

ALUMINIA

COATING MATERIAL.
GENERALLY GLASS
PASSIVATION LAYER.

RESISTIVE FILM.
GENERALLY RuO_2.

EXTENDED TOPSIDE
THICK-FILM CONDUCTOR.

THICKNESS

TERMINATION
LENGTH

BODY LENGTH

WIDTH

NOTE:
For land patterns, see Figures A-31, A-46, and A-48.
For vias, see Figures A-47 and A-49.

Chip Resistors

Figure A-5 covers standard EIA/IS 30 chip resistor outlines; Table A-4 gives the dimension values. In addition to the standard parts, there are a great variety of custom forms available for specific application needs.

Table A-4. Chip Resistor Dimensions

Size Code	Dimensions (mm/in)*			
	Body Length	*Width*	*Thickness*	*Termination Length*
RC0805	1.8–2.2 0.070–0.087	1.0–1.4 0.040–0.055	0.3–0.7 0.012–0.028	0.3–0.6 0.012–0.024
RC1206	3.0–3.4 0.118–0.134	1.4–1.8 0.055–0.070	0.4–0.7 0.016–0.028	0.4–0.7 0.016–0.028
RC1210	3.0–3.4 0.118–0.134	2.3–2.7 0.090–0.106	0.4–0.7 0.016–0.028	0.4–0.7 0.016–0.028

* Inch dimensions control and metric conversions are shown rounded to the nearest 0.1 mm.

Cylindrical Resistors

MELF, or other cylindrical resistor forms, are generally specified per the outline shown in Figure A-6. This outline conforms to EIAJ standards, and is essentially a JEDEC DO-35 without the leads.

Figure A-6 Cylindrical leadless resistor outline.

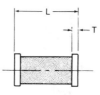

NOTES:
1. Terminations to be coated with tin/lead solder having a tin content of 58 to 68%. Solder coating 0.0075 mm (0.0003 inch) minimum thickness. Coating may be hot dipped or plated from solution. Plated coatings must be fused in a post-plating reflow operation.
2. For land patterns, see Figure A-32.
3. This outline also used for cylindrical diodes, etc.

CASE	DIMENSIONS (mm/inch)		
STYLE	L	D	T
MLL34	3.30-3.70 (0.130-0.146)	1.50-1.70 (0.059-0.067)	0.29-0.55 (0.011-0.022)
MLL41	4.80-5.20 (0.189-0.205)	2.44-2.54 (0.096-0.100)	0.35-0.51 (0.014-0.020)

Trimmer Potentiometers

Trimmer potentiometers, like their capacitor cousins, are as variable in their package outlines as in their electrical parameters. Figure A-7 shows several typical designs.

Resistor Networks

Resistor networks are generally in either a leadless or leaded ship carrier, and in SOIC-style packaging. The chip carriers are typically the same as the various standard 1.27-mm (0.050-inch) JEDEC outlines, as covered under *Actives* in Section A.1.2. Some SOIC styles are also in standard JEDEC-outline packages. However, many are wider than the narrow JEDEC part but not as wide as the SOLICs. Most special resistor packages conform to the outline drawing shown in Figure

Figure A-7
Typical trimmer
potentiometer
outlines.[6, 7, 8]

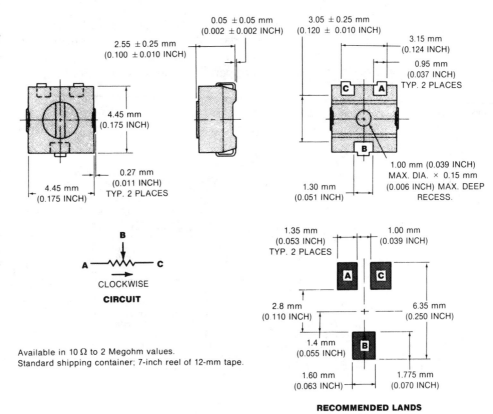

Available in 10 Ω to 2 Megohm values.
Standard shipping container; 7-inch reel of 12-mm tape.

(A) Bourns Model 3314; sealed body.

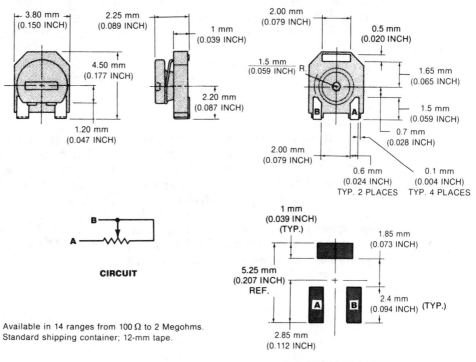

Available in 14 ranges from 100 Ω to 2 Megohms.
Standard shipping container; 12-mm tape.

(B) MuRata Erie RVG4F03A; open body.

A-8. But, a word of warning is in order. These special resistor network SOs may not be governed by any standard. Without standards control, manufacturers are free to choose package dimensions to suit their requirements. Before proceeding with a design for nonstandard parts, consult the vendor for the exact specifications and outline.

Figure A-8
Resistor networks in nonstandard SOIC packaging.

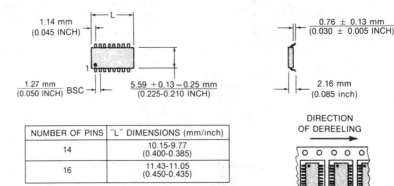

NUMBER OF PINS	"L" DIMENSIONS (mm/inch)
14	10.15-9.77 (0.400-0.385)
16	11.43-11.05 (0.450-0.435)

NOTES:
1. Terminations to be coated with tin/lead solder having a tin content of 58 to 68%. Solder coating 0.0075 mm (0.0003 inch) minimum thickness. Coating may be hot dipped or plated from solution. Plated coatings must be fused in a post-plating reflow operation.
2. For land recommendations, see Figure A-50.

PIN 1 ORIENTATION IN TAPE
(16-mm Tape/8-mm Pitch)

Thermistors
SMT thermistors are generally supplied in chip format, with dimensions taken from the ceramic chip-capacitor outlines, or with custom chip dimensions. Standard EIA monolithic capacitor dimensions are shown in **Figure A-1 and Table A-1**. For nonstandard parts, consult the device supplier for dimensions.

Tuning Coils and Transformers
In the absence of standards, a wide variety of alternatives are on the market. Consult the vendors for dimensional data.

Inductors
SMT inductors are supplied in open and protected cases. In the absence of a standard, many protected-case parts are modeled after tantalum brick-style components, as shown in Figures A-2 and A-3. Some of these "brick" inductors fit the tantalum-brick dimension scheme and others are dimensioned differently, but they follow the same basic form factor as the brick. Other protected parts follow their own unique molded construction. In addition, a wide variety of nonstandard parts, particularly of the open body construction, are offered for special applications. Consult the manufacturers for details, and use the SMTA component recommendations as a guideline for the specifications.

A.1.2 Actives

Diodes

Diodes may be packaged in MLL-34, SOD-80, or MLL-41 packaging, as shown earlier under MELF-style resistors. Larger diodes may be in the $\frac{1}{4}$-watt MELF resistor format. Both single and dual diodes are offered in SOT packages, such as shown in Figure A-9.

Transistors

Figures A-9A through A-9C detail the package outlines for common SMT transistor packages. Power transistors and diodes are often housed in lead-formed variants of the TO-220 through-hole package. Figure A-9D shows a typical SMT approach to packaging a power device.

Table A-5. Dimensions for Gull-Wing-Leaded SO and SOL IC Packages

Symbol	SO Package Dimensions (mm/inch)						
	SOIC 8	SOIC 14	SOIC 16	SOLIC 16	SOLIC 20	SOLIC 24	SOLIC 28
A	1.27 BSC 0.050	1.27 BSC 0.050	1.27 BSC 0.050	1.27 BSC 0.050	1.27 BSC 0.050	1.27 BSC 0.050	1.27 BSC 0.050
B	3.81 (REF.) 0.150	7.62 (REF.) 0.300	8.89 (REF.) 0.350	8.89 (REF.) 0.350	11.43 (REF.) 0.450	13.97 (REF.) 0.550	16.51 (REF.) 0.650
C (Nominal)	0.53 0.021	0.51 0.020	0.050 0.020	0.69 0.027	0.69 0.027	0.72 0.028	0.70 0.027
D	0.18–0.23 0.007–0.009	0.18–0.23 0.007–0.009	0.18–0.23 0.007–0.009	0.23–0.32 0.009–0.013	0.23–0.32 0.009–0.013	0.23–0.32 0.009–0.013	0.23–0.32 0.009–0.013
E (Nominal)	1.05 0.041	1.05 0.041	1.05 0.041	1.40 0.055	1.40 0.055	1.40 0.055	1.40 0.055
F	0.25–0.50 0.01–0.02	0.25–0.50 0.01–0.02	0.25–0.50 0.01–0.02	0.25–0.50 0.01–0.02	0.25–0.50 0.01–0.02	0.25–0.50 0.01–0.02	0.25–0.50 0.01–0.02
G	0–8°	0–8°	0–8°	0–8°	0–8°	0–8°	0–8°
H	1.3–1.75 0.053–0.069	1.3–1.75 0.053–0.069	1.3–1.75 0.053–0.069	2.4–2.6 0.094–0.102	2.4–2.6 0.094–0.102	2.4–2.6 0.094–0.102	2.4–2.6 0.094–0.102
J	0.5–1.15 0.02–0.045	0.5–1.15 0.02–0.045	0.5–1.15 0.02–0.045	0.58–1.25 0.023–0.049	0.58–1.25 0.023–0.049	0.58–1.25 0.023–0.049	0.58–1.25 0.023–0.049
L	4.75–5.0 0.188–0.197	8.53–8.74 0.336–0.344	9.78–10.00 0.385–0.394	10.11–10.41 0.398–0.413	12.6–13.0 0.496–0.512	15.21–15.6 0.599–0.614	17.7–18.11 0.697–0.713
O	5.8–6.2 0.228–0.244	5.8–6.2 0.228–0.244	5.8–6.2 0.228–0.244	10.0–10.6 0.394–0.419	10.0—10.6 0.394–0.419	10.0–10.6 0.394–0.419	10.0–10.6 0.394–0.419
P (Planarity)	±0.05 ±0.002	±0.05 ±0.002	±0.05 ±0.002	±0.05 ±0.002	±0.05 ±0.002	±0.05 ±0.002	±0.05 ±0.002
S	0.01–0.02 0.004–0.008	0.01–0.02 0.004–0.008	0.01–0.02 0.004–0.008	0.01–0.02 0.004–0.008	0.01–0.02 0.004–0.008	0.01–0.02 0.004–0.008	0.01–0.02 0.004–0.008
W	3.8–4.0 0.150–0.158	3.8–4.0 0.150–0.158	3.8–4.0 0.150–0.158	7.4–7.6 0.291–0.299	7.4-7.6 0.291–0.299	7.4–7.6 0.291–0.299	7.4–7.6 0.291–0.299
X	0.36–0.45 0.014–0.018	0.36–0.45 0.014–0.018	0.36–0.45 0.014–0.018	0.36–0.48 0.014–0.019	0.36–0.48 0.014–0.019	0.36–0.48 0.014–0.019	0.36–0.48 0.014—0.019
TAPE SIZE/ PITCH	12mm/ 8mm	16mm/ 8mm	16mm/ 8mm	16mm/ 12mm	24mm/ 12mm	24mm/ 12mm	24mm/ 12mm

Figure A-9 Typical SMT transistor package outlines.

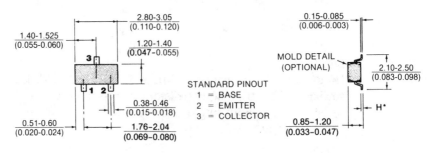

STANDARD PINOUT
1 = BASE
2 = EMITTER
3 = COLLECTOR

DIRECTION OF DEREELING

(8-mm Tape/4-mm Pitch)

MAX DIE SIZE: 0.76 mm (0.030 INCH) SQUARE

DIMENSION	LOW PROFILE	MEDIUM PROFILE	HIGH PROFILE
H*	0.01-0.10 mm (0.0004-0.004 inch)	0.08-0.13 mm (0.003-0.005 inch)	0.10-0.25 mm (0.004-0.010 inch)

*Standoff—bottom of seating plane to bottom of body

(A) SOT-23 (EIA TO-236) outline.

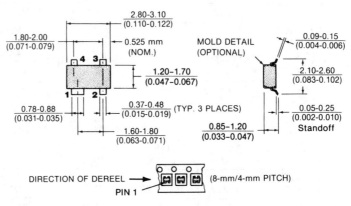

DIRECTION OF DEREEL →
PIN 1
(8-mm/4-mm PITCH)

(B) SOT-143 outline.

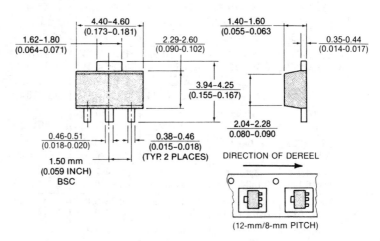

DIRECTION OF DEREEL

(12-mm/8-mm PITCH)

(C) SOT-89 (EIA TO-243) outline.

SOICs

The JEDEC standard gull-wing-leaded SOIC and SOLIC are covered in Figure A-10 and Table A-5.

SOJ ICs

The JEDEC standard "J"-leaded SO (SOJ IC) package outline is shown in Figure A-11.

PLCCs

The JEDEC postmolded and premolded 1.27-mm-pitch (0.050 inch) PLCC package outlines are shown in Figure A-12. The dimensions for the drawing are given in Table A-6.

Most PLCCs consumed today are of the postmolded variety. However, original JEDEC Type-A PLCC specifications were unclear on construction. Today, the MO-047 premolded part specification is applied to both premolded and postmolded packages. Note that the table of dimensions, Table A-6, lists specifica-

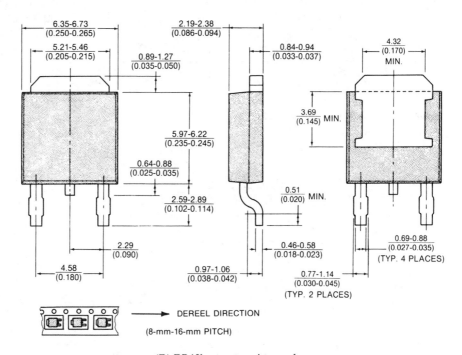

(D) DPAK power transistor package.

NOTES:
1. Values are in millimeters and inches.
2. For SOT-23 lands, see Figures A-34, A-51, and A-52.
3. For SOT-143 lands, see Figure A-35.
4. For SOT-89 lands, see Figures A-36, A-54, and A-55.
5. For DPAK lands, see Figure A-56.

6. Terminations to be coated with tin/lead solder having a tin content of 58 to 68%. Solder coating 0.0075 mm (0.0003 inch) minimum thickness. Coating may be hot dipped or plated from solution. Plated coatings must be fused in a post-plating reflow operation.

tions for only four postmolded-package pin counts. Some manufacturers, needing a postmolded package for other pin counts, use a dimensioning scheme like the table shows, but adjust the dimensions to suit the desired pin count. Others commonly follow the premolded dimension charts (Tables A-7 and A-8) even though they postmold the parts. The IPC-recommended land patterns for premolded and postmolded parts are not identical twins. Therefore, users should be sure to verify the dimension scheme followed on PLCCs before selecting land patterns, or they should resort to the alternate pattern given in Section A.3, which is suitable for either type.

Table A-6. Square Postmolded PLCC Dimensions (Type A)

Package Pin Count	Dimensions* (Millimeters/Inches)						Lands
	A	*B*	*C*	*D*	*E*	*H*	
28	11.94–12.95 0.470–0.510	11.43–11.582 0.450–0.456	2.29–3.05 0.090–0.120	See "A"	See "B"	4.19–4.57 0.165–0.180	For IPC lands specially designed for this part, see Figure A-38. For universal lands, see Figure A-58.
44	17.02–18.03 0.670–0.710	16.51–16.662 0.650–0.656	2.29–3.05 0.090–0.120	See "A"	See "B"	4.19–4.57 0.165–0.180	
52	19.56–20.57 0.770–0.810	19.05–19.202 0.750–0.756	2.29–3.30 0.090–0.130	See "A"	See "B"	4.19–5.08 0.165–0.200	
68	24.64–25.65 0.970–1.010	24.13–24.333 0.950–0.958	2.29–3.30 0.090–0.130	See "A"	See "B"	4.19–5.08 0.165–0.200	

* Dimensions in this table apply to Figure A-12.

Figure A-10
Gull-wing-leaded
SO and SOL IC
package outlines.

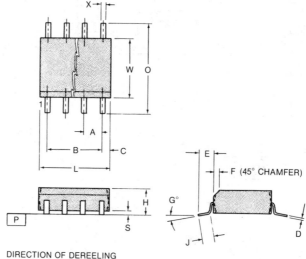

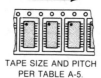

DIRECTION OF DEREELING

TAPE SIZE AND PITCH
PER TABLE A-5.

NOTES:
1. Terminations to be coated with tin/lead solder having a tin content of 58 to 68%. Solder coating 0.0075 mm (0.0003 inch) minimum thickness. Coating may be hot dipped or plated from solution. Plated coatings must be fused in a post-plating reflow operation.
2. For land patterns, see Figures A-37 and A-57.

Figure A-11
"J"-leaded SOJ
IC package
outline.

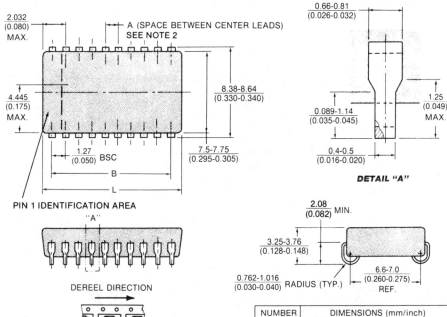

NUMBER	DIMENSIONS (mm/inch)		
OF PINS	A (See Note 2)	B	L
14	1.27 (0.050)	7.62 (0.300)	9.40-9.65 (0.370-0.380)
16	1.27 (0.050)	8.89 (0.350)	10.67-10.92 (0.420-0.430)
18	1.27 (0.050)	10.16 (0.400)	11.94-12.19 (0.470-0.480)
20	1.27 (0.050)	11.43 (0.450)	13.21-13.46 (0.520-0.530)
22	1.27 (0.050)	12.70 (0.500)	14.48-14.73 (0.570-0.580)
24	1.27 (0.050)	13.97 (0.550)	15.75-16.00 (0.620-0.630)
26	1.27 (0.050)	15.24 (0.600)	17.02-17.27 (0.670-0.680)
28	1.27 (0.050)	16.89 (0.650)	18.29-18.54 (0.720-0.730)

NOTES:
1. Terminations to be coated with tin/lead solder having a tin content of 58 to 68%. Solder coating 0.0075 mm (0.0003 inch) minimum thickness. Coating may be hot dipped or plated from solution. Plated coatings must be fused in a post-plating reflow operation.
2. For some applications requiring a large power bus, "A" is increased to 5.08 mm/0.200 inch or 3.81 mm/0.150 inch by omitting 3 pins for parts with an even number of 50-mil spaces per side, or 2 pins for parts with an odd number of 50-mil spaces per side.
3. For land patterns, see Figure A-37.
4. Dimensions are in millimeter and inches.

TapePaks

The TapePak package is unique in that it is shipped in a protective ring. After testing the device, the end user excises it from the ring and forms the leads for reflow attachment, as shown in Figure A-13. The dimensions for a TapePak package are given in Table A-9.

There are three main JEDEC-registered TapePak body sizes. The part shown in the graphics of Figure A-13, with a body that is 7.264 mm (0.286 inch) square is available with 32 pins on a 25-mil (0.635-mm) pitch, or 40 pins spaced at a 20-mil (0.508-mm) pitch. The medium-lead-count part has a body that is 12.83 mm (0.505 inch) square. It comes with 52 leads at 25 mil, or 84 pins at 20 mil. The high-pin-count part offers 100-, 132-, and 180-pin counts with pitches of 25, 20, and 15 mils (0.381 mm), respectively.

Table A-7. Square Premolded PLCC (JEDEC MO-047) Dimensions (Type A)

Package Pin Count	Dimensions* (Millimeters/Inches)						Lands
	A	B	C	D	E	H	
20	9.78–10.03 0.385–0.395	8.89–9.042 0.350–0.356	2.29–3.05 0.090–0.120	See "A"	See "B"	4.19–4.57 0.165–0.180	For IPC lands specially designed for this part, see Figure A-39. For universal Type-A PLCC lands, see Figure A-58.
28	12.32–12.57 0.485–0.495	11.43–11.582 0.450–0.456	2.29–3.05 0.090–0.120	See "A"	See "B"	4.19–4.57 0.165–0.180	
44	17.40–17.65 0.685–0.695	16.51–16.662 0.650–0.656	2.29–3.05 0.090–0.120	See "A"	See "B"	4.19–4.57 0.165–0.180	
52**	19.94–20.19 0.785–0.795	19.05–19.202 0.750–0.756	2.29–3.30 0.090–0.130	See "A"	See "B"	4.19–5.08 0.165–0.200	
68	25.02–25.27 0.985–0.995	24.13–24.333 0.950–0.958	2.29–3.30 0.090–0.130	See "A"	See "B"	4.19–5.08 0.165–0.200	
84	30.10–30.35 1.185–1.195	29.21–29.413 1.150–1.158	2.29–3.30 0.090–0.130	See "A"	See "B"	4.19–5.08 0.165–0.200	
100**	35.18–35.43 1.385–1.395	34.29–34.493 1.350–1.358	2.29—3.30 0.090–0.130	See "A"	See "B"	4.19–5.08 0.165–0.200	
124**	42.80–43.05 1.685–1.695	41.91–42.113 1.650–1.658	2.29–3.30 0.090–0.130	See "A"	See "B"	4.19–5.08 0.165–0.200	

* Dimensions in this table apply to Figure A-12.
** JEDEC qualified packages, but not widely available.

PQFPs and CQFPs

Figure A-14 covers the package outline for the Plastic Quad Flat Pack (PQFP) per JEDEC MO-069. Note 2 (in Figure A-14) details the variations of ceramic quad flat packs (CQFPs) from the plastic part outline shown. Basically, the ceramic package provides hermeticity. It is slightly smaller in body than the plastic device to provide additional space for the lead-form anvil without risk to the hermetic seals of the leads. And, the corner protrusions, or bumpers, are omitted because ceramic bumpers would be too fragile to serve any useful purpose. Otherwise, the two packages are identical.

CLLCCs

The ceramic leadless chip carrier was a very early entry in the SMT component arena. Thus, several variations developed before standards activity provided a clear guidance to package design.

Figure A-15 covers some common features of the four most-standard 50-mil lead-pitch packages (JEDEC Types A, B, C, and D). Figure A-16 details how these four packages vary. The dimensions for the device types shown in Figure A-16 are listed in Table A-12. Several additional specifications, mentioned in the notes of Figure A-16, cover JEDEC Types E and F rectangular parts.

Figure A-12 A
50-mil Type-A
PLCC package
outline.

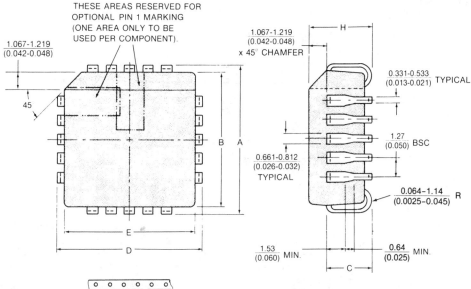

DEREEL DIRECTION

TAPE REEL DATA
Tape sizes shown in Table 2-11, Chapter 2.
Note that PLCC 44, PLCC 52, PLCC 68, and
PLCC 84 tape is sprocketed on both sides.

NOTES:
1. This drawing covers both premolded and
   postmolded devices. Lead form shown is
   for postmolded packages, and may vary
   with premolded.
2. Terminations to be coated with tin/lead
   solder having a tin content of 58 to 68%.
   Solder coating 0.0075 mm (0.0003 inch)
   minimum thickness. Coating may be hot
   dipped or plated from solution. Plated
   coatings must be fused in a post-plating
   reflow operation.
3. For dimensions, see Tables A-6, A-7, and A-8.
4. Dimensions are in millimeter and inches.

Table A-8. Rectangular† Premolded PLCC (JEDEC MO-052) Dimensions (Type A)

Package Pin Count	Dimensions* (Millimeters/Inches)						Lands
	A	*B*	*C*	*D*	*E*	*H*	
18	8.05–8.31 0.317–0.327	7.14–7.315 0.281–0.288	2.29–3.05 0.090–0.120	11.61-11.86 0.457–0.467	10.72–10.87 0.422–0.428	4.19–4.57 0.165–0.180	For IPC lands spe-cially de-signed for this part, see Figure A-39. For univer-sal Type-A PLCC lands, see Figure A-58.
18L	8.13–8.51 0.320–0.335	7.24–7.518 0.285–0.296	2.29–3.05 0.090–0.120	13.21–13.59 0.520–0.535	12.32–12.60 0.485–0.496	4.19–4.57 0.165–0.180	
22	8.13–8.51 0.320–0.335	7.24–7.518 0.285–0.296	2.29–3.05 0.090–0.120	13.21–13.59 0.520–0.535	12.32–12.60 0.485–0.496	4.19–4.57 0.165–0.180	
28	9.78–10.03 0.385–0.395	8.89–9.042 0.350–0.356	2.29–3.05 0.090–0.120	14.86–15.11 0.585–0.595	13.97–14.12 0.550–0.556	4.19–4.57 0.165–0.180	
32	12.32–12.57 0.485–0.495	11.43–11.582 0.450–0.456	2.29–3.05 0.090–0.120	14.86–15.11 0.585–0.595	13.97–14.12 0.550–0.556	4.19–4.57 0.165–0.180	

† Rectangular parts are not included in the SMT
  Association recommended components.

* Dimensions in this table apply to Figure A-12.

Figure A-13
TapePak package
outline.

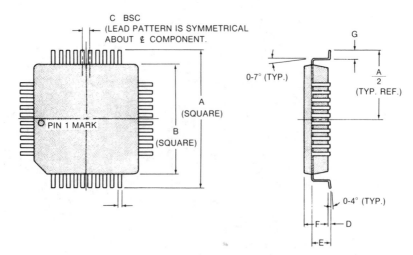

NOTES:
1. Dimensions are in millimeter and inches.
2. Recommended outline for formed and excised packages.
3. Refer to Table A-9 for package dimensions.
4. Terminations to be coated with tin/lead solder having a tin content of 58 to 68%. Solder coating 0.0075 mm (0.0003 inch) minimum thickness. Coating may be hot dipped or plated from solution. Plated coatings must be fused in a post-plating reflow operation.
5. For land patterns, see Figure A-63 for 20-mil lead pitch parts, and Figure A-65 for 10-mil pitch.

6. JEDEC MO-071 registration covers 32-, 40-, 52-, 84-, 100-, 132-, and 180-pin packages. There is a proposal to create a very high pin count family having a 30.6-mm (1.205-inch) square body and 168, 220, 284, 360, and 400 pins at 25-mil, 20-mil, 15-mil, 12-mil, and 10-mil lead pitches, respectively.
7. **For data and dimensions on unexcised part dimensions, test point dimensions, etc., consult the manufacturer. The unexcised part is shown in Figure 2-17 of Chapter 2.**

Figure A-14
Plastic and
ceramic quad flat
pack outlines.

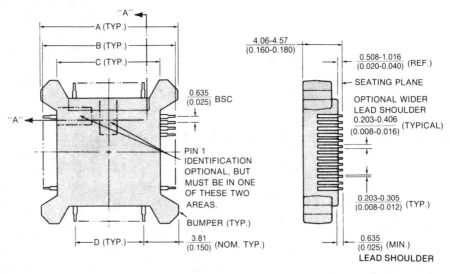

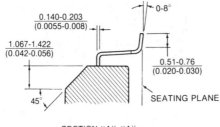

SECTION "A"–"A"

NOTES:
1. Terminations to be coated with tin/lead solder having a tin content of 58 to 68%. Solder coating 0.0075 mm (0.0003 inch) minimum thickness. Coating may be hot dipped or plated from solution. Plated coatings must be fused in a post-plating reflow operation.
2. Ceramic quad flat pack per this outline, except body width and length 1.27 mm (0.050 inch) smaller to allow room for lead-form anvil without risk of damage to glass seals, and except corner bumpers not included as ceramic bumpers would chip easily.
3. For land recommendations, see Figure A-40.
4. Values are in millimeter/inch. Inch dimensions control.

Table A-9. Dimensions for TapePak Package

Package Description		Dimensions*						
Body Size	Pin Count	A	B (Body Size)	C (Pitch)	D	E	F	G
7.25 mm / 0.286 inch SQUARE	32	8.99–9.30 / 0.354–0.366	7.26 / 0.286	0.635 / 0.025 BSC	0.05–0.25 / 0.002–0.010	1.14–1.35 / 0.045–0.053	1.55–1.75 / 0.061–0.069	0.048–0.635 / 0.019–0.025
	40	8.99–9.30 / 0.354–0.366	7.26 / 0.286	0.508 / 0.020 BSC	0.05–0.25 / 0.002–0.010	1.14–1.35 / 0.045–0.053	1.55–1.75 / 0.061–0.069	0.048–0.635 / 0.019–0.025
12.83 mm / 0.505 inch SQUARE	52	14.58–14.88 / 0.574–0.586	12.83 / 0.505	0.635 / 0.025 BSC	0.05–0.25 / 0.002–0.010	1.50–1.65 / 0.059–0.065	1.93–2.13 / 0.076–0.084	0.048–0.635 / 0.019–0.025
	84	14.58–14.88 / 0.574–0.586	12.83 / 0.505	0.058 / 0.020 BSC	0.05–0.25 / 0.002–0.010	1.50–1.65 / 0.059–0.065	1.93–2.13 / 0.076–0.084	0.048–0.635 / 0.019–0.025
20.45 mm / 0.805 inch SQUARE	100	22.76–23.06 / 0.896–0.908	20.45 / 0.805	0.635 / 0.025 BSC	0.28–0.48 / 0.011–0.019	1.73–1.88 / 0.068–0.074	1.93–2.13 / 0.076–0.084	0.76–0.91 / 0.030–0.036
	132	22.76–23.06 / 0.896–0.908	20.45 / 0.805	0.508 / 0.020 BSC	0.28–0.48 / 0.011–0.019	1.73–1.88 / 0.068–0.074	1.93–2.13 / 0.076–0.084	0.76–0.91 / 0.030–0.036
	180	22.76–23.06 / 0.896–0.908	20.45 / 0.805	0.381 / 0.015 BSC	0.28–0.48 / 0.011–0.019	1.73–1.88 / 0.068–0.074	1.93–2.13 / 0.076–0.084	0.76–0.91 / 0.030–0.036
	T.B.D. (In Design)		20.45 / 0.805	0.284 / 0.012 BSC				
	T.B.D. (In Design)		20.45 / 0.805	0.254 / 0.010 BSC				
30.6 mm / 1.205 inch SQUARE (Proposed)	168		30.61 / 1.205	0.635 / 0.025 BSC				
	220		30.61 / 1.205	0.508 / 0.020 BSC				
	284		30.61 / 1.205	0.381 / 0.015 BSC				
	360		30.61 / 1.205	0.284 / 0.012 BSC				
	400		30.61 / 1.205	0.254 / 0.010 BSC				

* Dimensions are millimeter/inch; inch dimensions control.

Figure A-17 covers the 40-mil CLLCCs under JEDEC MS-009. Tabular dimensions for the 40-mil part are shown in Table A-13.

CLDCCs

Figure A-18 covers the package outline for postleaded ceramic chip carriers under JEDEC MS-008. Table A-14 presents the tabular dimensions for the MS-008 part shown in Figure A-18.

As the drawing in Figure A-18 indicates, these specifications afford the manufacturer considerable latitude in selecting each detail of the lead configurations, an area critical to manufacturing yields and long-term reliability of the component. Some component manufacturers have exercised this freedom in favor of the end user's needs, providing top-brazed leads with high compliancy. Other manufacturers have taken advantage of the liberal specifications to cut their own production costs at the expense of device reliability. They have used stiff leads, often side brazed, to simplify meeting the coplanarity requirements. Such leads have too little compliance to absorb TCE stresses. The result is a higher field-failure rate of the solder joints. The moral, when selecting CLDCCs, is make sure that the leads will both meet coplanarity specifications and offer sufficient compliance to suit the design requirements.

JEDEC MS-007 specifically covers postleaded CCCs with side- brazed leads. Because side attachment reduces the available flexural element of the lead, we have not included this specification herein.

Preleaded ceramic chip carriers are also available. These have lead frames with hermetic glass-to-metal seals around the lead exit from the body. Preleaded CCCs in 50-mil centers are covered in JEDEC MS-044, as shown in Figure A-19. While this standard covers only 68- and 84-pin devices, many preleaded variants have been developed. Some are higher pin counts in 50-mil parts, and others have finer pitches. Fine pitches are offered in 0.8-mm (0.031-inch), 0.64-mm (0.025-inch), 0.5-mm (0.020-inch), and even 0.46-mm (0.018-inch) varieties.

Table A-10. Plastic and Ceramic Quad Flat Pack Dimensions

Pin Count	Dimensions (mm/inch)			
	A	B	C	D
52	15.16–15.32 0.597–0.603	14.61–14.86 0.575–0.585	11.35–11.51 0.447–0.453	7.62 BSC 0.300
68	17.70–17.86 0.697–0.703	17.15–17.40 0.675–0.685	13.89–14.05 0.547–0.553	10.16 BSC 0.400
84	20.24–20.40 0.797–0.803	19.69–19.94 0.775–0.785	16.43–16.59 0.647–0.653	12.70 BSC 0.500
100	22.78–22.94 0.897–0.903	22.23–22.48 0.875–0.885	18.97–19.13 0.747–0.753	15.24 BSC 0.600
132	27.86–28.02 1.097–1.103	27.31–27.56 1.075–1.085	24.05–24.21 0.947–0.953	20.32 BSC 0.800
164	32.94–33.10 1.297–1.303	32.39–32.64 1.275–1.285	29.13–29.29 1.147–1.153	25.40 BSC 1.000
196	38.02–38.18 1.497–1.503	37.47–37.72 1.475–1.485	34.21–34.37 1.347–1.353	30.48 BSC 1.200
244	45.64–45.80 1.797–1.803	45.09–45.34 1.775–1.785	41.83–41.99 1.647–1.653	38.10 BSC 1.500

Figure A-15 Common features of 1.27-mm-center (0.050-inch) leadless chip carriers under JEDEC MS-002 (Type A), JEDEC MS-003 (Type B), JEDEC MS-004 (Type C), and JEDEC MS-005 (Type D).

1. Metallization on sides in termination area, including castellation (where used).
2. Cavity down (Types A and D) metallized on top, with bottom optional.
3. Cavity up (Types B and C) metallized on bottom, with top optional.
4. Nontermination metallizations for heat transfer only, and must be clear of terminations.

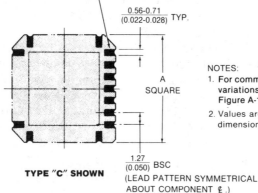

$\dfrac{0.56\text{-}0.71}{(0.022\text{-}0.028)}$ TYP.

A SQUARE

NOTES:
1. For common dimensions, see Table A-11. For variations between Types A, B, C, and D, see Figure A-16 and Table A-12.
2. Values are in millimeters and inches; inch dimensions control.

TYPE "C" SHOWN

$\dfrac{1.27}{(0.050)}$ BSC
(LEAD PATTERN SYMMETRICAL ABOUT COMPONENT ℄.)

Table A-11.
Dimensions for Standard 50-Mil Lead-Pitch Packages

Number Of Terminals	"A" Dimensions	
	Millimeters	*Inches*
16†*	7.4–7.8	0.291–0.307
20†	8.7–9.1	0.342–0.358
24†*	10.0–10.4	0.390–0.410
28	11.2–11.7	0.441–0.460
44	16.3–16.8	0.640–0.660
52*	18.8–19.3	0.740–0.760
68	23.9–24.4	0.940–0.960
84	28.8–29.6	1.135–1.165
100*	33.9–34.7	1.335–1.365
124*	41.3–42.0	1.625–1.655
156*	51.7–52.5	2.035–2.065

* Packages registered by JEDEC but not widely used as of this writing.
† Packages available in Type C only.

Note: Inch dimensions control. Metric dimensions are rounded to nearest 0.1 mm.

Figure A-16 Variations among common 50-mil leadless chip-carrier outlines.

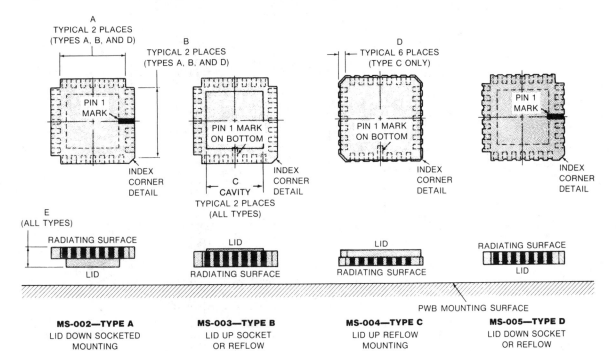

NOTES:
1. Unless otherwise indicated, dimensioned features are symmetrical about component X or Y axis.
2. Dimensions are tabulated in Table A-12.
3. In addition, there are Type E and F rectangular packages. Type E is available in 28 and 32 terminals. Type F comes in two 18-pin styles.

Table A-12.
JEDEC Leadless Chip-Carrier Dimensions for 50-Mil-Center Parts

Package	Dimensional Symbol	16	20	24	28	44	52	68	84	100	124	156
MS-002 / MS-003 / MS-005	A	N/A	N/A	N/A	8.76 – 9.01 / 0.345 – 0.355	13.82 – 14.12 / 0.544 – 0.556	16.34 – 16.69 / 0.643 – 0.657	21.39 – 21.79 / 0.842 – 0.858	26.47 – 26.87 / 1.042 – 1.058	31.47 – 32.03 / 1.239 – 1.261	39.09 – 39.65 / 1.539 – 1.561	49.25 – 49.81 / 1.939 – 1.961
	B	N/A	N/A	N/A	9.27 REF. / 0.365 REF.	14.38 REF. / 0.566 REF.	16.94 REF. / 0.667 REF.	22.05 REF. / 0.868 REF.	27.15 REF. / 1.069 REF.	32.13 REF. / 1.265 REF.	39.75 REF. / 1.565 REF.	49.91 REF. / 1.965 REF.
MS-002	C	N/A	N/A	N/A	7.06 – 7.37 / 0.278 – 0.290	8.9 – 12.4 / 0.350 – 0.490	9.8 – 15.0 / 0.386 – 0.590	11.6 – 20.1 / 0.458 – 0.790	13.5 – 25.1 / 0.530 – 0.990	15.3 – 30.2 / 0.602 – 1.190	18.0 – 37.8 / 0.710 – 1.490	20.9 – 48.0 / 0.822 – 1.890
	E	N/A	N/A	N/A	2.29 – 3.05 / 0.090 – 0.120	2.29 – 3.05 / 0.090 – 0.120	2.29 – 3.05 / 0.090 – 0.120	2.29 – 3.05 / 0.090 – 0.120	2.29 – 3.04 / 0.090 – 0.120	1.91 – 4.06 / 0.075 – 0.160	1.91 – 4.06 / 0.075 – 0.160	1.91 – 4.06 / 0.075 – 0.160
MS-003	C	N/A	N/A	N/A	7.06 – 8.12 / 0.278 – 0.320	8.9 – 13.2 / 0.350 – 0.520	9.8 – 15.7 / 0.386 – 0.620	11.6 – 20.8 / 0.458 – 0.820	13.5 – 25.9 / 0.530 – 1.020	15.3 – 31.0 / 0.602 – 1.220	18.0 – 38.6 / 0.710 – 1.520	20.9 – 48.8 / 0.822 – 1.920
	E	N/A	N/A	N/A	1.40 – 3.05 / 0.055 – 0.120	1.40 – 3.05 / 0.055 – 0.120	1.40 – 3.05 / 0.055 – 0.120	1.40 – 3.05 / 0.055 – 0.120	1.40 – 3.05 / 0.055 – 0.120	1.40 – 3.05 / 0.055 – 0.120	1.40 – 3.05 / 0.055 – 0.120	1.40 – 3.05 / 0.055 – 0.120
MS-004	C	6.53 – 7.82 / 0.257 – 0.308	7.80 – 9.09 / 0.307 – 0.358	9.02 – 10.41 / 0.355 – 0.410	10.31 – 11.63 / 0.406 – 0.458	12.57 – 14.22 / 0.495 – 0.560	12.57 – 14.22 / 0.495 – 0.560	12.6 – 21.9 / 0.495 – 0.862	12.6 – 27.1 / 0.495 – 1.066			
	D	1.02 / 0.040	1.02 / 0.040	1.02 / 0.040	1.02 / 0.040	1.02 / 0.040	1.02 / 0.040	1.02 / 0.040	1.02 / 0.040			
	E	1.63 – 2.54 / 0.064 – 0.100	1.63 – 2.54 / 0.064 – 0.100	1.63 – 2.54 / 0.064 – 0.100	1.63 – 2.54 / 0.064 – 0.100	1.75 – 3.05 / 0.069 – 0.120	2.08 – 3.05 / 0.082 – 0.120	2.08 – 3.05 / 0.082 – 0.120	2.08 – 3.05 / 0.082 – 0.120			
MS-005	C	N/A	N/A	N/A	7.07 – 7.37 / 0.278 – 0.290	8.9 – 12.4 / 0.350 – 0.490	9.8 – 15.0 / 0.386 – 0.590	11.6 – 20.1 / 0.458 – 0.790	13.5 – 25.1 / 0.530 – 0.990	15.3 – 30.2 / 0.602 – 1.190	18.0 – 37.8 / 0.710 – 1.490	20.9 – 48.0 / 0.822 – 1.890
	E	N/A	N/A	N/A	1.27 – 2.03 / 0.050 – 0.080	1.27 – 2.03 / 0.050 – 0.080	1.27 – 2.03 / 0.050 – 0.080	1.27 – 2.03 / 0.050 – 0.080	1.27 – 2.03 / 0.050 – 0.080	1.27 – 2.41 / 0.050 – 0.095	1.27 – 2.41 / 0.050 – 0.095	1.27 – 2.41 / 0.050 – 0.095

Number Of Terminations (Maximum)

Notes:
1. Dimensions in this table apply to Figure A-16.
2. Values are in millimeters and inches. Inch dimensions control.
3. N/A indicates that this is not a standard package.
4. For IPC standard lands for MS-003 and MS-004, see Figure A-41. For universal lands, see Figure A-58.

Figure A-17
JEDEC MS-009
40-mil leadless
chip carrier
devices.

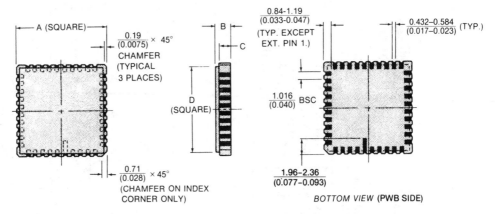

NOTES:
1. Unless otherwise indicated, dimensions shown centered are centered on the component axis.
2. For dimensions, see Table A-13.
3. For land patterns, see Figure A-62.
4. Dimensions are in millimeters and inches.

Figure A-18
JEDEC MS-008
50-mil-center
postleaded
ceramic chip
carrier.

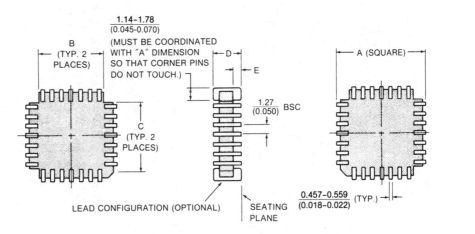

NOTES:
1. Lead pattern and A and B dimensions are centered about component center.
2. MS-007 also covers ceramic leaded carriers. We do not recommend these parts because their side-brazed leads offer less compliancy than top-brazed MS-008 parts.
3. Terminations to be coated with tin/lead solder having a tin content of 58 to 68%. Solder coating 0.0075 mm (0.0003 inch) minimum thickness. Coating may be hot dipped or plated from solution. Plated coatings must be fused in a post-plating reflow operation.
4. For dimensions, see Table A-14. For lands, see Figure A-58.
5. Values are in millimeter/inch.

Table A-13. JEDEC MS-009 40-Mil Leadless Chip-Carrier Dimensions

Dimensional Symbol	Number Of Terminations (Maximum)								
	16	20	24	32	40	48	64	84	96
A	5.84–6.22 0.230–0.245	8.26–8.64 0.325–0.340	8.76–9.14 0.345–0.360	10.54–10.92 0.415–0.430	12.07–12.50 0.475–0.492	14.07–14.53 0.554–0.572	18.08–18.62 0.712–0.733	23.14–23.75 0.911–0.935	26.19–26.80 1.031–1.055
B	1.02–2.41 0.040–0.095	1.40–2.79 0.055–0.110	1.40–2.79 0.055–0.110	1.40–2.79 0.055–0.110	1.40–2.79 0.055–0.110	1.40–3.18 0.055–0.125	1.40–3.18 0.055–0.125	1.40–3.18 0.055–0.125	1.78–3.43 0.070–0.135
C	0.76–1.52 0.030–0.060	1.14–1.91 0.045–0.075	1.14–1.91 0.045–0.075	1.14–1.91 0.045–0.075	1.14–1.91 0.045–0.075	1.14–2.29 0.045–0.090	1.14–2.29 0.045–0.090	1.14–2.29 0.045–0.090	1.52–2.54 0.060–0.100
D	5.33–5.72 0.210–0.225	7.49–7.87 0.295–0.310	7.87–8.38 0.310–0.330	9.40–10.16 0.370–0.400	11.05–11.68 0.435–0.460	12.70–13.59 0.500–0.535	15.24–15.88 0.600–0.625	16.38–17.91 0.645–0.705	16.38–17.91 0.645–0.705

Values are mm/inch. Inch dimensions control.

Table A-14. Dimensions for JEDEC MS-008 Postleaded Ceramic Chip Carriers

Dimensional Symbol	Number Of Terminations (Maximum)							
	28	44	52	68	84	100	124	156
A	11.81–12.83 0.465–0.505	16.84–17.96 0.663–0.707	19.35–20.52 0.762–0.808	24.38–25.65 0.960–1.010	29.41–30.78 1.158–1.212	34.52–35.84 1.359–1.411	42.14–43.46 1.659–1.711	52.30–53.62 2.059–2.111
B	8.76–9.02 0.345–0.355	13.82–14.12 0.544–0.556	16.33–16.69 0.643–0.657	21.39–21.79 0.842–0.858	26.47–26.87 1.042–1.058	31.47–32.03 1.239–1.261	39.09–39.65 1.539–1.561	49.25–49.81 1.939–1.961
C	9.27 0.365	14.38 0.566	16.94 0.667	22.05 0.868	27.15 1.069	32.13 1.265	39.75 1.565	49.91 1.965
D	2.67–4.95 0.105–0.195	2.67–4.95 0.105–0.195	2.67–4.95 0.105–0.195	2.67–4.95 0.105–0.195	2.67–4.95 0.105–0.195	2.29–5.46 0.090–0.215	2.29–5.46 0.090–0.215	2.29–5.46 0.090–0.215
E	0.89–1.27 0.035–0.050	0.89–1.27 0.035–0.050	0.89–1.27 0.035–0.050	0.89–1.27 0.035–0.050	0.89–1.27 0.035–0.050	0.89–1.27 0.035–0.050	0.89–1.27 0.035–0.050	0.89–1.27 0.035–0.050

Values are mm/inch. Inch dimensions control.

Figure A-19 Pre-leaded ceramic chip carrier per JEDEC MS-044.

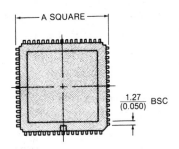
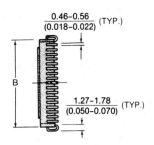

NOTES:
1. Terminations to be coated with tin/lead solder having a tin content of 58 to 68%. Solder coating 0.0075 mm (0.0003 inch) minimum thickness. Coating may be hot dipped or plated from solution. Plated coatings must be fused in a post-plating reflow operation.
2. Dimensions are in millimeter/inch.
3. For land patterns, see Figure A-42.

DIMENSIONAL SYMBOL	DIMENSIONS (mm/inch)	
	68 PIN	84 PIN
A	25.02-25.27 (0.985-0.995)	30.10-30.35 (1.185-1.195)
B	23.11-24.13 (0.910-0.950)	28.2-29.2 (1.110-1.150)

Figure A-20 Details of a typical surface-mount card edge connector.

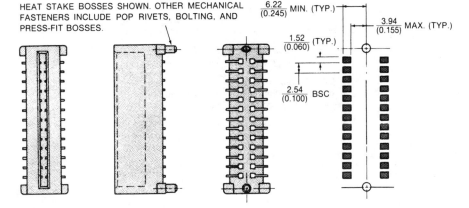

NOTES:
1. Typically available in up to 60 rows (120 contacts).
2. Many varieties are on the market. Dimensions here are for reference only. Check with vendor for outline and land recommendations.
3. Values are in millimeter/inch. Inch dimensions control.
4. Terminations to be coated with tin/lead solder having a tin content of 58 to 68%. Solder coating 0.0075 mm (0.0003 inch) minimum thickness. Coating may be hot dipped or plated from solution. Plated coatings must be fused in a post-plating reflow operation.

A.1.3 Other Components

Connectors

Figure A-20 covers the package outline of a typical SMT-style card-edge connector. Figure A-21 shows an approach to an SMT header. There are a multitude of application demands that impact the design of connectors. Because of this, standards are difficult to develop, and thus have moved slowly. Without standards, a wide range of outlines have been brought to the market. The configurations shown in Figures A-20 and A-21 are offered for reference only. But the general approach to lead form which is shown is quite common among SMT connectors. As these drawings suggest, SMT connectors are often modeled after IMC versions, and are given special lead forms for surface mounting. Generally, the housing material is also different from IMC types, in order to accommodate high-temperature exposure during reflow soldering.

Figure A-21
Details of a
typical SMT
header.

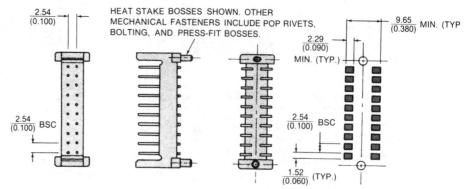

HEAT STAKE BOSSES SHOWN. OTHER
MECHANICAL FASTENERS INCLUDE POP RIVETS,
BOLTING, AND PRESS-FIT BOSSES.

NOTES:
1. Available in 10- to 100-pin configurations.
2. Many varieties are on the market. Dimensions here are for reference only. Check with vendor for outline and land recommendations.
3. Values are in millimeter/inch. Inch dimensions control.
4. Terminations to be coated with tin/lead solder having a tin content of 58 to 68%. Solder coating 0.0075 mm (0.0003 inch) minimum thickness. Coating may be hot dipped or plated from solution. Plated coatings must be fused in a post-plating reflow operation.

Figure A-22
Detail of a typical
SMT DIP socket.

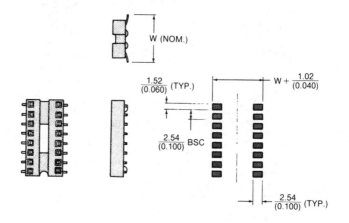

NOTES:
1. Terminations to be coated with tin/lead solder having a tin content of 58 to 68%. Solder coating 0.0075 mm (0.0003 inch) minimum thickness. Coating may be hot dipped or plated from solution. Plated coatings must be fused in a post-plating reflow operation.
2. Typical offerings include sockets for standard 300-mil DIP in 8, 14, 16, 18, and 20 pins; 400-mil DIP in 22 pins; and 600-mil DIP in 24, 28, and 40 pins.

3. Many varieties are on the market. Dimensions here are for reference only. Check with vendor for outline and land recommendations.
4. Some sockets are offered with features for mechanical anchoring to PWB.
5. Values are in millimeter/inch. Inch dimensions control.

Sockets

Like connectors, sockets are application-driven items which have proven difficult to standardize. And, like connectors, many variations of sockets have filled the standards vacuum. To bring some sense to the many offerings, we can divide them roughly, by application, into three basic families of SMT sockets.

First, there are sockets which allow the surface mounting of through-hole components. Figure A-22 shows a typical example of such a part. Some such sockets are gull winged as shown, while others have J or I leads. Some provide mechanical-attachment bosses or have provisions for screw attachment. The socket vendors can supply details of the outlines and the solder lands for a given device.

Second, there are sockets for reflow-solderable SMCs. The intent of these parts is primarily to facilitate device removal and reinstallation. A typical PLCC type is shown in Figure A-23. Note the two variations of lead depicted in Figure A-23. Most SMT sockets use one or the other. Surface-mountable SMC sockets are available for PLCCs, SOICs, CLLCCs, and other surface-mount components.

Figure A-23
Typical PLCC socket facilitates device removal/ replacement.

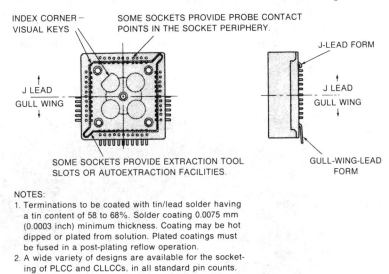

INDEX CORNER – VISUAL KEYS

SOME SOCKETS PROVIDE PROBE CONTACT POINTS IN THE SOCKET PERIPHERY.

J-LEAD FORM

J LEAD
GULL WING

J LEAD
GULL WING

SOME SOCKETS PROVIDE EXTRACTION TOOL SLOTS OR AUTOEXTRACTION FACILITIES.

GULL-WING-LEAD FORM

NOTES:
1. Terminations to be coated with tin/lead solder having a tin content of 58 to 68%. Solder coating 0.0075 mm (0.0003 inch) minimum thickness. Coating may be hot dipped or plated from solution. Plated coatings must be fused in a post-plating reflow operation.
2. A wide variety of designs are available for the socketing of PLCC and CLLCCs, in all standard pin counts.

Third, there are sockets designed for nonreflowable CLLCCs. These are similar to the PLCC socket shown in Figure A-23, except that they generally have separate hold-down covers which snap in over the chip carrier, securing it firmly into the socket.

Because there are so many types of sockets and no firm standards to guide us in deciding which to cover herein, we have not included dimensional data on sockets. Vendors can provide detailed information. Some sockets fit the same land pattern as the device they carry, and these provide a welcome measure of flexibility worth considering in the selection process.

Switches

As with the devices mentioned above, standards development has lagged far behind device-outline development in switches. Outwardly, most of today's SMT switches look like lead-prepped versions of through-hole parts. Figure A-24 shows a typical package outline for a DIP switch, and details several ways it may be lead prepped for surface mounting. As we mentioned in Chapter 2, though SMC switches look like their IMC cousins, they often differ substantially in materials and sealing details so as to withstand the SMT manufacturing environment.

Figure A-24 SMT
lead forms for
switches.

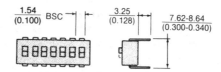

(A) Typical DIP switch.

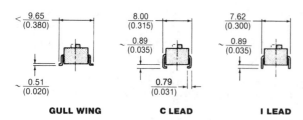

GULL WING   **C LEAD**   **I LEAD**

(B) SMT lead forms.

NOTES:
1. There are a wide variety of designs on the market.
   Dimensions presented here are for reference only.
   Check with vendor for the actual outline and land
   recommendations.
2. Values are in millimeter/inch. Inch dimensions control.
3. Terminations to be coated with tin/lead solder having
   a tin content of 58 to 68%. Solder coating 0.0075 mm
   (0.0003 inch) minimum thickness. Coating may be hot
   dipped or plated from solution. Plated coatings must
   be fused in a post-plating reflow operation.
4. For lands, see Figure A-43.

Figure A-25
Visual inspection
of "J" and gull-
wing lead forms.

Since switches present the designer with options of lead form, this is a good place to discuss the trade-offs involved in lead-form selection. Table A-15 lists the advantages and disadvantages for each. The comments regarding the visual inspection of J leads may seem suprising. However, referring to the photograph in Figure A-25, one can quickly and visually spot an open on a J-leaded part by the absence of reflected light. Gull wings will require a much closer look. However, the full solder joint and under-the-heel wetting is difficult to see on the J-lead part. Here, the gull-wing part has a distinct advantage.

Table A-15.
Advantages and
Disadvantages of
Lead Forms

Lead Form	Trade-Off Discussion	
	Advantages	*Disadvantages*
GULL WING	Strong solder joints. Full joint is visible.	Uses extra PWB area. Inspect only with close look.
C or J	Strong solder joints. Quick visual solder inspection. Minimizes PWB area.	Part of joint hidden from detailed visual inspection.
I	Quick visual solder inspection. Minimizes PWB area.	Weak solder joints.

Whatever lead form is used, if the parts are to be stored for any time, the lead ends as well as the lead sides should have a protective solder over-plate. Untreated beryllium-copper or Kovar surfaces quickly become unsolderable. And worse, the open end of a lead that was trimmed after solder coating provides a starting point for corrosion and surface blistering that may interfere with solderability elsewhere on the lead. This comment is doubly important for I-lead parts because the I lead inherently forms the weakest solder joint of the three types mentioned.

Crystals
Crystal standards work is proceeding. Surface-mount crystals are generally packaged in two alternatives. Many are in metal cans similar to IMC crystals, but with SMT lead forms. Others are in LLCC packages but use only designated pins on two sides of the package. Consult the device manufacturers for the exact outlines and recommended land patterns.

Relays
Relays are another class of devices that suffer from a paucity of standards and an abundance of varied packaging. Currently, most SMT relays, outwardly, look like IMC versions which are lead prepped for surface mounting. However, like switches, SMT relays often differ from IMC devices in both construction details and materials. Such differences are dictated by the harsh treatment of the components in the SMT assembly processes.

Jumpers
Chip jumpers are actually Zero-ohm resistors. They are controlled by the specifications governing chip resistors (see Chapter 1). The major difference between chip resistors and jumpers is that the Zero-ohm parts will carry substantially higher power levels than a resistor of the same body size.

DIPs

DIP packages may be lead prepped for surface mounting. Figure A-26 covers the package outline for a 16-pin DIP and shows several ways it can be customized for surface mounting. Other package pin counts follow the same customizing scheme.

Figure A-26
Lead prepping
DIPs for surface
mounting.

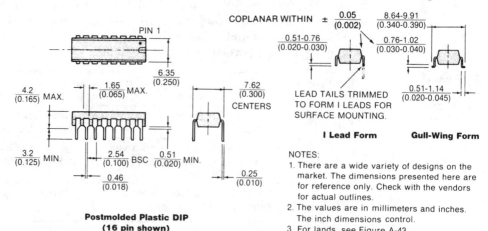

IPC-SM-782 lists the I lead as the preferred form for lead-prepped DIPs. The I form is preferred because the gull wing is more difficult to form within coplanarity specifications.[9] Also, the I lead is more space efficient than the gull wing. However, the gull-wing configuration, when properly formed per the specifications shown in Figure A-26, provides a stronger, more stress-resistant solder bond than does the I lead. And both parts fit the same land pattern, discounting concerns about real estate. Given appropriate tooling and process control to produce consistently coplanar feet, the gull wing is the better choice where high shear and tensile forces will be encountered. Note that "appropriate" includes the tooling which will protect the plastic body of the device from lead forming and trimming stress.

Whichever lead form is chosen, the parts should not be lead prepped and then stored. Lead trim-and-form operations should occur immediately before component assembly and soldering. This is because of concerns about solderability, as discussed earlier in the section on Switches.

A.2 ANSI/IPC-SM-782 Land Patterns

The following lands are suggested by the IPC for surface-mount components.[10] For reflow-soldered boards, the screened pattern of solder paste should be the same as the land patterns.

A.2.1 *SMT Passives*

Chip-component lands are determined by the size of the chip component. IPC-SM-782 states that the following formulae are to be used for land-geometry determination for nonstandard components, and are the basis of the geometries in the standard.

To calculate the land geometry of the device shown in Figure A-27, we need to determine the land width (W), the land length (L), and the gap between the lands (S). To do this, we calculate the land width by

$$W = W_c \text{ max} - K \qquad \text{(Eq. A-1)}$$

the land length by

$$L = H_c \text{ max} + T_c \text{ max} + K \text{ (for resistors)} \qquad \text{(Eq. A-2)}$$

and

$$L = H_c \text{ max} + T_c \text{ max} - K \text{ (for chip caps)}$$

and the gap between the lands by

$$S = L_c \text{ max} - 2T_c \text{ max} - K \qquad \text{(Eq. A-3)}$$

where,

W_c is the component width,
H_c is the component height,
L_c is the component length,
T_c is the termination distance,
K is a constant (0.25 mm/0.010 inch).

For tantalum bricks, L = Metallization Height (the termination height shown in Figure A-2).

Round all dimensions to a convenient number directed at accommodating expected board fabrication tolerances.

Figure A-27 IPC-SM-782 Standard land geometry nomenclature.

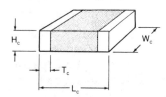

H_c = Component height,
L_c = Component length,
T_c = Termination distance,
W_c = Component width.

Figure A-28 Solder-joint quality determination.

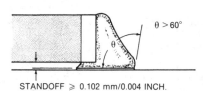

$\theta > 60°$

STANDOFF ⩾ 0.102 mm/0.004 INCH.

POOR DESIGN
LAND AND/OR PASTE PRINT THICKNESS SUPPLIED TOO MUCH SOLDER

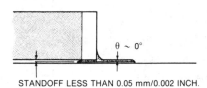

$\theta \sim 0°$

STANDOFF LESS THAN 0.05 mm/0.002 INCH.

POOR DESIGN
LAND AND/OR PASTE PRINT THICKNESS SUPPLIED TOO LITTLE SOLDER

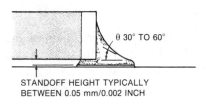

θ 30° TO 60°

STANDOFF HEIGHT TYPICALLY BETWEEN 0.05 mm/0.002 INCH AND 0.102 mm/0.004 INCH.

PROPER DESIGN
LAND AND/OR PASTE PRINT THICKNESS SUPPLIED RIGHT AMOUNT OF SOLDER

After the calculations, the result of each equation is rounded to the nearest convenient measurement; i.e., S = 0.062 (Equation A-3) would be rounded to 0.060 inch if the inch dimensions are to control, or converted to 1.57 mm and rounded up to 1.6 mm if metric dimensions are to control.

When using the formulae to determine a land for a nonstandard part, testing should be conducted to determine that the land geometry produced by the formulae yields good solder fillets and proper standoff height. "Good" solder fillets are determined by the solder wetting angle, as shown in Figure A-28.

Ceramic Chips

Figure A-29 shows the IPC-recommended land pattern for EIA standard-size ceramic-chip capacitors. Alternate reflow lands are shown in Figures A-46 and A-47. Optional flow-soldering optimized lands are covered in Figures A-48 and A-49.

Figure A-29
Lands for ceramic-chip capacitors per IPC-SM-782.

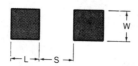

NOTES:
1. For component end terminations (per Figure A-1).
2. Component center at center of geometric pattern.
3. Where IPC-SM-782 inch conversions of metric dimensions were not exact, correct conversions, rounded to three places, have been substituted.
4. For alternate patterns, see Figures A-46 and A-47 for reflow soldering, and Figures A-48 and A-49 for flow soldering.

COMPONENT TYPE	DIMENSIONS (mm/inch)		
	WIDTH (W)	LENGTH (L)	SPACE (S)
0805 Capacitor	1.40 (0.055)	1.50 (0.059)	0.80 (0.031)
1206 Capacitor	1.60 (0.063)	1.60 (0.063)	1.80 (0.071)
1210 Capacitor	2.60 (0.102)	1.80 (0.071)	1.80 (0.071)
1812 Capacitor	3.20 (0.126)	1.80 (0.071)	3.20 (0.126)
1825 Capacitor	6.60 (0.260)	1.80 (0.071)	3.20 (0.126)

Tantalum and Aluminum Electrolytic Capacitors

Figure A-30 shows the IPC-recommended land pattern for EIA standard "brick" style electrolytic capacitors.

Figure A-30
Electrolytic capacitor land dimensions.

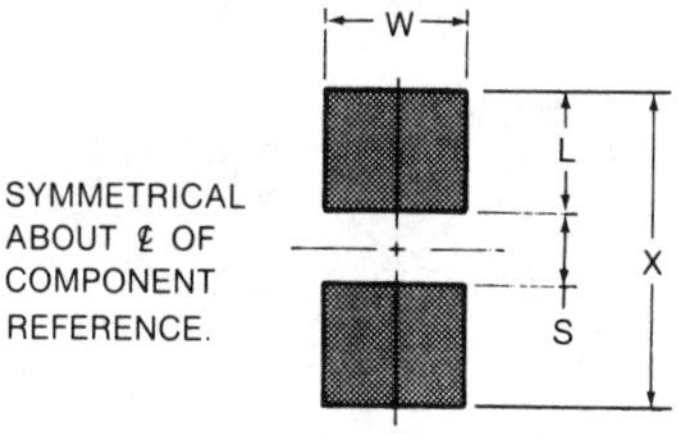

SYMMETRICAL ABOUT ℄ OF COMPONENT REFERENCE.

NOTES:
1. Values are in millimeters and inches.
2. All tolerances are ± 0.1 mm/ ± 0.004 inch.
3. For inch control, round inch dimensions to nearest 0.005 inch.
4. For components, per Figures A-2 and A-3.

COMPONENT DESIGNATION		DIMENSIONS (mm/inch)			
		L	S	W	X
Standard Capacitance	(A) 3216	2.0 (0.079)	0.8 (0.031)	1.4 (0.055)	4.8 (0.189)
	(B) 3528	2.0 (0.079)	1.2 (0.047)	2.4 (0.094)	5.2 (0.205)
	(C) 6032	2.4 (0.094)	3.2 (0.126)	2.4 (0.094)	8.0 (0.315)
	(D) 7243	2.4 (0.094)	4.2 (0.165)	2.6 (0.102)	**9.0 (0.354)**
Extended Range	3518	2.0 (0.079)	0.8 (0.031)	2.0 (0.079)	4.8 (0.189)
	3527	2.0 (0.079)	0.8 (0.031)	2.8 (0.110)	4.8 (0.189)
	7227	3.0 (0.118)	3.2 (0.126)	2.8 (0.110)	9.2 (0.362)
	7257	3.0 (0.118)	3.2 (0.126)	6.0 (0.236)	9.2 (0.362)

Variable Capacitors

Several of the many and varied outlines of trimming capacitors on today's market are shown in Figure A-4. For land-pattern recommendations for a specific part, consult the component manufacturer.

Chip Resistors

Figure A-31 shows the IPC-recommended land pattern for chip resistors. Alternate reflow-process optimized lands are shown in Figures A-46 and Figure A-47. Lands optimized for flow soldering are covered in Figures A-48 and A-49.

Figure A-31
Chip-resistor
land patterns.

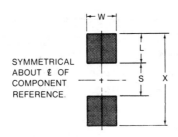

SYMMETRICAL
ABOUT ℓ OF
COMPONENT
REFERENCE.

COMPONENT	DIMENSIONS (mm/inch)			
DESIGNATION	L	S	W	X
RC0805	1.5 (0.059)	0.8 (0.031)	1.4 (0.055)	3.8 (0.150)
RC1206	1.6 (0.063)	1.8 (0.071)	1.6 (0.063)	5.0 (0.197)
RC1210	1.6 (0.063)	1.8 (0.071)	2.6 (0.102)	5.0 (0.197)

NOTES:
1. Values are in millimeters and inches.
2. All tolerances are ± 0.1 mm/ ± 0.004 inch.
3. For inch control, round inch dimensions to nearest 0.005 inch.

4. Use for components, per Figure A-5.
5. For alternate lands, reflow optimized, see Figures **A-46 and A-47**; for flow-solder optimized, see Figures **A-48 and A-49**.

Cylindrical Components

Figure A-32 shows the IPC-recommended land pattern for cylindrical parts under the ¼-watt MELF, and the MLL34, SOD80, and MLL41 designations. The notches shown on the inner faces of the lands are intended to help stabilize cylindrical parts during reflow. In some applications, the notches have been omitted. Without notches, the parts may be secured by adhesives or fixturing.

To notch or not to notch is discussed further under the MELF discussion in Section A.3.1.

Figure A-32
Cylindrical
component land
patterns.

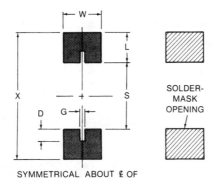

SOLDER-
MASK
OPENING

NOTES:
1. Notch (dimensions D and G) is optional to reduce swimming during reflow soldering. (See discussion of notch under MELFs **in Section A.3.1**)
2. Values are in millimeters and inches.
3. Tolerances are ± 0.1 mm/ ± 0.004 inch.
4. For inch control, round inch dimensions to nearest 0.005 inch.
5. Use for components outlined in Figure A-6.

SYMMETRICAL ABOUT ℓ OF
COMPONENT REFERENCE.

COMPONENT	DIMENSIONS (mm/inch)					
DESIGNATION	D	G	L	S	W	X
MLL-34	0.7 (0.028)	0.3 (0.012)	1.4 (0.055)	2.2 (0.087)	2.0 (0.079)	5.0 (0.197)
SOD-80	0.8 (0.031)	0.3 (0.012)	1.6 (0.063)	2.2 (0.087)	2.0 (0.079)	5.4 (0.213)
MLL-41	1.0 (0.039)	0.3 (0.012)	1.8 (0.071)	3.4 (0.134)	2.8 (0.110)	7.0 (0.276)
¼-Watt MELF	0.8 (0.031)	0.3 (0.012)	2.0 (0.079)	4.4 (0.173)	2.5 (0.098)	8.4 (0.331)

Trimmer Potentiometers

Figure A-7 shows the manufacturer's recommended land patterns for the two outlines covered herein. For other outlines (again, there are many), consult the component vendors.

Resistor Networks

Where resistor networks are in standardized JEDEC component outlines, such as the SOIC or CLLCC, standard SOIC or CLLCC land patterns are used. For nonstandard package outlines, consult the component vendor for land recommendations. Lands for one widely used nonstandard package are shown in Figure A-50.

Thermistors

Thermistors in EIA standard package sizes fit capacitor IPC land patterns, such as shown in Figure A-29. Or, use the process-tailored capacitor lands from Figures A-46 and A-47, or from Figures A-48 and A-49. Where nonstandard body sizes are required, the IPC chip-component formula in Figure A-27, or the customized land formulae in Figure A-46 or Figure A-48 may be used to determine the land geometry.

Tuning Coils and Transformers

The many and varied outlines on today's market prevent us from presenting a standard pattern herein for tuning coils and transformers. For land-pattern recommendations for a specific part, consult the component manufacturer.

Inductors

Figure A-33 shows the IPC-recommended land pattern for inductors in tantalum brick-style packaging (parts similar to tantalum bricks in form factor, but having a specialized inductor rather than the tantalum package dimensions). For open core and other packaging formats not covered herein, consult the manufacturer for package outline and land geometry information.

Figure A-33
Chip inductor lands for tantalum brick-style packages.

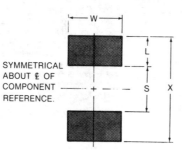

SYMMETRICAL ABOUT ₵ OF COMPONENT REFERENCE.

COMPONENT	DIMENSIONS (mm/inch)			
DESIGNATION	L	S	W	X
Size A	1.4 (0.055)	1.4 (0.055)	2.4 (0.094)	4.2 (0.165)
Size B	1.6 (0.063)	2.4 (0.094)	3.0 (0.118)	5.6 (0.220)

NOTES:
1. Values are in millimeters and inches.
2. Tolerances are ± 0.1 mm/ ± 0.004 inch.
3. For inch control, round inch dimensions to nearest 0.005 inch.
4. Use for component designations listed in chart.

A.2.2   *Actives*

Diodes

Diodes are packaged in both chip style and MELF packaging, using the lands per Figure A-32. Single and dual diodes are also available in SOT packages, using the lands patterns shown below. LEDs follow these same packaging directions.

Transistors

Figure A-34 shows the IPC-recommended land pattern for transistors in SOT-23 packaging, while Figure A-35 covers SOT-143 lands, and Figure A-36 presents land patterns for the SOT-89. Alternate SOT-23 lands are shown in Figures A-51, A-52, and A-53. SOT-89 CAD vias are shown in Figure A-53, and SOT-89 alternates are covered in Figure A-54 and Figure A-55. An additional land pattern for power transistors per the DPAK outline is presented in Figure A-56.

Figure A-34
SOT-23 land pattern dimensions.

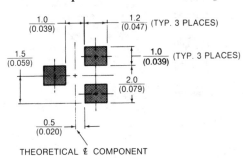

NOTES:
1. Values are in millimeters and inches.
2. Tolerances not stated in IPC-SM-782.
3. For inch control, round inch dimensions to nearest 0.005 inch.
4. Use for components covered under SOT-23 (Figure A-9A).
5. For lands optimized for reflow soldering, see Figures A-51 and A-52. Lands shown here are excellent for flow soldering. For CAD vias suitable for lands shown here, see Figure A-53.

Figure A-35
SOT-143 land pattern dimensions.

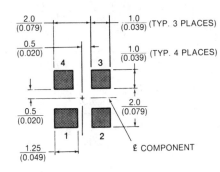

NOTES:
1. Values are in millimeters and inches.
2. Tolerances not stated in IPC-SM-782.
3. For inch control, round inch dimensions to nearest 0.005 inch.
4. Use for components covered under the SOT-143 designation in Figure A-9B.

Figure A-36
SOT-89 land pattern dimensions.

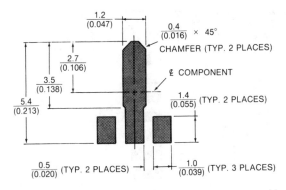

NOTES:
1. Values are in millimeters and inches.
2. Tolerances not stated in IPC-SM-782.
3. For inch control, round inch dimensions to nearest 0.005 inch.
4. Use for components covered under the SOT-89 designation in Figure A-9C.
5. For application-optimized alternate, see Figures A-54 and A-55.

SOICs

Figure A-37 shows the IPC-recommended land patterns for JEDEC standards SO, SOL, and SOJ ICs. CAD via recommendations for these land patterns are covered in Figure A-57. Figure A-50 displays lands for the nonstandard SOM IC (per the component outline given in Figure A-8). CAD vias for the SOM are also exhibited in Figure A-57.

Figure A-37 SO, SOL, and SOJ IC land pattern dimensions.

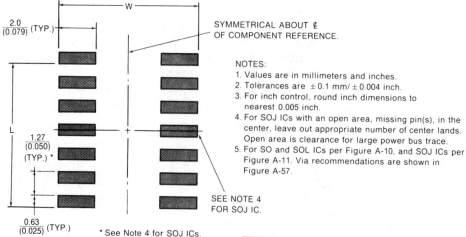

SYMMETRICAL ABOUT ₵ OF COMPONENT REFERENCE.

NOTES:
1. Values are in millimeters and inches.
2. Tolerances are ±0.1 mm/±0.004 inch.
3. For inch control, round inch dimensions to nearest 0.005 inch.
4. For SOJ ICs with an open area, missing pin(s), in the center, leave out appropriate number of center lands. Open area is clearance for large power bus trace.
5. For SO and SOL ICs per Figure A-10, and SOJ ICs per Figure A-11. Via recommendations are shown in Figure A-57.

SEE NOTE 4 FOR SOJ IC.

* See Note 4 for SOJ ICs.

SO AND SOL IC			
COMPONENT DESIGNATION	DIMENSIONS (mm/inch)		PACKAGE DRAWING
	L	W	
SO-8	3.8 (0.150)	7.6 (0.299)	See Figure A-10.
SO-14	**7.62 (0.300)**	7.6 (0.299)	
SO-16	8.9 (0.350)	7.6 (0.299)	
SOL-16	8.9 (0.350)	11.6 (0.457)	
SOL-20	**11.43 (0.450)**	11.6 (0.457)	
SOL-24	**13.97 (0.550)**	11.6 (0.457)	
SOL-28	16.5 (0.650)	11.6 (0.457)	

SOJ IC			
COMPONENT DESIGNATION	DIMENSIONS (mm/inch)		PACKAGE DRAWING
	L (Note 4)	W	
SOJ-14	7.62 (0.300)	8.89 (0.350)	See Figure A-11.
SOJ-16	8.89 (0.350)	8.89 (0.350)	
SOJ-18	10.16 (0.400)	8.89 (0.350)	
SOJ-20	11.43 (0.450)	8.89 (0.350)	
SOJ-22	12.70 (0.500)	8.89 (0.350)	
SOJ-24	13.97 (0.550)	8.89 (0.350)	
SOJ-26	15.24 (0.600)	8.89 (0.350)	
SOJ-28	16.51 (0.650)	8.89 (0.350)	

Figure A-38 JEDEC Standard postmolded PLCC lands.

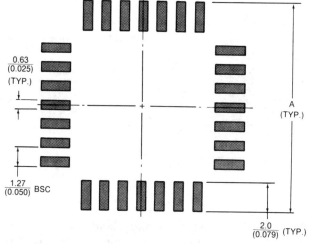

PIN COUNT	DIMENSION "A"
28	14.0 (0.551)
44	19.0 (0.748)
52	21.6 (0.850)
68	26.6 (1.047)

NOTES:
1. Values are in millimeters and inches.
2. Tolerances are ±0.1 mm/±0.004 inch.
3. For inch control, round inch dimensions to nearest 0.005 inch.
4. For CAD vias, see Figures A-59, A-60, and A-61.

5. For lands recommended by IPC for premolded standard parts, see Figure A-39. For universal Type-A PLCC lands, see Figure A-58.
6. For components, such as shown in Figure A-12 and dimensioned in Table A-6.

PLCCs

Figure A-38 shows IPC-recommended land patterns for JEDEC postmolded PLCCs. This figure covers lands specifically designed for components dimensioned in Table A-6 and illustrated in Figure A-12. Postmolded parts per this outline may also be placed on universal Type-A lands, as illustrated in Figure A-58, with vias per Figures A-59, A-60, and A-61.

Figure A-39 covers premolded packages in the square and rectangular formats. Parts fitting these lands are illustrated in Figure A-12 and dimensioned in Table A-7. Universal Type-A lands, per Figure A-58, may also be used for parts dimensioned per the premolded standards. Vias are shown in Figures A-59, A-60, and A-61.

Tables A-16 and A-17 provide the land dimensions for square packages per JEDEC MO-047 and rectangular packages per JEDEC MO-052, respectively.

Figure A-39 JEDEC MO-047 premolded PLCC square and rectangular lands.

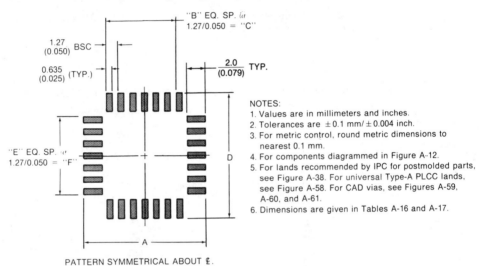

NOTES:
1. Values are in millimeters and inches.
2. Tolerances are ± 0.1 mm/ ± 0.004 inch.
3. For metric control, round metric dimensions to nearest 0.1 mm.
4. For components diagrammed in Figure A-12.
5. For lands recommended by IPC for postmolded parts, see Figure A-38. For universal Type-A PLCC lands, see Figure A-58. For CAD vias, see Figures A-59, A-60, and A-61.
6. Dimensions are given in Tables A-16 and A-17.

PATTERN SYMMETRICAL ABOUT ℄.

Table A-16. Premolded Square Packages per JEDEC MO-047

Pin Count	Dimensions*					
	A	B	C	D	E	F
20	$\frac{10.80}{0.425}$	4	$\frac{5.08}{0.200}$	Same as A	Same as B	Same as C
28	$\frac{13.34}{0.525}$	6	$\frac{7.62}{0.300}$	Same as A	Same as B	Same as C
44	$\frac{18.42}{0.725}$	10	$\frac{12.7}{0.500}$	Same as A	Same as B	Same as C
52	$\frac{20.96}{0.825}$	12	$\frac{15.24}{0.600}$	Same as A	Same as B	Same as C
68	$\frac{26.04}{1.025}$	16	$\frac{20.32}{0.800}$	Same as A	Same as B	Same as C
84	$\frac{31.12}{1.225}$	20	$\frac{25.40}{1.000}$	Same as A	Same as B	Same as C
100	$\frac{36.20}{1.425}$	24	$\frac{30.48}{1.200}$	Same as A	Same as B	Same as C
124	$\frac{43.82}{1.725}$	30	$\frac{38.10}{1.500}$	Same as A	Same as B	Same as C

* Values are millimeters/inches; tolerances are ±0.1 mm/±0.004 inch. For metric control, round metric dimensions to nearest 0.1 mm.

Table A-17. Pre-molded Rectangular Packages per JEDEC MO-052

Package Type	Dimensions*					
	A	B	C	D	E	F
18 PIN	9.02 / 0.355	3	3.81 / 0.150	12.57 / 0.495	4	5.08 / 0.200
18L PIN	9.40 / 0.370	3	3.81 / 0.150	14.61 / 0.575	4	5.08 / 0.200
22 PIN	9.40 / 0.370	3	3.81 / 0.150	14.61 / 0.575	6	7.62 / 0.300
28 PIN	10.80 / 0.425	4	5.08 / 0.200	16.00 / 0.630	8	10.16 / 0.400
32 PIN	13.46 / 0.530	6	7.62 / 0.300	16.00 / 0.630	8	10.16 / 0.400

* Values are millimeters/inches; tolerances are ±0.1 mm/±0.004 inch. For metric control, round metric dimensions to nearest 0.1 mm.

PQFPs and CQFPs

Figure A-40 shows the IPC-recommended land patterns for plastic and ceramic quad flat packs.

Figure A-40 Land patterns for JEDEC PQFPs and CQFPs.

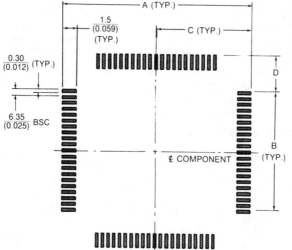

PIN COUNT	DIMENSIONS (mm/inch)			
	A	B	C	D
84	20.5 (0.807)	12.7 (0.500)	10.25 (0.404)	3.75 (0.148)
100	23.0 (0.906)	15.24 (0.600)	11.50 (0.453)	3.73 (0.147)
132	28.0 (1.102)	20.32 (0.800)	14.00 (0.551)	3.69 (0.145)
164	33.2 (1.307)	25.4 (1.000)	16.60 (0.654)	3.75 (0.148)
196	38.2 (1.504)	30.48 (1.200)	19.10 (0.752)	3.71 (0.146)
244	42.5 (1.673)	38.10 (1.500)	21.25 (0.837)	2.05 (0.081)

NOTES:
1. Values are in millimeters and inches.
2. Tolerances are ±0.1 mm/±0.004 inch.
3. For inch control, round inch dimensions to nearest 0.005 inch.
4. For components, such as shown in Figure A-14.

TapePaks

TapePak lands were not covered in the March, 1987, revision of IPC-SM-782. Lands for TapePak packages are covered under alternate patterns in Figure A-65.

CLLCCs

Figure A-41 shows the IPC-recommended land patterns for 50-mil-center square and rectangular reflow-mountable ceramic leadless chip carriers. Universal lands, per Figure A-58, may also be used for these CLLCCs; however, they require a slightly greater board real estate. Vias for CLLCCs are shown in Figures A-59, A-60, and A-61. CLLCCs with 40-mil lead pitch are covered in Figure A-62.

Figure A-41
Lands for JEDEC
50-mil CLLCCs.

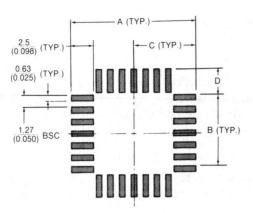

PIN	DIMENSIONS (mm/inch)			
COUNT	A	B	C	D
16	9.8 (0.386)	3.8 (0.150)	4.90 (0.193)	2.68 (0.106)
20	11.1 (0.437)	5.08 (0.2 00)	5.55 (0.219)	2.69 (0.106)
24	12.4 (0.488)	6.35 (0.250)	6.20 (0.244)	2.71 (0.107)
28	13.6 (0.535)	7.62 (0.300)	6.80 (0.268)	2.67 (0.105)
44	18.8 (0.740)	12.70 (0.500)	9.40 (0.370)	2.73 (0.107)
52	21.3 (0.839)	15.24 (0.600)	10.65 (0.419)	2.71 (0.107)
68	26.4 (1.039)	**20.32** **(0.800)**	13.20 (0.520)	2.73 (0.107)
84	31.5 (1.240)	25.4 (1.000)	15.75 (0.620)	2.73 (0.107)
100	36.5 (1.437)	**30.48** **(1.200)**	18.25 (0.719)	**2.70** **(0.106)**
124	44.2 (1.740)	**38.10** (1.500)	22.10 (0.870)	2.73 (0.107)
156	54.3 (2.138)	**48.26** **(1.900)**	27.15 (1.069)	**2.71** **(0.107)**

NOTES:
1. Values are in millimeters and inches.
2. Tolerances are ±0.1 mm/±0.004 inch.
3. For inch control, round inch dimensions to nearest 0.005 inch.
4. For those components covered in Figure A-15 and A-16, and Tables A-11 and A-12.
5. For universal lands, see Figure A-58; for vias, see Figures A-59, A-60, and A-61.

CLDCCs

Figure A-42 shows IPC-recommended land patterns for JEDEC MS-044 ceramic postleaded chip carriers. Universal Type-A lands, per Figure A-58, may also be used for parts dimensioned per MS-044. Vias for these CLDCCs are shown in Figures A-59, A-60 and A-61.

Figure A-42
JEDEC MS-044
Cerdip-type
CLDCC lands.

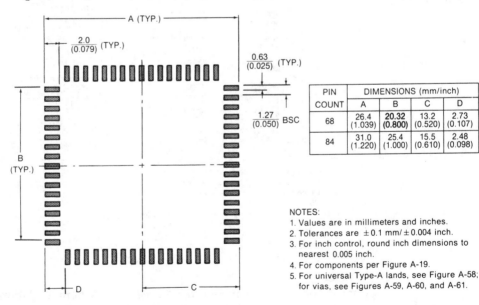

PIN	DIMENSIONS (mm/inch)			
COUNT	A	B	C	D
68	26.4 (1.039)	**20.32** **(0.800)**	13.2 (0.520)	2.73 (0.107)
84	31.0 (1.220)	25.4 (1.000)	15.5 (0.610)	2.48 (0.098)

NOTES:
1. Values are in millimeters and inches.
2. Tolerances are ±0.1 mm/±0.004 inch.
3. For inch control, round inch dimensions to nearest 0.005 inch.
4. For components per Figure A-19.
5. For universal Type-A lands, see Figure A-58; for vias, see Figures A-59, A-60, and A-61.

A.2.3   *Other Components*

Connectors
Figures A-20 and A-21 include the recommended land patterns for the typical SMT-style connectors they depict. Since nonstandardized connectors may be found in such a variety of outlines, consult the connector manufacturers for exact dimensional data on any given part.

Sockets
Figure A-22 shows a land pattern for a typical DIP socket. Again, see the socket vendor for the exact outline and land geometry information.

Switches
Standard DIP switches with custom surface-mount lead forms are mounted on a land pattern for the appropriate DIP lead form, as shown below. Lands for other package formats should be specified by the switch manufacturer.

DIPs
Figure A-43 shows the IPC-recommended land pattern for DIPs when they are lead prepped for surface mounting. The pattern is intended for the I lead form, but it will also work without modification for the gull-wing lead form. Vias for DIP land patterns are presented in Figure A-65.

Figure A-43
Land patterns for SMT lead-prepped DIPs.

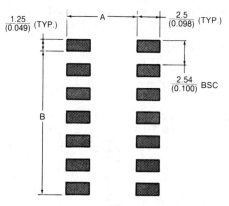

NOTES:
1. Values are in millimeters and inches.
2. Tolerances are ± 0.1 mm/ ± 0.004 inch.
3. For inch control, round inch dimensions to nearest 0.005 inch.
4. For components per Figure A-26. Also used for lead prepped parts, such as shown in Figure A-24.

DIP PACKAGE	DIMENSIONS (mm/inch) A	B
300 Mil, 8 Pin	7.7 (0.303)	7.62 (0.300)
300 Mil, 14 Pin	7.7 (0.303)	15.24 (0.600)
300 Mil, 16 Pin	7.7 (0.303)	17.78 (0.700)
300 Mil, 18 Pin	7.7 (0.303)	20.32 (0.800)
300 Mil, 20 Pin	7.7 (0.303)	22.86 (0.900)
400 Mil, 22 Pin	10.2 (0.402)	25.40 (1.000)
600 Mil, 24 Pin	15.2 (0.598)	27.94 (1.100)
600 Mil, 28 Pin	15.2 (0.598)	33.02 (1.300)

Crystals
Consult the crystal manufacturers for the land recommendations for crystals.

Relays
Consult the manufacturers for the land recommendations for relays.

Jumpers

Chip jumpers, packaged per standard chip-resistor sizes, will fit the same land patterns that the corresponding resistor size would require. See Figure A-31 for IPC standard chip-resistor lands, and Figures A-46 through A-49 for process customized patterns.

A.3 Land-Pattern Variations

The following lands differ from IPC recommendations in order to meet specific application objectives. For reflow-soldered boards, the screened pattern of solder paste should be the same as the land patterns.

A.3.1 SMT Passives

There are several potential difficulties with the formulae for standardized land geometries shown in Section A.2.1.

First, the formulae tie the extension of the lands past the ends of the components to the component thickness. This could result in a very large number of land patterns being required for a number of component-thickness variations. We could easily exceed the size of the aperture library available on typical photoplotting equipment. Of greater consequence, very tall components would require a very long land extension. Excessive extensions are undesirable because they contribute to component misalignment and because they take up valuable board real estate. Experience has shown that a fillet of sufficient strength is formed with a land extension of only about 0.65 mm (0.025 inch). Greater extension has little impact on solder-joint strength. In fact, too much solder can actually detract from solder-joint resistance to thermal cycling or board-flexure-induced stress, as shown in Figures A-44 and A-45. Also, past a certain point, greater extension has little impact on fillet formation.

Figure A-44 View of fillet formation vs. land extension.

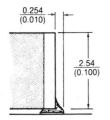 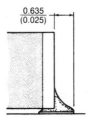 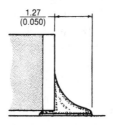 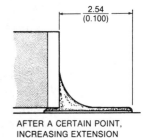

NOTE: Factors in Figures A-44A through A-44D apply to tall components ≥ 0.76 mm/0.030 inch thick.

AFTER A CERTAIN POINT, INCREASING EXTENSION DOES NOT ADD TO FILLET.

(A) Extension insufficient. Not enough solder for a strong joint.

(B) Joint approaching maximum available strength and resistance to cyclic stress.

(C) Joint at maximum strength but less resistant to cyclic stress than Figure A-44B.

(D) Joint no stronger than in Figure A-44C, but extra land extension wastes PWB real estate.

Figure A-45
Land extension
vs. solder-joint
strength.

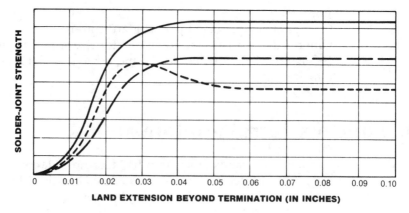

(A) Graph of strength vs. land size.

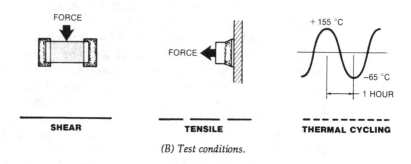

(B) Test conditions.

Note that the data illustrated in Figures A-44 and A-45 apply to chip components, not leaded parts.

Leadless chip carriers derive the bulk of their solder-joint reliability from the fillet outside the component body. Thus a minimum pad extension of 1.27 mm (0.050 inch) is necessary to produce the maximum stress-resistant CLLCC joint.

For J-leaded parts, both the heel and toe of the joint contribute about equally. A pad extension of 0.635 mm (0.025 inch) past the edges of the lead, inside and outside, is sufficient. SOICs and other gull-wing parts derive their solder-joint reliability primarily from the heel of the lead. Thus, the inner extension is critical for gull-wing forms.

Second, as discussed under chip capacitors and resistors later, reflow and wave-soldering process requirements differ. A land optimized for one process will not perform well in the other. A land that compromises between the needs of both processes will not give the consistent high yields and joint reliability that can be delivered by process-optimized lands.

Ceramic Chip Capacitors and Resistors

The standardized lands shown above are compromises between lands dimensioned specifically for reflow soldering and lands optimized for wave soldering.

For reflow soldering, lands should be designed to center the parts, using solder surface-tension forces. Also, it is vital that the lands be sized to minimize tombstoning.

In a study of land size vs. tombstoning defects, solder-paste manufacturer, Indium Corporation of America, found that the gap between the lands should essentially match the space between the terminations (bare body) on the component. Either reducing or increasing the gap from this ideal will increase tombstoning.[11] The reflow-optimized lands shown in Figure A-46, and dimensioned in Table A-18, are constructed along this model, with an allowance for basic manufacturing tolerances.

Figure A-46
Chip capacitor and resistor lands tailored for reflow-soldering manufacturing yields.

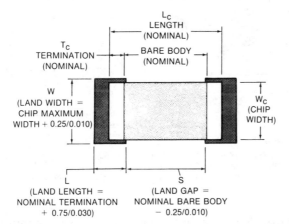

NOTES:
1. Values are in millimeters and inches.
2. All land dimensions calculated from English component dimensions (maximum or nominal as noted), rounded to the nearest 0.005 inch for inch control (or 0.05 mm for metric control).
3. For dimensions, see Table A-18. For component outline, see Figure A-1 for capacitors, and Figure A-5 for chip resistors.

Table A-18.
Standard Chip Lands Tailored for Reflow Soldering

Component Type	Dimensions (mm/inch)		
	Land Width	*Land Length*	*Land Gap*
0805 Capacitor	1.65 / 0.065	1.27 / 0.050	0.89 / 0.035
1206 Capacitor	2.03 / 0.080	1.27 / 0.050	1.78 / 0.070
1210 Capacitor	2.92 / 0.115	1.27 / 0.050	1.78 / 0.070
1812 Capacitor	3.68 / 0.145	1.27 / 0.050	3.18 / 0.125
1825 Capacitor	7.11 / 0.280	1.27 / 0.050	3.18 / 0.125
0805 Resistor	1.65 / 0.065	1.27 / 0.050	0.89 / 0.035
1206 Resistor	2.03 / 0.080	1.27 / 0.050	1.78 / 0.070
1210 Resistor	2.92 / 0.115	1.27 / 0.050	1.78 / 0.070

Note: Inch dimensions control.

Another custom land feature of interest is prelocated vias. The idea here is to include design-rule-conforming vias on each land. These vias are located as close as permitted to the land, while still remaining on grid. We arbitrarily place the component center on the grid, or centered between grids, as required to ensure grid conformance of the predesigned vias. This artifact is intended for CAD libraries. When a particular component is called from the library, its lands and associated vias are placed on the grid for testability. Vias may then be moved to a more remote location, where necessary. Figure A-47 shows automatic 50- and 100-mil-grid vias for reflow-soldered chips.

Figure A-47
Auto vias for reflow-soldered chip components.

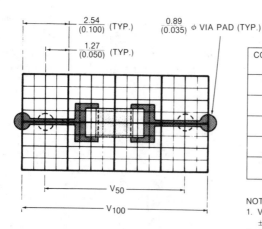

COMPONENT SIZE	DIMENSIONS (mm/inch)	
	V_{50} = 50-Mil Grid	V_{100} = 100-Mil Grid
0805	5.08 (0.200)	5.08 (0.200)
1206	6.35 (0.250)	7.62 (0.300)
1210	6.35 (0.250)	7.62 (0.300)
1812	7.62 (0.300)	7.62 (0.300)
1825	7.62 (0.300)	7.62 (0.300)

RULE (USING TEST GRID):
PLACE COMPONENT CENTER ON GRID OR HALF GRID. LOCATE VIA AT MINIMUM CLEARANCE FROM LAND, AND THEN SNAP OUT TO FIRST AVAILABLE TEST PROBE GRID.

NOTES:
1. Values are in millimeters and inches; tolerances are ± 0.05 mm/ ± 0.002 inch.
2. Assumes photo-imaged SMOBC or bare copper trace-pads-only surface layer; 0.035-inch via pad.
3. For metric control, round metric dimensions to correspond with metric test grid and comply with rules.
4. For lands per Figure A-46, and components per Figure A-1 or Figure A-5.

Flow-soldering defects differ from reflow soldering in several ways. Because parts are glued to the board, they couldn't tombstone if their life depended on it. Instead, flow defects generally involve the amount of solder applied to the joint. Sometimes, it's a feast. Excessive solder on a joint can short circuit to adjacent lands or terminations. This defect is called bridging. At other times, it's a famine. Too little solder (or none at all) results in an insufficient or even absent solder connection between the solder land and the device termination. Defects from insufficient solder are called "solder starved" joints. Missing connections are called skips, opens, or voids.

Using solder lands narrower than the component body will reduce bridging. Such lands would cause serious component misregistration (swimming) in reflow, but glued down parts stay put just fine.

Figure A-48
Chip capacitor and resistor lands tailored for flow-soldering manufacturing yields.

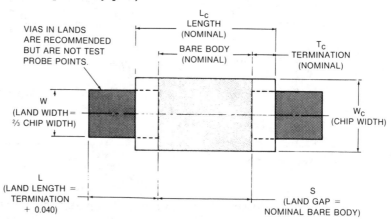

NOTES:
1. All land dimensions calculated from nominal English component dimensions and rounded to the nearest 0.005 inch.
2. For component per Figure A-1 or Figure A-5.

To limit opens, solder lands are extended well past the ends of the component. Thus, even if eddy currents in the molten solder cause a void around the device termination, solder will still contact the extended land. Solder-wetting forces will draw it in to the device termination, forming an acceptable joint. Figure A-48 shows lands dimensioned to suit this flow-soldering model. Tabular dimensions for the drawing are in Table A-19.

Table A-19. Standard Chip Lands Tailored for Flow Soldering.

Component Type	Dimensions (mm/inch)		
	Land Width	Land Length	Land Gap
0805 Capacitor	0.76 / 0.030	1.52 / 0.060	1.14 / 0.040
1206 Capacitor	1.02 / 0.040	1.52 / 0.060	2.03 / 0.080
1210 Capacitor	1.65 / 0.065	1.52 / 0.060	2.03 / 0.080
1812 Capacitor	2.16 / 0.085	1.52 / 0.060	3.43 / 0.135
1825 Capacitor	4.32 / 0.170	1.52 / 0.060	3.43 / 0.135
0805 Resistor	0.76 / 0.030	1.52 / 0.060	1.14 / 0.045
1206 Resistor	1.02 / 0.040	1.52 / 0.060	2.03 / 0.080
1210 Resistor	1.65 / 0.065	1.52 / 0.060	2.03 / 0.080

Table A-20. Dimensions for Automatic CAD-Located VIAs

Component Size	Dimensions* (50-mil Test Grid)		Dimensions* (100-mil Test Grid)	
	OV	LV	OV	LV
0805	5.08 / 0.200	3.81 / 0.150	5.08 / 0.200	—
1206	6.35 / 0.250	5.08 / 0.200	7.62 / 0.300	5.08 / 0.200
1210	6.35 / 0.250	5.08 / 0.200	7.62 / 0.300	5.08 / 0.200
1812	7.62 / 0.300	6.35 / 0.250	7.62 / 0.300	—
1825	7.62 / 0.300	6.35 / 0.250	7.62 / 0.300	—

* Values are in millimeters and inches.

Figure A-49 shows the automatic CAD-located vias for flow-soldered chip capacitors and resistors. Note that the flow-solder process allows, and even encourages, the location of vias much closer than would be usual in reflow soldering. This is discussed under via design in Chapter 5, Section 5. Table A-20 gives the necessary dimensions.

Cylindrical Resistors

The IPC specifications for MELFs show a locating notch centered on the inside of the land. The purpose of the notch is to reduce swimming and roll-off of cylindrical parts during soldering. The specification states that the notch is optional, and that, without the notch, parts may be secured in some other way.

Figure A-49
Auto vias for flow-soldered chips. (For standard chip dimensions, see Figure A-20.)

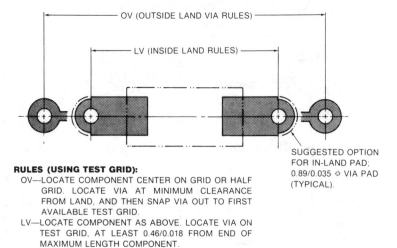

RULES (USING TEST GRID):

OV—LOCATE COMPONENT CENTER ON GRID OR HALF GRID. LOCATE VIA AT MINIMUM CLEARANCE FROM LAND, AND THEN SNAP VIA OUT TO FIRST AVAILABLE TEST GRID.

LV—LOCATE COMPONENT AS ABOVE. LOCATE VIA ON TEST GRID, AT LEAST 0.46/0.018 FROM END OF MAXIMUM LENGTH COMPONENT.

NOTES:

1. Values are in millimeters and inches; tolerances are ±0.05 mm/ ±0.002 inch.
2. Where in-land vias are used, all other design rules re: vias in-lands, must be met.
3. Assumes photo-imaged solder mask or no mask.
4. For metric control, round metric dimensions to correspond with metric test grid and comply with rules.
5. For lands, per Figure A-48, and components, per Figure A-1 or Figure A-5.

Figure A-50
Lands for common nonstandard SOIC resistor network.

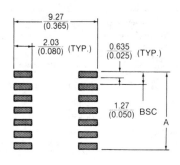

PACKAGE DESIGNATION	DIMENSION "A"
SOM-8	3.81 (0.150)
SOM-14	7.62 (0.300)
SOM-16	8.89 (0.350)

NOTES:

1. For SOIC-like nonstandard packages, per Figure A-8.
2. Values are in millimeters and inches. Tolerances are ±0.1 mm/ ±0.004 inch.
3. For metric control, round the metric dimensions to nearest 0.05 mm.

Some SMT manufacturers do glue or fixture MELFs for reflow. Some find that unrestrained reflow with the notches works well. Other users have reported no motion problem in the unrestrained reflow of parts without a notch. Several of these no-notch advocates have large-volume manufacturing operations, where, if reflow problems were going to occur, they would have been noticed. Finally, there is also a battery of manufacturers who have had trouble with round parts, with or without notches in the lands. The difference is in the process control regarding the parts and board solderability, the surface contour of the board lands, warpage of boards, and the reflow process. Regarding process, vapor-phase reflow is more likely to upset cylindrical parts than is IR reflow.

Figure A-51
SOT-23 tolerance
analysis—a land
design challenge.

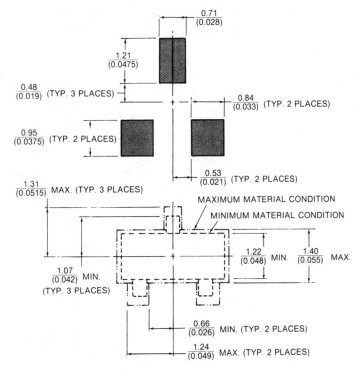

NOTES:
1. Values are in millimeters and inches. Inch dimensions control.
2. Use components covered under SOT-23, shown in Figure A-9.

Resistor Networks

A common nonstandard SOIC-like package, used for resistor networks, was illustrated in Figure A-8. Figure A-50 is a customized SOIC land pattern for this medium body SO.

A.3.2 Actives

Transistors

The SOT-23 land pattern recommended in IPC-SM-782 was intended as a compromise between lands optimized for reflow soldering and those tailored for wave soldering. In fact, it is close to an optimum for wave soldering, and may be a poor choice for Zero-defect reflow. The following discussion of the derivation of a reflow-optimized land pattern may prove useful in guiding nonstandard-component land derivation, as well.

For components where coplanarity problems may exist, we size lands to present enough solder to bridge the gaps produced by nonplanar leads. (Remember, in reflow soldering, your only control of the solder to the joint is in the solder-paste print-area width, length, and solder deposit thickness. Most chip components dictate a wet-paste print thickness of ≤0.010 inch.) Since the SOT-23 is a three-lead device, it always exhibits perfect coplanarity as long as its leads extend past the bottom of its body. Thus, we are free to reduce the land size to the minimums required by the component. Smaller lands conserve board space, and also produce very stress-resistant joints. (Remember, overly stiff joints fracture or break the components when stressed, and too much solder on leaded components stiffens the leads, reducing their flexural ability.)

Figure A-52
SOT-23 reflow-
optimized lands.

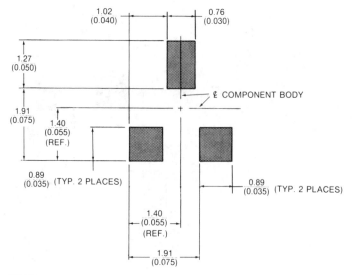

NOTES:
1. Values are in millimeters and inches. Inch dimensions control.
2. Use for components covered under SOT-23 in Figure A-9A.
   See Figure A-53 for standard vias.

Figure A-51 shows a tolerance analysis of the JEDEC dimensions for the SOT-23 at both maximum and minimum material conditions. Since the lead of the SOT-23 is relatively thin, only 0.085 to 0.152 mm (0.003 to 0.006 inch) thick, we know that a land extending at least 0.13 mm (0.005 inch) beyond a component lead will roughly equal the lead height and will produce a strong solder fillet with minimum excess extension. Therefore, we can predict a stress-resistant optimum reflow joint by setting a 0.005-inch extension around the limits of the tolerance pattern in Figure A-51. However, the SOT-23 inherently sees unequal wetting forces during reflow soldering, so we have extended the outer end of the single leg land in Figure A-51 an extra 0.010 inch to help balance forces. In practice, this modification reduces reflow defects common with SOT-23s.

Figure A-52 shows a conversion of the theoretical numbers we established in Figure A-51, changing them to more manageable dimensions. The rounding has been in favor of the convenient inch dimensions, but could just as easily be slanted to favor the convenient metrics where metric dimensions will control.

Figure A-53 shows CAD auto vias for SOTs. These can be used with either the lands shown in Figure A-52 or the IPC-recommended (flow) lands from Figure A-34.

SOT-89s
The SOT-89 center tab is the die flag as well as part of the device termination. The SOT-89 relies on heat-sinking through its center metal tab for a substantial portion of its dissipation. The standard land, with a width of 1.2 mm (0.048 inch), leaves up to 33% of this thermal path unsoldered, thus limiting its effectiveness in dissipation. The graphics of a tolerance analysis was used to provide a full thermal path, and the resulting land pattern is shown in Figure A-54. This alternate pattern is

recommended for any high-dissipation applications of SOT-89s. Figure A-55 shows CAD auto vias for SOT-89 lands.

Figure A-53
CAD auto vias
for SOT-23 and
SOT-143 devices.

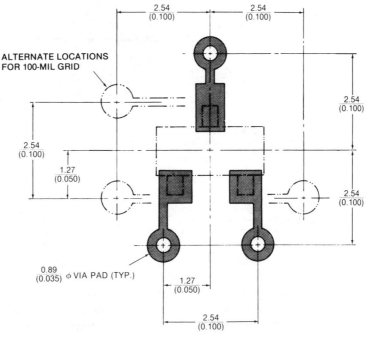

NOTES:
1. Values are in millimeters and inches; tolerances are ± 0.05 mm/ ± 0.002 inch.
2. For metric control, round the metric dimensions to correspond with metric test grid and comply with design rules.
3. Assumes photo-imaged SMOBC or bare copper and pads-only surface layer.
4. Used for lands per Figure A-52.
5. For SOT-143, use 2-via pattern on both sides.

Figure A-54
Lands to maxi-
mize the dissipa-
tion from SOT-
89s.

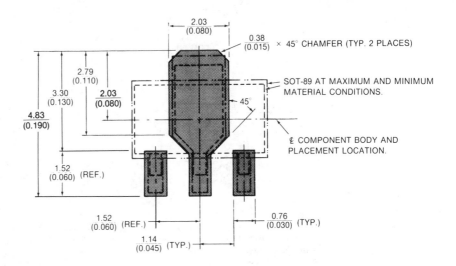

NOTES:
1. Values are in millimeters and inches; tolerances are ± 0.05 mm/ ± 0.002 inch.
2. For metric control, round the metric dimensions to correspond with metric test grid and comply with design rules.
3. Used for components covered under SOT-89 in Figure A-9C. See Figure A-55 for standard via locations.
4. Not recommended for flow soldering because of difficulty in assuring full solder bonding between the center tab of the component and the heak-sink solder land.

Figure A-55
CAD auto vias
for SOT-89
transistors.

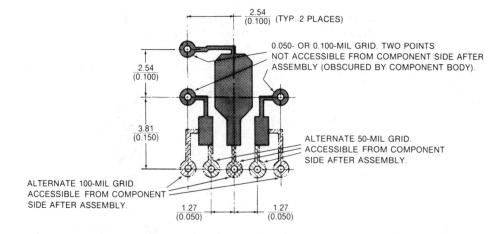

NOTES:
1. Values are in millimeters and inches; tolerances are ± 0.05 mm/ ± 0.002 inch.
2. Assumes photo-imaged SMOBC or bare copper and pads-only surface layer.
3. For metric control, round the metric dimensions to correspond with metric test grid and comply with design rules.
4. Use vias for lands, per Figure A-54.

DPAKs

Figure A-56 shows a land recommendation for the DPAK power transistor and diode package. Similar land patterns, amended for variations in device dimensions, may be used for other SMT parts which are made by lead prepping TO-220 devices and other IMC packages.

Figure A-56
Lands and CAD
auto vias for
DPAK power
transistors.

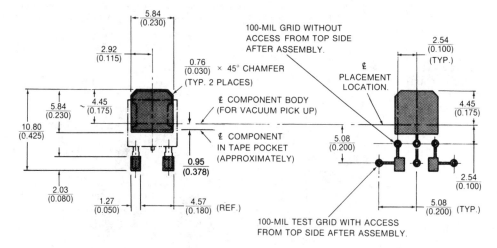

NOTES:
1. Values are in millimeters and inches; tolerances are ± 0.05 mm/ ± 0.002 inch.
2. Assumes photo-imaged SMOBC or bare copper and pads-only surface layer.
3. For metric control, round the metric dimensions to correspond with metric test grid and comply with design rules.
4. Used for components, per Figure A-9D.

SOICs

Figure A-57 shows 50- and 100-mil-grid CAD auto vias for SO and SOL ICs, and gives via recommendations for the nonstandard SOM (medium body) device.

Figure A-57
CAD auto vias
for SO, SOL, and
SOM ICs.

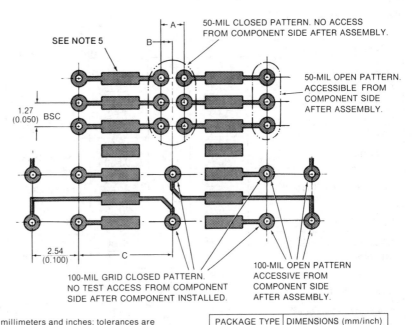

NOTES:
1. Values are in millimeters and inches; tolerances are ± 0.05 mm/ ± 0.002 inch.
2. Assumes photo-imaged SMOBC or bare copper and pads-only surface layer.
3. For metric control, round the metric dimensions to correspond with metric test grid and comply with design rules.
4. Use for SO and SOL ICs per Figure A-10, and on land patterns per Figure A-37. For SOM ICs per Figure A-8, and on lands per Figure A-50.
5. For very-high-density design, lands of 0.5 mm (0.020 inch) wide × 1.9 mm (0.075 inch) long, with rounded ends, may be substituted for IPC standard lands.

PACKAGE TYPE	DIMENSIONS (mm/inch)		
BODY SIZE	A	B	C
SO IC, 150 Mil	1.27 (0.050)	0.635 (0.025)	5.08 (0.200)
SOM IC, 220 Mil	2.54 (0.100)	1.27 (0.050)	7.62 (0.300)
SOL IC, 300 Mil	5.08 (0.200)	2.54 (0.100)	7.62 (0.300)

Figure A-58
Universal lands
for Type-A
PLCCs (pre-
molded and
postmolded).

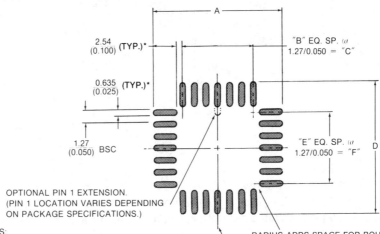

OPTIONAL PIN 1 EXTENSION.
(PIN 1 LOCATION VARIES DEPENDING ON PACKAGE SPECIFICATIONS.)

RADIUS ADDS SPACE FOR ROUTES BETWEEN CORNER LANDS. WHERE RADII ARE DIFFICULT TO PRODUCE, 0.4/0.015 × 45° CHAMFERS MAY BE USED ON CORNER LANDS ONLY.

SYMMETRICAL ABOUT ₵
(EXCEPT OPTIONAL PIN 1 EXTENSION)

NOTES:
1. Values are in millimeters and inches. Tolerances are ± 0.05 mm/ ± 0.002 inch.
2. Assumes photo-imaged SMOBC or bare copper and pads-only surface layer.
3. For metric control, round the metric dimensions to nearest 0.05 mm.
4. May be used for 50-mil CLLCCs per JEDEC MS-002 through MS-005, where PWB space permits. However, this pattern requires more space than lands, per Figure A-41.
5. Used for components, per Figure A-12. Also may be used for postleaded CLLCCs per Figure A-18. For CAD auto vias, see Figure A-59.

* For very-high-density PWBs, use a land of 0.020 × 0.075 inch and adjust A and B dimensions to suit component.

Table A-21. Dimensions for Figure A-58

Package Type(s)	Dimensional Data					
	A	B	C	D	E	F
20 Square (MO-047)	11.43 / 0.450	4	5.08 / 0.200	Same as A	Same as B	Same as C
28 Square (All Type A)	13.97 / 0.550	6	7.62 / 0.300	Same as A	Same as B	Same as C
44 Square (All Type A)	19.05 / 0.750	10	12.70 / 0.500	Same as A	Same as B	Same as C
52 Square (All Type A)	21.59 / 0.850	12	15.24 / 0.600	Same as A	Same as B	Same as C
68 Square (All Type A)	26.67 / 1.050	16	20.32 / 0.800	Same as A	Same as B	Same as C
84 Square (MO-047 andMS-008)	31.75 / 1.250	20	25.40 / 1.000	Same as A	Same as B	Same as C
100 Square (MO-047 and MS-008)	36.83 / 1.450	24	30.48 / 1.200	Same as A	Same as B	Same as C
124 Square (MO-047 and MS-008)	44.45 / 1.750	30	38.10 / 1.500	Same as A	Same as B	Same as C
156 Square (MS-008)	54.61 / 2.150	38	48.26 / 1.900	Same as A	Same as B	Same as C
18 Rectangular (MO-052)	10.16 / 0.400	3	3.81 / 0.150	12.70 / 0.500	4	5.08 / 0.200
18L Rectangular (MO-052)	10.16 / 0.400	3	3.81 / 0.150	14.86 / 0.585	4	5.08 / 0.200
22 Retangular (MO-052)	10.16 / 0.400	3	3.81 / 0.150	14.86 / 0.585	6	7.62 / 0.300
28 Rectangular (MO-052)	11.43 / 0.450	4	5.08 / 0.200	16.51 / 0.650	8	10.16 / 0.400
32 Rectangular (MO-052)	13.97 / 0.550	6	7.62 / 0.300	16.51 / 0.650	8	10.16 / 0.400

PLCCs

Figure A-58 shows a universal land pattern for Type-A PLCCs. Figure A-59 covers 50- and 100-mil-grid CAD auto vias for the JEDEC 50-mil PLCCs and CLLCCs on all appropriate land patterns herein. The dimensions are listed in Tables A-21 and A-22, respectively.

Figure A-60 presents a high-density 50-mil test-grid alternative to the via arrangement shown in Figure A-59. The vias shown in Figure A-60 are not accessible from the component side after assembly.

Table A-22. Dimensions for Figure A-59

Pin Count	Dimension "A"—Various Carrier Outlines				
	Type-A PLCC (Postmolded)	Premolded PLCC	50-mil CLLCC	CLDCC (JEDEC MS-044)	Type-A Universal (See Note 5)
16	N/A	N/A	$\frac{6.35}{0.250}$	N/A	N/A
20	N/A	$\frac{6.35}{0.250}$	$\frac{6.35}{0.250}$	N/A	$\frac{7.62}{0.300}$
24	N/A	N/A	$\frac{7.62}{0.300}$	N/A	N/A
28	$\frac{8.89}{0.350}$	$\frac{7.62}{0.300}$	$\frac{7.62}{0.300}$	N/A	$\frac{8.89}{0.350}$
44	$\frac{11.43}{0.450}$	$\frac{10.16}{0.400}$	$\frac{10.16}{0.400}$	N/A	$\frac{11.43}{0.450}$
52	$\frac{12.70}{0.500}$	$\frac{11.43}{0.450}$	$\frac{11.43}{0.450}$	N/A	$\frac{12.70}{0.500}$
68	$\frac{15.24}{0.600}$	$\frac{13.97}{0.550}$	$\frac{13.97}{0.550}$	$\frac{13.97}{0.550}$	$\frac{15.24}{0.600}$
84	N/A	$\frac{16.51}{0.650}$	$\frac{16.51}{0.650}$	$\frac{16.51}{0.650}$	$\frac{17.78}{0.700}$
100	N/A	$\frac{19.05}{0.750}$	$\frac{19.50}{0.750}$	N/A	$\frac{20.32}{0.800}$
124	N/A	$\frac{22.86}{0.900}$	$\frac{22.86}{0.900}$	N/A	$\frac{24.13}{0.950}$
156	N/A	N/A	$\frac{27.94}{1.100}$	N/A	$\frac{29.21}{1.150}$

Figure A-59
CAD auto vias
for PLCCs.

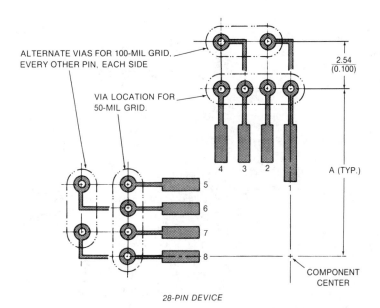

ALTERNATE VIAS FOR 100-MIL GRID, EVERY OTHER PIN, EACH SIDE

VIA LOCATION FOR 50-MIL GRID.

$\frac{2.54}{(0.100)}$

A (TYP.)

COMPONENT CENTER

28-PIN DEVICE

NOTES:
1. Values are in millimeters and inches; tolerances are ± 0.05 mm/ ± 0.002 inch.
2. For metric control, round the metric dimensions to correspond with metric test grid and comply with design rules.
3. Assumes photo-imaged SMOBC or bare copper and pads-only surface layer.
4. Use for components listed in Table A-21, per appropriate IPC recommendation.
5. Use for Type-A universal lands, per Figure A-58. **Dimensions A to suit A or D dimension for rectangular parts.**

Figure A-60
Modified via layout for dense applications. (28-pin PLCC lands are shown—premolded or Type-A postmolded.)

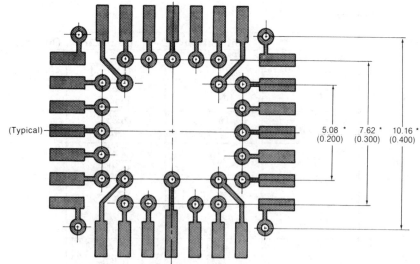

(Typical)

5.08 * (0.200)   7.62 * (0.300)   10.16 * (0.400)

* Dimensions shown for reference. Adjust to suit component pin count and land pattern used.

28-PIN PLCC LANDS

NOTES:
1. Vias are not accessible for in-circuit test from component side.
2. Used for components, per Figure A-12.

Figure A-61 shows another alternative, using filled or capped vias to allow via location within the solder lands. Filled and capped vias are discussed further in Chapter 5, Section 5. This design may be used for trace-length control in high-frequency design, or for space considerations. However, using standard test fixtures, these vias are not component-side accessible after assembly.

Figure A-61
Modified via layout for VHF/UHF design using filled or capped vias.

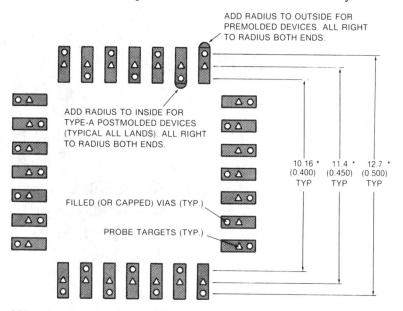

ADD RADIUS TO OUTSIDE FOR PREMOLDED DEVICES. ALL RIGHT TO RADIUS BOTH ENDS.

ADD RADIUS TO INSIDE FOR TYPE-A POSTMOLDED DEVICES (TYPICAL ALL LANDS). ALL RIGHT TO RADIUS BOTH ENDS.

10.16 * (0.400) TYP   11.4 * (0.450) TYP   12.7 * (0.500) TYP

FILLED (OR CAPPED) VIAS (TYP.)

PROBE TARGETS (TYP.)

* Dimensions shown for reference. Adjust to suit component pin count and land pattern used.

NOTES:
1. With appropriate lands, may be used with all styles of chip carriers having 50-mil lead pitch, per Figure A-12.
2. Vias are not accessible from component side after assembly.

CLLCCs

Figure A-62 presents land patterns and 50- and 100-mil-grid CAD auto vias for 40-mil-center CLLCCs. The pertinent dimensions are given in Table A-23.

Figure A-62
Land patterns
and CAD auto
vias for 40-mil
center
CLLCCs.

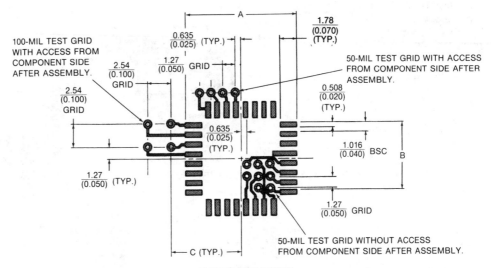

32-PIN DEVICE SHOWN

NOTES:
1. Values are in millimeters and inches; tolerances are ±0.05 mm/±0.002 inch.
2. Assumes photo-imaged SMOBC or bare copper and pads-only surface layer.
3. For metric control, round the metric dimensions to correspond with metric test grid and comply with design rules.
4. Test via pattern varies dependent on pin count. Component X and Y center lines may be on grid or may be centered between grid lines, whichever yields most space-efficient pattern **(if CAD permits).**
5. Features are symmetrical about ₵, except Pin 1 feature and vias.
6. Use for components per Figure A-17, and dimensions per Table A-6 through A-8.

Table A-23.
Dimensional
Data for Figure
A-62

Pin Count	Dimensions		
	A	*B*	*C*
16	$\dfrac{7.24}{0.285}$	$\dfrac{3.05}{0.120}$	$\dfrac{5.08}{0.200}$
20	$\dfrac{9.78}{0.385}$	$\dfrac{4.06}{0.160}$	$\dfrac{6.35}{0.250}$
24	$\dfrac{10.29}{0.405}$	$\dfrac{5.08}{0.200}$	$\dfrac{6.35}{0.250}$
32	$\dfrac{12.07}{0.475}$	$\dfrac{7.11}{0.280}$	$\dfrac{7.62}{0.300}$
40	$\dfrac{13.59}{0.535}$	$\dfrac{9.14}{0.360}$	$\dfrac{7.62}{0.300}$
48	$\dfrac{15.62}{0.615}$	$\dfrac{11.18}{0.440}$	$\dfrac{8.89}{0.350}$
64	$\dfrac{19.69}{0.775}$	$\dfrac{15.24}{0.600}$	$\dfrac{11.43}{0.450}$
84	$\dfrac{24.77}{0.975}$	$\dfrac{20.32}{0.800}$	$\dfrac{13.97}{0.550}$
96	$\dfrac{27.81}{1.095}$	$\dfrac{23.37}{0.920}$	$\dfrac{15.24}{0.600}$

Figure A-63
Land recommen-
dations and
suggested CAD
auto vias for
TapePaks with
20-mil and 25-mil
pitch.

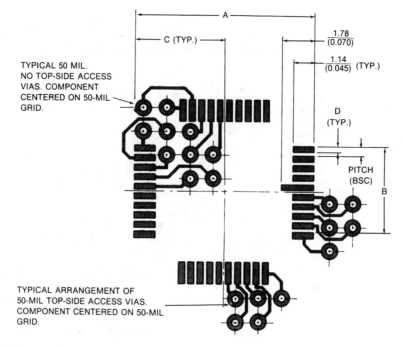

TYPICAL 50 MIL.
NO TOP-SIDE ACCESS
VIAS. COMPONENT
CENTERED ON 50-MIL
GRID.

TYPICAL ARRANGEMENT OF
50-MIL TOP-SIDE ACCESS VIAS.
COMPONENT CENTERED ON 50-MIL
GRID.

NOTES:
1. Values are in millimeters and inches; tolerances are ±0.05 mm/ ±0.002 inch.
2. For metric control, round the metric dimensions to correspond with metric test grid and comply with design rules.
3. Assumes photo-imaged SMOBC or bare copper and pads-only surface layer.
4. Use for components with 20- and 25-mil pitch, per Figure A-13.

Table A-24
Dimensional
Data for Figure
A-63

Pitch	Pin Count	Dimensions (mm/inch)			
		A	B	C	D
0.653 / 0.025	32	9.78 / 0.385	4.57 / 0.180	4.89 / 0.1925	0.38 / 0.015
	52	15.24 / 0.600	7.62 / 0.300	7.62 / 0.300	0.38 / 0.015
	100	23.50 / 0.925	15.24 / 0.600	11.75 / 0.4625	0.38 / 0.015
	168*	—*	26.16* / 1.030	—*	0.38* / 0.015
0.508 / 0.020	40	9.78 / 0.385	4.57 / 0.180	4.89 / 0.1925	0.30 / 0.012
	84	15.24 / 0.600	10.16 / 0.400	7.62 / 0.300	0.30 / 0.012
	132	23.50 / 0.925	16.26 / 0.640	11.75 / 0.4625	0.30 / 0.012
	220*	—*	27.43* / 1.080	—*	0.30* / 0.012

* Proposed future package design. Final
configuration and lead count may change.

Figure A-64
Land recommendations and CAD auto vias for 180-pin TapePaks.

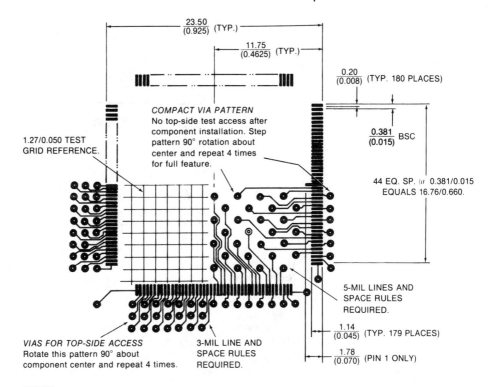

NOTES:
1. Values are in millimeters and inches; tolerances are ± 0.05 mm/ ± 0.002 inch.
2. Assumes photo-imaged SMOBC or bare copper traces and pads-only surface layer.
3. For metric control, impose the metric test grid.
4. Use for 180-pin part per Figure A-13.

TapePaks

Lands and vias for the 20-mil and 25-mil-pitch TapePak are shown in Figure A-63. Lands and vias for 15-mil 180- pin devices are shown in Figure A-64. These are not variants of the IPC-SM-782 rules as these packages were not covered in the March 1987 revision of IPC land-pattern specifications, which were available to us as of the publication of this work. The dimensions shown are the recommendations of National Semiconductor, the developer of the package. These dimensions are the result of both in-house and user research on mounting methods and solder-joint integrity.

A.3.3   Other Components

Switches

CAD auto vias for SMT lead-formed DIP switches are shown in Figure A-65. The same pattern is applicable to I-lead, gull-wing, and J-lead forms.

DIPs

Figure A-65 shows 50- and 100-mil-grid CAD auto vias for SMT lead-formed DIP lands. Table A-25 gives the dimensional data.

Figure A-65
CAD auto vias
for SMT lead-
formed DIPs.

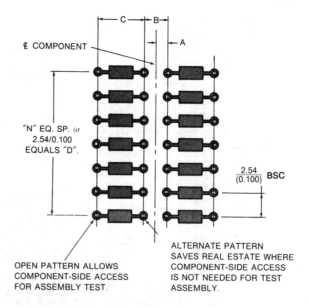

NOTES:
1. Values are in millimeters and inches. Tolerances are ±0.05 mm/±0.002 inch. Inch dimensions control. For metric control, round the metric dimensions to nearest 0.005 mm.
2. May be used with 50- or 100-mil test grids.
3. Use for components per Figure A-26, and similar lead forms, such as Figure A-24. Used for land patterns, per Figure A-43.

Table A-25.
Dimensional
Data for Figure
A-65

Dip Package	Dimensional Data (mm/inch)				
	A	B	C	D	N
300 Mil, 8 Pin	1.27 / 0.050	2.54 / 0.100	5.08 / 0.200	7.62 / 0.300	3
300 Mil, 14 Pin	1.27 / 0.050	2.54 / 0.100	5.08 / 0.200	15.24 / 0.600	6
300 Mil, 16 Pin	1.27 / 0.050	2.54 / 0.100	5.08 / 0.200	17.78 / 0.700	7
300 Mil, 18 Pin	1.27 / 0.050	2.54 / 0.100	5.08 / 0.200	20.32 / 0.800	8
300 Mil, 20 Pin	1.27 / 0.050	2.54 / 0.100	5.08 / 0.200	22.86 / 0.900	9
400 Mil, 22 Pin	2.54 / 0.100	5.08 / 0.200	5.08 / 0.200	25.40 / 1.000	10
600 Mil, 24 Pin	5.08 / 0.200	10.16 / 0.400	5.08 / 0.200	27.94 / 1.100	11
600 Mil, 28 Pin	5.08 / 0.200	10.16 / 0.400	5.08 / 0.200	33.02 / 1.300	13

A.4 References

1. *EIA-PDP-100, Parts Division Publication 100. Registered and Standard Mechanical Outlines for Electronic Parts*, Electronic Industries Association, Washington, DC, 1988.

2. *Surface Mount Technology Component Standardization Proposals—Commercial Grade Components*, SMTA, Edina, MN, 1986.

3. *EIA Specification RS-228, Fixed Electrolytic Tantalum Capacitors*, EIA, Washington, DC.

4. *Surface Mounted Components Catalog and Applications Manual*, Catalog No. 61-07A, MuRata Erie North America, Smyrna, GA, 1987, pg. 7.

5. "Ceramic Dielectric Trimmer Capacitors," *Sprague-Goodman Engineering Bulletin SG-305A*, Sprague-Goodman Electronics, Garden City Park, NY, 1987.

6. *The Trimmer Primer IV—A Look Below the Surface of SMT*, Bourns Trimpot, Riverside, CA, 1986.

7. *Bourns Model 3314 SMD 4mm Sealed Single-Turn Product Bulletin*, Bourns Trimpot, Riverside, CA, 1987.

8. *Trimming Potentiometers*, Catalog No. 61-09, MuRata Erie North America, Smyrna, GA, 1987, pg. 15.

9. Gray, Foster; Prasad, Ray; et al, "Surface Mount Land Patterns (Configurations and Design Rules)," *ANSI/IPC-SM-782*, IPC, Lincolnwood, IL, 1987, pg. 13.

10. Ibid.

11. Dr. Lee, Ning-Cheng, and Evans, Gregory, "Solder Paste: Meeting the SMT Challenge," *Screen Image Technology for Electronics*, June 1987, pg. 36.

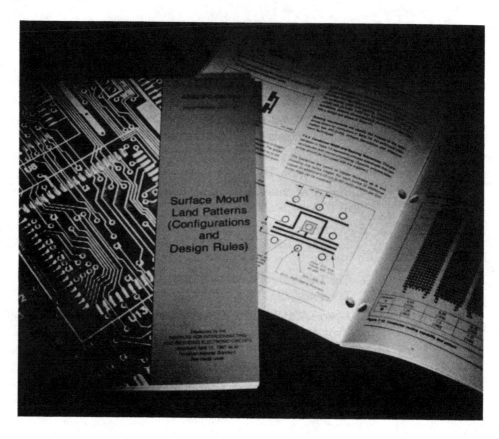

Figure B IPC-SM-782 helps clear the confusion. *(Photograph courtesy IPC.)*

B SMT Standards

Not long ago, there were crowds of professional SMT gainsayers. At the mere mention of SMT, they would throw up their hands and shout "Not till Standards are in place!" Now, SMT Standards work is well along. But are these same chronic critics stilled by the progress? No. They've just shifted gears. Now, when they hear SMT, they throw up their hands and shout, "Not till there's some order in this confusing batch of standards!" Well, the Standards document shown in Figure B should win a Nobel prize for reducing confusion. There is still work to be done. I hope this book and others like it are able to contribute. This chapter will show that needed standards are in place. I don't know if we'll ever win over the died-in-the-wool gainsayer.

We've cataloged some of the critical standards in this appendix, but for up-to-the-minute information, contact the organizations shown in Table B-1.

Table B-1.
Sources of Infor-
mation on Stan-
dards

Organization	Address	Phone
American National Standards Institute (ANSI)	1430 Broadway, New York, NY 10018	(212) 354-3300
ASTM	1916 Race Street, Philadelphia, PA 19103	(215) 299-5599
British Standards Institute (BSI)	2 Park Street, London W1A 12BS, UK	(441) 629-9000
Canadian Standards Association	178 Rexsdale Blvd., Rexsdale, Ontario, CANADA	(416) 747-4000
Defense General Supply Center (Department of Defense)	1. DGSC-SSC, Richmond, VA 23297-5000	(804) 275-3900
	2. Naval Publications and Forms Center, 5801 Tabor Avenue, Philadelphia, PA 19120	(215) 697-2000
Electronic Overstress and ESD Association	P.O. Box 298,Westmoreland, NY 13490	
EIA and JEDEC	2001 Eye Street N.W., Washington, DC 20006	(202) 457-4900
EIAJ (Japan)	2-2, Marunouchi 3 Chome, Chiyoda-Ku, Tokyo 100, JAPAN	(03) 213-1071
IEC	P.O. Box 131, 1211 Geneva 20, SWITZERLAND	(022) 340150
IEEE	345 East 47th Street, New York, NY 10017	(212) 705-7900
IPC	7380 N. Lincoln Avenue, Lincolnwood, IL 60646	(312) 677-2850
SMEMA	71 West Street, Medfield, MA 02052	(617) 359-7928
SMTA	5200 Willson Road, Suite 107, Edina, MN 55424	(612) 920-SMTA

B.1 American Society of Testing and Materials (ASTM)

Work is underway on Tape Automated Bonding Standards. Contact the ASTM for their current catalog of testing standards.

B.2 Electronics Overstress and ESD Association

No information was available as of this writing.

B.3 Electronics Industries Association (EIA)

The EIA and EIAJ (Japan's Electronics Industries Association) are working to bring the U.S. and Japanese component-packaging programs together under a single standard. The two groups have an agreement to circulate new standards proposals between groups for comment before any adoption. This activity should lead to world standards for new packaging, but the resistance to change of existing packages makes any retroactive action less likely.[1]

Three major EIA component standard works contain the bulk of data needed by SMT component users. These are on Components, Assembly, and Quality. For a detailed listing of the many individual offerings, contact EIA.

B.3.1 Components

Specification Number	Subject Matter
EIA-JEDEC-95	Outlines for Semiconductor Devices
EIA-PDP-100	Outlines for Passive Devices
EIA-RS-481A	Taping of SMCs for Auto Placement

B.3.2 Assembly

Specification Number	Subject Matter
EIA-CB-11	Guidelines for Surface Mount of MLC Capacitors

B.3.3 Quality

Specification Number	Subject Matter
EIA-IS-46	Test Procedure for SMC Resistance to VPS

B.4 Electronics Industries Association of Japan (EIAJ)

Please note the comments about any future joint activity, as listed under the EIA above. EIAJ has published close to 500 technical standards, and through cooperation with IEC, EIAJ is working on international standards.

B.5 International Electrotechnical Committee (IEC)

The IEC has a wide range of published standards covering all facets of electronic assembly—from components through processes and QC/test issues. The organization also publishes the *IEC Bulletin,* which reports on current activity. Contact IEC for further information on either the standards or the *IEC Bulletin.*

B.6 Institute for Packaging and Production of Printed Circuits (IPC)

The following is a listing of current IPC standards.

B.6.1 Assembly

Specification Number	Subject Matter
ANSI/IPC-BP-421	General Specifications for Rigid PWB Backplanes with Press Fit Contacts
ANSI/IPC-CM-770C	Printed Board Component Mounting Guidelines
ANSI/IPC-SM-782	SM Land Patterns (Configuration and Design Rules)
IPC-CM-78	Guidelines for SM and Interconnecting Chip Carriers

B.6.2 Cleaning

Specification Number	Subject Matter
ANSI/IPC-SC-60	Post Solder Solvent Cleaning Handbook
ANSI/IPC-SC-62	Post Solder Aqueous Cleaning Handbook

B.6.3 Coatings

Specification Number	Subject Matter
IPC-CC-830	Quality and Performance of Electrical Insulating Compounds for PWAs (Conformal Coating)

B.6.4 Design

Specification Number	Subject Matter
ANSI/IPC-SM-782	SM Land Patterns (Configuration and Design Rules)
IPC-D-249	Design Standard for Flex Single- and Double-Sided PWBs

IPC-D-319	Design Standard for Rigid 1- and 2-Sided PWBs
IPC-D-330	Printed Wiring Design Guide
IPC-MC-323	Design Standard for Metal-Core PWBs

B.6.5 General

Specification Number	Subject Matter
IPC-T-50	Terms and Definitions

B.6.6 Quality Control

Specification Number	Subject Matter
ANSI/IPC-A-610	Acceptability of Printed Board Assemblies
IPC-AI-641	User's Guide for Auto Solder Joint Inspection System
IPC-R-700	Modification and Repair of PWBs and PWAs

B.6.7 Soldering

Specification Number	Subject Matter
ANSI/IPC-S-804	Solderability and Test Methods for PWBs
ANSI/IPC-S-805	Solderability Tests for Component Leads and Terminations
ANSI/IPC-S-815	General Requirements for Soldering Electrical Interconnections
IPC-SF-818	General Requirements for Electronic Soldering Fluxes
IPC-SP-819	General Requirements for Electronic-Grade Solder Paste

B.7 Joint Electronic Device Engineering Council (JEDEC)

JEDEC standards are too numerous to list for each and every individual package. The EIA/JEDEC components' blanket standards, which are listed, are two important starting points in component outline data. They are also listed above under EIA, the parent organization of JEDEC.

B.7.1 Components

Specification Number	Subject Matter
EIA-JEDEC-95	Outlines for Semiconductor Devices
EIA-PDP-100	Outlines for Passive Devices

B.8 Military and DoD Specifications and Standards

The following Standards, Specifications, etc., are presented in alphabetical order by subject matter. Each specification has been assigned a relevance level from A to D. Level-A specifications contain specific mention of SMT and are an essential part of a specifications library for military SMT design. Many commercial users also turn to MIL SPECS for guidance. Level-B specifications are basic electronics design or manufacturing specifications which should form a part of either through-hole or SMT libraries. Level-C specifications cover specific topics which may be of interest in certain applications. Level-D specifications are nonmilitary specifications which are not necessarily listed elsewhere herein, but which are of potential interest to both military and commerical electronics manufacturers. Certain issues of some D-level specifications are DoD-adopted, as listed in DoDISS, and are specified by listed MIL SPECS and Federal specifications and standards.

B.8.1 Cleaning

Subject Matter	Specification Number	Level
Alcohol, Isopropyl	F/S TT-I-735a	B
Flux Residue, Verification and Inspection	MIL-P-28809	B
Hazardous Air Contaminants, Personnel Exposure Limits	BUMED 6270.3	B
Trichloroethane, 1,1,1, Inhibited (Methyl Chloroform) Vapor Degreasing	MIL-T-81533A	B
Trichloroethane, 1,1,1, Inhibited (Methyl Chloroform)	F/S O-T-620C	C
Trichlorotrifluoroethane Solvent Cleaning Compound	MIL-C-81302C	C

B.8.2 *Components*

Subject Matter	Specification Number	Level
Capacitor Chip, Multilayer, Unencapsulated	MIL-C-55681B	A
Capacitor, Packaging of	MIL-C-39028	B
Capacitors, Tantalum Electrolytic	MIL-C-39003	B
Clips, Edge, Hybrids and Chip Carrier, Dimensional Characteristics	EIA-STD-RS507	A & D
Connectors, Electrical Modular	MIL-C-28754	B
Designations, Electronics Parts and Equipment Reference	ANSI-Y-32.16 & IEEE-STD-200	D
Diodes and Transistors	MIL-C-15500	A
ESD Continuing Program Protection, Electrical Parts Assemblies and Equipment	DoD-STD-1686	B
Leads, Electrical Components	MIL-STD-1276C	C
Marking, Electrical and Electronic Parts	MIL-STD-1285	B
Marking, Identity of U.S. Military Property	MIL-STD-130	B
Marking, Shipment and Storage	MIL-STD-129	B
Packing and Packaging of Parts and Equipment, Procedures for	MIL-STD-794	B
Parts, Materials and Processes for Electronic Equipment	MIL-P-11268	B
Parts, Materials and Processes for Space and Launch Vehicles	MIL-STD-1547	B
Preservation, Methods of	MIL-P-116	B
Resistors, Chip, Solderability Requirements	MIL-R-55342C	A
Resistors, Fixed, Composite and Film	MIL-R-39008	B
Resistors, Preparation for Delivery of	MIL-R-39032	B
Semiconductor Devices, Specifications, General	MIL-S-19500	B
Semiconductor Devices, Standard, Lists	MIL-STD-701	B
Semiconductor Devices, Test Methods	MIL-STD-750C	B
Test Methods, Electronic and Electrical Component Parts	MIL-STD-202	B
Transistors and Diodes	MIL-C-15500	A

B.8.3 *Design Requirements*

Subject Matter	Specification Number	Level
Chip Carriers, Guidelines for Surface Mounting	IPC-STD-CM-78	A & D
Communications and Navigation Equipment, Interior, Electronic, Shipboard	MIL-I-983	C
Communications and Navigation Equipment, Interior, Electronic, Shipboard	MIL-E-16400	C
(Dendritic) Metal Growth, Avoidance of	IPC-TR-476	D
Design Specifications, Thick Film Multilayer Interconnection Board	NAVSEA 6228508	A & C
Electronic Equipment, Aerospace, General Requirements for	MIL-E-8983	C
Electronic Equipment, Airborne, General Requirements for	MIL-E-5400	C
Electronic Equipment, Fire Control, Navy, General Requirements for	MIL-F-18870	C
Electronic Equipment, Ground, General Requirements for	MIL-E-4158	C
Electronic Equipment, Surface Guided Missile, General Requirements	MIL-E-11991	C
Equipment Parts, Electronic, Selected Standards	MIL-STD-242	B
Equipment Requirements Standard General	MIL-STD-454	A
Interchangeability, Definitions, Models, Levels, Terms	MIL-STD-280	B
Microcircuits, General Specifications for	MIL-M-38510	C
Modules, Design Requirements for Standard	MIL-STD-1389C	B
Modules, Electronics, General Specifications for	NAVSEA 5516562	A
Modules, General Specifications for	MIL-M-28787	B
Modules, Standard Electrical, Requirements for Employing	MIL-STD-1378	B
PWBs, Flexible and Rigid Flexible, Electrical Design Requirements	MIL-STD-2118	A & C
PWBs for Electrical Equipment	MIL-STD-275E	A
Radiation, Gamma, Sealed Sources, Protection Against	NBS HDBK 73	D

Radiation, Ionizing, External Source, Permissible Dose	NBS HDBK 59	D
Radiation, Medical X-Ray, Protection, to 3 MegaVolts	NBS HDBK 76	D

NOTE: The following important additions to the Design Requirement specifications was released in June, 1987.

Soldering Techniques, High Quality and High Reliability	DoD-STD-2000-1	A
Solder Assembly, Component Mounting for High Quality and High Reliability	DoD-STD-2000-2	A
Soldering Techniques, Criteria for High Quality and High Reliability	DoD-STD-2000-3	A
Soldering Requirements, General Purpose for Electrical Equipment	DoD-STD-2000-4	B
Terms and Definitions for Interconnecting and Packaging Electrical Circuits	DoD-STD-2000-5	A

B.8.4 *Documentation*

Subject Matter	Specification Number	Level
Abbreviations for Use on Drawings, Specifications, and Standards	MIL-STD-12	B
CAD Preparation of Product Definition Data (Including Drawings)	ANSI-Y-14.26.3	D
Certification Boards, Flex and Rigid-Flex, Master Drawings	IPC-B-29/50884	D
Classified Information, DoD Industries Manual for Safeguarding Security	DoD-5220.22-M	B
Designing Electronics Parts and Equipment Reference (IEEE-STD-200)	ANSI-Y-32.16	D
Diagrams, Electrical and Electronic	ANSI-Y-14.15	D
Diagrams, Interconnection	ANSI-Y-14.15a & b	D
Digital Description of PWBs	IPC-D-350	D
Dimensioning and Tolerances for Engineering Drawings	ANSI-Y-14.5	D
Drawing, Engineering and Associated Lists	DoD-D-1000	B
Drawing Practices, Engineering	DoD-STD-100C	B
Drawing Sheet Size and Format, Engineering Drawings	ANSI-Y-14.1	D

Subject Matter	Specification Number	Level
Drawings, Undimensional, Reproducibles	MIL-D-8510B	B
Item Name Directory, Federal	HANDBOOK H6	B
Manufacturers, Federal Supply Code for	HDBK H4-2	B
Marking, Identification of U.S. Military Property	MIL-STD-130	B
Marking for Shipment and Storage	MIL-STD-129	B
Metric Practice Guide	ANSI-Z-210.1 & ASTM E380	D
Metric System, Application in New Design	DoD-STD-1476	B
Reference Designations for Electrical and Electronic Parts	ANSI-Y-32.14 & IEEE STD 200	D
Repair Parts, Accessories, Kits, Mechanical Packaging of	MIL-STD-196	B
Specifications, Military, Preparation of	MIL-STD-961	B
Symbols, Electrical Wiring Equipment, Ship's Wiring	MIL-STD-15-2	C
Symbols, Graphic, for Logic Diagrams	MIL-STD-806	B
Symbols, Graphic, for Logic Diagrams (Two-state Development)	ANSI-Y-32.14 & IEEE STD 91	D
Symbols, Graphic, for Electrical and Electronic Diagrams	ANSI-Y-32.2 & IEEE STD 315	D
Symbols, Graphic, for Electronic Diagrams	ANSI-Y-32.16 & IEEE STD 200	D
Symbols, Letter, for Scientific Units	ANSI/IEEE STD 260-1978	D
Symbols, Letter, Quantities in Electrical, Science, and Engineering	ANSI-Y-10.5 & IEEE STD 280	D
Symbols, Mechanical	MIL-STD-17-1	B
Symbols, Mechanical, Aeronautical, and Spacecraft	MIL-STD-17-2	C
Terms and Definitions, Electronic Circuits, Packaging and Interconnection	ANSI/IPC-T-50C	D
Terms and Definitions, Microelectronics	MIL-STD-1313	C
Terms and Definitions, Printed Wiring and PCBs	MIL-STD-429	B
Terms and Definitions, Welding	AWS A3.0	D

B.8.5 *Hybrids*

Subject Matter	Specification Number	Level
Board, Thick Film Multilayer Interconnection, Design Specifications	NAVSEA 6228508	A & C
Board, Thick Film Multilayer Interconnection, Performance Specifications	NAVSEA 6228507	A & C
Clips, Edge, Hybrids and Chip Carrier, Dimensional Characteristics	EIA-STD-RS507	D
Methods, Test, Microelectronics	MIL-STD-883	C
Quality Assurance, Custom Hybrid Manufacturing Line Process Certification	MIL-STD-1772	A & C
Specifications, General, for Microcircuits	MIL-M-38510	C
Specification Guidelines, Standard, Hybrid Microelectronics	ISHM SPA 001	D
Specifications Parameters to be Controlled, Microelectronics	MIL-STD-1331	C
Terms and Definitions, Microelectronics	MIL-STD-1313	C

B.8.6 *Manufacturing*

Subject Matter	Specification Number	Level
Chip Carriers, Guidelines for Surface Mounting	IPC-STD-CM-78	A & D
Equipment Requirements, Standard General	MIL-STD-454	B
Hazardous Air Contaminants, Personnel Exposure Limits	BUMED 6270.3	B
Marking, Electrical and Electronic Parts	MIL-STD-1285	B
Marking, Identification of U.S. Military Property	MIL-STD-130	B
Marking, Shipment and Storage	MIL-STD-129	B
Microcircuits, General Specifications for	MIL-M-38510	C
Modules, Design Requirements for Standard	MIL-STD-1389	B
Modules, Electronic, General Specifications for	NAVSEA 5516562	A
Modules, General Specifications for	MIL-M-28787	B

Subject Matter	Specification Number	Level
Packing and Packaging of Parts and Equipment, Procedures for	MIL-STD-794	B
Parts, Materials and Processes for Electronic Equipment	MIL-P-11268	B
Preservation, Methods of	MIL-P-116	B
Process Specifications and Procedures for Preparing and Soldering Electronics	NAVY WS-6536E	C

NOTE: *The following important additions to the Manufacturing specifications were released in June, 1987.*

Subject Matter	Specification Number	Level
Soldering Techniques, High Quality and High Reliability	DoD-STD-2000-1	A
Solder Assembly, Component Mounting for High Quality and High Reliability	DoD-STD-2000-2	A
Soldering Techniques, Criteria for High Quality and High Reliability	DoD-STD-2000-3	A
Soldering Requirements, General Purpose for Electrical Equipment	DoD-STD-2000-4	B
Terms and Definitions for Interconnecting and Packaging Electrical Circuits	DoD-STD-2000-5	A

B.8.7 *Materials*

Subject Matter	Specification Number	Level
Alcohol, Isopropyl	F/S TT-I-735a	B
Aluminum Alloy	F/S QQ-A-250	C
Boxes, Fiberboard, Wood-Cleated	F/S PPP-B-591	C
Boxes, Folding, Paperboard	F/S PPP-B-566	C
Boxes, Wood, Cleated, Plywood	F/S PPP-B-601	C
Boxes, Wood, Nailed and Lock-Corner	F/S PPP-B-621	C
Boxes, Fiberboard	F/S PPP-B-636	C
Brass, Leaded and Nonleaded	F/S-QQ-B-626	C
Bronze, Phosphor	F/S-QQ-B-750	C
Coatings, Inorganic and Surface Treatment, Metal Surfaces	MIL-S-5002	B
Coating, Insulating, PWBs	MIL-I-46058C	B
Copper Flat Products	F/S-QQ-C-576	B
Film, Sensitized, Diazotype, Roll and Sheet	F/S-L-F-340	C

Subject Matter	Specification Number	Level
Flux, Soldering, Liquid (Resin Paste)	MIL-F-14256D	B
Hazardous Air Contaminants, Personnel Exposure Limits	BUMED 6270.3	B
Nickel Bar, Flat Wire, and Strip	MIL-N-46025	C
Nickel, Electrodeposited Plating	F/S QQ-N-290A	C
Nickel, Electroless, Military Specifications for Coatings	MIL-C-26074C	C
Nickel Rod and Wire, Round, for Electronic Use	MIL-N-46026	C
Plastic Sheet, Laminated, Metal Clad, for PWBs	MIL-P-13949F	B
Resin, Phenolic Laminating	MIL-R-9299C	B
Solder, Tin, Tin-Lead, and Lead Alloy	F/S QQ-S-571E	B
Steel Bar, Alloy, Hot Rolled and Cold Finished	MIL-QQ-S-624	B
Treatment, Surface and Inorganic Coatings, Metal Surfaces	MIL-S-5002	B
Trichloroethane, 1,1,1, Inhibited (Methyl Chloroform)	MIL-T-81533A	B
Trichloroethane, 1,1,1, Inhibited (Methyl Chloroform)	F/S O-T-620C	C
Trichlorotrifluoroethane Solvent Cleaning Compound	MIL-C-81302C	C

B.8.8 *Plating*

Subject Matter	Specification Number	Level
Copper Plating	MIL-C-14550	B
Gold, Electrodeposited Plating, Military Specifications for	MIL-G-45204	C
Hazardous Air Contaminants, Personnel Exposure Limits	BUMED 6270.3	C
Nickel, Electrodeposited Plating	F/S QQ-N-290A	C
Nickel, Electroless, Military Specifications for Coatings	MIL-C-26074C	C
Tin Plating, Electrodeposited or Hot Dipped	MIL-T-10727	B
Tin-Lead, Electrodeposited Plating	MIL-P-81728	B

B.8.9   Printed-Wiring Boards

Subject Matter	Specification Number	Level
Board, Thick Film Multilayer Interconnecting Design Specifications	NAVSEA 6228508	A & C
Board, Thick Film Multilayer Interconnecting, Performance Specifications	NAVSEA 6228507	A & C
Certification Boards, Flex and Rigid-Flex, Master Drawings	IPC-B-29/50884	D
Coating, Insulating, PWBs	MIL-I-46058C	B
Digital Description of PWBs	IPC-D-350	D
Flexible and Rigid Flexible PCBs	MIL-P-50884C	D
Hazardous Air Contaminants, Personnel Exposure Limits	BUMED 6270.3	B
Multilayer PWBs for Electrical Equipment	MIL-STD-1495	B
Plastic Sheet, Laminated, Metal Clad, for PWBs	MIL-P-13949F	B
Preservation, Methods of	MIL-P-116	B
PWBs, Flexible and Rigid Flexible, Electrical Design Requirements	MIL-STD-2118	A & C
PWBs for Electrical Equipment	MIL-STD-275E	A
Resin, Phenolic Laminating	MIL-R-9299C	B
Specification, General, PWBs	MIL-P-55110D	B

B.8.10   Quality Assurance

Subject Matter	Specification Number	Level
Calibration System Requirements	MIL-STD-45662	B
(Dendritic) Metal Growth, Avoidance of	IPC-TR-476	D
ESD Control Program, Protection, Electrical Parts Assemblies and Equipment	DoD-STD-1686	B
Flux Residue, Verification and Inspection	MIL-P-28809	B
Inspection by Attribute, Sampling Procedures and Tables for	MIL-STD-105D	B
Inspection System Requirements	MIL-I-45208A	B
Methods, Test, Environmental	MIL-STD-810	B

Subject Matter	Specification Number	Level
Process Specifications and Procedures for Preparing and Soldering Electronics	NAVY WS-6536E	C
Quality Program Requirements	MIL-Q-9858	B
Radiation, Gamma, Sealed Sources, Protection Against	NBS HDBK 73	D
Radiation, Ionizing, External Source, Permissible Dose	NBS HDBK 59	D
Radiation, Medical X-Ray, Protection, to 3 MegaVolts,	NBS HDBK 76	D
Reliability Prediction of Electronic Equipment	MIL-STD-217	B

NOTE: The following important additions to the Quality Assurance specifications were released in June, 1987.

Subject Matter	Specification Number	Level
Soldering Techniques, High Quality and High Reliability	DoD-STD-2000-1	A
Solder Assembly, Component Mounting for High Quality and High Reliability	DoD-STD-2000-2	A
Soldering Techniques, Criteria for High Quality and High Reliability	DoD-STD-2000-3	A
Soldering Requirements, General Purpose for Electrical Equipment	DoD-STD-2000-4	B
Terms and Definitions for Interconnecting and Packaging Electrical Circuits	DoD-STD-2000-5	A

B.8.11   Soldering

Subject Matter	Specification Number	Level
Flux Residue, Verification and Inspection	MIL-P-28809	B
Flux, Soldering, Liquid (Resin Paste)	MIL-F-14256	B
Hazardous Air Contaminants, Personnel Exposure Limits	BUMED 6270.3	B
Process Specifications and Procedures for Preparing and Soldering Electronics	NAVY WS-6536E	C
Solderability Requirements, Chip Capacitors	MIL-C-55681B	A
Solderability Requirements, Chip Resistors	MIL-R-55342C	A

Subject Matter	Specification Number	Level
Solder, Tin, Tin-Lead, and Lead Alloy	F/S QQ-S-571E	B

NOTE: *The following important additions to the Soldering specifications were released in June, 1987.*

Subject Matter	Specification Number	Level
Soldering Techniques, High Quality and High Reliability	DoD-STD-2000-1	A
Solder Assembly, Component Mounting for High Quality and High Reliability	DoD-STD-2000-2	A
Soldering Techniques, Criteria for High Quality and High Reliability	DoD-STD-2000-3	A
Soldering Requirements, General Purpose for Electrical Equipment	DoD-STD-2000-4	B
Terms and Definitions for Interconnecting and Packaging Electrical Circuits	DoD-STD-2000-5	A

B.8.12   Testing

Subject Matter	Specification Number	Level
Calibration System Requirements	MIL-C-45662	B
Inspection by Attribute, Sampling Procedures and Tables for	MIL-D-105D	B
Methods, Test, Electronic and Electrical Components	MIL-STD-202	B
Methods, Test, Environmental	MIL-STD-810	B
Methods, Test, Microelectronics	MIL-STD-883	C
Methods, Test, Semiconductor Devices	MIL-STD-750C	B
SMT, Design for Test Access Guidelines	SMTA Test Guide	D
Test Equipment, Electronic and Electrical, General Specifications for	MIL-T-28880	B

B.9   Surface Mount Equipment Manufacturer's Association (SMEMA)

SMEMA does not set standards, but has issued guidelines for interface-ability of manufacturing equipment. These guidelines have been adopted by equipment manufacturers who are member companies of SMEMA. Thus, SMEMA equipment will interface electrically and mechanically in a straightforward manner for a pass through the manufacturing line.

B.10 Surface Mount Technology Association (SMTA)

The SMTA does not set standards, but has issued guidelines on the following topics.

B.10.1 Components

Specification Number	Subject Matter
Component Guide	SMT Preferred Packages for Commercial Design

B.10.2 Testing

Specification Number	Subject Matter
Testability Design Recommendations	SMT Design Guidelines for In-circuit Probe-ability

B.11 Reference

"Two EIAs Talk Standardization," *Electronic Buyers' News*, March 28, 1988, pg. 19.

Figure C. The surface-mount directory. *(Courtesy Info-Mation, New Britain, PA.)*

C SMT Bibliography

Happily, there are a few items to mention in this chapter. Had this book been written several years ago, the "additional reading" cupboard would have been nearly bare. But one major published work, which served to assist SMT designers, was around for years before this writing. The *Surface Mount Directory*, seen in Figure C, still is Number One in size and thoroughness of coverage of its subject matter. Details on the directory are presented in Section C.3.1.

This appendix is intended to both supplement the references listed in the body of the book, and to simplify accessing those referenced works which are pertinent and likely to be readily available. In Section C.1, we'll list nonprofit organizations which may be able to provide information on surface mount. Section C.2 will cover the more readily available referenced works. Section C.3 will list additional information resources not referenced herein.

C.1 Information Resources

In Table C-1 are listed the addresses and phone numbers for some of the nonprofit information resources. The U.S. Navy EMPF is a research facility with a focus on improving process yields for surface mounting and IMC assembly. It offers help to both defense and nondefense industries. The Electronic Overstress and ESD Association is concerned with increasing the knowledge of EOS/ESD problems and solutions. IEPS stands for International Electronics Packaging Society, a technical society focused as its name implies. ISHM–I/SMT is the SMT branch of the International Society for Hybrid Microelectronics. MEPPE is the Micro Electronic Packaging and Processing Engineering Group. NTCμ stands for National Training Center for Microelectronics. Operating from the Northampton County Area

Community College in Bethlehem, PA, NTCµ organizes courses and seminars on SMT. The Surface Mount Club is a British engineering organization interested in SMT. The Surface Mount Council was established jointly by the EIA and IPC to guide standardization of surface mounting. SMEMA, the Surface Mount Equipment Manufacturer's Association, is chartered to promote standard interfaces, which allow equipment from disparate vendors to be connected in-line. SMTA is the Surface Mount Technology Association, a technical society focusing on the information needs of SMT users. STACK is a German-based association with its U.S. headquarters located in Minneapolis, MN. The acronym is from *STAndard Computer Komponenten Gmbh.*

Table C-1.
Sources of Information on SMT

Organization	Address	Phone
Electronic Manufacturing Productivity Center	1417 N. Norma Street, Ridgecrest, CA 93555	(619) 446-7706
ElectronicOverstress and ESD Association	P.O. Box 298, Westmoreland, NY 13490	
IEPS	114 North Hale Street, Wheaton, IL 60187	(312) 260-1044
I^2MT (International Institute for Manufacturing	P.O. Box 549, MCV Station, Richmond, VA 23298-0001	(804) 786-6582
ISHM–I/SMT	P.O. Box 2698, Reston, VA 22090	(703) 471-0066
MEPPE	999 Commercial Street, #12, Palo Alto, CA 94303	(415) 493-6444
NTCµ	3835 Green Pond Road, Bethlehem, PA 18017	(215) 861-5486
SMC (Surface Mount Club)	British Overseas Trade Board, 1 Victoria Street, London SW1H OET	
SMC (Surface Mount Council)	Contact through EIA or IPC. (See Appendix B, Table B-1 for the addresses.)	(312) 677-2850
		(202) 457-4981
SMEMA	71 West Street, Medfield, MA 02052	(617) 359-7928
SMTA	5200 Willson Road, Suite 107, Edina, MN 55424	(612) 920-SMTA
STACK	5775 Wayzatta Blvd., #700, Minneapolis, MN 55416	(612) 593-0315

C.2 Referenced Works

At the end of each chapter, there is a complete listing of referenced works for that chapter. Below are items chosen from among the referenced works. These are listed because they have been particularly useful in our library, are readily available, and are reasonably current as of this writing. We have listed very few of the referenced magazine articles. Periodical references were limited not because of quality, but because they may not be widely available.

Together with the specifications referenced in Appendix B, the entries below provide a shopping list for the construction of a strong SMT background library.

Add this data to the periodicals and basic reference works listed herein and you will have a comprehensive data bank comparable to the best available today.

The material is separated into seven categories: component data, design information, SMT manufacturing issues, market reports, SMT overviews, soldering technology, and test and QA data. Under each heading, we have listed sources from books, commercial pamphlets and application notes, magazine articles, specifications and standards, and technical papers from published conference proceedings separately, to simplify accessing them.

C.2.1 Component Data

The following is a collection of information resources covering specific issues and general selection criteria for surface mount components.

Commercial Pamphlets and Application Notes
Consult the following for additional reading on components.

Chapter	Reference	Publication
5	4*	*Application and Performance of Gigabit 40 I/O Carriers*, GigaBit Logic, Inc., Newbury Park, CA, 1986.
5	5*	*Guidelines for the Use of Digital GaAs ICs*, GigaBit Logic, Inc., Newbury Park, CA, 1987.
7	7	*SMD Component and Substrate Solderability*, Signetics Corp., Sunnyvale, CA, 1986, pg. 2.
7	5	*SMD Reliability Data*, Signetics Corp., Sunnyvale, CA, 1986, pg. 23.
5	15	*SMD Thermal Data*, Signetics Corp., Sunnyvale, CA, 1987.
2	2	*Surface Mounted Components Catalog and Applications Manual*, MuRata Erie North America, Inc., 1987, pg. 30.
5	6*	*Thermal Management of PicoLogic and NanoRAM GaAs Digital IC Families*, GigaBit Logic, Inc., Newbury Park, CA, 1988.
2	9	*The Trimmer Primer IV*, Bourns Trimpot, Riverside, CA, 1986.

---

* Special, additional reading, high-frequency design references are given at end of Chapter 5 text.

Magazine and Newsletter Articles
Consult the following for additional reading on components.

Chapter	Reference	Publication
6	17	Andrews, J. L., Schroeder, J. E., Gingerich, B. L., Kolanski, W. A., Koga, R., and Diehl, S. E., "Single Event Upset Error Immune CMOS RAM," *IEEE Transactions on Nuclear Science*, Vol. NS-29, December 1982, pg. 2040–2043.

2	16	Braden, J. S., "New Surface Mountable Packages for VLSI Devices," *MEPPE Journal*, MEPPE, Palo Alto, CA, Vol. 1, No. 2, May 1987.
1	11	Bruno, Terry, "Update—Surface Mount Aluminum Electrolytics," *SMTA Newsletter*, SMTA, Edina, MN, November 1986.
9	18	Gillespie, T., "Semiconductor Seal Testing," *Journal of ISHM—Europe*, No. 15, January 1988, pg. 46.
4	9	Leibson, Steven H., "EDN's Hands-On SMT Project—Part 2: Selecting the Surface-Mount Components," *EDN*, June 11, 1987, pg. 165. (See also Part 1, *EDN*, May 30, 1987 issue.)
9	10	Pedder, D. J., "Flip Chip Solder Bonding for Microelectronic Applications," *Journal of ISHM—Europe*, No. 15, January 1988, pg. 4.

Specifications, Standards, and Guidelines
Consult the following for additional reading on components.

Chapter	Reference	Publication
—	—	*EIA-PDP-100, Parts Division Publication 100 Registered and Standard Mechanical Outlines for Electronic Parts*, Electronic Industries Association, Washington, DC, 1988.
—	—	*EIA Specification RS-228, Fixed Electrolytic Tantalum Capacitors*, EIA, Washington, DC.
—	—	*Surface Mount Technology Component Standardization Proposals—Commercial Grade Components*, SMTA, Edina, MN, 1986.

Technical Papers from Published Proceedings
Consult the following for additional reading on components.

Chapter	Reference	Publication
2	15	Derfiny, Dennis and Dody, Glen, "On Optimizing the PLCC Leadform," *Proceedings of NEPCON West 1987*, Cahner Exposition Group, Des Plaines, IL, February 1987, pg. 251.
1	4	Moore, Robert, and Weaver, Bruce, "Electrical and Mechanical Considerations in the Design of a Leadless Chip Carrier for High Performance Applications," *IEPS Surface Mount Technology Compendium of Technical Articles Presented at the 1st, 2nd, and 3rd Annual Conference*, International Electronics Packaging Society, Inc., Glen Ellyn, IL, 1984, pg. 529.

C.2.2 *Design Information*

In this section, we've listed referenced information resources on the design of SMT and mixed-technology boards.

Books

Consult the following for additional reading on SMT design.

Chapter	Reference	Publication
6	9	Oberg, Erik, et al; Ed. Henry H. Ryffel, *Machinery's Handbook, 22nd Edition*, Industrial Press, New York, NY, 1984.
2	4	*ANSI/IPC-SM-782, SMT Land Patterns*, Institute for Interconnecting and Packaging Electronics Circuits, Lincolnwood, IL, 1987, pg. 4.
5	8*	Davidson, C. W., *Transmission Lines for Communication*, John Wiley & Sons, New York, NY, 1978.
5	3*	*FAST Applications Handbook*, National Semiconductor, Digital, South Portland, ME, 1987.
6	5	Harper, Charles A., Editor, *Handbook of Wiring, Cabling, and Interconnecting for Electronics*, McGraw-Hill Book Co., New York, NY, 1972, pg. 8-18.
3	13	Hollomon, James K. Jr., *Focus on SMT Design*, Anatrek, Chesapeake, VA, 1988, pg. 3.2.4.2.
6	4	*Hybrid Microcircuit Design Guide*, ISHM-1402/IPC-H-855, ISHM, Reston, VA, and IPC, Evanston, IL, 1982.
5	1*	*Motorola MECL System Design Handbook*. Motorola Semiconductor Products, Inc., Phoenix, AZ, 1986.
6	13	*Rad-Hard/Hi-Rel CICD Data Book*, Harris Custom Integrated Circuit Division, Palm Bay, FL, 1987, pg. 1-8.
5	7*	Ramu, S., and Whinnery, T. Van Duzer, *Fields and Waves in Communications Electronics*, John Wiley & Sons, New York, NY, 1984.
2	18	Schoenthaler, D., Hymes, L., and Miller, W., Chairmen IPC Component Mounting Subcommittee, *Printed Board Component Mounting, ANSI/IPC CM-770C*, The Institute for Interconnecting and Packaging Electronic Circuits, Lincolnwood, IL, March 1987, pg. 37.

---

* Special, additional reading, high-frequency design references are given at end of Chapter 5 text.

Magazine and Newsletter Articles
Consult the following for additional reading on SMT design.

Chapter	Reference	Publication
9	15	De Mey, G., and Van Schoor, L., "Thermal Analysis of Hybrids with Mounted Components," *Journal of ISHM—Europe*, No. 15, January 1988, pg. 28.
6	15	Gauthier, Michael K., and Dantas, Armando Roberto V., "Radiation-Hard Analog-to-Digital Converters for Space and Strategic Applications," *JPL Publication 85-84*, NASA Jet Propulsion Laboratory, Pasadena, CA, pg. 5-3.
5	18	Leibson, Steven H., "EDN's Hands-On Project—Part 3: CAD and Surface Mount Technology," *EDN*, June 25, 1987, pg. 212.

Specifications, Standards, and Guidelines
Consult the following for additional reading on SMT design.

Chapter	Reference	Publication
7	16	*DoD-STD-2000, Military Standard—Soldering Technology, High Quality/High Reliability*, June 10, 1986, pg. 10.
1	2	Kolias, John T., "Packaging Impact on System Performance," *IEPS Surface Mount Technology Compendium of Technical Articles Presented at the 1st, 2nd, and 3rd Annual Conference*, International Electronics Packaging Society, Inc., Glen Ellyn, IL, 1984, pg. 363.
6	18	Srour, J., "Basic Mechanisms of Radiation Effects on Electronic Materials, Devices, and Integrated Circuits," *Proceedings of the IEEE Annual Conference on Nuclear and Space Radiation Effects*, 1982.

C.2.3. *Manufacturing of SMT*

Books
Consult the following for additional reading on SMT manufacturing.

Chapter	Reference	Publication
3	2	Hutchins, Charles L., Ph.D., *Surface Mount Technology—How to Get Started*, D. Brown Associates, New Britain, PA, September 1986.
7	19	Mullen, Jerry, *How to Use Surface Mount Technology*, Texas Instruments Incorporated, Dallas, TX, 1984, pg. 2-8.

Magazine and Newsletter Articles
Consult the following for additional reading on SMT manufacturing.

Chapter	Reference	Publication
5	6	Leibson, Steven H., "EDN's Hands-On Project—Part 4: Assembling the Surface Mount Project Board," *EDN*, July 9, 1987, pg. 75.

Specifications, Standards, and Guidelines
Consult the following for additional reading on SMT manufacturing.

Chapter	Reference	Publication
7	16	*DoD-STD-2000, Military Standard—Soldering Technology, High Quality/High Reliability*, June 10, 1986, pg. 10.

C.2.4 *Market Reports*

Books
Consult the following for additional reading on SMT manufacturing.

Chapter	Reference	Publication
4	1	Hollomon, James K., *Advanced Manufacturing Technology in the Upper Midwest—A Research Report*, Anatrek, Norfolk, VA, 1988.

C.2.5 *Overviews on SMT*

Books
Consult the following for overviews of SMT.

Chapter	Reference	Publication
9	3	Balde, John W., Caswell, Greg, et al, *Surface Mount Technology*, International Society for Hybrid Microelectronics, Reston, VA, 1984, pg. 3.
7	19	Mullen, Jerry, *How to Use Surface Mount Technology*, Texas Instruments Incorporated, Dallas, TX, 1984, pg. 2-8.

C.2.6 *Soldering Technology*

Books
Consult the following for additional reading on SMT soldering technology.

Chapter	Reference	Publication
2	1	Manko, H. H., *Solders and Soldering*, McGraw-Hill, 1979, pg. 119.

Magazine and Newsletter Articles

Consult the following for additional reading on SMT soldering technology.

Chapter	Reference	Publication
3	1	Carbonneau, Richard A., "Infrared Vs. Vapor Phase," *Circuits Manufacturing*, September 1986, pg. 27.

Specifications, Standards, and Guidelines

Consult the following for additional reading on SMT soldering technology.

Chapter	Reference	Publication
7	18	*ANSI/IPC-SF-819, Solder Paste for Surface Mounting Applications*, Institute for Interconnecting and Packaging Electronic Circuits, Lincolnwood, IL.
7	16	*DoD-STD-2000, Military Standard—Soldering Technology, High Quality/High Reliability*, June 10, 1986, pg. 10.
7	13	*IPC-SM-840A, Qualification and Performance of Permanent Polymer Coating (Solder Mask) for Printed Wiring Boards*, Institute for Interconnecting and Packaging Electronic Circuits, Lincolnwood, IL.

Technical Papers from Published Proceedings

Consult the following for additional reading on SMT soldering technology.

Chapter	Reference	Publication
8	7	Haimovich, Joseph, Ph.D., "Intermetallic Compound Growth in Tin and Tin-Lead Platings Over Nickel and Its Effects on Solderability," *12th Annual Electronics Manufacturing Seminar Proceedings*, Naval Weapons Center, China Lake, CA, February 1988, pg. 51.
7	20	Rall, Dieter, and Hollomon, James, "Advances in Reflow Soldering Process Control," *Proceedings of NEPCON West 1987*, Cahners Exposition Group, Des Plaines, IL, February 1987, pg. 393.
3	6	Sedrick, A. V., "Design Guidelines for Achieving High First-Time Solder Yields with Mixed Component Technology," *Proceedings of NEPCON West 1986*, Cahners Exposition Group, Des Plaines, IL, February 1986, pg. 797.

C.2.7 *Test and QA Data*

Magazine and Newsletter Articles
Consult the following for additional reading on QC/QA and test for SMT.

Chapter	Reference	Publication
4	8	Leibson, Steven H., "EDN's Hands-On SMT Project—Part 5: Automated Testing of SMT PC Boards," *EDN*, July 23, 1987, pg. 76.

Specifications, Standards, and Guidelines
Consult the following for additional reading on QC/QA and test for SMT.

Chapter	Reference	Publication
7	16	*DoD-STD-2000, Military Standard—Soldering Technology, High Quality/High Reliability*, June 10, 1986, pg. 10.
8	12	*MIL-HDBK-217D, Reliability Prediction of Electronic Equipment.*
7	21	*MIL-STD-810, Military Standard Environmental Test Methods and Engineering Guidelines*, July 19, 1983, pg. 7.
5	2	"Testability Guidelines," SMTA, Edina, MN, 1987.

Technical Papers from Published Proceedings
Consult the following for additional reading on QC/QA and test for SMT.

Chapter	Reference	Publication
3	9	Archer, Wesley L., Cabelka, Tim D., and Nalazek, Jeffrey J., "Quantitative Determination of Rosin Residues on Cleaned Electronic Assemblies," *11th Annual Electronics Manufacturing Seminar Proceedings*, Electronics Manufacturing Programs Office, China Lake Naval Weapons Center, China Lake, CA, pg. 19.
3	16	Hroundas, George, "PCB Test Strategies for Manufacturing Yield Improvement," *Proceedings of NEPCON West 1987*, Cahners Exposition Group, Des Plaines, IL, February 1987, pg. 911.
7	11	Ibid., pg. 913.
7	9	Lish, Earl F., and Weber, John O., "Solderability Testing of Leaded and Leadless SMDs by Means of a Modified Wetting Balance," *11th Annual Electronics Manufacturing Seminar Proceedings*, Naval Weapons Center, China Lake, CA, February 1987, pg. 283.

| 3 | 17 | Lussier, Paul V., "Developing a Strategic PCA Test Plan," *Proceedings of NEPCON West 1987*, Cahners Exposition Group, Des Plaines, IL, February 1987, pg. 920. |
| 3 | 8 | MacLeod, Norman, "MIL-P-28809: Testing in the Presence of SMDs," *Proceedings of Circuit Expo West 1986*, World Wide Convention Management, Libertyville, IL, 1986, pg. 77. |

C.3 Additional Reading

Some additional works are worthy of note and worthy of inclusion in a well-stocked SMT library. We'll cover them below in a separate list of books and periodicals. Undoubtedly, there are additional works and forthcoming works that are worthy of mention. If you note an omission, please write to us through the publisher and point out our error. We will include additional material in the next revision to the text.

C.3.1 *Books of Interest on SMT*

We have not made an effort to provide a comprehensive list. And, at the rate that new material is appearing, there would be little point. Check R. R. Bowker's *Books in Print* for additional titles. This listing is available through most public libraries, and includes all books currently in print which have an International Standard Book Number (ISBN) assigned.

1. *A Scientific Guide to SMT*, C. Lea. A highly recommended comprehensive guide to the science of the SMT process. There are 14 chapters covering an SMT overview, components, substrates, design and assembly, wave soldering, reflow soldering, nonsolder attachment, and solderability. Particularly commendable is the scientific discussion of the physical mechanisms of the soldering process. Contains 569 pages, well illustrated. Published by Electro Chemical Publications Ltd., 8 Barns St., Ayr, KA7 1XA, Scotland. Not available through U.S. distributors.

2. *Surface Mount Directory*. An exhaustive listing of components, including 1200 outline drawings, and a listing of contract services and consultants, equipment, materials, supplies, and CAD/CAM hardware and software. A required item for libraries where SMT component and supply specifications are an issue. Contains over 4000 pages, illustrated. Published by Info-Mation, Inc., 904 Town Center, New Britain, PA 18901.

3. *Surface Mounted Assemblies*, J. Pawling, Ed. A concise discussion of all aspects of SMT. Includes SMT beginnings, substrates, passive and active devices, circuit layout, footprints, component placement and attachment, and post-attachment processes. Contains 227 pages, illustrated. Published by Electro Chemical Publications Ltd., 8 Barns St., Ayr, KA7 1XA, Scotland. Not available through U.S. distributors.

4. Surface Mount Technology, Charles Henry Mangin and S. McCleland. This book provides an overview of SMT told primarily through six case

studies which cover SMT start-ups, including Iomega, Hewlett-Packard, Philips, Rank Zerox, and Texas Instruments. Contains 10 chapters, 219 pages, illustrated. Published by CEERIS International, POB 939, Old Lyme, CT 06371-0939.

C.3.2 *Periodicals Covering SMT*

The following is a list of general interest magazines which regularly devote space to the coverage of SMT. We have not performed an exhaustive search for magazines, and apologize to any publishers who may feel they should be included in this listing. Again, please write to us through the publisher and ask to be added to the next update. The reader is encouraged to contact associates in related job classifications and industry segments for additional periodicals that would be of value.

1. *Assembly Engineering*. Contains articles and advertising of interest to manufacturing engineers in all industries, including electronics. Published monthly by Hitchcock Publishing Co., 25W550 Geneva RD., Wheaton, IL 60188. Distributed free to qualified subscribers.

2. *Brazing and Soldering*. Includes articles of interest to electronics process engineers and research scientists. Contains minimal advertising. Published biannually, with paid subscriptions, by Wela Publications Ltd., 8 Barns Street, Ayr, KA7 1XA, Scotland, in conjunction with the British Association for Brazing and Soldering.

3. *California CircuitTree*. Consists of articles and advertising of interest to PWB manufacturers. Published monthly, and distributed free to qualified subscribers, by Raymond A. Rasmussen/Barry R. Matties, 620 E. Taylor Avenue, Sunnyvale, CA 94086.

4. *Circuits Manufacturing*. Contains articles and advertising of interest to engineers and managers in the electronics assembly businesses. Published monthly, and distributed free to qualified subscribers, by Miller Freeman Publications, Inc., 500 Howard Street, San Francisco, CA 94150.

5. *Circuit World*. Contains transactions of the Institute of Circuit Technology, Oxford, UK. Quarterly publication, with paid subscriptions; contains minimal advertising. Published by Wela Publications Ltd., 8 Barns Street, Ayr, KA7 1XA, Scotland.

6. *Connection Technology*. Consists of articles and advertising applicable to connector engineering and the applications of interconnect technology. Published monthly by Lake Publishing Corp., P.O. Box 159, Libertyville, IL 60048-0159. Distributed free to qualified subscribers.

7. *EDN*. Articles and advertising that is of interest to electronics system designers. Published 48 times a year (biweekly with additional monthly issues). Distributed free to qualified subscribers. Published by Cahners Publishing Co., 275 Washington Street, Newton, MA 02158-1630.

8. *Electronic Business*. Has articles and advertising that is of interest to the managers of electronics businesses. Published twice monthly (except for December), and distributed free to qualified subscribers. Published by Cahners Publishing Co., 275 Washington Street, Newton, MA 02158-1630.

9. Electronic Manufacturing. Contains articles and advertising applicable to engineers and managers of electronics manufacturing businesses. Published monthly by Lake Publishing Corp., P.O. Box 159, Libertyville, IL 60048-0159. Distributed free to qualified subscribers.

10. *Electronic Packaging and Production.* Includes articles and advertising of interest to electronics manufacturing and process engineers and system designers. Published monthly by Cahners Publishing Co., 275 Washington Street, Newton, MA 02158-1630. Distributed free to qualified subscribers.

11. *Electronics Purchasing.* Contains articles and advertising of interest to purchasing and materiel managers in electronics manufacturing. Published monthly by Cahners Publishing Co., 275 Washington Street, Newton, MA 02158-1630. Distributed free to qualified subscribers.

12. *Electronics Week.* Contains articles and advertising of interest to engineers and managers in electronics manufacturing. Published monthly by McGraw-Hill, Inc., 1221 Avenue of the Americas, New York, NY 10020. Distributed free to qualified subscribers.

13. *Evaluation Engineering.* Has articles and advertising that is of interest to QC, QA, and test engineers and managers. Published monthly by A. Verner Nelson Associates, 2504 N. Tatiami Trail, Nokomis, FL 33555-3479. Distributed free to qualified subscribers.

14. *Hybrid Circuits.* The journal of ISHM Europe (Hamburg, W. Germany; Delft, The Netherlands; Verrieres le Buisson, France; Lyngby, Denmark; Pavia, Italy; and Bracknell, Berkshire, UK). Contains minimal advertising, and issued biannually (paid subscriptions). Published by Wela Publications Ltd., 8 Barns Street, Ayr, KA7 1XA, Scotland.

15. *Hybrid Circuit Technology.* Has articles and advertising applicable to engineers and managers of hybrid-circuit-manufacturing businesses. Includes SMT data. Published monthly by Lake Publishing Corp., P.O. Box 159, Libertyville, IL 60048-0159. Distributed free to qualified subscribers.

16. *IEEE Transactions on Components, Hybrids, and Manufacturing Technology.* The journal of IEEE Components, Hybrid, and Manufacturing Technology Proceedings. Published by the Institute of Electrical and Electronic Engineering, 345 E. 47th Street, New York, NY 10017.

17. *Journal of Electronic Engineering.* A Japanese electronics engineering journal. Published by Dempa Publications, Inc., 11-15 Higashi Gotanda, Shina Gawa-Ku, Tokyo 141 JAPAN. The U.S. office is at 380 Madison Avenue, New York, NY 10017.

18. *Journal of SMT Standards* . Contains articles and advertising of interest to SMT engineers and managers concerned with standardization. Distributed free to qualified subscribers. Published quarterly by Info-Mation, Inc., 904 Town Center, New Britain, PA 18901.

19. *Printed Circuit Assembly.* Has articles and advertising of interest to PWB assemblers. Distributed free to qualified subscribers. Published monthly by PMS Industries, 1790 Hembree Road, Alpharetta, GA, 30201.

20. *Printed Circuit Design.* Contains articles and advertising of interest to PWB designers. Distributed free to qualified subscribers. Published monthly by PMS Industries, 1790 Hembree Road, Alpharetta, GA 30201.

21. *Printed Circuit Fabrication*. Contains articles and advertising of interest to PWB manufacturers. Distributed free to qualified subscribers. Published monthly by PMS Industries, 1790 Hembree Road, Alpharetta, GA 30201.

22. *Quality*. Contains articles and advertising that is of interest to QC, QA, and test engineers and managers in the electronics industry. Distributed free to qualified subscribers. Published monthly by Hitchcock Publishing Co., 25W550 Geneva Road, Wheaton, IL 60188.

23. *SAMPE Journal*. Contains articles, SAMPE papers, and advertising of interest to materials engineers and electronics designers. Published bimonthly by the Society for the Advancement of Materials and Process Engineering, P.O. Box 2459, Covina, CA 91722-2459. Distributed free to SAMPE members, with paid subscriptions for others.

24. *Semiconductor International*. Has articles and advertising that is of interest to engineers and managers in semiconductor manufacturing. Distributed free to qualified subscribers. Published monthly by Cahners Publishing Co., 275 Washington Street, Newton, MA 02158-1630.

25. *SITE (Screen Imaging Technology for Electronics)*. Contains articles and advertising of interest to engineers and manufacturing managers who are involved in screening technology. Distributed free to qualified subscribers. Published bimonthly by Signs of the Times Publishing Co., 407 Gilbert Avenue, Cincinnati, OH 45202.

26. *Solid State Technology*. Furnishes articles and advertising of interest to those engineers and managers involved in designing or applying semiconductors, or semiconductor packaging. Distributed free to qualified subscribers. Published monthly by PennWell Publishing Co., 14 Vanderventer Avenue, Port Washington, NY 11050.

27. *Surface Mount Technology*. Contains articles and advertising that is applicable to engineers and managers of SMT manufacturing businesses. Distributed free to qualified subscribers. Published monthly by Lake Publishing Corp., P.O. Box 159, Libertyville, IL 60048-0159.

28. *Test & Measurement World*. Has articles and advertising of interest to QC, QA, and test engineers and managers in electronics manufacturing. Distributed free to qualified subscribers. Published monthly by Cahners Publishing Co., 275 Washington Street, Newton, MA 02158-1630.

29. *VLSI System Design*. Contains articles and advertising of interest to designers and users of VLSI systems. Distributed free to qualified subscribers. Published monthly by CMP Publications, 600 Community Drive, Manhasset, NY 11030.

Figure D-1 IPC recommended test coupons.

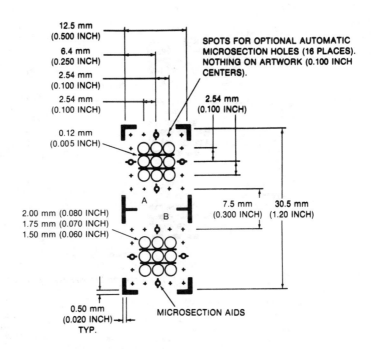

12.5 mm (0.500 INCH)

6.4 mm (0.250 INCH)

2.54 mm (0.100 INCH)

2.54 mm (0.100 INCH)

SPOTS FOR OPTIONAL AUTOMATIC MICROSECTION HOLES (16 PLACES). NOTHING ON ARTWORK (0.100 INCH CENTERS).

2.54 mm (0.100 INCH)

0.12 mm (0.005 INCH)

A

B

2.00 mm (0.080 INCH)
1.75 mm (0.070 INCH)
1.50 mm (0.060 INCH)

7.5 mm (0.300 INCH)

30.5 mm (1.20 INCH)

0.50 mm (0.020 INCH) TYP.

MICROSECTION AIDS

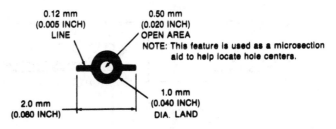

0.12 mm (0.005 INCH) LINE

0.50 mm (0.020 INCH) OPEN AREA

NOTE: This feature is used as a microsection aid to help locate hole centers.

2.0 mm (0.080 INCH)

1.0 mm (0.040 INCH) DIA. LAND

(A) Test coupon A and B.

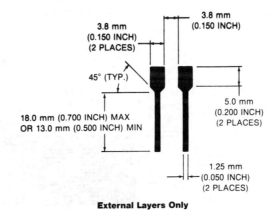

3.8 mm (0.150 INCH) (2 PLACES)

3.8 mm (0.150 INCH)

45° (TYP.)

18.0 mm (0.700 INCH) MAX OR 13.0 mm (0.500 INCH) MIN

5.0 mm (0.200 INCH) (2 PLACES)

1.25 mm (0.050 INCH) (2 PLACES)

External Layers Only

(B) Test coupon C.

Figure D-1 IPC recommended test coupons. (cont.)

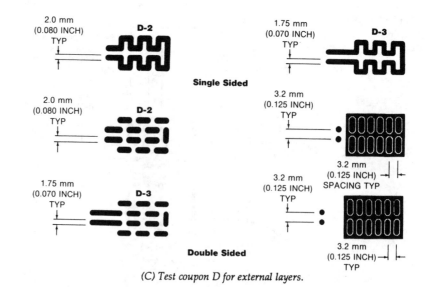

(C) *Test coupon D for external layers.*

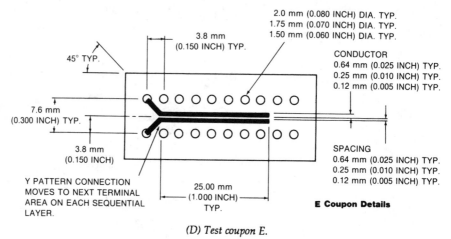

(D) *Test coupon E.*

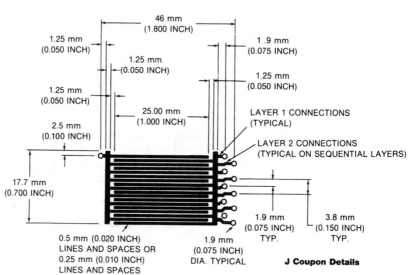

(E) *Test coupon J.*

D SMT Test Board Patterns

Test patterns on coupons can yield valuable insight into the accuracy and degree of the process control of PWB manufacturing and assembly processes. Such patterns will testify to PWB under-/over-etching, trace adhesion, cleaning, insulation resistance, solderability, etc. Coupons are particularly valuable when standard patterns are used, because a large bank of heuristic data has already been assembled, allowing a prompt accurate interpretation of the test observations.

IPC-recommended test coupons are shown in Figure D-1. By incorporating coupons on all board or panel perimeters, you can develop meaningful test data on the board and assembly quality. That is, test data can be correlated to previous data because the data pertain to exactly the same test pattern(s). To evaluate the test results, obtain copies of IPC-A-600C, *Acceptability of Printed Boards—Guidelines*, and IPC-A-610, *Acceptability of Printed Board Assemblies.*

Additional IPC offerings for PWB quality control and assurance are listed below:

Specification Number	Subject Matter
IPC-BP-421	Backplanes, Rigid PWB with Press-Fit Contacts.
IPC-D-249	Design Standard, 1- and 2-Sided Flexible PWBs.
IPC-D-319	Design Standard, 1- and 2-Sided Rigid PWBs.
IPC-D-949	Design Standard, Rigid MLBs.
IPC-FC-241B*	Dielectric, Flexible, Metal-Clad, Flex PWBs.
IPC-FC-231B*	Dielectric, Flexible PWBs.
IPC-D-300G	Dimensions and Tolerances, PWBs.
IPC-FC-232B*	Film, Adhesive-Coated Dielectric, Flex PWB Cover.
IPC-AM-372	Film, Copper, Electroless, for Additive Boards.

Specification Number	Subject Matter
IPC-FC-233A*	Film, Flexible Adhesive Bonding.
IPC-FC-150E*	Foil, Copper, for Printed-Wiring Applications.
IPC-HM-860	Hybrid Multilayer Performance Specifications.
IPC-L-112	Laminate, Composite, Foil-Clad Polymeric Standards.
IPC-L-125	Laminate, Plastic Sheet, Metal-Clad, High-Frequency PWB.
IPC-L-115A	Laminate, Plastic Sheet, Metal-Clad, High-Temperature PWB.
IPC-L-130	Laminate, Thin, Metal-Clad, General-Purpose MLBs.
IPC-L-108A	Laminate, Thin, Metal-Clad, High-Temperature MLB Specifications.
IPC-RF-245	Multilayer Boards, Rigid Flex.
IPC-L-109A	Prepreg, Resin and Glass Cloth, High-Temperature MLB Specifications.
IPC-D-352*	PWB Database Description, Electronic Design.
IPC-CC-830	PWB Dielectric Qualification and Performance.
IPC-D-351*	PWB Documentation Digital Form.
IPC-D-325	PWB Documentation, End Product.
IPC-MC-324	PWB, Metal-Core, Performance Specifications.
IPC-NC-349*	PWB NC Formatting for Drilling and Routing.
IPC-D-350B*	PWB NC Formatting for End Product Description.
IPC-D-322	PWB Sizes, Standard Panel Guidelines.
IPC-SD-320B	PWB Specifications, End Product, Rigid 1- and 2-Sided.
IPC-ML-950C	PWB Specifications, MLB Performance.
IPC-FC-250A	PWB Standard, Flexible 1- and 2-Sided with Interconnects.
IPC-S-804A	Solderability Test Methods for PWBs.
IPC-SM-840A	Solder Mask Qualification and Performance for PWBs.
IPC-AM-361	Substrate Specification, Rigid, for Additive Process.

*Specifications approved for U.S. Department of Defense use.

In addition, the IPC-G-400 guidelines manual provides a convenient collection of currently approved specifications on accept/reject criteria for PWBs and assemblies, artwork and design generation, press-fit backplanes, modification and repair, and component mounting, including surface mounting. IPC-D-330 is a compendium of design guidelines and specifications. And, IPC-TM-650 and IPC-MI-660 gather test methods and incoming inspection techniques, respectively.

All IPC specifications and compendiums may be obtained from the Institute for Interconnecting and Packaging Electronic Circuits (IPC), 7380 Lincoln Avenue, Lincolnwood, IL 60646, USA. Their telephone number is (312) 677-2850, their TWX number is (510) 601-6005, and their FAX number is (312) 677-9570.

We'd like to close this section with a special note of thanks to the IPC and the International Electronic Packaging Society (IEPS) for their generous support of this work through the provision of data and specifications.

E Glossary of SMT Terms

In this glossary, we will not attempt to define common dictionary words except where they carry a unique connotation for surface mounting. Also, because of space limitations, we will not define words in common usage in through-hole assembly or general electronic engineering. Such words are amply defined elsewhere. We will concentrate on defining words peculiar to SMT and SMT-related areas of hybrid technology.

Abrasive Trimming—The trimming of material from a component, such as a resistor, to bring it to a specified value, where such trimming is performed by means of abrasive cutting. Abrasive dust is generally directed through a nozzle and at the part, in a miniature air blast.

ACPI—An acronym for Automated Component Placement and Insertion. Also, the name of a committee dealing with related issues in the Joint Electron Device Engineering Council (JEDEC).

Active—A component which changes its basic character under applied voltage. For example, semiconductors, which are conductive only when certain conditions of applied voltage are met, are ~.

Active Trim—The trimming of a component while monitoring a remote electrical parameter which the trimmed component controls.

Air Gap—The minimum space between conductors on a PWB.

Air Knife—A high-pressure air flow directed through a elongated narrow-slot-style nozzle and used to sweep excess solder and solder bridges off flow-soldered boards. Particularly useful in removing bridges and icicles during the flow soldering of SMCs.

Alumina—Polycrystalline or amorphous aluminum oxide, Al_2O_3. Commonly used as a hybrid substrate and material for leadless chip carriers. *See also* Sapphire.

Angle of Attack—The angle that the squeegee of a screen printer forms with the screen.

Annular Ring—That portion of conductive material completely surrounding a hole in a PWB. *See also* Pad.

ANSI—An acronym for the American National Standards Institute located at 1430 Broadway, New York, NY 10018 USA.

Artwork—An accurately scaled representation of the desired pattern for a set of features on a PWB. ...solder screen ~, first layer conductor ~.

ASIC—An acronym for Application Specific Integrated Circuit, a custom IC developed for a specific user application.

Aspect Ratio—The ratio of the length to the width of a feature, component, etc.

"A" Stage—The low molecular weight stage of a "resign" in which it is readily soluble and fusible.

ASTM—An acronym for the American Society of Testing and Materials, which is located at 1916 Race Street, Philadelphia, PA 19103 USA.

ATE—An acronym for "automated test equipment."

Automatic Placement—The automated assembly of SMT boards using pick and place machinery. An adjective phrase to describe equipment that performs this process.

Auto Route—A CAD term referring to the capability of the CAD system to automatically interconnect components given inputs of a net list and a component placement database.

Axial—A term describing a discrete device having leads coming out of its ends along the longitudinal axis; an ~ component.

Azeotropic—An adjective used to describe a liquid made up of two or more distinct fluid compounds, but having a constant boiling point at a given composition.

Back Bonding—The bonding of dies with their noncircuitry side toward the substrate, as in chip and wire techniques. Antonym: Face Bonding.

Backplane—A PWB assembly fitted with multiple connectors and conductors for interconnecting several other PWB assemblies.

Bake—1. A baking process used in SMT to dry volatiles from solder paste, activate flux, and firmly fix components prior to reflow soldering. 2. To heat for the purpose of performing a bake. Sometimes called Bake-Out.

Bed of Nails—A test probing device which has spring-loaded test pins in an equidistant-spaced matrix. Boards or assemblies that are to be tested will have test points spaced on exact multiples of the matrix spacing.

Beryllia—A ceramic of beryllium oxide, BeO. Beryllia is highly thermally conductive. It is used as a substrate where its thermal conductivity is useful. The dust of BeO is toxic when inhaled.

Beryllium Copper—An alloy of beryllium and copper noted for good electrical and thermal conductivity and high wear resistance. It is used in lead frames of SMCs to promote dissipation.

BIT—An acronym for *Binary digIT*. A bit is considered the indivisible limit of smallness in binary information. It has the value of either 1 or 0.

Bleed—In screen or stencil printing, the act of ink diffusing into adjacent areas beyond the limits of the desired print.

Bond—1. The interconnect formed on the surface of a bare semiconductor chip, or the interconnect between the wire leading from such a chip and a bonding pad on a board. This generally refers to the thermal compression or ultrasonic bonding process in chip and wire hybrids. 2. The connection between the bottom side of a semiconductor die and a circuit board. Chips may be attached by a eutectic, epoxy, or solder die ~.

Bonding Pad—A metallized site for the attachment of a wire through a wire bond.

Bonding Wire Crossover—A potential defect where the routing of bond wires from I/O bonding pads on an IC to its external I/O bonding pads forces bond wires to crisscross, raising a risk of short circuiting.

Breakaway—*See* Snapoff.

Breakout—Portion(s) of a substrate scribed, notched, or otherwise mechanically processed, so that they may be broken apart after assembly operations.

Bridge—1. A soldering process defect. An unwanted solder connection formed between closely spaced leads or metallized features on a PWB assembly. Antonym: Open. 2. The act of forming such a defect.

BSI—An acronym for the British Standards Institute, which is located at 2 Park Street, London W1A-12BS, UK.

"B" Stage—The medium molecular weight condition of a "resin" where it is less soluble than in the "A"Stage, but is plastic and fusible. "Prepregs" are resin/reinforcement compositions with B-Stage resin.

Bump—1. In TAB technology, a thermocompression or reflow-bondable hemispherical metal deposit. Bumps may be grown on TAB tape (as in bumped-tape technology) or on the IC die (as in bumped-chip technology). Bumped-die technology, with gold bumps grown on the wafer, is the common approach as of this writing. 2. In flip-chip technology, the name applied to the solder hemispheres grown on the bonding sites of IC dies to facilitate their reflow attachment to substrates.

Buried Via—A via not opening to either outer surface of a MLB.

Burn-In—The process of electrically stressing a device or assembly (usually at a high ambient temperature) for a sufficient time to eliminate failure due to component infant mortality. A large percentage of device failures generally occur during early life. Devices which do not succumb to infant mortality tend to live to their predicted life span. Thus, by eliminating infant mortalities from a population, the apparent reliability of the population may be significantly raised. Sophisticated burn-in tests log real-time anomalous event information to the nodal level during test.

Butt Lead—An SMT lead form; also called an I lead. The lead extends from the component center and turns toward the component base like a standard DIP lead, but is cut to extend just below the device body.

CAD—An acronym for "computer-aided design."

CAE—An acronym for "computer-aided engineering."

CAM—An acronym for "computer-aided manufacturing."

Castellation—A semicircular cutout centered on the metallization on the periphery of a leadless chip carrier.

CAT—An acronym for "computer-aided testing."

CCC—An acronym for "ceramic chip carrier." May be leaded or leadless. *See also* Ceramic Chip Carrier.

Centering—Describes a device used for precisely registering the center of a component so that it may be accurately placed or handled by an automated machine.

Centipoise—A unit of measure of the coefficient of dynamic viscosity. One centipoise equals one one-hundredth of 1 dyne sec/cm^2.

Ceramic Chip Carrier (CCC)—A ceramic IC package with leads on four sides. May be leadless or leaded.

Ceramic Quad Flat Pack (CQFP)—A fine-pitch hermetic surface-mount semiconductor package with a gull-wing lead form and peripheral I/O on all four sides.

Chase—The frame which supports a screen or stencil.

Chip—1. A single substrate on which all the active and passive circuit elements have been fabricated in situ, using one or all of the semiconductor techniques of diffusion, passivation, masking, photoresist, and epitaxial growth. *See also* IC. 2. An adjective used to describe a small uncased discrete component; i.e., *Replace a large radial capacitor with a tiny ~*.

Chip and Wire—A hybrid interconnect method employing faceup-bonded IC devices which are interconnected to the substrate by wire bonding.

Chip Capacitor—A small uncased rectangular or square discrete device which introduces capacitance into an electronic circuit.

Chip Carrier—A high-density packaging technique often used for ICs. Its input and output terminals are around the perimeter of the device. Chip carriers are available in leadless and leaded formats.

Chip-on-Board—Chip and wire technique applied to traditional PWB materials in lieu of hybrid substrates.

Chip Resistor—A small ceramic substrate-based resistor, generally from 3.2 mm (125 mils) to as little as 0.5 mm (20 mils) in length. The ceramic chip is an inert substrate with the resistor formed (generally from ruthenium oxide) on the surface. Chip resistors are small, have extremely low shunt or parasitic capacitance, no inductance, and are stable.

Chip Shooter—A slang term for automatic placement equipment that is optimized for the very-high-speed placement of passive chips.

CIM—An acronym for "computer integrated manufacturing."

Clam Shell Fixture—A test fixture which closes like a clam's shell to allow the probing of two sides of a PWB assembly simultaneously.

CLDCC—The preferred acronym for "ceramic leaded chip carrier."

Clinch—In electronics assembly, to bend the inserted leads in order to mechanically secure the leaded components in their holes before soldering.

CLLCC—The preferred acronym for "ceramic leadless chip carrier."

COB—An acronym for "chip-on-board."

Code 39—The most commonly used bar-code standard for the identification of components, assemblies, etc. It provides adequate field area and the alphanumeric data capability required for the application.

Coefficient of Thermal Expansion (CTE)—A variant of "Thermal Coefficient of Expansion."

Cold—When applied to solder joints in electronics, this denotes the condition also called "upset joint," where two or more pieces being joined by solder have moved during the solder solidification process. Such joints are reliability concerns. They are recognized by a grainy rough surface finish.

Cold-Crack—A metallurgical term for flaws which develop in certain alloys and metals upon low-temperature exposure.

Cold Work—The embrittlement resulting from repeated flexural stresses to a metal or alloy.

Component—The preferred term for a discrete electrical element.

Component Side—For IMC boards, the side of the board bearing the components, and the opposite of the "solder side," which travels through the molten solder during the flow-soldering process.

Computer Integrated Manufacturing (CIM)—The high-level computer intergration of a manufacturing operation. CIM generally is taken to include the computer integration of computer-aided engineering (CAE), computer-aided design (CAD), computer-aided manufacturing (CAM), computer-aided test (CAT), and includes test-program generation from CAE/CAD and CAE/CAD feedback and upgrading from CAM/CAT.

Condensation Soldering—*See* Vapor-Phase Soldering.

Conduction Reflow—A reflow-soldering process, whereby assemblies are conductively heated for soldering by being passed over heated platen(s) with an intimate thermal contact to the heated surface(s).

Contact Printing—Screen-printing terminology for printing with no snapoff. This is commonly done in stencil printing.

Controlled Collapse—In flip-chip technology, controlling the reduction in standoff of a chip during reflow. This may be done by the growth of high-melting-point pedestals within the solder bumps, etc.

Convection Reflow—Reflow soldering in a convection oven.

Coordinatograph—A high-precision drafting machine used to cut scribe-coat films in the preparation of IC or microcircuit original artwork.

Coplanar—Lying in a single plane. In SMCs, the coplanarity specification defines the maximum variance of any lead from a single plane at the points where the leads contact the substrate for mounting.

Copper/Invar/Copper—Also called Copper-Clad Invar. A laminated metallic material composed of two outer layers of copper over an inner layer of Invar. Used to form heat-sink, TCE-restraint, stiffening, and power/ground layers in organic laminate MLBs. In proper proportions, Copper/Invar/Copper can match the TCE of an MLB to the TCE of the leadless ceramic components mounted on it.

Copper/Molybdenum/Copper—Also called Copper-Clad Moly. A laminated metallic material composed of two outer layers of copper over an inner layer of Molybdenum, which is noted for its high strength-to-weight ratio. Used to form heat-sink, TCE-restraint, stiffening, and power/ground layers in organic laminate MLBs. In proper proportions, copper-clad molybdenum can match the TCE of an MLB to the TCE of the leadless ceramic components mounted on it.

Core—In PWBs, a central layer or layers of a different material than that used elsewhere in the board laminate. May serve as power, ground, heat-sink, TCE-restraint, and stiffening layer(s), or as a combination of these functions.

Coupon—A small test board fabricated with a production run of boards under matching manufacturing and control conditions. Test coupons generally have an intricate pattern and/or small holes suitable for testing the PWB pattern and/or the via quality.

CQFP—An acronym for "Ceramic Quad Flat Pack."

Cream—A solder formulation of tiny solder spheres or powder, a liquid flux, and a thixatropic vehicle, plus, in some instances, a flux activator. Solder creams may be designed for application by screening, stencil printing, or dispensing. *Synonymous with* Ink, Paste.

Crosstalk—An unwanted signal produced in an interconnect line by inductive and/or capacitive coupling with an adjacent line(s).

CSA—An acronym for the Canadian Standards Associations, located at 178 Rexsdale Blvd., Rexsdale, Ontario, CANADA.

"C" Stage—The high-molecular-weight, fully cured state of a resin. In the C Stage, the resin is low in solubility and is infusible.

CTE—An acronym for "Coefficient of Thermal Expansion." *See also* Thermal Coefficient of Expansion.

Dam—A dielectric barrier printed or fabricated over a conductor trace to prevent unwanted solder migration along the conductor and away from the intended reflow sites.

Definition—1. Screening terminology for the sharpness of the demarcation line between a printed pattern and the adjacent substrate area. 2. Photolithographic terminology for the sharpness of demarcation between patterns and background in film masters, masks, or etched patterns.

Delid—The action of removing the hermetically sealed lid from a hybrid circuit package.

Dendrite—A galvanic growth of circuit metallization between conductors of opposite charges. Often, the dendrite takes on an appearance like the branches of a tree.

Dendritic—Of, or pertaining to, dendrites or their growth.

Detritus—Loose material produced by spatter from a trimming process, and adhering to a substrate or assembly after trimming.

Dewet—In soldering, the action of molten solder pulling back from an area where it has been applied because of insufficient wetting forces in the application area.

Dewetting—A solder defect. The condition where solder lands, metallizations, or leads have dewetted.

Differential Scanning Calorimetry (DSC)—A quantitative test for the cure state of a resin by determining the glass transistion temperature (Tg) of a sample.

DIP—An acronym for "dual in-line package."

Direct Chip Assembly—Assembly methods involving the direct mounting of semiconductor chips on substrates. Some examples are COB, Flip Chip, and Chip and Wire.

Direct Emulsion—In screening terminology, an emulsion coating applied directly to screen material as a liquid, as opposed to solid emulsions transferred to the screen from a film backing.

Direct-Metal Mask—A mask produced by etching a desired positive pattern into a metal-foil sheet. One application for such masks is stencil printing.

Discrete—1. A component, such as a resistor, capacitor, transistor, etc., which has an individual identity, as opposed to an in situ component of a hybrid. 2. A component forming a single circuit element as opposed to an integrated circuit composes of multiple circuit elements. 3. Used as an adjective, it means of, or pertaining to, such a component.

DMA—An acronym for "dynamic mechanical analysis."

DoD—An acronym for the U.S. Department of Defense. Also in common usage as DOD.

Double-Side Populated—An adjective which describes a board (usually primarily SMT) which has components located on both sides.

DPAK—An SM transistor package with a large, metal, heat-sink area on its bottom, similar in format to the TO-220 package, with a surface-mount lead form. Used primarily for high-power transistors and fast-switching rectifiers.

DRAM—An acronym for "dynamic RAM." A read/write memory that must be clocked or refreshed at regular intervals to maintain its memory integrity.

Drawbridge—A term for a solder defect in which a small chip component raises one termination, producing an open, during reflow soldering. The defect is caused by unbalanced forces of solder wetting and surface tension acting on the two ends of the chip. When this imbalance is great enough to overcome gravity, the component begins to stand on one end like a drawbridge opening. The term is used for parts in any stage of erection great enough to produce a sol-

der open on the raising end. Also, the act of drawbridging. *Synonymous with* Pop Wheelies, Tombstone (preferred), Manhattan Effect.

DSC—An acronym for "differential scanning calorimetry."

Dual In-Line Package (DIP)—An insertion-mountable package, having a row of extended I/O pins on two opposing sides. The extended pins are typically mounted through-hole in a PWB. The format is generally used to package ICs.

Dual Wave—An SMT flow-soldering approach, using two solder waves. The first is a high-pressure turbulent wave to throw solder on all metallizations, reducing the likelihood of solder skips. The second is a low-pressure, laminar flow wave designed to remelt and remove any bridges and icicles left by the turbulent wave.

Dynamic Mechanical Analysis (DMA)—A quantitative test method for resin materials involving the measurement of the change in stiffness of laminates and prepregs with changes in temperature.

Dynamic Printing Force—Screening terminology for the fluid force which operates on a thixatropic paste, causing it to act as a liquid and flow through a screen or stencil, wetting the workpiece to be printed.

EIA—An acronym for the Electronics Industries Association, located at 2001 Eye Street, Washington, DC 20006 USA.

EIAJ—An acronym for the Electronics Industries Association of Japan, located at 2-2, Marunouchi 3-Chome, Chiyoda-ku, Tokyo 100, JAPAN.

Electromagnetic Interference (EMI)— Unwanted signals induced in a circuit by magnetic fields.

Electromagnetic Pulse (EMP)—A brief high-power electromagnetic radiation. EMPs are generated over a brief period during a nuclear or thermonuclear explosion.

EMI—An acronym for "electromagnetic interference."

EMP—An acronym for "electromagnetic pulse."

EMPF—An acronym for the Electronic Manufacturing Productivity Facility, a cooperative electronics and SMT manufacturing research project of the U.S. Navy and several American electronics corporations. It's located at 1417 Norma Street, Ridgecrest, CA 93555 USA.

Eutectic—A metallurgical term for an alloy composition which has the lowest melting point of that alloy series. Unlike other alloy compositions in the series, the eutectic has no plastic or mushy phase. Its liquidus and solidus phases coincide, making eutectic solders useful for soldering to reduce the likelihood of upset solder joints. In the tin/lead series, the eutectic alloy is approximately 63% Sn, 37% Pb.

Exponential Failures—Failures that occur at an exponentially increasing rate. Also called "wear-out" failures.

Extender—A material added to a plastic, paste, etc., to reduce the amount of the primary material required.

Face Bonding—The bonding of dies with their circuitry side toward the substrate, as in flip-chip mounting. *Antonym of* Back Bonding.

Fillet—The concave-shaped interfacial connection formed by solder between a device termination and the solder land on a PWB.

Film—Hybrid terminology for a coating deposited by thick-film (screen print), or thin-film (evaporation, or sputtering) processes. *See also* Thick Film, Thin Film.

Final Seal—In hybrid processing, the lid sealing operation that covers the circuit assembly such that further processing cannot be done to the circuit without disassembling the package (delidding).

Fine Leak—In hermetically sealed packages, a leak passing less than 10^{-5} cm³/sec. at a pressure differential of one atmosphere.

Fire—To bake, at high temperatures, a green ceramic or printed thick-film pattern in order to alter its basic properties and bring it to a condition suitable for use.

First-Pass Yield—The percentage of the product conforming to specifications as it leaves the manufacturing line and arrives at final test. This percentage is generally expressed as failures in parts per million (PPM) for high-yield processes and a percentage of conforming parts for low-yield lines (greater than 1000 PPM nonconforming assemblies, or less than 99.9% conforming).

Fissure—In ceramics, a crack in a dielectric in the area of metallizations, which is usually due to stresses produced in an incorrect firing cycle. Also, to crack the dielectric in such a fashion.

Flag—An area of a component lead frame used for support of the semiconductor die.

Flat Pack—1. A semiconductor package having leads on two or four sides. The lead-frame material is very thin and fragile. Flat packs are typically shipped with the lead frame untrimmed and unformed. The parts are commonly trimmed and formed into gull-wing format for surface mounting. 2. A style of hybrid package with a thin Z-axis dimension and two rows of ribbon-wire leads, one extending from each long side. Before forming, the leads extend in a single plane. Such packages are designed for gull-wing lead forming to make a surface-mountable hybrid circuit.

Flexible—When applied to machines or production lines, this indicates that the system can readily accommodate a range of assembly specifications. In the most extreme usage, "flexible automation," it implies that equipment has the capability to automatically recognize and adapt to a wide range of assembly tasks and changes in workpiece specifications. However, when used in advertising or sales claims, this word appears to have lost any precise meaning.

Flip Chip—An IC die prepared for reflow face bonding by the growth of solder hemispheres on the bond pads of the chip. More broadly, the word is used to refer to the full technology of mounting such devices.

Flood—Screening terminology for the nonprinting stroke of a screen printer. This stroke is used to drag paste over the screen before the print stroke. In some printers, the flood stroke is accomplished by a separate flood bar. In others, the flooding is provided by a stroke of the print squeegee, where the normal printing pressure is not applied.

Flow—A soldering method involving the immersion of metallizations in molten solder to produce the desired solder joints. (Wave soldering in the dominant flow-soldering method used in IMC PWB assembly.)

Footprint—1. The land pattern for an SMC. 2. The outline that a component body or chip will occupy when placed on an assembly.

Frit—Powered glass used as a component in thick-film pastes to give adhesion to the composition after firing, and also, used singly as an overglaze for the protection of underlying materials.

GaAs—An acronym for Gallium Arsenide. A composite semiconducting material used in lieu of silicon in very-high-speed discrete devices and IC applications.

Gang Placement Equipment—Automatic P&P machinery which places a large number of components in each single machine cycle. Also called *simultaneous placement equipment*.

Glassivation—Semiconductor protection caused by the firing of a pyrolytic glass-frit deposition over the semiconductor face. A die passivation method.

Glass Phase—In hybrid firing, the part of the cycle where the glass-frit binder is in its molten phase.

Glass-to-Metal Seal—A hermetic dielectric seal, such as that formed between a lead which passes through a package wall and the wall of the package. The seal is made by filling the space between the lead and package with glass frit and firing it to oxide layers on the metals.

Glass Transistion Temperature (Tg)—In polymer technology, the temperature above which the thermal expansion rate of the polymer increases dramatically and changes from a linear to an asymptotic curve.

Glaze—1. To produce a glassy dielectric layer by firing a thick-film dielectic ink on a substrate. 2. In PWB drilling, to cause a glassy coating to form on the barrel of the drilled hole due to improper drilling rotational or feed speed, or due to a dull bit. Glazed holes must be etched to produce a rougher surface for plating adhesion prior to through plating. 3. The glassy surface produced in either of these two processes.

Glue Chip—A slang term for simple merchant market logic ICs. The term derives from the fact that such logic is often used to attach VLSI parts together to form a system.

Green—1. Ceramic technology, of or pertaining to a ceramic material in the unfired state. 2. The color of early epoxy-glass compositions, which were green due to the presence of impurities. Now applied as a name for the coloring material added to pure epoxy (which is milky pink) to produce the customary green hue.

Gross Leak—A leak in a hermetically sealed package which passes more than 10^{-5} cm^3/sec. at a pressure differential of one atmosphere.

Gull Wing—A lead form where the leads exit a package body from a plane near or on the Z-axis center of the part, turn downward to just below the body of the device, and then turn outward to form feet for reflow-solder mounting. So named for the lead appearance, which resembles a gull's wing bent downward in flight.

Hairpin Mounting—A high-density through-hole style of mounting axial components, where one lead is left straight and the other lead is bent 180° to conserve square-mounting area. This mounting method is not allowed for high-reliability assemblies because the hairpin leads are prone to damage from impact, shock, and vibration.

Halo—In ceramic technology, the term for a defect in which glassy halos form around conductors during the firing process. Generally avoided by proper firing profiles and correct material combinations.

Hidden Via—A via in an MLB, which exits through one surface layer but does not come through the opposite surface layer.

Hot-Crack—A metallurgical term for the cracking of a metal or alloy upon freezing. It may occur because of stress developed in the alloy due to unequal cooling. The phenomenon is sometimes accompanied by "hotshortness."

Hotshortness—The brittleness exhibited by some alloys and metals at elevated temperatures.

Hybrid—A type of microcircuit that consists of elements which are fabricated in situ directly on the substrate material in combination with discrete add-on components.

Icicle—A solder stalactite which forms on circuit metallizations or component terminations. Icicles are generally produced by an improperly controlled or designed immersion-soldering operation. Also, the formation of such a defect.

IEC—An acronym for the International Electrotechnical Commission, whose address is: P.O. Box 131, 1211 Geneva 20, SWITZERLAND.

IEEE—An acronym for the Institute of Electrical and Electronics Engineers, located at 345 E. 47th Street, New York, NY 10017 USA.

IEPS—An acronym for the International Electronics Packaging Society, located at 114 North Hale Street, Wheaton, IL 60187 USA.

I^{2}MT—An acronym for International Institute of Manufacturing Technology. An organization conducting SMT and advanced technology continuing education and research. Headquarters are at Box 549, MCV Station, Richmond, VA 23296-0001.

ILB—An acronym for "inner-lead bond."

IMC—An acronym for "insertion-mounted component."

Immersion Soldering—Soldering effected by the dipping of parts to be joined in a molten solder bath, wave, or jet.

Indirect Emulsion Screen—In screening terminology, a screen on which the emulsion is applied as a separate sheet or film pressed into the screen mesh, as opposed to a direct emulsion which is applied as a liquid.

Infant Mortality—The failure of an electronic device or system very early in its expected lifetime. Devices which do not succumb to infant mortality tend to live for their predicted lifespan, but a large percentage of device failures generally occur during their early life. Thus, by eliminating infant mortalities from a population, the apparent reliability of the population may be significantly raised.

Infrared—Of, pertaining to, or being in the band of radiation which has wavelengths from just longer than visible light to just shorter than microwaves. Generally used in SMT to describe a furnace or soldering process which uses such radiation as one of its major heating mechanisms.

Ink—A screen-printable thick-film paste.

In-Line Placement—Placement equipment which passes boards through a number of assembly stations in-line with each other. At each station, one component is added in simple in-line systems. More complex systems may add several or many components at each station.

Inner-Lead Bond—A TAB bond to a bonding pad on an IC (an inner lead), as opposed to bonds on the circuit or interconnect structure (outer lead bonds).

Insertion-Mount Component (IMC)—A leaded component designed for mounting to a printed-wiring board by insertion of its leads through holes in the board, and then the soldering of those leads to circuitry on the substrate surface(s).

Integrated Circuit—A monolithic microcircuit consisting of elements inseparably associated and formed in situ on or within a single substrate (such as silicon).

Interfacial Bond—An interconnection made between conductors on two separate conducting layers of a substrate.

Intermetallic—Of, or pertaining to, a compound formed by two or more metals where the compound has a definite composition, its own unique crystal structure, and properties widely divergent from its constituent materials. Intermetallics are often refractory and brittle. Intermetallic formation in solder joints raises concern about the predictability of the solder joint's reliability.

Invar—A trademark of the Carpenter Technology Corporation. A metal alloy consisting of 64% iron and 36% nickel. Its notable properties are high strength, good thermal conductivity, and a low TCE.

I/O—The abbreviation for input/output. Generally applied to the outer interconnects of an IC or a PWB assembly, it refers to those interconnects through which a given device or assembly communicates with related devices or assemblies in an electronic system.

IPC—An acronym for the Institute for Interconnecting and Packaging Electronic Circuits (IPC), located at 7380 North Lincoln Avenue, Lincolnwood, IL 60646 USA.

IR—An acronym for "infrared."

ISHM—An acronym for the International Society for Hybrid Micro-electronics, whose address is: P.O. Box 2698, Reston, VA 22090 USA.

I/SMT—An acronym for the Interconnect/Surface Mount Technology division of ISHM. *See also* ISHM.

ISO—An acronym for the International Standards Organization.

JEDEC—An acronym for the Joint Electron Device Engineering Council, which is the component standardization arm of the EIA, Washington, DC. *See also* EIA.

J Lead—A lead form commonly used on PLCCs. The lead departs from the package body at about the Z-axis center of the body, turns immediately toward the bottom of the component, and is formed in a radius (resembling the base of the letter J), turning under the package body. It then terminates in a protective pocket under the body.

K—Symbol for the dielectric constant of an insulating material. *See also* K factor.

Kevlar—A trademark of New England Ropes, Inc. Also, a tradename for a high-strength polymer. In SMT, Kevlar is used as a reinforcing and TCE restraining fibre in polymer PWB materials. Kevlar's TCE restraint is the result of the fiber's property of becoming shorter and fatter when heated. Thus, the warp and woof of woven Kevlar cloth restrains a PWB's X and Y expansion, but amplifies its Z-axis expansion.

K Factor—A term for the value of thermal conductivity of a material. *See also* K.

Kirkendall Voids—These are voids which occur along the interface between two dissimilar materials that have different diffusion rates. The higher diffusion-rate material develops these voids as it diffuses into the adjacent substance faster than it is replaced by that substance.

Land—The preferred term for a metallized area used for the attachment of a surface-mounted component termination to a substrate. Also called a *pad*, which is not preferred because of a potential confusion with via pads.

Laser Reflow—Refers to soldering using a laser, with the laser generally emitting in the infrared wavelengths, to impart a short burst of precisely located heat to reflow a solder joint.

LCC—An early acronym for "Leadless Chip Carrier." Not favored now because of the confusion with "Leaded Chip Carrier."

L Cut—An L-shaped notch made in a resistive material in a resistor-trimming operation. The first part of the trim is made perpendicular to the resistor longitudinal axis to rapidly increase the resistance. As the desired value is approached, the trimmer turns the cut 90° so that the rate of resistance change slows and the final value can be more precisely controlled.

LDCC—The preferred acronym for "Leaded Chip Carrier."

Leach—The action of one material diffusing into another. In soldering, leaching applies to the diffusion of metals from the device and/or PWB metallizations into solution in molten solder.

Leaded Chip Carrier (LDCC)—A chip carrier with leads. The leads may be in the form of an integral lead frame or may be post attached to a leadless chip carrier.

Lead Frame—A metal interconnect structure used to provide I/O terminals for an electronic package or substrate.

Leadless Chip Carrier (LLCC)—A chip carrier, usually ceramic, with no external lead wires. I/O connections are through metallizations around the periphery at the bottom and/or the sides, and/or the top of the device. Also uses the less precise acronym of *LCC*.

Leadless Pad Grid Array (LPGA)—A package format used primarily for high-pin-count ICs. The leadless SMT package provides I/O connections through metallized pads arrayed in a matrix on its base.

Lid—A cover for a hermetic package, used with hybrid circuits and chip carriers for mechanical, and/or hermetic protection of uncased dice. Also, means "To install such a cover."

LID—An acronym for "Leadless Inverted Device." An open ceramic carrier for small semiconductor dice. The LID is designed for surface mounting by reflow attachment of four bottom-side I/O metallizations.

Line Definition—A term for the precision of the line width of a screen printing process. The value is obtained by dividing twice the line edge variation measurement by the line width. The value is typically expressed as a percent.

LLCC—The preferred acronym for "Leadless Chip Carrier."

LPGA—An acronym for "Leadless Pad Grid Array." Also, an acronym for the *Ladies Professional Golf Association.*

Machine Vision—Optical pattern recognition used as a sensor to facilitate the computer guidance of automated equipment.

Manhattan Effect—A term for a solder defect in which a small chip component raises one termination, producing an open, during reflow soldering. The defect is caused by unbalanced forces of solder wetting and surface tension acting on the two ends of the chip. When this imbalance is great enough to overcome gravity, the component begins to stand on one end. The term is best used for parts standing straight up, but is sometimes applied to parts in any stage of erection which produces a solder open on the raising end. *Synonymous with* Tombstone (preferred), Drawbridge, Pop Wheelies.

Margin—The distance from the edge of a substrate to the beginning of the features of the substrate.

Mask—A thin metal sheet with etched features. Masks are used in lieu of screens for stencil printing, and in vapor deposition to define the thin-film deposition areas. Also, means "To cover or protect with a mask."

Matrix Tray—A shipping container specifically designed to protect relatively fragile components, such as bare IC dies. The matrix tray is named for its egg-crate matrix of pockets for part storage. Also called a *Waffle Pack*.

Megahertz (MHz)—A term meaning one million hertz (one million cycles per second).

MELF—An acronym for "Metal Electrode Face (Bonded)." A cylindrical package for two-pin devices. Similar to a DO-35 package, but with end metallizations instead of lead wires. The part takes its name from the common practice of face-bonding metal caps over the ends to form terminations.

Mesh Size—A measurement of the size of the openings in a screen's cloth; expressed as the number of openings per linear inch.

Metal-to-Glass Seal—*See* Glass-to-Metal Seal.

Migration—The motion of ions from a material into an adjacent material where the concentration of ions is lower. Often, migration is an undesirable condition, such as in dendritic growth, leaching, or the migration of plasticizers from a polymer.

Misregistration—The offset error between the actual and theoretical X, Y, and theta dimensions of the tooling features to the working features of a substrate.

MLB—An acronym for "multilayer board."

MLC—An acronym for "monolithic ceramic capacitor."

MLL-34—A cylindrical MELF-type package often used for diodes, usually about 1.5 mm (0.059 inch) in diameter by 3.5 mm (0.138 inch) long. Very similar to, and sometimes interchanged with, the SOD-80 package.

MLL-41—A cylindrical MELF-type package often used for diodes, usually about 2.5 mm (0.098 inch) in diameter by 5 mm (0.197 inch) long.

Monolithic Ceramic Capacitor—A term sometimes used for a chip capacitor laminated from a ceramic multilayer construction.

Morphology—As applied in engineering, the science and study of the form, size, and shape of things. For example, in SMT, Quality Control is concerned with the morphology of solder pastes.

Multilayer Board (MLB)—A composite wiring board, generally laminated from multiple layers, with each having a dielectric and at least one conductive layer. Successive MLB layers are generally interconnected by through-plated vias.

Multiple-Circuit Layout—A circuit layout, or CAD design of an array of identical circuits, used to fill a substrate that is greater than twice the size of a single circuit.

NASA—An acronym for the National Aeronautics and Space Administration, the U.S. governmental space agency. Its address is: NASA Scientific and Technical Information Facility, Technology Utilization Office, P.O. Box 8757, Baltimore, MD 21240-0757 USA.

NASA-JPL—An acronym for the NASA Jet Propulsion Laboratory at the California Institute of Technology, 4800 Oak Grove Drive, Pasadena, CA 91109 USA.

Neck Down—A term for the reduction of the width of conductors only in those areas where they closely approach adjacent features. Using relatively large lines, and necking down the traces only where necessary, can improve PWB manufacturing yields, thus reducing cost without sacrificing high density.

Noise—Any signal, particularly a random persistent disturbance, that obscures the intended circuit signals.

Ohms/Square—The unit of measure of sheet resistance. It is the resistance exhibited by a square pattern of the resistive material when printed at a given thickness. The area of the square pattern has no bearing on the measured resistance, but it controls the dissipation ability of the film resistor.

OLB—An acronym for "outer lead bond."

Open—A solder defect in either the reflow or flow-soldering process that occurs when insufficient solder is applied to a joint in effecting a connection (see *Skip*), when poor wetting causes solder to fail to make a connection (see *Tombstone*), or when a previously formed electrical connection fails and approaches infinite resistance. An antonym is: *Bridge*.

Orange Peeling—A defect of solder masks printed over tin or tin/lead-plated traces on a PWB. The defect occurs when the high temperatures of the PWB surface during SMT soldering operations reflow the metals under the solder mask. When this occurs, the mask takes on a rippled surface texture. In extreme cases, the mask may crack or open. The presence of "orange peeling" indicates the reflow of underlying metals, which points to another serious potential defect, the capillary migration of solder from reflow lands through traces. Such capillary migration may remove enough solder to cause solder-starved joints or open connections.

Outer Lead Bond (OLB)—A TAB bond connecting the TAB lead frame to a substrate, or interconnect structure as opposed to the "Inner Lead Bonds," which connect the TAB lead frame to the IC.

Outer Lead Bonder—A machine for the outer lead bonding of TAB devices.

Overglaze—A glassy coating applied over circuitry or over components to provide mechanical and electrical protection. Also, "To apply such a coating."

P&P—An acronym for Pick & Place. *See* Placement Machine.

Package—1. A housing or protective covering for an electronic component. 2. A housing for a circuit assembly (usually a hybrid circuit). 3. A PWB assembly. 4. A housing for circuit board(s) or an electronic system. 5. To design or assemble a circuit assembly.

Packaging—1. The act of putting an item(s) in a housing or protective covering. 2. The act of designing an electronic circuit-board assembly—a package. 3. The material forming a protective package.

Pad—1. An annular ring around a via. Sometimes imprecisely used to mean the land for a surface-mount device. 2. A bonding site.

Pad Cap Layer—A surface layer of an MLB having only SMC lands, short or nonexistent connecting traces, and vias to internal routing layers.

Pads-Only Layer—*See* Pad Cap Layer.

Parasitic—Refers to capacitance, inductance, or resistance not designed into a circuit, but produced by circuit interconnections. Parasitics cause losses extraneous to those designed into circuit components.

Parasitics—Parasitic losses in a circuit.

Passive—1. A circuit element which does not change its basic characteristics when an electrical signal is applied to it. Resistors, capacitors, inductors, and transformers are some of the passives. 2. Of, pertaining to, or describing a circuit element which performs its intended electrical function without changing its basic character to do so.

Pass Through—A type of assembly line equipment where the workpieces are automatically transported from an entry point on one end of the machine to an exit point on the other end. Also, a line made up of such machines, and linked to automatically pass workpieces through the line.

Paste—1. A formulation of tiny solder spheres or powder, a liquid flux, a thixatropic vehicle, and, in some instances, a flux activator. Solder pastes may be designed for application by screening, stencil printing, or dispensing. *Synonymous with* Cream. 2. Any screenable material for thick-film processing. *Synonymous with* Ink.

Paste Blending—The mixing of two or more inks to obtain a paste with properties between those of the original two. Resistor pastes are sometimes blended to achieve an ohms/square value between the values of the two original materials.

PCB—An acronym for "printed-circuit board."

PFIB—An acronym for "Perfluoro Isobutelyne," a highly toxic decomposition product of perfluoronated hydrocarbon fluids (such as those used for vapor-phase soldering). PFIB is generated when perfluoronated hydrocarbon fluids are subjected to extremes of high-temperature stress.

PGA—An acronym for a "pin grid array." *See also* LPGA.

P/IA—An acronym for "printed interconnect assembly." An IPC-recommended term for a PWB assembly.

Pick & Place Machine—*See* Placement Machine.

Pin—1. A lead of an electronic component. 2. A small post forming an interconnect facility when inserted in a PWB.

Pin Grid Array (PGA)—An IMC package used for high-pin-count devices. The pin grid array has a matrix of I/O pins extending from its base. *See also* LPGA.

Pin Transfer—A printing method employing pin(s) to pick up, and transfer to a printing site, a paste or ink. Used in SMT processing for adhesive application, and occasionally for solder pastes.

P/IS—An acronym for "printed interconnect structure." The IPC-recommended term for a bare PWB.

Pit—A solder defect. An opening or cavity in the surface of a solidified solder joint. It is usually caused by entrapped gas or volatiles bubbling up through the solder as it is solidifying. It may indicate the hidden presence of voids. *See also* Void.

Placement—The act of locating an SMC on its correct solder lands on a PWB.

Placement Machine—An automated or semiautomated machine used for the assembly of SMT components onto PWBs. Often called a "Pick & Place" machine because it functions by picking components from specified delivery mechanisms and placing them at programmed locations on a PWB assembly.

Plastic Leaded Chip Carrier (PLDCC or PLCC)—A plastic IC chip-carrier package having J-leaded SMT I/O on four sides. Both premolded and postmolded parts are available in this package format.

Plastic Quad Flat Pack (PQFP)—A fine-pitch postmolded plastic surface-mount semiconductor package with a gull-wing lead form, peripheral I/O on all four sides, and is distinguished by extended corners which form bumpers for the protection of the leads during part handling and transport.

Plated-Through Hole (PTH)—A hole in a PWB with a plated metallization on its inner walls, which allows the interconnection of circuit metallizations on the opposite sides of two-sided boards and/or an interconnection with the inner-layer metallizations of an MLB.

PLCC—An acronym for "Plastic Leaded Chip Carrier."

PLDCC—The preferred acronym for plastic leaded chip carrier.

Poise—The unit of measure of the coefficient of dynamic viscosity. One poise equals 1 dyne-sec/cm^2.

Pop Wheelies—The physical action that creates a solder defect in which a small chip component raises one termination, thereby producing an open, during reflow soldering. The defect is caused by the unbalanced forces of solder wetting and surface tension acting on the two ends of the chip. When this imbalance is great enough to overcome gravity, the component begins to stand on one end like a drawbridge opening. The term is used for parts in any stage of erection which is sufficient to produce a solder open on the raising end. *Synonymous with* Tombstone (preferred), Drawbridge, Manhattan Effect.

PPM—An acronym for "parts per million."

PQFP—An acronym for "Plastic Quad Flat Pack."

Preform—A small preshaped piece of material, such as solder, for placement at a soldering site. Some solder preforms also have a layer of flux preapplied. Epoxy and hot-melt adhesives may also be applied as preforms.

Printed-Circuit Board (PCB)—The precise term for a substrate containing both interconnect wiring traces and in situ components (circuitry). Often used when referring to printed-wiring boards as well, but

printed-wiring boards are actually printed interconnect structures only, having no in situ circuitry.

Printed-Wiring Board (PWB)—The preferred term for a wiring board fabricated by photolithographic print-and-etch techniques, and which contains only printed interconnect structure(s) with no components (circuits) formed in situ. Also, sometimes called a Printed-Circuit Board (PCB), which is not preferred, since the board is not actually an electronic circuit but merely the wiring for a circuit assembly.

Process Control—1. The science of improving manufacturing yields by bringing the variability of each critical element of a manufacturing process under tight control. 2. The measurement and control of a given variable in a manufacturing process.

Process Quality—The percentage of product which conforms to specifications as it leaves the manufacturing line and arrives at final test.

Profile—1. The time-temperature relationship experienced by a workpiece in a process such as linear furnace soldering, with given furnace settings. The X–Y plot of such a time-temperature relationship. 2. To adjust and correctly set the time-temperature relationship of a piece of equipment.

Propagation Delay—The time delay between the input and output of a signal.

PTH—An acronym for "plated-through hole."

Pull Test—A test for the force required to pull a component from its mounting lands with the force acting only in the +Z vector.

Push-Off Test—A test for the force required to remove a component from its mounting lands by a single axis push that is acting only in the X or the Y axis.

PWB—An acronym for "printed-wiring board."

Quad Flat Pack—Any flat pack with a peripheral I/O on all four sides.

Quad Pack—Literally, a semiconductor package having I/O on all four sides at its periphery. Generally usage does not include leadless components as quad packs. Also, packages with a distinct name, such as PLCCs, PQFPs, CQPFs, etc., are not called quad packs. It is the generic name for four-sided I/O-leaded semiconductor packages which do not have some other specific name.

Quality—The conformance to specifications by a new device or system.

Radial—Of or pertaining to an electronic component lead form in which the leads extend from the outer radius of the device tangentially. Also, such a device.

Radio-Frequency Interference (RFI)—Unwanted (interference) signals received in a circuit from external or internal radio-frequency noise sources.

RAM—An acronym for "random-access memory."

Random-Access Memory (RAM)—A memory in which data may be read or written in any selected binary address location.

Real Estate—Refers to the space on a printed-wiring board; its square area.

Reference Edge—The term for the edge or edges of a substrate which are to locate it for manufacturing processes.

Reflow—A type of soldering that involves the melting of previously applied solder to form device solder bonds on a circuit assembly.

Reliability—The continued conformance to specifications by a device or system over a specified extended period of time.

Repair—The work done to a circuit assembly to cause it to operate, but not necessarily to bring it into complete conformance with its original specifications. Also, the act of performing such work. *See also* Rework.

Resist—A dielectric coating used to mask selected areas of a PWB and prevent solder from adhering to metallizations. Resists may be temporary, allowing the selective solder coating of PWB features during board manufacture, or permanent, as in the case of solder masks.

Resistor Drift—A measure of the change in resistance of a resistor over an extended period. It is generally expressed as a percentage of change in resistance over 1000 hours.

Resistor Overlap—In hybrid circuitry, the overlap distance of the resistor print and its conductor trace.

Reverse Reading—In photolithography, a photo tool in which the desired pattern is seen as a mirror image when viewed from the emulsion side of the film.

Rework—Work done to a circuit assembly to bring it into complete conformance with its original specifications. Also, the act of performing such work.

RFI—An acronym for "radio-frequency interference."

Right Reading—In photolithography, a photo tool in which the pattern is seen as the required circuit image when viewed from the emulsion side of the film.

Riser—In hybrid MLBs, a via formed by thick-film print and fire techniques.

Runner—In a PWB layout, an individual conductor. *Synonymous with* Trace (preferred), Track.

Sapphire—The monocrystalline form of aluminum oxide (Al_2O_3). Used as a substrate material, where its highly polishable surface finish, low camber, radiation hardness, and/or uniform lattice structure is needed. Also used for high-speed silicon-on-sapphire MOS devices. *See also* Alumina.

Scallop Marks—Screening terminology for a defect where the edges of a printed pattern are jagged. The defect is seen when printing pressures are incorrect or the screen emulsion is too thin.

Scavenge—Also called "leaching." An action that causes one material to diffuse into another. In soldering, leaching applies to the diffusion of metals from the device and/or PWB metallizations into solution in molten solder.

Score—Hybrid terminology for the process of scribing break lines on ceramic, silicon, glass, or sapphire snapstrates.

Screen—The image-defining mask used in screen printing. Also, the act of using such a mask to print.

Screenability—A measure (often subjective) of the quality of print definition in a screen printing operation.

Scribe Coat—A two-layer polymer film with a clear base and a semiopaque cover that is used for preparation of semiconductor and microcircuit artwork. Circuit patterns, in an enlarged scale, are defined on the film by cutting and stripping away unwanted portions of the opaque cover, using a coordinatorgraph.

Self-Passivating—A term used to describe materials which inherently form a passivating overglaze when fired in a thick-film process. Some resistor inks are self-passivating.

Sequencer—A machine designed for extracting leaded components from shipping container carrier tape reels and placing them on an assembly tape reel in the sequence in which they will be inserted in a PWB assembly.

Sequential Placement—A class of placement equipment which programmably places components sequentially, one at a time.

Sequential Simultaneous Placement—A class of placement equipment which programmably places components sequentially, and which places more than one component during each machine cycle (sequence).

Serpentine Cut—A serpentine-shaped or sinusoidal trim cut made in a resistor.

Shadowing—A term for reduced heating in infrared ovens on areas of a workpiece when the adjacent objects on the assembly shield those areas from IR radiation. Now largely eliminated by oven design improvements, whereby a greater percentage of the total energy transfer is by conduction and convection than it was in earlier furnace designs.

Shear Rate—The relative rate of flow of a viscous fluid.

Sheet Resistance—The measured resistance of a square pattern of controlled-thickness film of resistive material, expressed in ohms/square. *See also* Ohms/Square.

Silicon Monoxide—A material that is sometimes vapor deposited as a passivating layer for thin-film circuits.

Silver Chromate Paper Test—A qualitative test for the presence of organic halides. Used in SMT soldering to test RMA and other mildly activated fluxes to determine that no organic halides are present in the activators. *See also* Silver Nitrate Test.

Silver Nitrate Test—A qualitative test for the presence of organic halides. Used in SMT soldering to test RMA and other mildly activated fluxes to determine that no organic halides are present in the activators. *See also* Silver Chromate Paper Test.

Simultaneous Placement—The placement of a full complement of SMCs on an assembly, or the placement of a large number of SMCs simultaneously. Also, a class of placement machines which operate in this fashion. Generally, "simultaneous placement" refers to inflexible dedicated automation. Also called "Gang Placement."

Single In-Line Package (SIP)—A package format that is often used for resistor networks or ICs. The package is named for its single line of I/O pins extending from one side. SIP leads are designed for insertion mounting in PWBs.

Single-Layer TAB Tape—In TAB technology, a TAB tape consisting only of a metal foil which is etched to produce repetitive lead-frame patterns and sprocket holes for handling.

SIP—An acronym for "single in-line package."

Skip—An open-circuit defect in flow soldering, where insufficient solder to make a desired solder connection is applied to a solder joint. *Antonym:* Bridge. *See also* Open.

Slump—To spread after being deposited. This applies to viscous fluids and thixatropic pastes. The slumping of screened ink before or during firing is a defect which blurs screened image definition.

SM—An acronym for Surface Mount, Surface Mounted, and Surface Mounting. See definition under *Surface-Mount Technology.*

SMA—An acronym for "surface-mount assembly."

SMC—An acronym for "surface-mount(ed) component."

SMD—An acronym for "surface-mount device," SMD is a trademark of Philips. See definition under *Surface-Mounted Component.*

SMOBC—An acronym for "Solder Mask Over Bare Copper."

SMT—An acronym for "surface-mount technology."

Snapback—The return of the screen to its at-rest position after a squeegee stroke.

Snap-Off—A measure of the distance between the top of the substrate and the bottom of the undeflected screen when they are mounted on a screen printer. In printing, the squeegee deflects the resilient screen causing it to contact the board. After the squeegee stroke, the screen snapback lifts the screen from the substrate and returns it to its original at-rest clearance.

Snapstrate—A multiple-circuit substrate, which is scribed so that it may be broken (snapped) into individual assemblies after batch assembly processing.

SO—An acronym for "small-outline." Refer to SOB, SOD, SOT, SOIC, and SOJ IC.

SOB—1. An acronym for "small-outline bridge." A four-leg SOT-143-like package with an oversized body for housing a full-wave bridge rectifier. 2. The suppressed crying sound made by engineers in reaction to the proliferation of electronics acronyms.

SOD—An acronym for "small-outline diode"; a MELF-type package.

SOIC—An acronym for "small-outline IC," a package format having I/O on two sides. The leads exit the component body around its center line and are formed into a gull-wing shape for surface mounting. The JEDEC standard SOIC body is nominally 3.8 mm (0.150 inch) wide, and the package length varies dependent on the number of 1.27-mm (0.050-inch) center leads per side.

SOJ IC—An acronym for "small-outline J-leaded IC." An IC package similar to the SOIC, except that its body is wider and its leads turn under the body in J-lead form rather than turning out in the "gull-wing" fashion.

Solder—1. A low melting point metal or alloy used to join metal(s) having a higher melting point. For SMT, the tin/lead alloys are the most common solders. 2. To join several metals using solder.

Solder Ball—A solder defect where solder forms in small spheres outside the solder-joint area. Solder balls are difficult or impossible to remove from under low-clearance components, but can move while in service, causing short circuits and catastrophic failures. Therefore, solder balling is considered a serious defect that is to be completely avoided. Solder balls may result from the use of overly oxidized solder paste or the violent outgassing of volatiles from improperly baked solder paste during rapid reflow heating.

Solder Bridge—A short-circuit defect produced when solder crosses a dielectric space and electrically joins two leads or metallizations which are supposed to be electrically isolated.

Solder Bumps—Solder hemispheres grown on, or added to, a semiconductor die to facilitate interconnection of the chip. *See* Flip Chip.

Solder Mask Over Bare Copper—A PWB manufacturing technique that involves the application of the solder mask over bare copper traces, with subsequent operations to apply a tin/lead overplate to the solder lands left exposed by the mask. The SMOBC technique eliminates solder-filled capillaries under the mask, thus improving reflow solder yields.

Solder Paste—A solder formulation of tiny solder spheres or powder, a liquid flux, and a thixatropic vehicle, plus, in some instances, a flux activator. Solder pastes may be designed for application by screening, stencil printing, or dispensing.

Solder Side—In IMC boards, the side of the board opposite the components. This is the side of the PWB that is immersed in solder during the flow soldering of through-hole component leads.

SOL IC—An oxymoronic acronym for Small-Outline Large IC, a package format similar to the SOIC, but with a wider body that is capable of housing a larger die. The JEDEC standard SOL IC body is nominally 7.6 mm (0.300 inch) wide, and package length varies dependent on the number of 1.27-mm (0.050-inch) center leads per side.

SOM IC—An acronym for Small-Outline Medium IC. A nonstandard package that is sometimes used to house resistor networks with substrates too large to fit in an SOIC.

The package is like the SOIC, but it has a nominal body width of 5.6 mm (0.220 inch).

SOS—1. An acronym for "silicon-on-sapphire," a high-speed device in MOS IC technology. 2. An acronym for "silicon-on-silicon," a technology that makes use of a silicon substrate.

SOT-23—An acronym for Small-Outline Transistor (style 23), it uses a JEDEC TO-236 package. A small-signal transistor package. Package is also used for diodes, dual diodes, LEDs, etc. The package has three leads and can house dies up to about 0.6 mm (0.025 inch) square. Its maximum dissipation in free air at 25 °C is 200 milliwatts.

SOT-89—An acronym for Small-Outline Transistor (style 89), it uses a JEDEC TO-243 package. A small-outline transistor package capable of handling higher-power applications than the SOT-23. Suitable for dies up to about 1.5 mm (0.060 inch) square. The package can dissipate a maximum of 500 milliwatts in 25 °C free air.

SOT-143—An acronym for Small-Outline Transistor (style 143). A small-signal transistor package, that is also used for dual diodes, GaAs FETs, RF transistors, etc. The package has four leads and can house dies up to about 0.6 mm (0.025 inch) square. It dissipates a maximum of 200 milliwatts in free air at 25 °C.

Space—In PWB layouts, the minimum distance between the adjacent edges of two conductors (not the center-to-center distance).

SRAM—An acronym for "static RAM," a memory device which does require periodic clocking or refresh, to maintain memory integrity. *Contrast with* DRAM.

Stair-Step—Term used to describe a screen printing defect where the print edges show the distinct pattern of the screen mesh. This defect is produced by inadequate printing pressure or by the use of screens with too thin an emulsion.

Standoff—In SMT, the distance between the top of the substrate and the bottom of an SMC mounted on it. An adequate standoff height is critical to the cleaning of surface-mount assemblies.

Stencil—Screening terminology for a metal mask that is used in lieu of a screen for printing. For solder printing, stencils are more expensive and require longer lead times than screens. However, stencils last considerably longer, are easier to clean and maintain, and are capable of reproducing finer printed patterns than screens.

Step and Repeat—A process wherein a pattern, or a sequence of automated machine motions, is repeated one or more times at equal increments in the X and/or Y axis. For instance, a small circuit pattern may be stepped and repeated multiply on a large substrate.

Stick—A plastic shipping container for components. Sticks also may be used as a component delivery mechanism for many types of automated processing equipment.

String—1. The name for the phenomenon evidenced when internal cohesive forces of a dispensed fluid or gel cause a part of the dispensed droplet to draw up as a tail, following the withdrawing dispensing tool. Excessive stringing must be avoided, because such tails, after breaking their cohesive connection with the departing tool, may fall into areas of the assembly where they will be a quality or reliability problem. 2. The action of stringing.

Substrate—In an electronics assembly, the supporting material, with or without its printed interconnections, upon which components are mounted or fabricated to produce a circuit assembly. A multilayer board is spoken of as a single substrate, even though it is laminated from several layers of composite materials.

Surface-Mounted Component—A component designed for planar mounting as opposed to through-hole or chip-and-wire mounting. Variants are standard SMT components, flip chips, and TAB formats.

Surface-Mount Technology—The science of planar mounting components on a PWB or on hybrid-printed interconnect substrates.

Surfactant—A contraction of *SURFace ACTive AgeNT*. A material which reduces the surface tension of liquids.

Swim—The act of X–Y motion of components during reflow soldering. Swimming is the phenomenon seen when a part displaces more than its own weight in solder, thus floating during the reflow process. Floating parts tend to move to equalize the forces of wetting and solder surface tension. With good wettability and proper design, these forces cause parts to center in the lands.

TAB—An acronym for "tape automated bonding."

Tape—1. A carrier mechanism for shipping, storing, and automatically feeding SMCs to assembly processes. Carrier tape has pockets cut or embossed in it. Components are retained in the pockets by means of cover tape. The carrier tape has peripheral sprocket holes to facilitate the precise feeding of pocketed components to an assembly machine. Cover tape is peeled back to expose one component after another as they are required for assembly. 2. The act of installing components in carrier tape. NOTE: There are numerous other electronics usages of the word *Tape*, derived from its standard dictionary definitions. The above definitions and that shown under *Tape Automated Bonding* are listed herein as they are unique to the SMT process.

Tape and Reel—A term to distinguish carrier tape from other forms of tape used in electronics (axial and radial components are commonly taped for automatic insertion, but such tape is not called *Tape and Reel*, or *Carrier Tape*). The phrase derives from the reel storage system used for the shipping and handling of carrier tapes.

Tape Automated Bonding—A high-density approach to interconnecting IC dies to substrates or leadframes. A tape of either metal foil or a foil/polymer composite is used to carry and provide an interconnect to the die. Sprocket holes are formed in the periphery of the tape to facilitate precise automated handling. The metal foil is etched to form a leadframe. Inner leads of the tape leadframe are thermocompression bonded to the IC die. At final assembly, the chip and leadframe portion are excised from the outer carrier portion of the tape. To interconnect the TAB chip to its higher-level assembly, the outer leads are thermocompression or solder bonded to the lands of the higher-level interconnect structure. *See* TAB.

TCE—An acronym for "Thermal Coefficient of Expansion."

Termination—A lead or metallization of a component, which provides a point to interconnect an I/O line of the component to other circuitry.

Tg—An acronym for "Transistion Temperature, Glass." In polymer technology, the temperature above which the thermal expansion rate of the polymer increases dramatically and changes from a linear curve to an asymptotic curve.

TGA—An acronym for "thermogravimetric analysis."

Theoretical Origin—In an artwork or pattern generated from an artwork, that point from which all features are measured in the X and Y offset.

Thermal Coefficient of Expansion—Also called *Coefficient of Thermal Expansion (CTE)*. A measure of the coefficient of expansion of a material under temperature increases. Expressed in parts per million (PPM)/°C. This is critical in SMT where leadless parts have no flexural element to absorb the stress generated by the disparate expansion of substrate and component during temperature cycling. *See* TCE.

Thermal Stress—1. Stress generated by disparate expansion rates between separate bonded bodies in an assembly during temperature cycling. 2. Stress developed within a homogeneous body due to temperature differentials within the body during temperature cycling. (This sense of the phrase is also referred to as "thermal shock.") 3. Electrical and mechanical stress produced by operation at elevated ambient temperatures.

Thermocompression—A type of bonding process in which pressure and elevated temperatures are used to join two materials by interdiffusion across their boundary.

Thermogravimetric   Analysis—A quantitative analytical test used to determine the cure state of a polymer resin by observing the effect of temperature changes on the sample's weight. *See also* TGA.

Thermomechanical   Analysis—A quantitative analytical test used on polymer PWBs to determine their dimensional expansion curve in relationship to temperature. TMA can be used to pinpoint the Tg of a polymer material. *See also* TMA.

Thick Film—A film generated on a substrate surface by a screening or stencil-printing process. Such films are generally greater than 100 microns (0.004 inch) thick. Typical SMT thick films include conductive, resistive, and dielectric films.

Thin Film—A film accreted on a substrate surface, generally by a vapor-deposition or sputtering process. Such films are generally less than 5 microns (0.0002 inch) thick, and, more usually, under 0.3 microns (0.00001 inch). Typical SMT thin films include conductive, resistive, and dielectric materials.

Thixatropic—A term used to describe the property that certain gels and liquids have to liquefy when stirred or placed under dynamic pressure, but to return to the previous viscous state when static. Thixatropic agents allow solder paste to flow through a screen under squeegee pressure, but stand without slumping on the substrate after printing.

Three-Layer TAB Tape—In TAB technology, a tape comprised of a base polymer layer (generally polyimide) with necessary handling features, such as sprocket holes, an adhesive reinforcing layer running the full width of the tape, and a metal-foil layer etched to form the TAB leadframes. Generally, the metal foil of three-layer tapes covers only the center portion of the tape and is just wide enough to encompass the needed leadframe features.

Through Hole—1. A via hole extending from one side to the other in a PWB. 2. The technology and practice of mounting leaded components by insertion in holes through PWBs.

Through-Hole Component—A leaded component designed for mounting to a printed-wiring board by the insertion of its leads through holes in the board, and then, the soldering of those leads to circuitry on the substrate surface(s). *See also* Insertion-Mount Component (IMC).

TMA—An acronym for "thermomechanical analysis."

Tombstone—The preferred term for a solder defect in which a small chip component raises one termination, producing an open, during reflow soldering. The defect is caused by unbalanced forces of solder wetting and surface tension acting on the two ends of the chip. When this imbalance is great enough to overcome gravity, the component begins to stand on one end. The term is best used for parts standing straight up, but is sometimes applied to parts in any stage of erection which produces a solder open on the raising end. Also, the act of tombstoning. *Synonymous with* Drawbridge, Pop Wheelies, Manhattan Effect.

Top Hat—A descriptive term for a type of thick-film resistor geometry resembling a top hat. Top-hat resistors have a projection from one or both sides (like the brim of a top hat). The projection(s) allow a trim notch to be cut into the center of the resistor, creating a serpentine geometry and thereby increasing resistance. Top hats are recommended where high aspect ratios are required.

Trace—In PWB layout, an individual conductor. Also, the width of an individual conductor. *Synonymous with* Runner (not preferred), and Track.

Trace and Space Rules—In PWB layout, the minimum allowable trace and space dimensions.

Track—In PWB layout, an individual conductor. *Synonymous with* Runner (not preferred), and Trace (preferred).

Trim—To remove material (cut a notch) in a film resistor or other component to adjust the value of the component. Laser and abrasive trimming are two typical processes.

Tube—A plastic shipping container for components. Tubes also may be used as a component delivery mechanism for many types of automated processing equipment. *Synonymous with* Stick.

Two-Layer TAB Tape—In TAB technology, a tape made up of a base layer of polymer film (such as polyimide) and a metal-foil layer etched to form the TAB leadframe. Sprocket holes and handling features are fabricated in the periphery of the composite tape to permit the precise indexing of the material in assembly operations.

Type 1 SMT—A surface-mount assembly style that employs surface mounting exclusively, and is characterized by the reflow soldering of all solder connections. Type 1 SMT may be used on either a single side or both sides of a board.

Type 2 SMT—A surface-mount assembly style that employs SMCs mixed with through-hole components on the same side of a board assembly, and is characterized by the reflow soldering of SMC solder connections and the flow soldering of IMCs where they are mixed on the same side. Type 2 SMT may involve double-sided population using any style of assembly, but is still considered Type 2 as long as one side mixes SMCs and IMCs.

Type 3 SMT—A surface-mount assembly method characterized by having IMCs on the classical component side of the board, and SMCs glued to the bottom side of the board. After assembly, the board is positioned with the SMCs down and is flow soldered, simultaneously forming through-hole and SMC solder bonds.

Ultraviolet Cure—A curing process used to activate the cross linking of certain epoxy adhesives by exposing the material to strong ultraviolet radiation. Also used to cure certain silk-screened inks, such as those used on some PWB silk-screen legends.

Underglaze—1. A glass or ceramic glaze that is fired onto a substrate under the area where a resistor will be screened and fired. 2. To apply such a glaze.

Upset—When applied to solder joints in electronics, this term denotes the condition that is also called a "cold joint," where two or more pieces being joined by solder moved during the plastic phase of the solder-solidification process. Such joints are reliability concerns. They are recognized by a grainy, rough, surface finish.

Vapor-Phase Soldering (VPS)—A reflow-soldering method, which uses a vapor blanket above a boiling fluid to conductively heat a workpiece. The heating is rapid, since vapor condensing on the relatively cool workpiece yields its latent heat of vaporization to the workpiece. For typical SMT solder requirements, high-boiling-temperature perfluorinated fluids are used. The most commonly used material boils at 215 °C.

VCD—An acronym for "variable center distance," a style of insertion machine capable of programmably adjusting to form-and-insert axials at a variety of lead-center distances.

Via—1. In PWBs, an interlayer connection in a two-sided or multilayered board, provided by a through-plated hole. 2. In hybrids, an interlayer connection provided by printing an upper-layer conductive trace over an opening in the separating dielectric layer. The opening in the dielectric is arranged to expose the lower-layer conductor that is to be interconnected. Also called a *Riser*.

Void—A type of solder defect. A cavity inside a solidified solder joint. Generally caused by entrapped gases or volatiles which are unable to escape as the solder solidifies. *See also* Pit.

VPS—An acronym for "vapor-phase soldering."

Waffle Pack—A matrix tray with "egg crate"-style pockets for the protection of delicate components during shipment, storage, and handling.

Warp and Woof—In screen technology, a description of the lay of the fibers that make up the screen mesh. The warp fibers run parallel to the length of the fabric, and are warped alternatively upward and then downward in the weaving process. The woof fibers run cross-wise to the fabric's length (or parallel to the length of the weaving machine).

Wave Soldering—An immersion soldering method. Workpieces are moved across a wave or waves of molten solder at a height such that the bottom of the board is momentarily immersed. Leaded components are on the board's top-side with their leads inserted through metallized holes in the board. Any SMCs that are to be wave soldered are glued to the bottom (solder) side. Both the leads extending through the PTHs and the bottom-side SMCs are simultaneously soldered in passing through the solder immersion(s). SMT wave soldering generally requires equipment with special features in order to reduce solder defects on the attachment of surface-mount components.

Wetting Agent—A material which reduces the surface tension of liquids. In SMT, a wetting agent improves a cleaning fluid's ability to penetrate under low standoff components.

Wick—A term used to describe flow through a capillary, as when solder flows away from lands through capillaries which are produced by reflowed metal on masked conductor traces. Wicking is likely to produce defects due to the solder starving of SMC joints.

WIP—An acronym for "Work In Process."

Wire Bond—1. An interconnection between components, or between a component and leads of a circuit. A wire bond is made by means of a fine wire bonded to the points to be interconnected. 2. The act of making such interconnections.

Work In Process—Defines the goods involved as raw material, subassemblies, and assemblies in a manufacturing line. A large amount of WIP increases inventory amounts, reduces inventory turns, and generally reduces manufacturing yield while driving costs upward.

"X" Axis—For drafting, normally taken as the up-and-down length of the paper as it is viewed. For automated equipment, convention places the X axis parallel to the flow of workpieces through the machine.

"Y" Axis—For drafting, normally taken as the length across the paper as it is viewed. For automated equipment, convention places the Y axis perpendicular to the flow of workpieces through the machine.

"Z" Axis—For automated equipment, the Z axis is an axis perpendicular to the plane of the workpieces.

Zero Zero—The theoretical origin point of an artwork or the pattern generated from an artwork. That point from which all features are measured in the X and Y offset.

ZIP—An acronym for "zigzag in-line package." An IMC package format in which leads extend from only one side of the package body, and alternate (zigzag) between two parallel rows, which are generally spaced 2.54 mm (0.100 inch) apart.

Index

A

Accessibility, 143

Active components, 54-69, 383-397, 406-411, 419-431
CLLCCs, 388-391, 410, 427-428
CLDCCs, 391-397, 411
diodes, 54-55, 383, 406
DPAKs, 422
ICs, 57-68
PLCCs, 385-387, 409-410, 424-426
PQFPs and CQFPs, 388, 410
SOICs, 385, 407-408, 422-424
SOJ ICs, 385
SOT-89s, 420-421
TapePaks, 387, 410, 429
transistors, 55-57, 383, 407, 419-420

Adhesives, 81, 97-100, 105, 228-229, 232-238, 240-241, 331
application, 97-100, 232-238
conductive, 331
curing, 81, 105, 240-241
design, 98-100
dispensing, 98
hybrid circuits, 331
quality assurance, 228-229, 232-238
reliability, 228-229
screening and stenciling, 98
single part, 228-229
two part, 229
thermostat, 229
transfer printing, 98
U-V activated, 229

Aerospace applications, 29-30, 371

Aluminum electrolytic capacitors, 45-47, 378, 404

American Society of Testing and Materials (ASTM), 434

ANSI/IPC-SM-782 land patterns, 402-413
actives, 406-411
other components, 412-413
SMT passives, 402-406

Applications, SMT, 10-12, 20-35, 188-190, 313-315, 361-362
cost-driven, 362
disadvantages, 11
size-driven, 361
strategies, 188-190

Aqueous cleaners, 108, 244

Aspect ratio, via, 162

Assembly burn-in testing, 110

Assembly clearance, 144

Assembly flow, 145-147

Assembly integrity tests, 108-111, 248-251
laser solder inspection, 249
mockup tests, 248-249

x-ray analysis, 249

Assembly-level packaging, 185-207
application-determined strategies, 188-190
board flexure, 196-200
environment, 200-207
interconnections, 194-196
patches, 192-194
real estate, 186-188, 190-192
system size, 185-186

Assembly standards, 435, 436

Attachment media , 78-81, 105, 238, 352
application, 238, 352
curing, 80-81, 105
dispensing, 78-79
(See Adhesives, Solder)

Automated Optical Inspection (AOI), 222, 237-240

Automated vision systems, 289-292

Automatic component placement, 8, 19, 79-80, 101, 112-114

Automation, 112-114
factory, 114
hand assembly, 112-113
islands, 113
line, 113

Automotive applications, 20-22, 313-315

B

Bare copper boards, 160

Bare copper traces, 160

Batch cleaners, 107-108, 244

Bed of nails testing, 221-222

Benefits of SMT, 3-35
examples, 20-35

Board-level interconnects, 194-196

Boards, 9-10, 167-168, 187, 192-200, 273-277, 254, 238, 280-282, 467-469
daughter boards, 193-194
design constraints, 273-274
flexure, 196-200, 254
interconnects, 194-196
panelization, 274
PC, cost savings, 9-10
registration, 238
shape, 274
size, 273-274
standards, 280-282
stiffeners, 196-197
TCE, 167-168
test board patterns, 467-469
through-hole, 192-194

Books about SMT, 460-461

Brick style capacitors, 45-47

Built-in test, 257-258

Buried resistors, 192

Burn-in testing, 110, 219, 248

C

CAD for component placement, 103-105

standards, 71, 278-280, 435, 439, 449
switches, 70, 399-401
testing, 87, 269, 304-305
thermal considerations, 167-172
thermistors, 72
variable capacitors, 71
(*See Surface mount components*)
Computer applications, 22-25, 315-316
Computer Integrated Manufacturing, 7-8
Conductive adhesives, 331
Conductor form factors, 178-179
Conductors, transmission line, 178-179
Cone of visibility, 153
Conformal coatings, 331
Connectors, 69-70, 194-196, 200, 204, 333,
    343-347, 397-398, 412
  board-level, 194-196
  cycling, 204
  hybrids, 333, 343-347
  long, 200
Constraints, design, 268-277
Consumer applications, 25-27, 316-317
Contract assembly, 27-29
Convective reflow stations, 112
Coplanarity, 231, 278-279
Copper boards, 168
Copper/Invar/Copper cores, 168
Copper/Molybdenum/Copper cores, 168
Corrosion, 206-207
Cost issues and reductions, 9-10, 293, 362
CQFP packages, 388, 410
Credibility, 294
  design, 294
Crossovers, 333, 342
Crosstalk, 6
Crystals, 70-71, 401, 412
Curing of attachment media, 80-81, 105,
    240-241, 354
  design, 105
  (*See Adhesive, Solder*)
Cylindrical components, 50-51, 380, 405,
    417-418
  resistors, 50-51

D

Database for design and manufacturing,
    277-278
Daughter boards, 193-194
Decoupling capacitors, 189-190
Delivery, 294-295
  design, 294-295
Dense assemblies, 85
Density, board, 5, 84, 156-158, 182, 313
  hybrid circuits, 313
  levels, 156-158
  testing, 84
Design, 85-86, 96-97, 98-100, 103-107,
    113-115, 119-138, 173-182, 257-264,
    267-295, 332-343, 363, 366-367, 436,
    440-441, 455-456

adhesive application, 98-100
capital investment, 133-134
constraints, 268-277
component placement, 103-105
credibility, 294
delivery, 294
environment considerations, 134-136, 363
equipment, 268-273
features, 295
feedback, 136-137
form, fit, and function, 129-133
high density, 182
hybrid circuits, 332-343
information, 455-456
inspection, 288-292
line automation, 113-114
nonmanufacturing issues, 292-295
manufacturing, 133, 267-295
military applications, 366-367
price, 295
reliability, 134, 262-264, 295
repairability, 260-262
reviewing (*See Reviewing of designs*)
service, 295
simultaneous engineering approach, 120-126
soldering process, 96-97, 106-107, 173-175
standards, 436, 440-441
system performance, 119-138
testability, 85-86, 129, 257-262
transmission lines, 175-182
volume consideration of, 114-115
Destructive physical analysis (DPA), 11, 219,
    222, 253-257, 368-369
  board flexure, 254
  environmental aging, 254-256
  microsectioning, 254
  military applications, 368-369
  solder strength, 253
  thermal cycling, 256-257
Device polarity and orientation, 279-280
Dielectric constant of boards, 177
Dielectric layers for hybrid circuits, 330
Dielectric separation, 177-178
Differential scanning calorimetry, 272
Diodes, 54-55, 383, 406
  2-pin packages, 54-55
DIP modules, 187, 212-213, 322-324, 344,
    402, 412, 429-430
  lead frames, 344
  compared with SMCs, 212-214
Disadvantages of SMT, 10-13
Discrete components, 325-326
  resistors, 325
  capacitors, 325
  transistors, 325
  inductors, 325
  thermistors, 326
  other, 326
Dispenser control, 232
Documentation of rework, 292

A Bell Atlantic Company

Howard W. Sams

Your Technology Connection to the Future!

Now You Can Visit Howard W. Sams & Company <u>On-Line</u>: http://www.hwsams.com

Gain Easy Access to:

- **The PROMPT Publications catalog, for information on our *Latest Book Releases*.**
- **The PHOTOFACT Annual Index.**
- **Information on Howard W. Sams' Latest Products.**
- ***AND MORE!***

PROMPT®

PUBLICATIONS

CALL 1-800-428-7267 TODAY FOR THE NAME OF YOUR NEAREST PROMPT PUBLICATIONS DISTRIBUTOR

Optoelectronics, Volume 1
The Introduction
Vaughn D. Martin

This book is the first in a three-part series on optoelectronics. It is the introductory self-teaching text and includes descriptions of basic concepts, photometrics, and optics.

Optoelectronics is an exciting technology which is useful, vitally important, rapidly emerging, and constantly evolving. This text will walk readers through the field at their own pace, and allow them to verify their progress. It provides a thorough understanding of optoelectronics, and bridges the gap between theories and more practical aspects and applications. Equations are used only when no other means of explanation can clearly illustrate a point. Topics covered in Optoelectronics, Volume 1, include terminology and concepts, measuring and testing, visible light-emitting sources, photocells, photodiodes, photomultipliers, LED secondary optics, and more.

Optoelectronics, Volume 2
Intermediate Study
Vaughn D. Martin

Optoelectronics, Volume 1, introduced you to the basic concepts of the field, as well as photometrics and optics. Now, Optoelectronics, Volume 2, presents you with an intermediate study in the practical aspects and uses of optoelectronics.

Written for experienced technicians and electronics students who want to broaden their knowledge of optoelectronics, Optoelectronics Volume 2, presents you with comprehensive information on radiometrics, color CRTs, and much more. This book also contains fascinating and easy-to-follow projects that will show you how to put your newly acquired optoelectronics knowledge to practical use. Equations are used only when no other means of explanation can clearly illustrate a point.

Electronic Theory
352 pages + Paperback + 8-1/2 x 11"
ISBN: 0-7906-1091-4 + Sams: 61091
$29.95 + January 1997

Electronic Theory
258 pages + Paperback + 8-1/2 x 11"
ISBN: 0-7906-1110-4 + Sams: 61110
$29.95 + March 1997

CALL 1-800-428-7267 TODAY FOR THE NAME OF
YOUR NEAREST PROMPT PUBLICATIONS DISTRIBUTOR

PUBLICATIONS

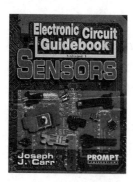

Electronic Circuit Guidebook
Volume 1, Sensors
Joseph J. Carr

Most sensors are inherently analog in nature, so their outputs are not usable by the digital computer. Even if the sensor is supposedly a digital output design, it is likely that an inherently analog process is paired with an analog-to-digital converter. In *Electronic Circuit Guidebook, Volume 1: Sensors*, you will find information you need about typical sensors, along with a large amount of information about analog sensor circuitry. Amplifier circuits are especially well covered, along with differential amplifiers, analog signal processing circuits and more. This book is intentionally kept practical in outlook. Some topics covered include electronics signals and noise, measurement, sensors and instruments, instrument design rules, sensor interfaces, analog amplifiers, and sensor resolution improvement techniques.

Electronic Circuit Guidebook
Volume 2, IC Timers
Joseph J. Carr

Timer circuits used to be a lot of trouble to build and tame for several reasons. One major reason was the fact that DC power supply variations would cause a frequency shift or slow drift. Part I of this book is organized to demonstrate the theory of how various timers work. This is done by way of an introduction to resistor-capacitor circuits, and in-depth chapters on various TTL and CMOS digital IC devices. Through simplified equations and detailed graphics, the information presented is perfect for both practicing technicians and enthusiastic hobbyists. Part II presents a variety of different circuits and projects. Some of the circuits include: analog audio frequency meter, one-second timer/flasher, relay and optoisolator drivers, two-phase digital clock and more. *Electronic Circuit Guidebook, Volume 2: IC Timers* will teach you enough that you will not only be able to rework and modify the circuits covered here, but also design a few of your own.

Electronic Theory
340 pages + Paperback + 7-3/8 x 9-1/4"
ISBN: 0-7906-1098-1 + Sams: 61098
$24.95 + April 1997

Electronics Technology
240 pages + Paperback + 7-3/8 x 9-1/4"
ISBN: 0-7906-1106-6 + Sams: 61106
$24.95 + August 1997

CALL 1-800-428-7267 TODAY FOR THE NAME OF
YOUR NEAREST PROMPT PUBLICATIONS DISTRIBUTOR

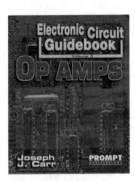

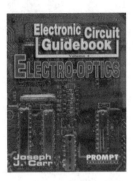

Electronic Circuit Guidebook
Volume 3, Op Amps
Joseph J. Carr

The operational amplifier is the most commonly used linear IC amplifier in the world. The range of applications for the op amp is truly awesome – it has become a mainstay of audio, communications, TV, broadcasting, instrumentation, control, and measurement circuits. Third in a series covering electronic instrumentation and circuitry, *Electronic Circuit Guidebook, Volume 3: Op Amps* is design to give you some insight into how practical linear IC amplifiers work in actual real-life circuits. Because of their widespread popularity, operational amplifiers figure heavily in this book, though other types of amplifiers are not overlooked. This book allows you to design and configure your own circuits, and is intended to be a practical workshop aid. Some of the topics covered in detail include linear IC amplifiers, ideal operational amplifiers, instrumentation amplifiers, isolation amplifiers, active analog filter circuits, waveform generators, and many more.

Electronics Technology
273 pages + Paperback + 7-3/8 x 9-1/4"
ISBN: 0-7906-1131-7 + Sams: 61131
$24.95 + August 1997

Electronic Circuit Guidebook
Volume 4, Electro-Optics
Joseph J. Carr

Electronic Circuit Guidebook, Volume 4: Electro-Optics is mostly about E-O sensors — those electronic transducers that convert light waves into a proportional voltage, current, or resistance. The coverage of the sensors is wide enough to allow you to understand the physics behind the theory of operation of the device, and also the circuits used to make these sensors into useful devices. This book examines the photoelectric effect, photoconductivity, photovoltaics, and PN junction photodiodes and phototransistors. Also examined is the operation of lenses, mirrors, prisms, and other optical elements keyed to light physics.

Electronic Circuit Guidebook, Volume 4: Electro-Optics is intended to teach the physics and operation of E-O devices, then proceed to circuits and methods for actual application of the devices in real situations.

Electronics Technology
416 pages + Paperback + 7-3/8 x 9-1/4"
ISBN: 0-7906-1132-5 + Sams: 61132
$29.95 + October 1997

CALL 1-800-428-7267 TODAY FOR THE NAME OF YOUR NEAREST PROMPT PUBLICATIONS DISTRIBUTOR

Semiconductor Cross Reference Book Fourth Edition

Howard W. Sams & Company

This newly revised and updated reference book is the most comprehensive guide to replacement data available for engineers, technicians, and those who work with semiconductors. With more than 490,000 part numbers, type numbers, and other identifying numbers listed, technicians will have no problem locating the replacement or substitution information needed. There is not another book on the market that can rival the breadth and reliability of information available in the fourth edition of the *Semiconductor Cross Reference Book*.

Professional Reference
688 pages - Paperback - 8-1/2 x 11"
ISBN: 0-7906-1080-9 - Sams: 61080
$24.95 ($33.95 Canada) - August 1996

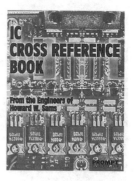

IC Cross Reference Book Second Edition

Howard W. Sams & Company

The engineering staff of Howard W. Sams & Company assembled the *IC Cross Reference Book* to help readers find replacements or substitutions for more than 35,000 ICs and modules. It is an easy-to-use cross reference guide and includes part numbers for the United States, Europe, and the Far East. This reference book was compiled from manufacturers' data and from the analysis of consumer electronics devices for PHOTOFACT® service data, which has been relied upon since 1946 by service technicians worldwide.

Professional Reference
192 pages - Paperback - 8-1/2 x 11"
ISBN: 0-7906-1096-5 - Sams: 61096
$19.95 ($26.99 Canada) - November 1996

CALL 1-800-428-7267 TODAY FOR THE NAME OF YOUR NEAREST PROMPT PUBLICATIONS DISTRIBUTOR

The Component Identifier and Source Book

Victor Meeldijk

Because interface designs are often reverse engineered using component data or block diagrams that list only part numbers, technicians are often forced to search for replacement parts armed only with manufacturer logos and part numbers.

This source book was written to assist technicians and system designers in identifying components from prefixes and logos, as well as find sources for various types of microcircuits and other components. There is not another book on the market that lists as many manufacturers of such diverse electronic components.

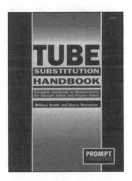

Tube Substitution Handbook

William Smith & Barry Buchanan

The most accurate, up-to-date guide available, the *Tube Substitution Handbook* is useful to antique radio buffs, old car enthusiasts, and collectors of vintage ham radio equipment. In addition, marine operators, microwave repair technicians, and TV and radio technicians will find the *Handbook* to be an invaluable reference tool.

The *Tube Substitution Handbook* is divided into three sections, each preceded by specific instructions. These sections are vacuum tubes, picture tubes, and tube basing diagrams.

Professional Reference
384 pages - Paperback - 8-1/2 x 11"
ISBN: 0-7906-1088-4 - Sams: 61088
$24.95 ($33.95 Canada) - November 1996

Professional Reference
149 pages - Paperback - 6 x 9"
ISBN: 0-7906-1036-1 - Sams: 61036
$16.95 ($22.99 Canada) - March 1995

CALL 1-800-428-7267 TODAY FOR THE NAME OF YOUR NEAREST PROMPT PUBLICATIONS DISTRIBUTOR

P PROMPT®
PUBLICATIONS

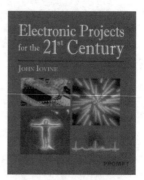

Basic Digital Electronics
Alvis J. Evans

Electronic Projects for the 21st Century
John Iovine

Explains digital system functions and how digital circuits are used to build them! Digital — what does it mean? Why is it that electronic systems are being designed using digital electronic circuits? Find the answer to these questions and more, as you learn the difference between analog and digital systems, the functions required to design digital systems, the circuits used to make decisions, code conversions, data selections, adding and subtracting, interfacing and storage, and the circuits that keep all operations in time and under control.

Learn about logic circuits, flip-flops, registers, multivibrators, counters, 3-state bus drivers, bidirectional line drivers and receivers, and more using easy-to-read, easy-to-understand explanations coupled with detailed illustrations.

If you are an electronics hobbyist with an interest in science, or are fascinated by the technologies of the future, you'll find Electronic Projects for the 21st Century a welcome addition to your electronics library. It's filled with nearly two dozen fun and useful electronics projects designed to let you use and experiment with the latest innovations in science and technology — innovations that will carry you and other electronics enthusiasts well into the 21st century!

Electronic Projects for the 21st Century contains the expert, hands-on guidance and detailed instructions you need to perform experiments that involve genetics, lasers, holography, Kirlian photography, and more. Among the projects are a lie detector, an ELF monitor, air pollution monitor, pinhole camera, laser power supply for holography, synthetic fuel, and an expansion cloud chamber.

Electronic Theory
192 pages + Paperback + 8-1/2 x 11"
ISBN: 0-7906-1118-X + Sams: 61118
$19.95 + April 1997

Electronic Projects
256 pages + Paperback + 7-3/8 x 9-1/4"
ISBN: 0-7906-1103-1 + Sams: 61103
$19.95 + June 1997

CALL 1-800-428-7267 TODAY FOR THE NAME OF YOUR NEAREST PROMPT PUBLICATIONS DISTRIBUTOR